Climate Change Management

Series Editor

Walter Leal Filho, International Climate Change Information and Research Programme, Hamburg University of Applied Sciences, Hamburg, Germany

The aim of this book series is to provide an authoritative source of information on climate change management, with an emphasis on projects, case studies and practical initiatives – all of which may help to address a problem with a global scope, but the impacts of which are mostly local. As the world actively seeks ways to cope with the effects of climate change and global warming, such as floods, droughts, rising sea levels and landscape changes, there is a vital need for reliable information and data to support the efforts pursued by local governments, NGOs and other organizations to address the problems associated with climate change. This series welcomes monographs and contributed volumes written for an academic and professional audience, as well as peer-reviewed conference proceedings. Relevant topics include but are not limited to water conservation, disaster prevention and management, and agriculture, as well as regional studies and documentation of trends. Thanks to its interdisciplinary focus, the series aims to concretely contribute to a better understanding of the state-of-the-art of climate change adaptation, and of the tools with which it can be implemented on the ground.

Notes on the quality assurance and peer review of this publication

Prior to publication, the quality of the works published in this series is double blind reviewed by external referees appointed by the editor. The referees are not aware of the author's name when performing the review; the referees' names are not disclosed.

Walter Leal Filho · Diogo Guedes Vidal ·
Maria Alzira Pimenta Dinis
Editors

Climate Change and Health Hazards

Addressing Hazards to Human and Environmental Health from a Changing Climate

Springer

Editors
Walter Leal Filho
Faculty of Life Sciences
Hamburg University of Applied Sciences
Hamburg, Germany

Maria Alzira Pimenta Dinis
UFP Energy, Environment and Health Research Unit (FP-ENAS)
University Fernando Pessoa (UFP)
Porto, Portugal

Diogo Guedes Vidal
Centre for Functional Ecology - Science for People & the Planet (CFE), TERRA Associate Laboratory, Department of Life Sciences (DCV)
University of Coimbra (UC)
Coimbra, Portugal

ISSN 1610-2002 ISSN 1610-2010 (electronic)
Climate Change Management
ISBN 978-3-031-26591-4 ISBN 978-3-031-26592-1 (eBook)
https://doi.org/10.1007/978-3-031-26592-1

This Springer imprint is published by the registered company Springer Nature Switzerland AG
The registered company address is: Gewerbestrasse 11, 6330 Cham, Switzerland

Preface

According to the Intergovernmental Panel on Climate Change (IPCC), human-induced climate change is known to be causing dangerous and widespread disruptions in nature, is affecting the lives of billions of people around the world, despite efforts to reduce the risks. Climate change is also negatively influencing health and is a mounting threat to our well-being and a healthy planet.

The world is also facing significant climate hazards over the next two decades, with global warming expected to soon reach 1.5°C. Even temporarily exceeding this warming level will result in additional severe impacts, some of which may be irreversible. There is therefore a perceived need for publications that may foster a greater understanding of how climate change connects to human health, and the role played by hazards in this context. It is against this background that this book is being prepared. It contains a set of papers that explore the links between climate change, health, and hazards and demonstrate how they interact. It emphasizes the urgency of immediate and more ambitious action to address climate risks.

Produced by the European School of Sustainability Science and Research (ESSSR) and the Inter-University Sustainable Development Research Programme (IUSDRP), this book aims to provide a significant contribution to a more comprehensive and interdisciplinary understanding, through a state-of-the-art description of approaches, methods, initiatives, and projects from universities, health institutes, research organizations, and civil society across the world, regarding climate change and health. Under the title "Climate Change and Health Hazards: Addressing Hazards to Human and Environmental Health from a Changing Climate", this publication is expected to fill in a gap in relation to integrative publications on the connections between climate change, health, and health hazards, considered cross-cutting issues.

The book is structured into 3 parts:

Part 1—Theoretical Approaches to Climate Change Impacts on Human and Environmental Health
Part 2—Climate Change Impacts on Human and Environmental Health
Part 3—Climate Resilience and Adaptation to Climate Change

These, in turn, address the following aspects:

- Transdisciplinary theoretical approaches to climate change impact on humans and non-humans lives
- Theoretical discussions on the impacts of climate change on mental health
- Assessing different impacts of climate change on tourism, agriculture, water, vector-borne diseases, disasters, and gender
- Global north and global south perspectives of climate change
- Actions and strategies of resilience and adaptation to climate change impacts

A distinctive feature of this book is that it not only presents a wide range of philosophies, approaches, methods, and analyses about climate change impacts on human and environmental health but also documents and disseminates a polyphonic discussion, where human and non-humans voices are called to reflect on the biggest challenge that society is facing on the contemporaneity: a changing climate. In this regard, the presentation of case studies illustrates the main drivers and barriers involved in climate change, being faced in practice.

We would like to thank all authors and reviewers for making available their experience in the chapters included in this book and their willingness to share their ideas. By providing their inputs, the authors have made a strong contribution to a debate that needs to be fruitful and reach a depth far beyond what conferences, workshops, or seminars may be able to offer.

Hamburg, Germany — Walter Leal Filho
Coimbra, Portugal — Diogo Guedes Vidal
Porto, Portugal — Maria Alzira Pimenta Dinis
Summer 2023

Contents

Part II Climate Change Impacts on Human and Environmental Health

Editors and Contributors

About the Editors

Walter Leal Filho holds the chairs of Climate Change Management at the Hamburg University of Applied Sciences (Germany) and Environment and Technology at Manchester Metropolitan University (UK). He directs the Research and Transfer Centre "Sustainability Development and Climate Change Management." His main research interests are in the fields of sustainable development and climate change, also including aspects of climate change and health. He has over 30 years of experience on climate change projects and directs the International Climate Change Information and Research Programme.

Diogo Guedes Vidal is an Environment and Nature Sociologist interested in all its interconnections with and health of human and non-human beings. Ph.D. in Ecology and Environmental Health with a project focused on the attempt to look beyond the greenness of public urban green spaces funded by the main Portuguese public body agency—FCT. Currently is researcher in the Project PHOENIX "The rise of citizens' voices for a greener Europe" (Horizon 2020; Grant agreement ID: 101037328) and Invited Assistant Researcher at Centre for Functional Ecology—Science for People & the Planet" (CFE) in the Department of Life Sciences of the University of Coimbra, Portugal.

Maria Alzira Pimenta Dinis, Ph.D., M.Sc. is Associate Professor with Habilitation at the Faculty of Science and Technology (FCT), University Fernando Pessoa (UFP), Porto, Portugal. She is graduated in Metallurgical Engineering and Master Degree in Materials Engineering at the Faculty of Engineering, University of Porto, Portugal. She holds a Ph.D. degree in Earth Sciences at FCT, UFP. Researcher at FP-ENAS, UFP Energy, Environment and Health Research Unit in environment-related topics. She is an experienced author, editor, and reviewer, with scientific outputs translated by a number of publications. Sustainability expertise.

Contributors

Roberto Ariel Abeldaño Zuñiga University of Sierra Sur, Oaxaca, Mexico

Khandaker Jafor Ahmed Walsh School of Foreign Service, Georgetown University, Washington, District of Columbia, United States

Diana Guadalupe Antonio Sánchez University of Sierra Sur, Oaxaca, Mexico

Eduardo Paz Barroso FP-I3ID, Faculty of Human and Social Sciences, University Fernando Pessoa (UFP), Porto, Portugal;
LabCom—Communication & Arts Research Unit, University of Beira Interior, Covilhã, Portugal

Girma Berhe Haramaya University, Institute of Technology, School of Water Resource and Environmental Engineering, Dire Dawa, Ethiopia

Juliane Boenecke Research and Transfer Centre "Sustainable Development and Climate Change Management", Hamburg University of Applied Sciences, Hamburg, Germany;
Department Health Sciences, Hamburg University of Applied Sciences, Hamburg, Germany

Clemens Buter PAHO/WHO Office of Barbados & the Eastern Caribbean Countries, Bridgetown, Barbados

Allison Margaret Byrne Universidade Fernando Pessoa, Porto, Portugal

Álvaro Campelo FP-I3ID, Faculty of Human and Social Sciences, University Fernando Pessoa (UFP), Porto, Portugal;
Fernando Pessoa Behaviour and Social Sciences Research Centre (FP-B2S), University Fernando Pessoa (UFP), Porto, Portugal;
Centre for Research in Anthropology (CRIA), Lisbon, Portugal

João Casqueira Cardoso Department of Political and Humanitarian Studies, University Fernando Pessoa/CEPESE, Porto, Portugal

Paulo José Ribeiro Cardoso Universidade Fernando Pessoa, Porto, Portugal;
Universidade Lusíada do Porto, Porto, Portugal

Sónia Maria Martins Caridade Psychology Research Center, School of Psychology, University of Minho, Braga, Portugal;
Interdisciplinary Center of Gender Issues, Higher Institute of Social and Political Sciences, University of Lisbon (ISCSP-ULisboa), Lisbon, Portugal

Arijana Cenov Department of Environmental Health, Faculty of Medicine, University of Rijeka, Rijeka, Croatia;
Department of Environmental Health, Teaching Institute of Public Health of Primorje-Gorski Kotar County, Rijeka, Croatia

Pedro Cunha FP-I3ID, Faculty of Human and Social Sciences, University Fernando Pessoa (UFP), Porto, Portugal;
Fernando Pessoa Behaviour and Social Sciences Research Centre (FP-B2S), University Fernando Pessoa (UFP), Porto, Portugal

Carla Alexandra Martins da Fonte Faculty of Human and Social Sciences, University Fernando Pessoa (UFP), Porto, Portugal

Tonderai Dangare Department of Electronics and Telecommunications Engineering, University of Zimbabwe, Harare, Zimbabwe

Justice Kofi Debrah Faculty of Science and Technology, University Fernando Pessoa (UFP), Porto, Portugal

Gabriela Narcizo de Lima Faculty of Letters, Geography Department, Porto University, Porto, Porto, Portugal

Maria Alzira Pimenta Dinis UFP Energy, Environment and Health Research Unit (FP-ENAS), University Fernando Pessoa (UFP), Porto, Portugal

Rui Estrada FP-I3ID, Faculty of Human and Social Sciences, University Fernando Pessoa (UFP), Porto, Portugal;
Center for Transdisciplinary Research «Culture, Space and Memory» (CITCEM), Faculty of Arts and Humanities, University of Porto (FLUP), Porto, Portugal

Maria Pia de Melo Alvim Ferraz Tavares Departamento de Engenharia Metalúrgica e de Materiais, Faculdade de Engenharia da Universidade do Porto, Universidade do Porto, Porto, Portugal;
Instituto de Engenharia Biomédica, Universidade do Porto, Porto, Portugal;
i3S, Instituto de Investigação e Inovação em Saúde da Universidade do Porto, Porto, Portugal

Diogo Cunha Ferreira CERIS, Instituto Superior Técnico, University of Lisbon. Av, Lisbon, Portugal

Mafalda S. A. Fonseca Master in Humanitarian Action, Cooperation and Development, University Fernando Pessoa, Porto, Portugal

Ana Gabriela Frazão-Nogueira FP-I3ID, Faculty of Human and Social Sciences, University Fernando Pessoa (UFP), Porto, Portugal;
ICNOVA—Instituto de Comunicação da NOVA, Faculdade de Ciências Sociais E Humanas, Universidade NOVA de Lisboa, Lisbon, Portugal

Gloria Svietlana Jólluskin García Universidade Fernando Pessoa, Porto, Portugal

Marin Glad Department of Environmental Health, Faculty of Medicine, University of Rijeka, Braće Branchetta 20, 51000 Rijeka, Croatia;
Department of Environmental Health, Teaching Institute of Public Health of Primorje-Gorski Kotar County, Krešimirova 52a, 51000 Rijeka, Croatia

Laurine Grzelak Master in Humanitarian Action, Cooperation and Development, University Fernando Pessoa, Porto, Portugal

Weronika Z. Gąsior NeuroLandscape, Warsaw, Poland

Nelly Victoria Gutiérrez Moguel Universidad de La Cañada, Oaxaca, Mexico

Shah Md Atiqul Haq Department of Sociology, Shahjalal University of Science and Technology, Sylhet, Bangladesh

Judith Harvey PAHO/WHO Office of Barbados & the Eastern Caribbean Countries, Bridgetown, Barbados

Johanna Heimfarth School of Public Health, Charité - Universitätsmedizin Berlin, Berlin, Germany

Umar Ibrahim Public and Environmental Health Department, Faculty of Basic Medical Sciences, College of Medicine and Allied Medical Sciences, Federal University Dutse (FUD), Dutse, Jigawa State, Nigeria

Deysi Ofelmina Jerez-Ramírez Research Institute in Risk Management and Climate Change, University of Sciences and Arts of Chiapas, Tuxtla Gutiérrez, Mexico

Walter Leal Filho Research and Transfer Centre "Sustainable Development and Climate Change Management", Hamburg University of Applied Sciences, Hamburg, Germany;
Department Health Sciences, Hamburg University of Applied Sciences, Hamburg, Germany

Marina Prieto Afonso Lencastre FP-I3ID, Faculty of Human and Social Sciences, University Fernando Pessoa (UFP), Porto, Portugal

Željko Linšak Department of Environmental Health, Faculty of Medicine, University of Rijeka, Rijeka, Croatia;
Department of Environmental Health, Teaching Institute of Public Health of Primorje-Gorski Kotar County, Rijeka, Croatia

Dijana Tomić Linšak Department of Environmental Health, Faculty of Medicine, University of Rijeka, Rijeka, Croatia;
Department of Environmental Health, Teaching Institute of Public Health of Primorje-Gorski Kotar County, Rijeka, Croatia

Hélder Lopes Lab2PT e Departamento de Geografia da Universidade Do Minho, Braga, Portugal;
Grupo de Climatología, Departamento de Geografía de la IdRA, Universidad de Barcelona, Barcelona, Spain

Dražen Lušić Department of Environmental Health, Faculty of Medicine, University of Rijeka, Rijeka, Croatia;
Center for Advanced Computing and Modelling, University of Rijeka, Rijeka, Croatia;
Faculty of Health Studies, University of Rijeka, Rijeka, Croatia

Darija Vukić Lušić Department of Environmental Health, Faculty of Medicine, University of Rijeka, Rijeka, Croatia;
Department of Environmental Health, Teaching Institute of Public Health of Primorje-Gorski Kotar County, Rijeka, Croatia;
Center for Advanced Computing and Modelling, University of Rijeka, Rijeka, Croatia

Daniel Maestro Department of Environmental Health, Institute for Public Health of the Federation of Bosnia and Herzegovina, Sarajevo, Bosnia and Herzegovina;
Department of Environmental Health, Faculty of Health Studies, University of Sarajevo, Sarajevo, Bosnia and Herzegovina

Nerma Maestro Department of Communal Hygiene, Institute for Public Health of Canton Sarajevo, Bosnia and Herzegovina, Sarajevo, Bosnia and Herzegovina

Susana Magalhães FP-I3ID, Faculty of Human and Social Sciences, University Fernando Pessoa (UFP), Porto, Portugal;
Instituto de Investigação e Inovação em Saúde (i3S), Universidade Do Porto, Porto, Portugal

Rui Leandro Maia UFP Energy, Environment and Health Research Unit (FP-ENAS), Universidade Fernando Pessoa, Porto, Portugal;
CITCEM—Centro de Investigação Transdisciplinar «Cultura, Espaço E Memória», Faculdade de Letras da Universidade Do Porto, Porto, Portugal

Javier Martín-Vide Grupo de Climatología, Departamento de Geografía de la IdRA, Universidad de Barcelona, Barcelona, Spain

Newton R. Matandirotya Department of Geosciences, Faculty of Science, Nelson Mandela University, Port Elizabeth, South Africa;
Centre for Climate Change Adaptation and Resilience, Kgotso Development Trust, Beitbridge, Zimbabwe

Teresa Letra Mateus CISAS—Centre for Research and Development in Agrifood Systems and Sustainability, Escola Superior Agrária, Instituto Politécnico de Viana Do Castelo, Viana Do Castelo, Portugal;
EpiUnit—Instituto de Saúde Pública da Universidade do Porto, Laboratory for Integrative and Translational Research in Population Health (ITR), Porto, Portugal;
Veterinary and Animal Research Centre (CECAV), Associate Laboratory for Animal and Veterinary Sciences (AL4AnimalS), UTAD, Vila Real, Portugal

Eduardo José Rocha Medeiros Instituto Universitário de Lisboa (ISCTE-IUL), DINÂMIA'CET - IUL, Lisboa, Portugal

Tanya Méndez-Matus Research Institute in Risk Management and Climate Change, University of Sciences and Arts of Chiapas, Tuxtla Gutiérrez, Mexico

Solange Moreira Escola Superior Agrária, Instituto Politécnico de Viana Do Castelo, Viana Do Castelo, Portugal

Alexandre Morais Nunes Centre for Public Administration and Public Policies, Institute of Social and Political Sciences, Universidade de Lisboa, Rua Almerindo Lessa, Lisbon, Portugal

Mina Margaret Ogbanga Indianapolis School of Social Work, Indiana University Purdue University-IUPUI, Indianapolis, IN, USA;
CEO Institute for Partnership and Development Studies of CEDSI Nigeria, Building Exceptional Organizations for Effective and Sustainable Project Delivery, UN Sustainable Development Solutions Network, Port Harcourt, Nigeria

Gisela Marta Teixeira de Sousa Oliveira UFP Energy, Environment and Health Research Unit (FP-ENAS), University Fernando Pessoa, Porto, Portugal

Agnieszka Olszewska-Guizzo NeuroLandscape, Warsaw, Poland

Mahfuza Parveen Department of Environmental Science and Disaster Management, Daffodil International University, Dhaka, Bangladesh

Miguel Alves Pereira INESC TEC, Porto, Portugal;
CEGIST, Instituto Superior Técnico, University of Lisbon, Lisbon, Portugal

Maja Radišić Faculty of Civil Engineering Rijeka, Rijeka, Croatia

Paula Remoaldo Lab2PT e Departamento de Geografia da Universidade Do Minho, Braga, Portugal

Vítor Ribeiro Lab2PT e Departamento de Geografia da Universidade Do Minho, Braga, Portugal;
Escola Superior de Educação Paula Frassinetti, ESEPF, Porto, Portugal

Josip Rubinić Faculty of Civil Engineering Rijeka, Rijeka, Croatia

Dianne Rudd Department of Geography, Environment and Population, School of Social Sciences, The University of Adelaide, Adelaide, South Australia, Australia

Bijoya Saha Department of Sociology, Shahjalal University of Science and Technology, Sylhet, Bangladesh

María Alejandra Sánchez Bandala University of Sierra Sur, Oaxaca, Mexico

Isabel Maria Sousa Lopes Silva Universidade Fernando Pessoa, Porto, Portugal

Yan Tan Department of Geography, Environment and Population, School of Social Sciences, The University of Adelaide, Adelaide, South Australia, Australia

Nazwa Tahsin NeuroLandscape, Warsaw, Poland

Derya Taser Research and Transfer Centre "Sustainable Development and Climate Change Management", Hamburg University of Applied Sciences, Hamburg, Germany

Nishat Tasneem Department of Sociology, Shahjalal University of Science and Technology, Sylhet, Bangladesh

Godfred Kwesi Teye Key Laboratory of Integrated Regulation and Resource Development on Shallow Lakes of Ministry of Education, College of Environment, Hohai University, Nanjing, China

Teresa Martinho Toldy FP-I3ID, Faculty of Human and Social Sciences, University Fernando Pessoa (UFP), Porto, Portugal;
Centre for Social Studies (CES), University of Coimbra, Coimbra, Portugal

Dana van Alphen PAHO/WHO Office of Barbados & the Eastern Caribbean Countries, Bridgetown, Barbados

Diogo Guedes Vidal Centre for Functional Ecology—Science for People & the Planet (CFE), TERRA Associate Laboratory, Department of Life Sciences, University of Coimbra, Coimbra, Portugal

Zujaja Wahaj Faculty of International Business and Management, National University of Sciences and Technology, Islamabad, Pakistan

Roberto Ariel Abeldaño Zuñiga University of Helsinki, Finland, Mexico

Part I
Theoretical Approaches to Climate Change Impacts on Human and Environmental Health

Chapter 1
Thinking Beyond Borders: How Can Humanities and Social Sciences Help to Deal with Climate Change Health Hazards in the Twenty-First Century? Outputs of the Transdisciplinary Network "Composing Worlds: Humanities, Well-being and Health in the 21st century"

Diogo Guedes Vidal, Marina Prieto Afonso Lencastre, Susana Magalhães, Eduardo Paz Barroso, Álvaro Campelo, Pedro Cunha, and Rui Estrada

Abstract Climate Change (CC) is the most noticeable evidence of the "Metamorphosis of the World" and it is changing regimes. The unpredictability of climate phenomena and their multidimensional impacts on human and environmental health has contributed to making our understanding of the world increasingly difficult. Furthermore, history proved that technoscience alone is not enough to deal with these issues. CC, diseases linked to poverty and human migration, health systems increasingly required by democratic societies, and artificial intelligence, are all technical issues, but also social, which require a broad understanding for creative solutions. A global call is now waking up for the humanities and social sciences visions to deal with the human, and non-human matters raised by CC, namely through the integration of health and sustainability areas. This chapter aims to present and discuss some of the first outputs of the transdisciplinary network "Composing worlds: humanities, well-being and health in the 21st century", which consists of a network of experts in the humanities and social and health sciences who think about the issues of well-being and health in contemporary technological societies. The methodology used in

Diogo Guedes Vidal and **Marina Prieto Afonso Lencastre** have equally contributed to this work.

D. G. Vidal (✉)
Centre for Functional Ecology—Science for People & the Planet (CFE), TERRA Associate Laboratory, Department of Life Sciences, University of Coimbra, Calçada Martim de Freitas, 3000-456 Coimbra, Portugal
e-mail: diogoguedesvidal@hotmail.com

M. P. A. Lencastre (✉) · S. Magalhães · E. P. Barroso · Á. Campelo · P. Cunha · R. Estrada
FP-I3ID, Faculty of Human and Social Sciences, University Fernando Pessoa (UFP), Praça Nove de Abril, 349, 4249-004 Porto, Portugal

W. Leal Filho et al. (eds.), *Climate Change and Health Hazards*, Climate Change Management, https://doi.org/10.1007/978-3-031-26592-1_1

the first phase of the project was a thematic analysis of an interview with an open answer script, built in a participatory manner by the experts' network. It was an exploratory research, which used text analysis to identify the key ideas original to each author, and the induction of the corresponding main themes. The themes were then organized by semantic correspondence groups or thematic clusters. Of the nine clusters that emerged, five of them may pave the way for a true discussion on the CC hazards in human and environmental health: (i) Public understanding and post-normal science; (ii) critical thinking and ethics in health; (iii) well-being, health, democracy and social justice; (iv) holistic (cross-disciplinary) approach to health and well-being; and (v) environment, health, sustainability and equity. It is expected that these clusters contribute to discussing how CC has already altered our way of being, living and thinking about the world, and to dealing with the challenges related to CC hazards on health and well-being in the twenty-first century.

e-mail: mlencast@ufp.edu.pt

S. Magalhães
e-mail: susanat@ufp.edu.pt

E. P. Barroso
e-mail: epb@ufp.edu.pt

Á. Campelo
e-mail: campelo@ufp.edu.pt

P. Cunha
e-mail: pcunha@ufp.edu.pt

R. Estrada
e-mail: restrada@ufp.edu.pt

S. Magalhães
Instituto de Investigação e Inovação em Saúde (i3S), Universidade Do Porto, 4200-135 Porto, Portugal

E. P. Barroso
LabCom—Communication & Arts Research Unit, University of Beira Interior, Rua Marquês D'Ávila E Bolama, 6201-001 Covilhã, Portugal

Á. Campelo · P. Cunha
Fernando Pessoa Behaviour and Social Sciences Research Centre (FP-B2S), University Fernando Pessoa (UFP), Praça Nove de Abril, 349, 4249-004 Porto, Portugal

Á. Campelo
Centre for Research in Anthropology (CRIA), Edifício ISCTE, Sala 2w2, 1649-026 Lisbon, Portugal

R. Estrada
Center for Transdisciplinary Research «Culture, Space and Memory» (CITCEM), Faculty of Arts and Humanities, University of Porto (FLUP), Via Panorâmica Edgar Cardoso S/N, 4150-564 Porto, Portugal

Introduction

According to the UK Royal Society of Medicine (2021), there is overwhelming evidence that Climate Change (CC) endangers human health. There are global problems, where the environmental and ecosystem changes cause shifts in the health of entire human populations that depend on adequate supplies of drinking water, food, absence of pollution or infections, as well as safety from floods or heat waves. And there are local problems like respiratory or autoimmune disorders, likely related to chemical air pollution, an increase in low birth weight babies and miscarriages and aggravation of cardiovascular symptoms. Excess cancer deaths related to CC are also reported. Forest fires cause air and water pollution, population displacement, mental health disorders and death, extreme temperatures cause heat exhaustion and hyperthermia, and changes in the patterns of disease-carrying insects or waterborne diseases cause new epidemics.

There are also indirect effects of CC on human and non-human health. Food scarcity results from altered agriculture due to drought or flood and mediated through modified societal systems, may cause violent conflict and population displacement. Rising sea levels and the unpredictability of its effects will disrupt major cities' functioning or even their existence. Evidence shows that poor groups will bear the most risk and, globally, the greatest burden will fall on poor countries and poor children and elderly. All of these new realities rise the feeling of strangeness in face of a new world. To figure out the paramount changes we are facing, the German sociologist Beck (2016) called this transformation "the metamorphosis of the world".

There are practical solutions for particular problems, and leading groups advocate that health should be a major concern in the CC debate. But the situation is paradoxical and very difficult to change. We know that healthcare professionals are a key resource in the management of CC-induced health hazards, but they represent, at the same time, an important carbon footprint due to their modes of action. So, this is a contradictory scenario that forces us to rethink the traditional healthcare system in the context of CC. Not only from a practical point of view but essentially from a conceptual one, for there is a need for new ideas to operationalize new practices. As famously stated by K. Lewin in the 1950's, nothing is more practical than a good theory; to become friendly again, the metamorphosis of the world implies a metamorphosis of the mind. This is the general aim of this chapter: to propose some theoretical, as well as practical frameworks to rethink CC hazards and health, drawing specifically on the results of the transversal project "Composing worlds: humanities, well-being and health in the 21st century" developed by a network of experts from three Portuguese universities.

Theoretical Framework

The world as we knew it will never be the same. We are experiencing a metamorphosis of social life and environmental dynamics which, according to Beck (2016), is contributing to make it more difficult to understand and decode. The world is facing a myriad of events that are challenging its balance and harmony such as terrorist attacks, nuclear disasters, breaches of privacy, economic crises, wars and climate changes. Technoscience is controlling the whole technic and economic processes, overlapping humanistic and democratic values, such as liberty, formerly fundamental in social-political ideals Society has been continuously understood as apart from nature (Aldeia and Alves 2019) but this duality only makes sense in Western societies hegemonic paradigm (Descola 2005; Bragança et al. 2021).

Beck (2016) states that the basic concepts and certainties of modernity are no longer constant, which expresses a "detraditionalization of life forms". In this sense, the author believes that Foucault and Pierre Bourdieu's theories are no longer able to understand the social reality that is metamorphosing, since words like "revolution", "evolution" and "transformation" cannot capture it. This new concept implies a much more radical transformation, in which the old certainties of modern society are disappearing and something entirely new is emerging. This new reality is defined as "a major shift to something different and imply a complete transformation into a different type, a different reality, a different way of being in the world, of seeing the world and of doing politics" (Beck 2016, p. 19). This is what he means by the strangeness of the world, in which humanity is trying to find some answers. This "strangeness" can be related to Latour's viewpoint (2016) that sates that we are living in a new, fast-paced world, full of possibilities and new artifacts, but for which we still seem to lack words and ideas. Beck (2016) explores the application of metamorphosis by linking it with the need to move beyond what he calls a "methodological nationalism" to a "cosmopolitism nationalism" since humanity is truly impregnated with global realism independently of our own will. To him, cities are part of a logic of "cosmopolitan risk communities", because instead of sharing common norms and values (traditionally associated with the concept of the community), they now share crises, risks and threats, that is, the "evils" of society. Its purpose is driven by the risk that is the anticipation of catastrophe. This is evident in the Portuguese case, where several cities are part of a wide project (ClimAdaPT.Local) that valued the integration of climatic dimension in the municipal planning processes (Mourato et al. 2018). Latour (2022) reflects on *cosmopolitics* as a project that refers to the possibility of placing a multiplicity of worlds in equal forms, a project that would guide the policy of scientific knowledge in a new direction than the one it followed until the present.

In this complex, confused and interdependent world, human society is claiming a clear project of humanity within a vulnerable planet. A cosmopolitan CC narrative may help pave the way to the diversity of values at stake, both human (Dias et al. 2020) and non-human (Martin et al. 2016). Latour (2017a) reflects on the fact that under the CC threat, which represents global changes (Santos 2012), the Western philosophical perspective is also experiencing a reverse: while the history of humanity stays paralyzed, the natural history is controlling the course of humanity's future. CC is transforming the world's geography, where its impacts are transversal but assume different forms. The unpredictability of climate phenomena may impact the vineyards in France (Pieri et al. 2012), increase gender violence in developing countries (Caridade et al. 2022), and also ameliorate croplands in Canada (Smith et al. 2013). Here, the biggest difference between these realities is resilience: the capacity to face adversities (Skoufias 2012; Islam and Winkel 2017) and profit from opportunities. While vineyard producers in France may easily change the target market and the wine produced, low-income communities dependent on family farming enter a cycle of poverty and disruption causing depression, anxiety, and violence (Costello et al. 2009).

Building up on Beck's (2016) and Latour's (2016) previous works, as well as from our former experience, we propose the following framework to understand the general dynamics between CC and health hazards. This framework tries to give an integrated picture of nature and culture, and is an update of Lencastre and Leal work (2006), meeting some of the requirements of the *sociogeobiology* and the cosmopolitics of the Anthropocene. Each point will be briefly discussed and will serve later for the final image synthesising the project:

i. **Co-evolution, change and the metamorphosis of the world**—Since the beginning, the Earth has undergone vast changes permeated by long periods of stability, in which species succeeded one another and transformed habitats and, more generally, terrestrial ecosystems and biomes. Associated with geological and climatological dynamics, they gave rise to the current situation that interacts with the effects of anthropic origin, generating the Anthropocene era (Steffen et al. 2011). Human cultures, in their diversity, contribute in different ways and with different weights to the global co-evolutionary framework. Modern cultures have significantly accentuated the rates of local and global change, contributing to many of the environmental problems we face. Modern societies are facing the externalities of their values and actions and there is mounting evidence for an urgent need to change ways of life and general politics. Moderns feel their world is undergoing a fundamental metamorphosis (Beck 2016) and this situation affects health and well-being. Ecoanxiety (Hickman et al. 2021), anomia and depression are health problems related to specific environmental catastrophes like forest fires, heat waves and drought that affect entire human and non-human populations, particularly the poorer and most exposed;
ii. **Non-linear complexity and relationality**—Environmental changes are traditionally described as complex systems that include a high set of variables in non-linear interaction, tending to end states described as "chaotic attractors" of a probabilistic outcome. When facing local problems needing urgent interventions, these systems are better described as relational imbroglios of humans and non-humans living in finite territories, asking for the concrete scientific inventories of their conditions of existence (Latour 2017b) to take decisions in post-normal settings (Funtowicz and Ravetz 1994). Return to the concrete

local allows for the pragmatic organization of communal cooperation in face of environmental hazards affecting health and well-being, away from unique central government liabilities;

iii. **Uncertainty and strangeness**—Human and non-human imbroglios imply levels of uncertainty in scientific, social, cultural and political knowledge. When systems uncertainty and decision stakes are high, controversy alongside necessary decisions make democratization and public knowledge of science a mandatory step. Post-normal science in environment and health includes those, human and non-human, who are affected by technoscientific decisions (Funtowicz and Ravetz 1994). To take practical decisions, local extended peer communities must deal with levels of uncertainty and socioecological strangeness (Beck 2016), value loading and a plurality of legitimate perspectives. This scenario applies to health and the environment, namely when communication of health hazards of subliminal perception (e.g. infection by the virus) implies public health politics that are not immediately palpable, or when genetically engineered seeds are proposed to local farmers;

iv. **Risk, precaution and reflexivity**—Complexity and uncertainty traditionally refer to the capacity of science to identify risks and precautions associated with decision-making. The ideas of simple causality, non-retroactivity and independence of variables that characterized the chains of determination of traditional science are abandoned. Ignorance often permeates scientific knowledge in predicting global physical, biological and social phenomena which, in turn, interfere with ecological, economic and political dynamics. The Anthropocene shows that human action affects geological stability and may have negative effects on ecological systems. The idea that modern life engendered and now permeates global and local territorial profiles is generally accepted. Awareness and reflexivity are the results of the current situation, even if the action does not immediately follow. One of the reasons for passive reflexivity is western nature-culture opposition. Individual and collective action seem insignificant in face of the magnitude of externalized natural hazards. But it should be clear that every cultural decision is immediately a natural one: we live in worlds of technocultural commonalities. Our actions, as well as non-human actions like infecting viruses or oxygen-producing bacteriae, have common effects that produce common worlds. How can we render compatible this new conceptualization of nature-culture imbroglios and western ideas of science, nature, society, democracy and individual freedom of choice in precautionary scenarios?;

v. **Regenerative landscaping and modes of co-existence**—With the ecological crisis of the twentieth century, the need for new perspectives on human development emerged, which consider and respect the dynamic balance of ecological and social systems. Thus, the concept of sustainable development, and the desirable tripartite sustainability—ecological, social and economic—emerged as one of the main concerns and one of the great challenges for the twenty-first century. Its goal is to guide human communities towards sustainable development, throughout time (concern for future generations), and space (attention to different regions of the planet), balancing ecological, economic and social requirements in their complexity and cultural and geographic regionalities. But sustainability did not fulfil its promises because its goals were centred on satisfying fundamental human needs, without compromising the possibility of future generations satisfying theirs, a view too dependent on economic interests and funding. As an alternative, regenerative landscaping is emerging as a new concept that describes the restoration and revitalization of local territories, urban and non-urban, based on their sources of energy, materials, and climatic and species profile. The idea is to create resilient and equitable ecological systems that integrate the needs of human and non-human populations. Regenerative design (Wahl 2016) dwells on biomimicry, biophilia (Lencastre and Farinha Marques 2021; Lencastre et al. 2022) and circular economics, as well as on rewilding and restorative justice ideas. It focuses on mapping relationships and promoting harmonious co-evolution through a co-creation approach where people are part of a collective social project, a path towards inclusive, fair and healthy territories (e.g. urban green spaces promoting respiratory health);

vi. **Territories and local human and non-human diversity**—Diversification seems to be one of the most important adaptive characteristics of resilient natural and cultural systems, being an important concept to describe at various levels: molecular, genetic, physiological, ethological, psychological, social, cultural and ecological. The different ecological qualities of cultures are evident when associated with local ways of life. Human miscegenation, based largely on current and future migrations, could increase diversity and constitute one of the most important axes of human demographic dynamics in the twenty-first century. Diversification appears associated with its opposite tendency, which is the homogenization operated by natural/cultural selection. Local territories are subject to species mobility, in search of life conditions, as well as human mobility. To describe local dynamics of nature/culture like the usage of water, the distributions of food, and the needs for energy or shelter, means to maintain a dynamic diversity lens that allows for a global perception of different relations to the territory. In terms of environmental mobility and health, it is important to consider how local populations adapt to entering new populations (e.g. migratory infectious diseases and local adaptive polymorphisms) and to leaving populations;

vii. **Cosmopolitics and multispecies equity**—Equity means a differential notion of justice that distributes wealth and material or symbolic resources in a way adapted to the local needs of human and non-human groups. In a globalised world with finite resources where different populations strive, there is a growing concern for sharing spaces with others, including other species. This concern requires a re-thinking of territories and their human and non-human inhabitants, entangled in complex relations and interest imbroglios. In the Anthropocene, territories where life dwells, seem sometimes disordered, sometimes strange and unpredictable, imperfectly attuned by new and old local stories, practical politics and social habits, as well as by species mobilities and compromises. Cosmopolitical thinking and action address these diverse stories and experiences, following a description of the plurality of the modes of existence to understand relations and create new strategic possibilities for action (Stenghers 2005). This concept raises fundamental questions for democracy: how to describe and ascribe needs and resources? Who distributes and who receives? This question dramatically arose in the context of Covid-19 with the problem of differential purchasing and distributing of the vaccine;

viii. **Modes of action and controversy**—One of the central characteristics of the current sciences is its controversial nature, i.e., subject to different reasoned opinions, coming from different valid actors. When publicly discussing controversial environmental health hazards like, for example, allergic reactions caused by industrial food chemicals, it is important to clearly describe what is at stake, namely the modes of action of the different agents involved in the dispute (the chemicals, health reactions, the industrials, citizens, local governance, other interested groups like non-human animals and plants, etc.). These entities occupy the same space and have unique ways of relating one to one another and to the other elements that compose the collective. Different facts and different values are interwoven, and deliberative procedures should clarify them first, to allow the discussion that leads to decision-making;

ix. **Inventory, diplomacy and action**—Decision-making, in a complex and uncertain collective of beings, makes the collaborative processes of consensus production one of the essential points of contemporary cosmopolitics. It must base itself on 'radical empiricism and irreductionism' (Latour and Muecke 2021, p. 12), opening the discussion to ontological and disciplinary pluralism. Latour proposes a new diplomacy that negotiates in the middle space between intervenients, closely looking at the description of things and avoiding abstract concepts like 'society' or 'nature'. The pragmatic inventory of beings should lead to a concrete description before proposing an explanation. Diplomatic mediation between modes of existence should lead to the description of a common world.

Due to the complexity of CC and its multidimensional impacts on the health and wellbeing of humans and non-humans, quantitative approaches may enhance and quantify the impacts but cannot deconstruct their meanings and associated discourses. Therefore, a reflexive approach based on participatory methods may be a contribution to a deeper understanding of this complex phenomenon. The methodological steps of the project "Composing Worlds" are described in the next section.

Methodology

The methodology used in this project is the interview with an open response script (Table 1.1), built-in a participatory way by the experts' network. The objective of the interview is to bring out personal and well-founded knowledge and sensitivities in the face of major issues involving the humanities, health and well-being in contemporary societies, including CC and health hazards. The present investigation is characterized by being inductive, realistic and at a semantic level; its goal is to identify original themes emerging in each author, and not the analysis of their historical, critical or psychological context.

This exploratory research uses thematic analysis with various levels of reading and text coding to identify the original key ideas for each author, and the induction of the corresponding main themes. We followed the methodological indications developed in Braun and Clarke (2006). The experts' answers to the interview were the subject of six levels of reading, analysis and synthesis:

1. Global reading and familiarization with the text of each expert;
2. Global reading highlighting the key ideas;
3. Global reading and analysis, highlighting and coding, by terms, the related key ideas;
4. Partial reading and analysis, organizing, throughout the text, the related terms into themes;
5. Return to the text and key ideas, reviewing and synthesizing them into 10 main themes;
6. The definitive review of the corpus of the key ideas and main themes by the experts.

The themes were later organized into correspondence groups, or clusters, in which they were brought together by semantic proximity. From this clustering, the transversal thematic axes were abstracted. The transversal thematic axes constitute vectors of interdisciplinary and transdisciplinary development for various training, research, social intervention and cultural extension activities.

Results and Discussion

Of the nine clusters that emerged, **five** of them may pave the way for a true discussion on the CC hazards in human and environmental health:

(i) **Public knowledge and post-normal science**

According to Latour's seminal work (1993), the idea of Science with a majuscule has been replaced by a more flexible and realistic idea of the sciences developing inside concrete networks of scientists, technologies, objects, peoples, and interests. In fact, for Latour, Science and Modernity engendered each other, as well as the idea

Table 1.1 Dimensions and questions of the interview script

Dimension	Question
Personal background	1. What is your subject area of initial training and how does it relate to your current professional activity and your main research interests? Please detail your answer
	2. Does your activity as a researcher and teacher relate to the area of well-being and/or health? If yes, how are these areas addressed? If not, what were the influences that, in your scientific and academic career, marked your interests? How can they contribute to your understanding of well-being and/or health?
Concepts definition and the utility of humanities and social sciences	3. This is a project about wellness and health. How do you define these concepts and what are, in your opinion, the areas of the humanities that can contribute most to their study? Please tell why.
	4. In your experience, and starting from your area of research, what are the major issues that, in the area of well-being and/or health, solicit your thought today? How can your research contribute to their understanding? 5. What are the themes in the area of well-being and/or health that you consider most embarrassing or, conversely, most promising for the future?
	5. What are the themes in the area of well-being and/or health that you consider most embarrassing or, conversely, most promising for the future?
Technological and ethical issues	6. Technologies for human well-being and/or health are rapidly changing and affect contemporary societies in different ways. How does it redefine the place of the human in this world increasingly marked by technologies?
	7. What is your opinion on the role of critical and ethical thinking at a time of great technological advances, environmental transformations, changes in social discourses and dialogues with other non-Western partners? How do you think this impacts on the search for truth in knowledge?
Current research experience related to well-being and health	8. How could your subject area participate in interdisciplinary projects related to health and/or well-being? If possible, can you give one or more examples, proposing the general designs of this research?

(continued)

Table 1.1 (continued)

Dimension	Question
	9. Do you have any other questions or important life events that shape your ideas about human well-being and/or health that you would like to develop?

of a time arrow inevitably leading to Progress. But today's environmental situations, particularly CC, show that this idea of progress leads to a messy state of affairs, where the traditionally separated concepts of Science, Nature and Culture are intertwined and inevitably produce externalities that affect the 'objects' and the 'subjects' of modernity. CC and its effects on health demand a more flexible and open conception of the sciences interacting with collectives of humans and non-humans that are affected by them.

This state of affairs is also recognizable in the context of the transmission of science between experts and laypeople. The linear model of transmission has been replaced by an idea of collective negotiated 'meanings' that arise during the social process that takes place at various levels, at different times and involving representatives of different groups and from different socio-cultural origins. These hybrid cultural spaces between sciences and collectives include specialists from different backgrounds, as well as interested people and non-human beings like animals, plants, geological elements or ecosystems, represented by human speakers. Latourian new diplomacy is based on radical empiricism, that is, a close description of what is, in a real place, or territory; and irreductionism, that is, the inclusion of everything relevant for a concrete inventory and explication of a collective, living in space (the opposite of scientific reduction); and, finally, consensus, that is, the negotiation in the middle space between intervenients, closely looking at the description of things and avoiding abstract concepts. In this hybrid space, critical and ethical thinking find their greatest relevance for applied science. This means that lab specialized science is incorporated into the real world, where real things operate and where living beings, including humans, strive. Diplomacy, in the context of this cosmopolitan conception of lived, earthly places, represents an effort of mediation and consensus for the making of a common, livable world.

Post-normal science (Funtowicz and Ravetz 1994), already addressed the issue of translating scientific content into collectives and public knowledge. Post-normal designation evolved from T. Kuhn's concept of normal science (Kuhn 1962). When faced with pressing applied science issues, where knowledge is uncertain and decision risks are high, post-normal science proposes that scientists, politicians, social partners and interested citizens form large communities of peers that will define the policies to follow. This is typically the case with CC hazards and their impact on human and non-human health; this was the case with public health decisions addressing the Covid-19 pandemic, and also with other environmental health issues (e.g. decarbonization and respiratory health). These public health problems are often characterized by uncertain chains of facts, institutional controversy, strong social and economic stakes and the need for urgent solutions.

In public and political spaces, scientific processes and results are mediated by cultural belongings and social cognition. This translation of sciences into public knowledge produces a modified language which all scientists are not able to handle. This condition makes the building of consensus difficult and complex. That is the reason why the extended community of peers must integrate mediators, and diplomatic translators of knowledge that focus above all on the quality of deliberative processes, rather than the quantitative results of applied sciences. Given the uncertainty and controversy of predictions, they cannot provide veto answers. The co-construction of post-normal health expertise based on science is also based on interests and values (e.g. economic considerations), aesthetics and social sensitivities (e.g. the preservation of special symbolic or natural health spaces), tradition (e.g. traditional lifestyles), and local debates have shown that there are no right answers or only one application of science in decision making. If we introduce ecological considerations about CC, including the consideration of other species and habitats, the problem of decision-making exponentially increases. With obvious exceptions like indigenous people or people living in poor countries, human modern way of life depends deeply on what needs to be changed if we want to stop global warming and all its planetary consequences. So, post-normal consensus depends on the assessment of health and CC risks, in the context of ecological sensitivities and information from different sources, where the credibility of the intervenients is as important as their knowledge.

In contemporary technological societies that face important CC local impacts on health, a careful inventory of local data is as critical for the correct exercise of preventive post-normal science, as is critical and ethical thinking. For that purpose, an extended peer community should include speakers for the non-humans, like non-human animals, but also natural elements like fresh water, earth, and plants. Deeply felt, poetical dimensions of human and non-human sensitivities should also be taken into account, because the complexity and novelty of CC emergent scenarios, and their impact on global health, means that they require radical new solutions. Entering these dimensions into diplomacy may lead to an emotion-laden, as well as a cognitive interpretation, and solution, of the novel situations.

The Covid-19 scenario showed that the infectiousness assessment, and the corresponding health decision-making, depended on a set of factors ranging from research on the zoonotic origin of the virus, the typification of its variants in particular human populations, the response capacity of concrete health systems, the local and global ecological and social processes of infection, the economic concerns of managers and politicians and, last but not least, to the type of media communication that influenced public opinion. What traditionally appeared as an object defined and contained by positive science, is now inscribed into a broader eco-socio-scientific context, open to uncertainty and risk, to values and controversy. Post-normal science proposes trans-disciplinary methodological frameworks and problem-solving strategies to include the uncertainty of natural systems and human and non-human interests and values in decision-making. Like we said earlier, to be able to operate in the real world.

(ii) **Critical thinking and ethics in health**

Assuming that health is now conceived as the *equilibrium* achieved among the different human dimensions (bio-psycho-social-spiritual) of each individual and being aware that this *equilibrium* interacts and is affected by other individuals, animals, plants and their shared environment,[1] it is clear that healthcare and health management requires critical thinking. In the current technological world, faced with the risk of depersonalization and "Undinge" (Han 2021)—a world where objects are replaced by information—the challenge is a keyword. One needs to challenge assumptions and context and to (re)imagine alternative paths to live in and with complexity. Moreover, health is a continuum between being healthy and being ill, where healthy people may live with diagnosed diseases for a long time and *ill* people may be healthier than in the so-called healthy world. Binary thinking and enormous (the pathology of modernity) (Weil et al. 2017) should therefore be deconstructed in the context of health and the two main tools to achieve this aim are critical reflection and ethical thinking, being both interrelated. Challenging assumptions and contexts imply a critical revision of knowledge, self and the world as well as a multilayered approach to health issues and education, focused not only on knowing what to do and how to think but also on knowing **how to be**.

Ageing populations need non-binary and critical approaches to their care since there is no longer right and wrong diagnosis (and treatment) but meaningful outcomes; treating chronic patients is more about disease management rather than technical expertise, and there has been a shift from *cure to care.*

Unpredictability and uncertainty seem to be at odds with the search for the right diagnosis and the best course of action/treatment based on the best scientific evidence that is the basis of *evidence-based medicine.* A patient's experience of disease (i.e. illness) coexists with social and scientific perspectives, adding more layers of complexity to decision making by both health professionals and individual patients. What is at stake here is the need to keep personhood in those who experience illness and in those who care. There is no person-centred care if the person is denied the possibility of Being (which is always rooted in values, preferences, meaningful places and people). *Phronesis,* as the knowledge that gives direction to action in a concrete situation and requires the task of deliberation, should be promoted in healthcare by setting up safe places for self-reflection, providing insights into who one is and not only what one knows or does:

> *Phronetic knowledge* describes an inward state of professionalism—it is something practitioners are, not what they have or do. This can only come about through critical self-reflection—making an honest assessment of strengths, limitations and professional motivation. Finally, phronesis is the ability to act in the concrete situation, which involves engagement with real cases and interaction with peers and other professionals. It can't be done

---

[1] One Health is defined by the Centers for Disease Control and Prevention (2022) as "a collaborative, multisectoral, and transdisciplinary approach — working at the local, regional, national, and global levels — with the goal of achieving optimal health outcomes recognizing the interconnection between people, animals, plants, and their shared environment."

theoretically, it can't even be prepared for, because it is action in a new situation. (Tyreman 2000)

Considering Paul Ricoeur's definition of ethics as aiming at the good life, with and for others, in just institutions, it is precisely the meaning of goodness and justice that healthcare professionals, patients, caregivers and policy makers are required to integrate into their deliberative and decisional making procedures. From the macro to the micro-level, from the abstract norms and generalizable scientific knowledge to the unique and singular narrative, care can only take place if it addresses what really matters to the suffering person.

(iii) **Wellbeing, health, democracy and social justice**

A recent study, entitled 'Climate anxiety in children and young people and their beliefs about government responses to climate change: a global survey' (Hickman et al. 2021), shows that young people suffer from eco-anxiety, fear the future, and blame governments for this dramatic situation: "According to our study, children and young people in countries around the world report climate anxiety and other distressing emotions and thoughts about climate change that impact their daily lives. This distress was associated with beliefs about inadequate governmental response and feelings of betrayal. A large proportion of children and young people around the world report emotional distress and a wide range of painful, complex emotions (sad, afraid, angry, powerless, helpless, guilty, ashamed, despair, hurt, grief, and depression). Similarly, large numbers report experiencing some functional impact and have pessimistic beliefs about the future (people have failed to care for the planet; the future is frightening; humanity is doomed; they won't have access to the same opportunities their parents had; things they value will be destroyed; security is threatened, and they are hesitant to have children)" (2021, p. 8). For example, young Portuguese respondents are particularly pessimistic, and the authors suggest an explanation: "Countries expressing more worry and a greater impact on functioning tended to be poorer, in the Global South, and more directly impacted by climate change; in the Global North, Portugal (which had dramatic increases in wildfires since 2017) showed the highest level of worry" (2021, p. 4).

As previously mentioned, according to the UK Royal Society of Medicine, there is solid evidence that climate change endangers the health of humans. In other words, young people have reason to be worried, anxious and, on the other hand, angry with most of the leaders who rule the world; this is not, therefore, a teenage or post-adolescent fad, but a dramatic and real situation.

The idea that climate change is of anthropogenic origin also raises no doubts in the scientific community. As the climatologist, Filipe Duarte Santos (2017): "(…) research published on precisely this subject concludes that 97% of climate scientists share the consensus that the observed global warming is anthropogenic". What then seems to be failing, despite the evidence of science on the subject, so that we can categorically reverse the announced disaster and at the same time restore the new generations' confidence in the future?

The answer is complex, since even in democratic societies—societies in which knowledge and the institutions that produce it should be respected, cherished and listened to—the emergence of populist leaders has hindered the clarification and decisions on the subject. It has not been enough, as we mentioned above, to reach the necessary consensus regarding climate or, more recently, Covid-19, that science, by and large, adheres to the new communicational reality. The mediators or translators of scientific knowledge can do little when they come up against denialist leaders who, in some cases, govern very powerful countries.

Take for example former US President Donald Trump: on 6 November 2012, he says: "The concept of global warming was created by and for the Chinese to make US manufacturing non-competitive" (Santos 2017). In 2018, the already elected president, in an interview conducted by CBS journalist Lesley Stahl, says the following regarding Hurricane Michael: "They say that we had hurricanes that were far worse than what we just had with Michael," Trump said. "Who says that? 'They say?'" Stahl pushed back, "Who says that? "They say". In response, Trump simply repeated his mantra: "People say, people say that…" (Rosenblum and Muirhead 2019, p. 108).

Summing up this point, we can say that the world today faces, let us admit it has always been so, imminent challenges—for example, climate change and the Covid-19 pandemic-, but also an unfortunately contemporary trend that is endangering the existence of a common language among humans and thus undermining the extraordinary democratic experiment. We refer specifically to the disdain for knowledge and experts. Now, for better or for worse, but certainly more for better (see the recent example of the Covid-19 vaccine), evidence, knowledge and reasoned and critical reflection are our decisive values and guarantee our individual and collective existence.

The relationship between nature and culture changed profoundly from the second half of the twentieth century onwards. Authors such as Edgar Morin warned us of the enormous risk of humans living apart from nature, a reality well illustrated by what he called the death of the notion of "insular man" (Morin, 1973). Today, in the age of the Anthropocene, there is a call to understand, as humans, that the entry into the political agenda of non-humans should give rise to a metamorphosis of mind. Therefore, it remains capitalism that must be subjected to scrutiny and discussion through the lens of the "Capitalocene" (Moore, 2017).

(iv) **Holistic (Transdisciplinarity) approach to health and well-being**

The decade of 1970 was a time of critique of the standard education and research model financed by the governments and the universities, and transdisciplinarity emerged with ethical concerns related to the linear applications of research to society. The problems were much too complex to be solved by the simple translation of lab science into the real world. 20 years later, there were new, highly complex global problems to be addressed, namely CC and urgent issues of environmental sustainability, biodiversity loss, pollution and poverty. The relations between science and technology, social policies, education and the role of the humanities and the arts were being equated in new forms, and transdisciplinarity appeared to be an interesting new way of producing theoretical, as well as practical knowledge.

A transdisciplinary approach to CC, health and well-being means that teaching, learning or researching take into account what is, simultaneously, inside, between, across and beyond the different disciplines that address CC, health and well-being. Transdisciplinarity is also concerned with the different partners involved in research and action, considering them as 'subjects' and not as 'objects'. In other words, what is sought is the integration of the different knowledges and viewpoints to achieve a deeper, and also larger understanding of the situations, as well as the possibility of a relevant action. Its goal is holistic in the sense that it tries to understand phenomena in a more comprehensive, synthetic framework, one of its imperatives being the unity of knowledge and the search for meaning.

There are two main trends in transdisciplinary tradition (Bernstein 2015). The first one is shaped by the *Charter of Transdisciplinarity* (1994a), where the views of B. Nicolescu, a Romanian physicist, are presented: transdisciplinarity recognizes the fundamental complexity of different levels and dimensions of reality and accepts the 'included middle', which means abandoning the excluded middle of Aristotelian logic that governs western thought for centuries (e.g. separate to understand). The second trend is known as the Mode 2 knowledge production (Gibbons et al. 1994b): transdisciplinarity means to address a concrete situation with several intervenients, from scientific academia to industry and local governments, to local institutions like NGOs, museums, architecture, humanities and the arts. Mode 2 is descriptive, analytic and applied, while Nicolescu's transdisciplinary approach is philosophical, metaphysical and even mystical. Mode 2 approach has recognizable affinities with post-normal science, Nicolescu's approach emphasizes the human life world and lived meanings in the phenomenological tradition. Both approaches are complementary to each other, the first one addressing real-world issues, the second developing the theoretical implications of transdisciplinary knowledge.

Transdisciplinarity is an interesting epistemological framework when it comes to explore or teach CC hazards and their impacts on human and non-human health. For example, in education, transdisciplinary projects in medical or psychological curricula have the potential to show more clearly the tacit values of the scientific disciplines: molecular and causal determination in systemic social situations, analytical materialism of CC impacts on disease, categorization and diagnostic universality, etc. These are intrinsic scientific values that can be rethought in the light of other values, like the humanities' values. Vignettes of visual arts or literature that address scientific content applied to difficult CC and health issues should allow students to deepen their understanding of the concrete effects of their practices.

Transversality in CC and health can also be addressed from the viewpoint of environmental humanities (EH). Born in the last years of the twentieth century, EH emphasized the need for a conceptually sensitive approach to the environment that enriches its vocabulary to include its different social, economical and ecological dimensions (Rose et al. 2012). The Anthropocene is the geological era where humans and nature are intimately intertwined. To understand hybrid phenomena like CC and its impact on health, it doesn't suffice to understand the direct effects isolated in lab or Big Data statistics; the findings from the humanities and the social sciences are important because they speak about cultural representations of health and climate, they reconstruct the ways CC came into being and how history and economics shaped the present situation, how norms, tradition and ethics guide our relationships to health and climate (Schmidt et al. 2020). The transversality of the humanities in environment and health allows us to rethink what it means to be human and how humans, but also non-humans, should be treated. As the concept of "One Health" (Brymer et al. 2019) portrays, there is growing evidence that health and well-being also concern a multitude of disciplines that deal with social, natural and urban ecosystems, as well as with plants and animals. The complexity of these interactions is easily understood when we tell stories that display them.

(v) **Environment, health, sustainability and equity**

Equity is a core issue in CC impacts on health. When not present, it is defined by Whitehead (1992) as an unfair and uneven distribution of healthcare; it holds a strong association with individuals' socioeconomic positions and societies' environmental sustainability (Vidal et al. 2018, 2019; Oliveira et al. 2019).

Anthropogenic and environmental factors are the main causes of CC, impacting human and ecosystem health disproportionately, resulting in situations of environmental injustice (Schlosberg 2007), environmental racism (Salas 2021), and climate injustice (Comim 2008). In fact, history has shown that vulnerable and marginalized social groups have been continuously exposed to low air quality, lack of access to green spaces, and consequently experience environmental poverty (Roberts et al. 2022). Therefore, altogether with CC impacts, which aggravate their conditions, these social groups have also the poorest health outcomes, generating a cycle hard to be broken.

Climate projections showed that extreme weather events caused by CC, especially extreme heat and cold waves, are expected to increase in the next years (Johnson et al. 2018). In this sense, there is also expected that these waves may negatively impact individuals' health conditions. However, vulnerable social groups—those with respiratory or circulatory health conditions, children, and the elderly—are likely to be more exposed to these impacts. This situation worsens when individuals experience energy poverty (Horta et al. 2019; Peduzzi et al. 2020), which is related to the incapacity to keep houses cold during the summer and warm during the winter. CC adaption and mitigation cannot be dissociated with health and environment equity, since policies must ensure the diversity of socio-cultural contexts, drivers, and barriers, which can undermine the sustainability efforts if ignored.

The four clusters, when combined, may help to reinforce the role of humanities and social sciences to deal with CC health hazards. Figure 1.1 presents the main interlinked outputs that arise from the project "Composing Worlds" and which can be applied to CC.

Due to the complexity and multidimensionality of CC, there is a need to co-create an extended and inclusive peer community that goes far beyond the merely human dimension. The impacts of CC are transversal to all beings so their rights must be respected and considered in the design of the actions to mitigate and adapt. In fact, as Beck (2016) stated, communities all over the world, humans and non-humans, are together facing the unpredictable impacts of climate phenomena. This is somewhat related to the "detraditionalization of life forms" which implies a shift of the dominant paradigm that puts society on the side of culture and non-humans on the

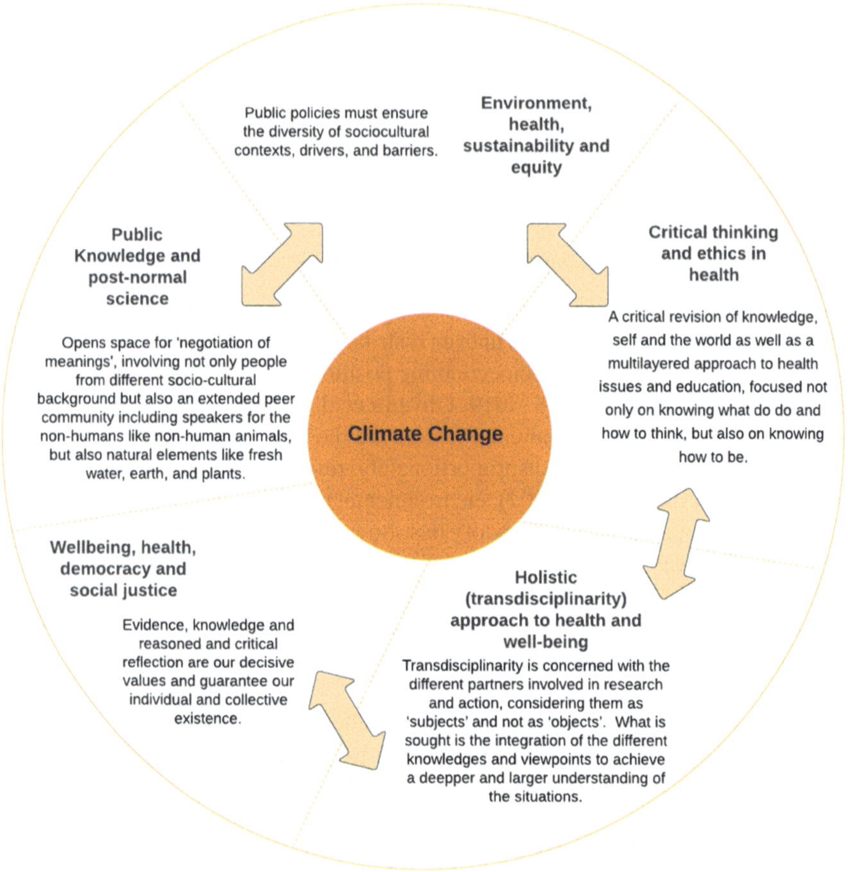

Fig. 1.1 Clusters emerging from the project "Composing Worlds" that may help do deal with Climate Change health hazards from humanities and social sciences viewpoint

side of nature. A new way of being part of the world, a common ecosystem, where the plurality of knowledges and beings are considered, may pave the way to deal with CC health hazards. The strangeness of the world that Latour (2016) describes implies critical thinking in health beyond the biomedical paradigm where individuals are at the centre. Also, democratic values, anchored in social justice and wellbeing for all, at all ages, are being challenged and thus a new narrative to humanity that places the possibilities of multiverses is much needed. This vulnerable planet, where environment, health, and sustainability are trying to thrive, is being undermined by the loss of the meaning of equity. Humanities and social sciences may pave the way to reinterpret the meanings of core concepts and thinking about our place on the planet and, more importantly, about humans' relations with nature.

Final Remarks

Technoscience proved not to be enough to deal with the complexity of global changes that emerged from CC. Therefore, this chapter is an attempt to propose some theoretical rethinking of CC health hazards through the lens of social sciences and humanities. Framed by a transdisciplinary network "Composing Worlds", it deeply explores the powerful connection between social values, democracy, health, well-being and sustainability, considering the need to move towards the integration of multiple voices and knowledge, from human and non-human, to deal with CC. Anchored in the concept of "Metamorphosis of the World", developed by Beck (2016), we argue that social sciences and humanities may not only theoretically but also practically deal with the health hazards that result from CC through the desconstruction of climate narratives and the complex socioecological phenomenon that humanity and others beings will face in a closer future. The clusters here identified should now be further explored empirically to disclose their full potential.

Acknowledgements This work was developed in the scope of the project "Composing worlds: humanities, well-being and health in the 21st century" (Compor mundos: humanidades, bem-estar e saúde no século XXI), supported by Fundação Ensino e Cultura Fernando Pessoa (FFP).

References

Aldeia J, Alves F (2019) Against the environment. Problems in society/nature relations. Front Sociol 4. https://doi.org/10.3389/fsoc.2019.00029

Beck U (2016) The metamorphosis of the world how climate change is transforming our concept of the world. Polity Press, Cambridge

Bernstein J (2015) Transdisciplinarity: a review of its origins, development, and current issues. J Res Pract 11:1–20

Bragança LS, Lopes MSB, Grossi G de B et al (2021) Natureza política e sustentabilidade: limites e horizontes [Political nature and sustainability: limits and horizons]. In: Lopes MSB, Bragança L (eds) Natureza política: Rupturas, aproximações e figurações possíveis. Agência de Iniciativas Cidadãs, Belo Horizonte, pp 29–51

Braun V, Clarke V (2006) Using thematic analysis in psychology. Qual Res Psychol 3:77–101. https://doi.org/10.1191/1478088706qp063oa

Brymer E, Freeman E, Richardson M (2019) Editorial: one health: the well-being impacts of human-nature relationships. Front Psychol 10. https://doi.org/10.3389/fpsyg.2019.01611

Caridade SMM, Vidal DG, Dinis MAP (2022) Climate change and gender-based violence: outcomes, challenges and future perspectives. In: Leal Filho W, Vidal DG, Dinis MAP, Dias RC (eds) Sustainable policies and practices in energy, environment and health research. Springer, Cham, pp 167–176

Centers for Disease Control and Prevention (2022) One health basics. https://www.cdc.gov/onehealth/basics/index.html. Accessed 7 Jun 2022

Comim F (2008) Climate injustice and development: a capability perspective. Development 51:344–349. https://doi.org/10.1057/dev.2008.36

Costello A, Abbas M, Allen A et al (2009) Managing the health effects of climate change. Lancet and University College London Institute for Global Health Commission. Lancet 373:1693–1733. https://doi.org/10.1016/S0140-6736(09)60935-1

Descola P (2005) Par-delà nature et culture [Beyond nature and culture]. Gallimard, Paris

Dias NMOC, Vidal DG, Sousa HFPE et al (2020) Exploring associations between attitudes towards climate change and motivational human values. Climate 8:1–21. https://doi.org/10.3390/cli8110135

Funtowicz SO, Ravetz JR (1994) Uncertainty, complexity and post-normal science. Environ Toxicol Chem 13:1881–1885. https://doi.org/10.1002/etc.5620131203

Gibbons M, Limoges C, Nowotny, H., Schwartzman S et al (1994a) The charter of transdisciplinarity. Encycl Relig Sci. https://inters.org/Freitas-Morin-Nicolescu-Transdisciplinarity. Accessed 8 June 2022

Gibbons M, Limoges C, Nowotny H et al (1994b) The new production of knowledge: The dynamics of science and research in contemporary societies. Sage, London

Han B-C (2021) Undinge: Umbrüche der Lebenswelt. Ullstein Verlag, Berlin

Hickman C, Marks E, Pihkala P et al (2021) Climate anxiety in children and young people and their beliefs about government responses to climate change: a global survey. Lancet Planet Heal 5:e863–e873. https://doi.org/10.1016/S2542-5196(21)00278-3

Horta A, Gouveia JP, Schmidt L et al (2019) Energy poverty in Portugal: Combining vulnerability mapping with household interviews. Energy Build 203. https://doi.org/10.1016/j.enbuild.2019.109423

Islam N, Winkel J (2017) Climate change and social inequalities. United Nations, New York

Johnson NC, Xie SP, Kosaka Y, Li X (2018) Increasing occurrence of cold and warm extremes during the recent global warming slowdown. Nat Commun 9. https://doi.org/10.1038/s41467-018-04040-y

Kuhn T (1962) The structure of scientific revolutions. The University of Chicago Press, Chicago

Latour B (2016) Reset modernity! MIT Press, Cambridge, MA

Latour B (2022) A cosmopolítica como pluralismo ontológico [Cosmopolitics as ontological pluralism]. Trab Antropol e Etnol 62:17–34

Latour B (2017a) Facing Gaia: eight lectures on the new climatic regime. Polity Press, Cambridge, MA

Latour B (2017b) Où atterrir ? Comment s'orienter en politique [Or land? Comment s'orienter in politics]. Ed. La Découverte, Paris

Latour B (1993) We have never been modern. Harvard University Press, Cambridge

Latour B, Muecke S (2021) Protective measures an exercise. Cult Polit 17:11–16. https://doi.org/10.1215/17432197-8797459

Lencastre MPA, Farinha Marques P (2021) Da Biofilia à Ecoterapia. A Importância dos Parques Urbanos para a Saúde Mental [From Biophilia to Ecotherapy. The Importance of Urban Parks for Mental Health]. Trab Antropol e Etnol 61:131–155

Lencastre MPA, Leal RM (2006) Narratividade educativa e literacias ambientais: O projecto TERRA [Educational narrative and environmental literacies: The TERRA project]. In: Many E, Guimarães S (eds) Como Abordar a Metodologia de Trabalho de Projecto. Areal Editores, Porto, pp 100–123

Lencastre MPA, Vidal DG, Estrada R et al (2022) The biophilia hypothesis explored: regenerative urban green spaces and well-being in a Portuguese sample. Int J Environ Stud 1–15. https://doi.org/10.1080/00207233.2022.2067411

Martin JL, Maris V, Simberloff DS (2016) The need to respect nature and its limits challenges society and conservation science. Proc Natl Acad Sci USA 113:6105–6112. https://doi.org/10.1073/pnas.1525003113

Morin, E (1973) Le paradigme perdu. La nature humaine [The Lost Paradigm. Human nature], Seuil, Paris

Moore JW (2017) The capitalocene Part I: On the nature & origins of our ecological crisis. J Peasant Stud 44(3):594–630. https://doi.org/10.1080/03066150.2016.1235036

Mourato J, Schmidt L, Ferrão J, Bussler A (2018) Reframing adaptation to climate change in Portugal: the case of ClimAdaPT.Local. In: Delicado A, Domingos N, Sousa L (eds) Changing societies: legacies and challenges. The diverse worlds of sustainability. Imprensa de Ciências Sociais, Lisboa, pp 153–177

Oliveira GM, Vidal DG, Ferraz MP et al (2019) Measuring health vulnerability: an interdisciplinary indicator applied to mainland Portugal. Int J Environ Res Public Health 16:1–18. https://doi.org/10.3390/ijerph16214121

Peduzzi E, Baldi MG, Pisoni E et al (2020) Impacts of a climate change initiative on air pollutant emissions: Insights from the Covenant of Mayors. Environ Int 145. https://doi.org/10.1016/j.envint.2020.106029

Pieri P, Lebon E, Brisson N (2012) Climate change impact on French vineyards as predicted by models. Acta Hortic 931:29–38. https://doi.org/10.17660/actahortic.2012.931.2

Roberts JD, Dickinson KL, Hendricks MD, Jennings V (2022) "I Can't Breathe": examining the legacy of American racism on determinants of health and the ongoing pursuit of environmental justice. Curr Environ Heal Reports. https://doi.org/10.1007/s40572-022-00343-x

Rose DB, van Dooren T, Chrulew M et al (2012) Thinking through the environment, unsettling the humanities. Environ Humanit 1:1–5. https://doi.org/10.1215/22011919-3609940

Rosenblum NL, Muirhead R (2019) A lot of people are saying: The new conspiracism and the assault on democracy. Princeton University Press, Princeton

Salas RN (2021) Environmental racism and climate change—missed diagnoses. N Engl J Med 385:967–969. https://doi.org/10.1056/nejmp2109160

Santos FD (2012) Alterações Globais: os Desafios e os Riscos Presentes e Futuros. Fundação Francisco Manuel dos Santos, Lisboa

Santos FD (2017) Pós-verdade e ciência nos EUA com o Sr. Donald Trump [Post-truth and science in the USA with Mr. Donald Trump]. Público

Schlosberg D (2007) Defining environmental justice: theories, movements, and nature. Oxford University Press, Oxford

Schmidt M, Soentgen J, Zapf H (2020) Environmental humanities: an emerging field of transdisciplinary research. GAIA Ecol Perspect Sci Soc 29:225–229. https://doi.org/10.14512/gaia.29.4.6

Skoufias E (2012) The poverty and welfare impacts of climate change. World Bank, Washington, D.C.

Smith WN, Grant BB, Desjardins RL et al (2013) Assessing the effects of climate change on crop production and GHG emissions in Canada. Agric Ecosyst Environ 179:139–150. https://doi.org/10.1016/j.agee.2013.08.015

Steffen W, Persson A, Deutsch L et al (2011) The anthropocene: from global change to planetary stewardship. Ambio 40:739–761. https://doi.org/10.1007/s13280-011-0185-x

Stenghers I (2005) The Cosmopolitical Proposal. In: Latour B, Weibel P (eds) Making things public: Atmospheres of democracy. MIT Press, Cambridge, pp 994–1003

Tyreman S (2000) Promoting critical thinking in health care: phronesis and criticality. Med Health Care Philos 3:117–124. https://doi.org/10.1023/A:1009973021449

Vidal DG, Maia RL, Oliveira GM et al (2019) Cities challenges in the contemporary societies: urban sustainability and environmental issues. Sociol LINE, Rev da Assoc Port Sociol 20:119–138. https://doi.org/10.30553/sociologiaonline.2019.20.6

Vidal DG, Pontes M, Barreira E et al (2018) Differential Mortality and Inequalities in Health Services Access in Mainland Portugal. Finisterra - Rev Port Geogr 53:53–70. https://doi.org/10.18055/Finis14118

Wahl D (2016) Designing regenerative cultures. Triarchy Press, Dorset

Weil P, Leloup JY, Crema R (2017) Normose: a patologia da normalidade [Normosis: the pathology of normality]. Vozes, Rio de Janeiro

Whitehead M (1992) The concepts and principles of equity and health. Int J Heal Serv 22:429–445. https://doi.org/10.2190/986L-LHQ6-2VTE-YRRN

Diogo Guedes Vidal is an Environment and Nature Sociologist interested in all its interconnections with and health of human and non-human beings. Ph.D. in Ecology and Environmental Health with a project focused on the attempt to look beyond the greenness of public urban green spaces funded by the main Portuguese public body agency—FCT. Currently is researcher in the Project PHOENIX "The rise of citizens' voices for a greener Europe" (Horizon 2020; Grant agreement ID: 101037328) and Invited Assistant Researcher at Centre for Functional Ecology—Science for People & the Planet" (CFE) in the Department of Life Sciences of the University of Coimbra, Portugal.

Marina Prieto Afonso Lencastre is Retired Full Professor (Jubilated) at the University of Porto and Full Professor at the University of Fernando Pessoa. She is a Specialist Psychotherapist and Supervisor of the Portuguese Society of Clinical Psychology, holding a Specialty in Clinical and Health Psychology and also in Advanced Specialty in Psychotherapy by the Portuguese Order of Psychologists. Additionally, she is an Accredited Trainer in Environmental Education.

Susana Magalhães is Ph.D. in Bioethics by the Portuguese Catholic University, is Lecturer at University Fernando Pessoa and the Head of the Unit for Responsible Conduct in Research at the Institute for Research and Innovation in Health of the University of Porto (i3S). Her research interests are Narrative Medicine and Research Ethics/Integrity. She is a Certified Virt2UETrainer and member of the steering group of the Portuguese Reproducibility Network. Coordinator of a Group of Study and Reflection on Narrative Medicine (Grupo de Estudos e Reflexão em Medicina Narrativa—GERMEN, https://www.facebook.com/GermenMedicinaNarrativa/) and member of the European Society of Narrative Medicine (EUNAMES).

Eduardo Paz Barroso is Full Professor of Communication Sciences in the Faculty of Human and Social Sciences of the University Fernando Pessoa. He is Researcher at LabCom, a research unit in communication and arts at the University of Beira Interior. Among being a lecturer, cultural manager, and programmer, he has been appointed to several public charges in Portugal in the cultural and artistic sector. Author of about a hundred works, such as books, book chapters, and articles in national and international journals.

Álvaro Campelo is Ph.D. in Anthropology and Associate Professor at the Fernando Pessoa University. He conducts research in the area of cultural heritage, socio-anthropology of health, cultural landscape and uses of space. He is also a researcher at CRIA.

Pedro Cunha is Psychologist and Sociologist. Full Professor at the Fernando Pessoa University (Portugal) and guest lecturer at the Porto Business School at the University of Porto. He has Certificate of Conflict Mediator and Family Mediator. For 25 years, his research interests have been primarily focused on conflict management, negotiation and mediation in different areas (namely family, schools, health, and organizations). He has published several books in the area and in Portuguese and foreign specialty magazines.

Rui Estrada holds a Ph.D. in Theory of Literature by the Faculdade de Letras of the University of Lisboa. Is Full Professor at Universidade Fernando Pessoa and Researcher at CITCEM (Faculdade de Letras, University of Porto). He won the Pen Club Award Essay in 2002 and published four books, edited six and is author of several papers and book chapters (national and international).

Chapter 2
Climate Change and Health: Essay on the Limits of Naturalism

Marina Prieto Afonso Lencastre, Diogo Guedes Vidal, Eduardo Paz Barroso, Álvaro Campelo, Teresa Martinho Toldy, Susana Magalhães, Ana Gabriela Frazão-Nogueira, and Rui Estrada

Abstract One of the main problems of the twenty-first century is climate change and its impacts on global health. Nonetheless, there is a certain optimism that is rooted in the new world opened up by the possibilities of technoscience, namely in the control of health and climatic and environmental hazards. But on the horizon of these possibilities, there are increasingly clear signs that technological and scientific naturalism alone is running out of capacity to contain its own challenges. Naturalism has problems in resolving climate change impacts on health, founding ethics and epistemology, and defining relations with non-humans. In an era that is largely dependent on the actions that result from our thinking, there is a call to imagine multiverses or different versions of what naturalism seems to present as a single reality. This essay aims to explore the limits of naturalism to address climate change health hazards.

M. P. A. Lencastre · E. P. Barroso · Á. Campelo · T. M. Toldy · S. Magalhães · A. G. Frazão-Nogueira · R. Estrada
FP-I3ID, Faculty of Human and Social Sciences, University Fernando Pessoa (UFP), Praça Nove de Abril, 349, 4249-004 Porto, Portugal
e-mail: mlencast@ufp.edu.pt

E. P. Barroso
e-mail: epb@ufp.edu.pt

Á. Campelo
e-mail: campelo@ufp.edu.pt

T. M. Toldy
e-mail: toldy@ufp.edu.pt

S. Magalhães
e-mail: susanat@ufp.edu.pt

W. Leal Filho et al. (eds.), *Climate Change and Health Hazards*, Climate Change Management, https://doi.org/10.1007/978-3-031-26592-1_2

Introduction

Since the end of the last century, but especially in the first decades of this century, we have been witnessing vast changes in social life, in human and non-human environments—by non-humans we mean all organisms, their habitats and ecosystems, shared or not with humans, that live in the Anthropocene era—climate change (CC), and health systems, but also in concepts and theories that are slowly beginning to emerge. Threats such as global warming and the massive extinction of species, human population growth, poverty in specific regions of the world, an increase in pandemics resulting from the abuse of natural ecosystems, and rapid processes of globalization, are parallel, in the social area, with an exponential availability of resources and information, with intelligent technologies that allow us to process an immense amount of data, predict natural and anthropic events, and carry out biological and physical analyzes on previously unimaginable time scales. We have effective means of

A. G. Frazão-Nogueira
e-mail: ana@ufp.edu.pt

R. Estrada
e-mail: restrada@ufp.edu.pt

M. P. A. Lencastre · Á. Campelo
Fernando Pessoa Behaviour and Social Sciences Research Centre (FP-B2S), University Fernando Pessoa (UFP), Praça Nove de Abril, 349, 4249-004 Porto, Portugal

D. G. Vidal (✉)
Centre for Functional Ecology—Science for People & the Planet (CFE), TERRA Associate Laboratory, Department of Life Sciences, University of Coimbra, Calçada Martim de Freitas, 3000-456 Coimbra, Portugal

E. P. Barroso
LabCom–Communication & Arts Research Unit, University of Beira Interior, Rua Marquês D'Ávila E Bolama, 6201-001 Covilhã, Portugal

Á. Campelo
Centre for Research in Anthropology (CRIA), Edifício ISCTE, Sala 2w2, 1649-026 Lisbon, Portugal

T. M. Toldy
Centre for Social Studies (CES), University of Coimbra, Colégio de S. Jerónimo, Apartado 3087, 3000-995 Coimbra, Portugal

S. Magalhães
Instituto de Investigação e Inovação em Saúde (i3S), Universidade Do Porto, 4200-135 Porto, Portugal

A. G. Frazão-Nogueira
ICNOVA—Instituto de Comunicação da NOVA, Faculdade de Ciências Sociais E Humanas, Universidade NOVA de Lisboa, Lisbon, Portugal

R. Estrada
Center for Transdisciplinary Research «Culture, Space and Memory» (CITCEM), Faculty of Arts and Humanities of the University of Porto (FLUP), Via Panorâmica Edgar Cardoso S/N, 4150-564 Porto, Portugal

prevention and treatment of health hazards; we are developing ever more rapid forms of globalization and communication. In general, in the Anthropocene era (Steffen et al. 2011), the conditions for human life expectancy have largely improved (Pinker 2018) and, globally, we have never done so well.

The improvement of human life conditions is due to, to a large extent, the possibilities of technoscience and the ways in which it interferes with social and environmental situations. Recent changes in social discourse and recent conceptual optimism—as is the case for transhumanism (Bostrom 2005)—are rooted in the new world opened by these technological possibilities. In fact, more than ever, technologies are present in all social, political, and environmental settings: without leaving home, we can organize far distant travels, buy objects from the other side of the planet, consume conferences and professional courses, take care of health and well-being. We can change our appearance and embody various digital *personae*, we can interact with intelligent systems as if they were people. We can influence the electoral systems of our own, and other countries, we can manipulate information into fake news, and we can reduce our social life to digital networks. Environmental technologies allow us to try to stop the melting of glaciers by geoengineering, change rainfall patterns by cloudseeding (Farahat and Abuelgasim 2022), capture CO_2 and develop renewable energies, monitor migratory macrofauna and control microfauna, turn plastic waste into reusable clothing and build islands out of trash. In the Anthropocene, human industriousness seems inexhaustible, and naturalistic technoscientific ingenuity seems endless.

Science, and its applied social counterpart, which is technology, appear inextricably associated and ready to answer the most unlikely human demands. But this technological optimism is not yet accompanied by a true reflection on the implications and interferences of technologies in sensitive areas such as the human and non-human environments, CC, and its effects on social life and health. Nowadays, there is a predominance of quick consumption and applied ethics, the one that deliberates on the uses of objects and techniques, and there is an absence of fundamental ethics, one that thinks about important questions such as what it means to have a good life in a finite planet, the practical qualities of human and non-human existences, the transforming effects of technoscience, its irreversible impacts on natural and artificial environments, CC, and the general orientation it gives to the processes of humanization, in diverse social and cultural contexts.

It seems that the dominant mentality prefers things that work rather than what qualifies as a good life (Barroso and Estrada 2018). Old questions such as 'truth', 'beauty', and 'goodness' have given way to a diversity of utopian or dystopian projections that speak more about the personal futures of those who project, than about true visions of real societies dwelling within concrete territories (Latour 2017). As Nussbaum (2019) states: "(…) nations around the world will soon produce generations of useful machines, instead of complete citizens who can think for themselves, criticize tradition and understand the meaning of sufferings and achievements of others. The future of world democracy remains uncertain" (2019, p. 38). A little further back in her book, Nussbaum details this idea of the useful machine, quoting the republican Marco Rubio: "we need more welders and fewer philosophers" (2019, p. 22).

In many ways, we are living in a new, fast-paced world (Virilio 1998), full of new artifacts and new possibilities, but for which we still seem to lack the words (Latour 2016) and, to a large extent, the ideas. We live in a confused, complex, interdependent world, a world in metamorphosis, as stated by Beck (2016), which is in need of a clear project of humanity inhabiting, with non-humans, a vulnerable planet and mortal existences. This dominant mentality, which is not aware of the precariousness and vulnerability of the present, is the same that remains indifferent to the risks of the unregulated use of instruments and industries that technology provides.

Based on these assumptions, this essay aims to explore the limits of naturalism to address climate change health hazards. The essay starts by discussing the interconnections between naturalism, CC, and health and ends with a reflection on the limits of naturalism before drawing some final remarks on the topic discussed.

Naturalism, Climate Change, and Health

Naturalism and Climate Change

The Intergovernmental Panel on Climate Change (2018) states that if an immediate and important decrease in greenhouse emissions does not take place, the goal of a 1.5 °C global temperature rise will be overcome, and result in increasing heat waves and long warmer, and short colder seasons. This CC will have critical effects on agriculture and health.

There is a series of diseases related to CC. A higher risk of waterborne diseases like malaria or dengue is occurring, due to high temperatures and the intensifying of rainfall and associated flooding. Inversely, intense drought, woodfire and animal diseases and death take place in other regions of the planet. Drinking fresh water is being affected, both in quality and quantity. Major cities are already witnessing rising temperatures and small coastal floodings, as well as increasing air pollution due to CO_2 emissions. It is expected that cardiovascular, lung, cancer, and immune diseases will grow, as well as mental health problems. As CC becomes more extreme, eco-anxiety and stress are increasing in children and the elderly. Loss of biodiversity and of natural ecosystems, alongside with the exploration of wild species, intensify disease spreading between animals and humans, as confirmed by the COVID-19 pandemic. Other problems like microplastics have contaminated our food chains, and eventually, they have entered our bloodstreams. Waste recycling is not as effective as thought, but consumption continues to grow, expanding polluting waste disposals, particularly in the poorer countries. Human environmental migration due to CC is occurring, mainly from the southern to the northern countries, putting even more pressure on their already crowded health systems. These find themselves in a paradoxical situation: being dedicated to mitigate CC diseases, they too massively contribute to the CC emissions (The Royal Society of Medicine 2021). So, there is a structural problem to be addressed, and policies and governments, both global and local, are struggling with solutions, within a general naturalistic paradigm.

Naturalism is a concept introduced by French anthropologist P. Descola (2005). It generally states that, in Western culture, mainly Europe, from the seventeenth century onwards, body and mind were treated as separate realities, the body belonging to the realm of the physical things (like other animals, plants, and physical matter), and the spirit belonging to the realm of subjectivity. For naturalism, differently from other cultural traditions, physical things can be studied and manipulated by science and reason, subjectivity cannot because it is where thought and reason belong. So, the naturalist paradigm relies on science and technology to deal with objective reality. It states that thought is independent of material reality, and that culture and nature are two different spheres. Thus, what results from human subjectivity such as language, the arts, and creativity, cannot be the object of the naturalist approach. But the body, its anatomy, physiology, or the mechanics of movement, they can be studied objectively, insofar as they resemble what happens with other animals. The material and objective properties of the body are attributed to non-humans as well, for humans and non-humans share a similar physicality, and a different subjectivity. It is naturalism that invents the engineering of the body and of nature and, more recently, also of the mind. In fact, the advances in biological and psychological sciences in the second half of the twentieth century promoted the expansion of the physicalist conception of the bodies to the study of subjectivity and the spirit. The development of neurosciences and, more recently, microscopic technoscience linked to brain health (genomics, nanotechnologies applied to the brain, etc.) are the result of a naturalist approach to subjectivity.

Naturalism is thus pervasive in Western culture, and we can see clearly how it acts in the areas of health and CC. Diagnostic machines, tests, smartphone applications, and others, translate physiological, psychological and environmental entities into objects that are externalized in relation to people. Health applications can even propose partial interventions and solitary recovery from bodily or mental disorders. The consequence is that non-organic interlocutors of the body and mind, such as the various systems of objectification with which we fill our daily lives and, to an increasing extent, our subjectivity, also mechanize the human condition itself. The extreme naturalist metaphor that conceives the body/mind as a machine appears explored in futuristic philosophies such as transhumanism, and by ideas about the post-human, which collapse the symbolic, and the sacred itself, to the banalest register of the concrete, which is its physicality (Delio 2012).

Solutions to CC offered by naturalism imply the usage of sciences and technologies in different settings. One example is curbing greenhouse gas emissions by carbon removal directly from the atmosphere, with an engineered system, that compresses and stores CO_2 underground (1POINTFIVE 2022). Climate-smart agriculture using genetic technologies for food production innovation, household energy efficiency and clean energy technologies, are some of the solutions proposed by engineers and other scientists. Climate repair is another area of research that should diminish, in the future, the damages of pollution, change local rainfall patterns and control elevating temperatures. But there are important problems with technological solutions. The first one regards the continuation of a 'do and waste' logic that is based on the availability of natural resources for the construction of ever more sophisticated apparatuses, and an exponential increase in dangerous waste. The second major problem is the accentuation of a culture of innovation that doesn't value what is already there and should be requalified and reused. Another major problem is the accentuation of cultural inequalities, because traditional solutions are not sufficiently valued and technology, and waste, are exported and followed by external experts. Research and experience have shown that creating dependency from the outside is a quick way for local community culture to fall apart. One significant consequence of technology applied to societies is citizen passivity in face of problems like CC, because the implicit message is that technology will solve it all. The same is true for health. A compliant and, at the same time, aggressive and demanding democracy results from this unsustainable culture. This, and other related behaviors, capture us in unending paradoxical situations that should be carefully analysed to be overcome. A major shift would be the reintegration of humans into nature, and the recognition of nature/culture as a functional whole. Another, deeper shift, would be to overcome the typical utilitarian relations that humans have with non-humans. The empathic recognition of animal and plant sensitivities, as well as the value of their particular behaviors, could inspire new ways of relating to them, to their habitats and, in turn, to their relations to us. This would not be a sentimentalist shift, but rather a cultural and symbolic one, weaving new respectful meanings and practices. Citing Descola's work with the Achuar, a tribe of the Upper Amazon, Stuart-Smith (2020) describes how gardening women quietly sing to the plants they are tending, because they believe the plants have something resembling a soul, and that Nukui, the spirit mother of all plants dwells under the soil. Chanting to Nukui means calling nourishment into existence and assuring its continuing propagation. This is a 'horticultural mothering' (Descola 2005) that also works for the women, because they find a mothering to themselves, within the magic garden. We, in the West, need to reenchant the world with meaning and treat it kindly. For that purpose, we need to feel that we are part, and agents of concrete places, and that our actions affect finely tuned wholes that exist for a very long time.

Naturalism is closely related to Western history and culture and has anthropocentrism at its core. This is visible in the economy and legal systems that consider humans at its center, and nature as an external resource; it is also tangible in religion, the arts and monuments and, generally, in the representations of nature and culture (Schmidt et al. 2020). Some authors propose that one of the ways to mitigate CC would be by framing it mainly as a human health issue, alongside a human environmental issue (Rossa-Roccor et al. 2021). Being framed as a human health issue, CC becomes personal and might put more pressure on policymakers. But this strategy implies strong ecosystem and cultural limitations, by not considering the very same structural conditions that started CC in the first place. As stated by Beck (2016), we witnessed a metamorphosis of the world caused by CC; we need a metamorphosis of the mind, that challenges our ideas and cultural practices themselves.

Naturalism and Health

Until the beginning of this century, the limits of the human circumstance were guided by fragility and death, but today it is believed that science will solve these questions, turning humans into a-corporeal and a-mortal. With technoscience constantly updating, humans have taken the places of the gods. Dying is no longer inevitable, and it became a technical problem (Harari 2018). Biological reproduction disaggregates genetic affiliation from physiological affiliation, psychological affiliation, and social affiliation, redefining new parenting and family relationships, which are in some cases merely virtual. In the clinical setting, the traditional position between the caregiver and the person asking to be cared for has been reversed by the growing personal responsibility in face of biomedicine and biopsychology. Traditional diagnostic and treatment processes, the face-to-face relationship between the patient and her doctor or therapist, were transformed into procedural protocols, with the participation of the patient herself (Turnpenny et al. 2011). There is disproportionate inflation of technical processes to the detriment of a balanced perspective that recognizes us as wholly human. In Ancient Greece, for example, the learning and use of a *techne* implied a general framework of polymathy, where critical thinking and argumentative skills were essential (Pinto 2005). The passage from Gorgias (Platão 1973), who had a doctor brother, in Plato's dialogue of the same name, says: "Many times I accompanied my brother and other doctors to the homes of patients who did not want to take medicine or undergo treatment with iron or iron fire. Now, when the doctor proved incapable of persuading the patient, I did it, with no more resources than rhetoric" (456b). What Gorgias means, in this passage, by 'rhetorical' is precisely what is beyond the mere technical procedure: face-to-face, attention to vulnerability (Campelo 2020), in short, the treatment of the human person and not the organ.

José Cardoso Pires' book *De Profundis, Valsa Lenta*, written in 1997, includes a text from his friend and neurosurgeon João Lobo Antunes—"A letter to a 'new' friend"—that shows the relevance of a holistic approach to medicine. Toldy and Estrada (2017, p. 7) in their text about this book and this letter, ask themselves: "What is the point of a letter written by a neurosurgeon, written by João Lobo Antunes?" And they answer this question, saying: "We think that this letter makes sense precisely because it was written as a kind of a gloss of Cardoso Pires' sentence: 'Science is not a shrine of technologies' (Pires 1997, pp. 68–69). João Lobo Antunes recognizes the 'unique state of a wounded humanity' (…) as the essence of each and every disease" (Toldy and Estrada 2017, p. 7).

Health sciences are now characterized by the 4 p's—personalized, predictive, preventive, and participatory, they emphasize the technoscientific aspects of therapies and not the affiliative and larger symbolic human needs, which are essential for healing (Lencastre 2015). Today there are "tools" to solve a number of difficult problems, including suffering and death, and some of these "tools" for the mental or the physical are accessible in customized applications on smartphones, and other digital devices. The drama, but also the art of human encounters, are removed from the clinical setting. In a path of solitary authorship, well-being and health are increasingly individual choices that can go as far as the physical improvement of the mind—micro-implants to improve brain function—and the metonymic transformation of the body—by the hypertrophy of some part of the body that is socially valued.

For the World Health Organization, health is a state of complete physical, mental, and social well-being and not merely the absence of disease or the good functioning of organs. Well-being is defined as a state of harmony between the body and the mind in a given physical and social environment, and this concept is generically related to the evolution of ideas about what constitutes bodily and mental health. One of the most important aspects that are emphasized since the 1970s is the complexity and interdependence of the various physical, biological, social, and environmental factors of health. However, even if the WHO recognizes, through the concept of One Health (Brymer et al. 2019; Vidal et al. 2022), the integrated nature of these factors, medical, psychological, environmental and social practices appear mostly divided in Western societies.

We have biochemical and surgical interventions on the body, psychosocial and psychotherapeutic interventions of different types on the mind, CC interventions on ecosystems, and festive events celebrating the social. In the West, technoscientific modernity has not generated a theory and a practice embedded in an emotionally significant cosmology that makes these different interventions coherent and supported by a solid symbolic fabric, capable of motivating people's actions. This state of affairs implies that social activities, CC interventions, and therapies themselves are often perceived as fragmented and even intrusive, aimed at promoting some aspect of social life, capturing emissions and conserving one or more species in an ecosystem, healing an organ or a thought that is sick, but not considering the whole integrated reality.

All cultures have tacit or explicit conceptions about the relationships between environment, climate, and health, and all have developed concepts and practical ways to deal with them (Sussman 2004). Non-Western cultures generally maintained an integrated approach to the environment, to social and individual health, proposing holistic theories about the relationships between body and mind, spirit and nature (Kellert 1991; Riviere 1992; Watanabe 1974). In case of illness or natural disaster, they aim to restore a feeling of coherence and symbolic appropriation in relation with the whole (Gielen et al. 2004; Singer and Erickson 2011). These societies are generally organized into collectivist social systems, and the sacred and the profane are held together through complex systems of meaning by which humans and non-humans interact. Even populous countries, such as China or India, that are subject to rapid processes of globalization and economic, political, and scientific westernization, maintain an integrated view of people, nature, and the wider cosmos.

The same applied to the West until the seventeenth century, when cartesian mechanistic thinking, typical of our material action on physical things, imposed itself on the study of natural phenomena. From then on, and for almost three centuries, the study of physical phenomena and the study of human phenomena became independent, promoting what Ph. Descola (2005) designates scientific naturalism. The objective study of nature imposed a mandatory reductionist methodology that treated as insignificant the inter-relational aspects of humans, animals and plants, and the cultural and evolutionary processes that result from them.

The Limits of Naturalism

Naturalism corresponds to a way of thinking and disposing of reality that has punctuated Western culture during the last three centuries (Descola 2005). But today, considering the magnitude of the problems, namely CC, and some of the excesses of technoscience, this naturalist solidity seems to be volatilizing for many of us. It is with perplexity that we witness today the absence of secure semantic supports on which to compose human and non-human worlds. The absence of a strong version of what it means to be human, a clear understanding of the non-human, the limits of experimentation and of technological achievements, are questions of science, and society at large. The ideas we have don't seem to adapt to the new questions that arise in a technological and CC scenario. Ethical ignorance is often the result of this state of affairs and, at best, the precautionary principle works. Ancient wisdom pales in face of the complexity of contemporary issues that involve the individuals and the collectives. How can we think and promote our common future, which in many cases is already here? What ecological purposes should we value, what processes of humanization are we to set in motion, how can we choose without being overwhelmed by the force of scientific evidence, by the anonymity of large numbers, by the social imaginary of risk?

On the horizon of these questions, there are increasingly clear signs that naturalism is exhausting its capacity to contain contemporary problems. Regarding nature, civic movements linked to different environmental ethics (Callicott and Frodeman 2009), impose the recognition of the status of the subject, and the respective rights, to animals and also to natural places such as lakes and rivers (Viaene 2022), as was the case in New Zealand and Ecuador (Pecharroman 2018). Ethology has revised behavioral mechanicism to include the interiority and subjective life of animals (Bekoff 2015), and some claim the possibility of an internal and relational dynamic for plants as well (Goodall and Hudson 2015; Wohlleben 2017). Following the seminal work of J. Goodall on Gombe chimpanzees, de Waal (2009), an important primatologist who specialized in pygmy chimpanzees, showed the relevance of empathy, altruism, and cooperation in non-human social life, and in its evolutionary processes. Darwin's (1872) early observation on the importance of sympathy for animal social life is now being fully recognized. Pinker took up these ideas in his 2012 work on the good part of human nature, showing how they interfere with social institutions, education, and social transformation in order to converge towards the reduction of different types of violence, at a global level. The very concept of the Anthropocene (Steffen et al. 2011), by recognizing the human role in the gestation of a new geological era, seems to reintroduce the old relationships of each one with the whole, as in pre-modern analogical ontologies. Only, in this case, the relationships of the collectives with the whole are material and causal, instead of analogous, that is, based on humans and non-humans being related to each other by metaphor or metonymy. Perhaps, in our culture, the papal *Encyclical Letter Laudato Sí* (2015), with its proposal for integral ecology, is the document that best approaches a symbolic and spiritual revaluation of the whole. There, CC and the notions of individual well-being and health acquire a relational dimension with the well-being and health of others, human and non-human, in an invitation to 'cultivate and guard the world's garden'.

Naturalism also presents some problems in health conceptions. Contrary to the abstract idea of the universal body, subject to the deterministic laws typical of naturalist thinking, recent studies point to the importance of cultural aspects in the expression of physical and mental pathologies (Levin and Browner 2005; Lewis-Fernández and Kirmayer 2019; Sussman 2004). The diagnoses are culturally contextualized, as are the treatments. Another important aspect is the recognition of the therapeutic value of the placebo effect (Horin et al. 2014), which points to the unity of the body-mind-spirit, and questions its analytical fragmentation and partial interventions. The importance of human relationships in health may be one of the reasons for the growing interest in traditional therapies in the West (Gielen et al. 2004). These therapies consider the person, living within a personal, relational, cultural, and environmental context. Suffering is inscribed in chains of meaning that transcend individual responsibility and relate it to the realities that are shared by all. Ethics in health is another aspect that should be considered: although the ethical considerations associated with research and intervention are mandatory, they do not yet have the integrated characteristics that make them truly a system of general reasoning for practical decisions. Ethical decisions are often taken on a case-by-case basis, and not within a global vision of values for contemporary collectives. For individuals, ethical neutrality of institutions is often resolved with their own moral or spiritual values, in order to help with difficult choices, and to lend deeper meanings to lonely experiences of suffering (Eersel 2017).

Naturalism also seems to present difficulties in epistemology. In its recent versions, epistemology considers the virtues of those who know (Greco 2003). It is not enough to be objective, it is necessary to be honest in the presentation of data, humble in its scope and meaning, open to refutation, and capable of reviewing paradigms. The post-normal conception of science shows the importance of people's participation in science-based decisions because, in contemporary CC and health scenarios, the uncertainty of the facts is often high, there is a great diversity of values in dispute, the risks of decisions are big and action is urgent. To overcome these characteristics of contemporary science, Funtowicz and Ravetz (1993) propose that technoscientific deliberation be carried out within extended communities of peers that include specialists from other areas, such as the humanities, lay people affected by science decisions, representatives of local institutions, and representatives of non-humans (Latour 2017).

For Descola (2018), at a conference given at the Louis Vuitton Foundation, the contemporary West may be progressively abandoning naturalism, which intends to reduce the diversity to its simple elements and artificially recompose the whole from the parts, to become more interested in new forms of analogism, which maintains diversity and weaves relationships between the entities. This 'metastable analogy of differences' is recognizable in the author's analysis of contemporary visual arts and raises important questions for future forms of knowledge. These should be desirably non-anthropocentric and respectful of the myriad of differences existing in the real world, capable of creating an order that does not reduce this complexity, and truly addresses CC as a nature/culture phenomenon. De-anthropologizing naturalism and recognizing that we are governed by nature, and not the other way around, is one of the imperatives of today.

According to Descola (2005), analogism prevailed in the West until the seventeenth century. Nature, body, spirit, and social life were related. Nowadays, the departmentalization of reality may be, once again, losing out in favor of greater integration, facilitated by the systemic technologies, globalization, and its corollary, cooperation, which is related with the themes of systems thinking and interdependence that have been extensively addressed by E. Morin (2000). But, above all, by the search for knowledge of shared interdependence of multiple interiorities and physicalities.

We will have to discover the relationships that unite contemporary collectives in concrete territories (Latour 2017) and find the ways that sustain the ones that thrive. Discovering relationships implies to identify the entities that are related; naming allows description, highlights entities, and makes relational life much more interesting. How to compose this common world?

Imagination is a source of intuition and knowledge essential for this task. What do contemporary performing arts, visual arts, literature, philosophy, landscaping, environment and social sciences, psychology, and medicine tell us about the present, CC and health? What new entities, new relationships, and new collectives appear on the horizon of these disciplines? One of the characteristics of imagination is to think about reality from different points of view, describing multiverses, different versions of society, environment, CC and health. In a global world that is largely dependent on our actions, several descriptions on how CC and health may be approached, and not mainly by technology, may help. Here, other cultures' ecological traditions are important to this task, as well as knowing how other animals thrive in different natural and human ecosystems. Returning to Nussbaum: "These abilities are associated with the imagination and the arts: the ability to think critically; the ability to transcend local loyalties and approach world problems as a 'world citizen'; and, finally, the ability to compassionately imagine other people's [and non-human] problems" (Nussbaum 2019, p. 43).

Conclusions

The mitigation of CC and its impacts on health need more than technological solutions to pervade culture. Imagination, particularly in the humanities, consists of an invaluable contribution to the task of thinking about the contemporary situations.

It helps us in the description, in the precision of languages and in the search for meanings that allow to think about the multiple relationships that are weaved today between the concepts inherited by tradition, and the new "technonatural" realities, like CC and health. Nature/culture declines today, as it always has, in new combinations, new oppositions, new hybridizations and relationships that metabolize the old realities into new ones. We need to recognize them in order to build our common future.

This essay has some limitations that need to be pointed out: firstly, it is an introductory discussion on the topic, being of narrative reflection of a generalist nature whose objective is precisely to provoke reflection and critical thinking; secondly, it is not an empirical work, so in the future it would be important to define hypotheses and conditions for empirical research.

We know that cultures make the world speak, and we also know that human knowledge results from the ways we dwell in particular places and times. This is a complex ecological, cultural, and social process, but it is also, above all, a relational process, the common landscape where the formulas of the future are rehearsed.

Acknowledgements This paper is based on the Introduction to the network project "Composing worlds: humanities, well-being and health in the 21st century" (Compor mundos: humanidades, bem-estar e saúde no século XXI), supported by Fundação Ensino e Cultura Fernando Pessoa (FFP).

References

1POINTFIVE (2022) 1PointFive is developing practical decarbonization solutions at scale. https://www.1pointfive.com/about. Accessed 22 Jun 2022

Barroso EP, Estrada R (2018) De Hípias Menor a Trump: das virtudes do erro (e da mentira) ao erro da pós-verdade [From Hippias minor to Trump: from the virtues of error (and lies) to post-truth error]. Estud em Comun: 301–309. https://doi.org/10.20287/ec.n26.v1.a17

Beck U (2016) The metamorphosis of the world how climate change is transforming our concept of the world. Polity Press, Cambridge

Bekoff M (2015) A vida emocional dos animais [The emotional life of animals]. Texto Editores, Lisboa

Bostrom N (2005) A history of transhumanist thought. J Evol Technol 14:1–25

Brymer E, Freeman E, Richardson M (2019) Editorial: One health: The well-being impacts of human-nature relationships. Front Psychol 10. https://doi.org/10.3389/fpsyg.2019.01611

Callicott B, Frodeman R (eds) (2009) Encyclopedia of climate change ethics and philosophy. Macmillan Gale, USA

Campelo A (2020) Como se faz o corpo. A construção da antropologia: da antropologia física à antropologia da saúde. In: Jorge VO (ed) (Coord,). Modos de Fazer / Ways of Making, Porto: Universidade do Porto—CITCEM, pp 137–158

Darwin C (1872) The expression of the emotions in man and animals. John Murray, London

Delio I (2012) Transhumanism or ultrahumanism? Teilhard de chardin on technology, religion and evolution. Theol Sci 10:153–166. https://doi.org/10.1080/14746700.2012.669948

Descola P (2005) Par-delà nature et culture [Beyond nature and culture]. Gallimard, Paris

Descola P (2018) Conversation: Philippe Descola. In: Fond. Louis Vuitt. https://www.youtube.com/watch?v=sTtBDfTLkhg&t=4464s

van Eersel P (2017) À la recherche de la vie intérieure [In search of inner life]. Albin Michel, Paris
Farahat A, Abuelgasim A (2022) Effect of cloud seeding on aerosol properties and particulate matter variability in the United Arab Emirates. Int J Environ Sci Technol 19:951–968. https://doi.org/10.1007/s13762-020-03057-5
Funtowicz SO, Ravetz JR (1993) Science for the post-normal age. Futures 25:739–755. https://doi.org/10.1016/0016-3287(93)90022-L
Gielen UP, Fish JM, Draguns JG (2004) Handbook of culture, therapy, and healing. Lawrence Erlbaum Associates Publishers, Mahwah
Goodall J, Hudson G (2015) Seeds of hope: wisdom and wonder from the world of plants. Grand Central Publishing, New York
Greco J (2003) Virtues in epistemology. In: Moser P (ed) The Oxford handbook of epistemology. Oxford University Press, New York, pp 287–312
Harari YN (2018) Sapiens. História Breve da Humanidade [Sapies. Brief history of humanity]. Elsinore, Amadora
Horin A, Lee K, Colloca L (2014) Placebo effects in therapeutic outcomes. Curr Clin Pharmacol 9:116–122. https://doi.org/10.2174/1574884708999140101145528
IPCC (2018) The Intergovernmental Panel on Climate Change. In: Glob. Warm. 1.5 °C. https://www.ipcc.ch/sr15/. Accessed 21 Sep 2020
Kellert S (1991) Japanese perceptions of wildlife. Conserv Biol 5:297–308. https://doi.org/10.1111/j.1523-1739.1991.tb00141.x
Latour B (2016) Reset modernity! MIT Press, Cambridge, MA
Latour B (2017) Où atterrir ? Comment s'orienter en politique [Where to land? How to orient yourself in politics]. Ed. La Découverte, Paris
Lencastre MPA (2015) Ecologias e concepções culturais de saúde, doença e cura: Reflexões iniciais [Ecologies and cultural conceptions of health, illness and healing: initial reflections]. Sensos IO—Rev do Cent Investig e Inovação em Educ 5:169–184
Levin BW, Browner CH (2005) The social production of health: critical contributions from evolutionary, biological, and cultural anthropology. Soc Sci Med 61:745–750. https://doi.org/10.1016/j.socscimed.2004.08.048
Lewis-Fernández R, Kirmayer LJ (2019) Cultural concepts of distress and psychiatric disorders: understanding symptom experience and expression in context. Transcult Psychiatry 56:786–803. https://doi.org/10.1177/1363461519861795
Morin E (2000) Les Sept savoirs nécessaires à l'éducation du futur [The seven knowledges necessary for the education of the future]. Seuil, Paris
Nussbaum MC (2019) Sem Fins Lucrativos. Porque Precisa a Democracia das Humanidades [Non-profit. Why humanities need democracy]. Edições 70, Lisboa
Papa Francisco (2015) Laudato Sí. Sobre o cuidado da casa comum [Laudato Si. About the care of the common home]. Ed. Paulus, Apelação
Pecharroman LC (2018) Rights of nature: rivers that can stand in court. Resources 7. https://doi.org/10.3390/resources7010013
Pinker S (2018) Enlightment now. The case for reason, science, humanism and progress. Penguin Random House LLC, New York
Pinto MJ V (2005) Introdução. Os Sofistas e a Sofística [Introduction. The sophists and sophistry]. In: Sofistas. Testemunhos e Fragmentos. IN-CM, Lisboa, pp 11–39
Pires JC (1997) De profundis, valsa lenta [From deep, slow waltz]. Relógio D'Água, Lisboa
Platão (1973) Górgias. Verbo, Lisboa
Riviere C (1992) Attitudes africaines face à l'environnement [African attitudes towards the environment]. ANTHROPOS 87:365–378
Rossa-Roccor V, Giang A, Kershaw P (2021) Framing climate change as a human health issue: enough to tip the scale in climate policy? Lancet Planet Heal 5:e553–e559. https://doi.org/10.1016/S2542-5196(21)00113-3

Schmidt M, Soentgen J, Zapf H (2020) Environmental humanities: an emerging field of transdisciplinary research. GAIA Ecol Perspect Sci Soc 29:225–229. https://doi.org/10.14512/gaia.29.4.6
Singer M, Erickson PI (2011) A companion to medical anthropology. Wiley-Blackwell Ed, United Kingdom
Steffen W, Persson A, Deutsch L et al (2011) The anthropocene: from global change to planetary stewardship. Ambio 40:739–761. https://doi.org/10.1007/s13280-011-0185-x
Stuart-Smith S (2020) The well gardened mind. Rediscovering nature in the modern world. Harper Collins Publishers, London
Sussman LK (2004) The role of culture in definitions, interpretations, and management of illness. In: Gielen UP, Fish JM, Draguns JG (eds) Handbook of culture, therapy, and healing. Lawrence Erlbaum Associates Publishers, Mahwah, pp 37–65
The Royal Society of Medicine (2021) Confronting the health challenges of climate change. https://www.rsm.ac.uk/latest-news/2021/confronting-the-health-challenges-of-climate-change/. Accessed 8 Apr 2022
Toldy T, Estrada R (2017) A vulnerabilidade como lugar ético: sobre De Profundis, Valsa Lenta, de José Cardoso Pires [Vulnerability as an ethical place: on De Profundis, Valsa Lenta, by José Cardoso Pires]. Didaskalia 47. https://doi.org/10.34632/didaskalia.2017.3272
Turnpenny J, Jones M, Lorenzoni I (2011) Where now for post-normal science?: A critical review of its development, definitions, and uses. Sci Technol Hum Values 36:287–306. https://doi.org/10.1177/0162243910385789
Viaene L (2022) Can rights of nature save us from the anthropocene catastrophe? Some critical reflections from the field. Asian J Law Soc 1–20. https://doi.org/10.1017/als.2022.2
Vidal DG, Oliveira GM, Pontes M et al (2022) The influence of social and economic environment on health. In: Prata JC, Isabel A (eds) One health. Elsevier, Amsterdam, pp 205–229
Virilio P (1998) La bombe informatique, Galillé, Paris
De Waal WF (2009) The age of empathy: nature's lessons for a kinder society. Harmony, USA
Watanabe M (1974) The conception of nature in Japanese culture. Science (80) 183:279–282. https://doi.org/10.1126/science.183.4122.279
Wohlleben P (2017) A vida secreta das plantas [The secret life of plants]. Sextante, Rio de Janeiro

Marina Prieto Afonso Lencastre is Retired Full Professor (Jubilated) at the University of Porto and Full Professor at the University of Fernando Pessoa. She is a Specialist Psychotherapist and Supervisor of the Portuguese Society of Clinical Psychology, holding a Specialty in Clinical and Health Psychology and also in Advanced Specialty in Psychotherapy by the Portuguese Order of Psychologists. Additionally, she is an Accredited Trainer in Environmental Education.

Diogo Guedes Vidal is an Environment and Nature Sociologist interested in all its interconnections with and health of human and non-human beings. Holds Ph.D. in Ecology and Environmental Health with a project focused on the attempt to look beyond the greenness of public urban green spaces funded by the main Portuguese public body agency—FCT. Currently is Researcher in the Project PHOENIX "The rise of citizens' voices for a greener Europe" (Horizon 2020; Grant agreement ID: 101037328) and Invited Assistant Researcher at Centre for Functional Ecology – Science for People & the Planet" (CFE) in the Department of Life Sciences of at the University of Coimbra, Portugal.

Eduardo Paz Barroso is Full Professor of Communication Sciences in the Faculty of Human and Social Sciences of the University Fernando Pessoa. He is Researcher at LabCom, a research unit in communication and arts at the University of Beira Interior. Among being a lecturer, cultural manager, and programmer, he has been appointed to several public charges in Portugal in the cultural and artistic sector. Author of about a hundred works, such as books, book chapters, and articles in national and international journals.

Álvaro Campelo is Ph.D. in Anthropology and Associate Professor at the Fernando Pessoa University. He conducts research in the area of cultural heritage, socio-anthropology of health, cultural landscape and uses of space. He is also a researcher at CRIA.

Teresa Martinho Toldy is Ph.D. in Theology (feminist theology) at the Philosophisch-Theologische Hochschule Sankt eorgen (Frankfurt/Germany), Master in Theology (Catholic University, Lisbon) and 1st grade in Theology at the same University. Holds Postdoctorate in CES. She is Full Professor at the Fernando Pessoa University (Porto/Portugal); Teaches in the field of Ethics, Gender Studies and Citizenship; Vice-President of the Ethics Commission of the same University; Researcher at the Center for Social Studies; Coordinator of POLICREDOS (Observatory on Religion in Public Space); President of the Portuguese Association of Feminist Theologies and former Vice-President of the Portuguese Association on Women's Studies (2009–2014). Her Fields of specialization are religion; feminist studies. Her other fields of interest are: Citizenship. Publishes in the field of religion and feminist studies.

Susana Magalhães is Ph.D. in Bioethics by the Portuguese Catholic University, is Lecturer at University Fernando Pessoa and the Head of the Unit for Responsible Conduct in Research at the Institute for Research and Innovation in Health of the University of Porto (i3S). Her research interests are Narrative Medicine and Research Ethics/Integrity. She is a Certified Virt2UETrainer and member of the steering group of the Portuguese Reproducibility Network. She is Coordinator of a Group of Study and Reflection on Narrative Medicine (Grupo de Estudos e Reflexão em Medicina Narrativa—GERMEN, https://www.facebook.com/GermenMedicinaNarrativa/) and member of the European Society of Narrative Medicine (EUNAMES).

Ana Gabriela Frazão-Nogueira is Ph.D. in Audiovisual Communication and Journalism at the University of Santiago de Compostela (2016) is responsible for the Press, Radio and Hipermedia Laboratories of the Fernando Pessoa University, in Porto, also providing teaching support in the Radio Laboratory and Practice in Radio, Press and Hipermedia. Among others, Ana Nogueira holds a Diploma in Advanced Studies in the Program of Communication and Journalism of Doctorate at USC and attended the postgraduate course in Marketing and Strategic Com her curriculum also includes training courses in Creative Writing; Teaching and Multimedia, Multimedia Infographics; Film Script and Animation; Voiceover and Dubbing; Techniques of Journalism and Image Design, among others short courses. Ana is also an integrated member of the Investigation Group for Media & Journalism of the ICNova, an external collaborator of Grupo Novos Medios, University of Santiago de Compostela and member of the research group Cultura y Comunicaci6n Interactiva, Universidade of Coruña.

Rui Estrada holds a Ph.D. in Theory of Literature by the Faculdade de Letras of the University of Lisboa. Is Full Professor at Universidade Fernando Pessoa and Researcher at CITCEM (Faculdade de Letras, University of Porto). He won the Pen Club Award Essay in 2002 and published four books, edited six and is author of several papers and book chapters (national and international).

Chapter 3
Mental Health, Well-Being and Climate Change: Scope and Challenges

Carla Alexandra Martins da Fonte, Sónia Maria Martins Caridade, and Maria Alzira Pimenta Dinis

Abstract Climate change (CC) manifests itself through extreme weather conditions and is gaining more and more attention from the scientific community and civil society, considering the social implications in various domains of daily life. Undeniable social, economic and environmental consequences have been disclosed worldwide, in addition to the significant psychosocial impacts that the resulting weather events pose to mental health and well-being of people. Post-traumatic stress disorder (PTSD), vicarious trauma, anxiety, substance abuse, depression, survivor guilt, recovery fatigue, and suicidal ideation, are some of the mental health problems linked with the occurrence of catastrophic weather events. Other threats involve financial and relationship stress and the consequent potentiation of the risks of violence, in particular gender-based violence (GBV) and aggression, disproportionately affecting those who are most marginalized. Paradoxically, and in response to CC, the research shows that these same circumstances may also inspire individuals to more positive and adaptive behaviours, such as altruism, optimism and compassion, promoting a life meaning and individual development, motivated by CC post-traumatic growth. Based on an expert-driven literature review, this chapter intends to assess and discuss the relevant literature describing how CC impacts on the mental health and well-being

C. A. M. da Fonte (✉)
Faculty of Human and Social Sciences, University Fernando Pessoa (UFP), Praça 9 de Abril 349, 4249-004 Porto, Portugal
e-mail: cfonte@ufp.edu.pt

S. M. M. Caridade
Psychology Research Center, School of Psychology, University of Minho, Campus de Gualtar, 4710-057 Braga, Portugal
e-mail: scaridade@psi.uminho.pt

Interdisciplinary Center of Gender Issues, Higher Institute of Social and Political Sciences, University of Lisbon (ISCSP-ULisboa), Lisbon, Portugal

M. A. P. Dinis
UFP Energy, Environment and Health Research Unit (FP-ENAS), University Fernando Pessoa (UFP), Praça 9 de Abril 349, 4249-004 Porto, Portugal
e-mail: madinis@ufp.edu.pt

W. Leal Filho et al. (eds.), *Climate Change and Health Hazards*, Climate Change Management, https://doi.org/10.1007/978-3-031-26592-1_3

of individuals and societies, as well as the hazards associated with increasing problems of violence and aggression. Likewise, it also aims to analyse and discuss specific responses to the CC that can contribute to positive changes in learning, growth and behaviour throughout life. The promotion of pro-environmental behaviour through education is considered as an important strategy in this context.

Introduction

Often presented as an environmental problem that integrates the global political agenda, climate change (CC) constitutes an important socioecological problem (Clayton 2019). It has been argued that the impact of CC on mental health has repercussions in Europe and the rest of the world, with particularly severe effects in CC hotspots, such as the Mediterranean region (Massazza et al. 2022). In addition to being a global problem, it is also documented that CC affects different population groups directly exposed to it, with particular impact on groups that show greater vulnerability in terms of their geographical conditions, as well as the deficiency of access to resources, information and protection (Cianconi et al. 2020). Considered a relevant threat to sustainable development (Basha et al. 2017; Filho et al. 2022a, 2022b), CC poses many hazards, potentially compromising the global well-being of populations (Filho et al. 2022a, 2022b), but also gender equality. The broadly systemic health risk associated with CC has been debated, as it can contribute to increase people's social and psychological stress (Johnson 2020). CC can manifest itself in different ways, such as natural disasters or extreme weather events, e.g., desertification, droughts, forest fires, floods and cyclones, resulting in significant implications in several areas of life. Social, economic and environmental consequences are inevitable, with a particular impact on those being most marginalized. Their contributions to the increase of gender-based violence (GBV) have also been discussed (Caridade et al. 2022). The World Health Organization "estimates an increase of 250,000 excess deaths per year between 2030 and 2050 due to the well understood impacts of climate change" (WHO 2015b). The often-unobserved consequences of CC include the mental health effects, so more interest must be addressed in relation to analyse this subject. A wide range of new concepts and terminologies emerge in literature focused on the connection between CC and mental health, examples being: ecopsychology, ecoemotions, ecoanger, ecoanxiety, ecoguilt, ecological grief (Cianconi et al. 2020).

To ensure high standards of health in every way, the Lancet Commission on Health and CC was created in 2015, in order to map the impacts of CC and the required policy responses. Multidisciplinary and international in nature and with great collaboration between academic centres in Europe and China, this commission seeks to address CC as the greatest global health opportunity of the twenty-first century. In this sense, it has presented several recommendations to accelerate worldwide action in the next 5 years, for example: (i) greater investment in public health and climate change research, monitoring and surveillance; (ii) mobilize funding for the development of climate-resilient health systems worldwide; (iii) stimulate and support a transition to cities that promote healthy lifestyles for populations and the planet; and (iv) increase access to renewable energy in low-income countries and middle-income countries (Watts et al. 2017). Such recommendations are fundamental for moving towards global sustainable development, better public health, and achieving long-term economic growth. This is also critical to meeting and accelerating the implementation of the sustainable development goals of the UN 2030 Agenda, namely poverty reduction (SDG1); ending hunger (SDG2); better health outcomes (SDG3); improving the level of education (SDG4) resulting in the decrease of early marriage and increasing the likelihood of having a prosperous life; as well as in drinking water and sanitation (SDG6); and the promotion of climate action (SDG13) (United Nations 2015, 2021).

This chapter is based in an expert-driven literature review methodology, unveiled to explore the connections between CC and mental health and well-being. Based on a selection of relevant publications carried out on the Scopus and Web of Science databases, also including Google Scholar, and focusing the key terms "climate change", "mental health", "social inequalities", "gender violence" and "positive change" and "well-being", carefully analysed, it is intended to analyse and discuss the relevant literature on both the direct and indirect effects of CC on the mental health and well-being of individuals. The role of the CC in expanding social inequalities and the greater vulnerability of certain social groups, such as women, children, the elderly, people with disadvantaged socio-economic status, people with disabilities or even pre-existing health conditions, to climate hazards is also reviewed, as well as the risks associated with growing problems of violence, especially in the increase of GBV, e.g., domestic violence and violence against women. Finally, and based on a psychosocial approach informed by trauma and resilience, it is aimed to reflect on how CC can catalyse responses to learning, growth and positive change. Promoting pro-environmental behaviour through education (Filho et al. 2021a), in particular relating CC (Filho et al. 2021b), appears, therefore, as a fundamental strategy to increase well-being and to respond to an important dimension of mental health of populations.

Impact of Climate Change on Mental Health and Well-Being

The research on mental health has been changing through the years. The concept of mental health as referring only to the absence of symptoms has been questioned by some authors (Huppert et al. 2006; Keyes 2005, 2013). In this sense, it has been argued that mental health also involves the individual's ability to feel well-being and to deal with and actively contribute to one's life and the life of the community where the individual is inserted (Carvalho and Matos 2019). The World Health Organization (2008) defines mental health as "a state of well-being in which the individual realizes his or her own abilities, can cope with the normal stress of life, can work productively and fruitfully and is able to make a contribution to his or her community". This definition demonstrates that mental health is more than the absence of disease, and may involve different emotional, social and psychological aspects that will help to determine the level of mental health of a person at a given moment (Carvalho and Matos 2019; WHO 2008, 2015a). Keyes (2002, 2013), in turn, conceptualizes mental health as involving a set of positive emotions and positive overall functioning. This author proposed a model of mental health that aims to assess it through subjective well-being (e.g., happiness and life satisfaction); psychological well-being (e.g., positive affect, avowed happiness or avowed life satisfaction), self-acceptance, and social well-being (e.g., social actualization). To operationalize mental health, Keyes (2005) sums it up on a continuum model, two continua model, that involves the "languishing, flourishing and moderate mental health" concepts. The first one indicates a negative mental health interfering with the individual´s functioning. The second model preconizes the existence of "emotional well-being" and "positive functioning" and the third concept means that those who are not "flourishing", not "languishing" are on a "moderate mental health" state. So, the "flourishing" is related to the presence of mental health, meaning the absence of pathology and the existence of high levels of "psychological, emotional and social well-being". According to Keyes (2013), a person can have a mental illness and simultaneously present high levels of well-being. The non-existence of psychopathology is neither needed nor enough to guarantee individual useful, prosperous, and realized lives. So, although mental health and mental illness are two distinct components, they are strongly connected (Fonte 2019). Therefore, Keyes conceived the "two continua model of mental health" and illness, confirming that mental health and mental illness are not poles of the same continuum but two axes that become distinct, but correlated. Accordingly, mental health must be viewed as a complete state of well-being (Fonte et al. 2020; Keyes and Annas 2009; Westerhof and Keyes 2010).

Multiple factors could endanger people's mental health. In the first line are individual factors, such as behavioural characteristics, personality, resilience, strategies and coping to deal with life experiences and early adversities, among others. Social, family and contextual factors are another set of factors that, when interacting with individuals, can help to understand individual differences in the way people manage their mental health (APA 2015). Regarding resilience, some debate exists about the importance of distinguishing between resilience as a skill, a prerequisite for recovery, and resilience as a process, contemplating different mechanisms underlying recovery, which should also be considered to better understand the response of individuals to climate events (Bogardi and Fekete 2018).

CC can be considered as an important risk factor that can compromise the mental health of populations, and can also lead to mental illness (Massazza et al. 2022; Swim et al. 2010). The growing recognition of the impact of CC on mental health, in terms of mental illness, may occur through different mechanisms. Specifically, the psychological impacts of CC have been in the recent years recognized as a complexity phenomenon, one of the biggest tasks challenging humanity today. Recent literature review carried out until now (Cianconi et al. 2020; Charlson et al. 2021; Doherty and Clayton 2011) shows that the effects of CC can be direct or indirect, short-term or long-term, especially in the face of more severe and prolonged climatic events, or with the potential to be transmitted to future generations (Cianconi et al. 2020).

The direct effects of psychological impacts refer to severe or traumatic effects of extreme weather events and also a changed environment, which are more frequent and intense under a changing climate (Sanson et al. 2019; Watts et al. 2017). In this sense, in terms of children and adolescents but also in terms of adult population, some studies indicated that trauma from natural disasters can trigger post-traumatic stress disorder (PTSD) (Massazza et al. 2022; Swim et al. 2010) and often comorbidity with depression and anxiety symptoms (Albrecht et al. 2007; Barkin et al. 2021; Hanna and Oliva 2016), with the three groups being equally exposed to these effects. Literature also reports an increase of problematic externalizing behaviours, such as aggression and mood disorders (Younan et al. 2018). Also, higher ambient temperatures, flooding and hurricanes, are linked with more emergency room visits for mental health conditions and an increase of suicide rates, which as well occurs on adults (Azuma et al. 2014; Charlson et al. 2021; Dumont et al. 2020; Fernandez et al. 2015; North and Pfefferbaum 2013). A recent scoping review conducted by To et al. (2021) summarizes the results from 254 studies that have examined the impact of wildfires on mental health. The results indicated an expanded rate of PTSD, anxiety and depression, CC post-wildfire, found in children and adolescents, as in adults. Also, Cunsolo et al. (2014), build upon existing research, outlined that CC could unfavourably impact mental health and well-being in the remote region of Circumpolar North.

The indirect effects of psychological impacts refer to "threats to emotional well-being" based on the analysis "of impacts and concern or uncertainty about future risks", as a result of damages of social and physical infrastructure, food and conflict (Doherty and Clayton 2011; Hayes et al. 2018; Sanson et al. 2019). Young people are more distressed and concerned about CC caused by the environmental degradation, reporting stronger emotional reactions, even for those who have not yet been directly affected (Massazza et al. 2022; van Nieuwenhuizen et al. 2021). Also, some literature have discussed the correlation between "eco-emotions", i.e., "ecoanxiety, eco-anger and eco-depression", pro-climate activism, and well-being. The "ecoanxiety" is related to the fact that people are frequently surrounded by problems associated with CC (Albercht et al. 2007; Samantha et al. 2021). One another climate hazard that indirectly affects mental health is drought, that can affect food and water supplies and consequently affect the mental health. Drought can also lead to forced migration, conflict and the almost certain probability of experiencing anxiety, stress and trauma (Obrien et al. 2014).

Climate Change, Inequality and Violence

Furthermore, the impacts of CC on mental health, its role in enhancing social inequalities and the greater vulnerability of certain social groups to climate risks is also known and debated (Caridade et al. 2022; Filho et al. 2022a). In fact, it has been shown that the most marginalized populations in social and environmental aspects, and which are therefore the target of greater discrimination at different levels, e.g., gender, age, race, socioeconomic status, disability or even with pre-existing health conditions, which do not infrequently intersect, are more vulnerable to climate risks (Islam and Winkel 2017). CC is also perceived as an important threat to a sustainable development (Basha et al. 2017), and therefore challenging efforts to achieve global sustainable development and social equity. As an example, Christenson et al. (2014) found that in countries with low human development indices, there is greater exposure to the impacts of CC (about 50%), when compared to what is seen in countries with high development indices (which is 14% more). It is also known that if there is no decrease in the CO_2 emissions, this uneven exposure resulting from the main impacts of CC, e.g., droughts and floods, could increase between 9 and 17% in 2030. In fact, it is in developing countries that the damage caused by floods is more significant, considering the worst structural conditions in which people live, with houses being built with fragile materials, for example. In this sense, it is important to consider that not all people are at the same level, whether in material, social, economic and mental terms, to recover after extreme events, and that needs to be considered when addressing this topic (Orr et al. 2020).

Likewise, and given the role that women assume more often in these developing countries, relegated to the performance of domestic tasks, they end up being more exposed to the impacts of CC (Filho et al. 2022a; Memon 2020), also constituting a group of greater vulnerability to the experience of sexual and domestic violence, with severe and harmful consequences to their reproductive and sexual health during any type of disaster, either natural or not (FRA 2014; WHO 2015b). Rezaeian (2013) also mentions gender differences in terms of the adverse impact of climate events, which is especially reflected in women experiencing greater inequality in this field. Likewise, Memon (2020) reports that women experience and react to the impact of a disaster in different ways, when compared to men. It is also known that women's greater vulnerability to CC is increased in developing countries. This is explained by the fact that they experience greater social inequalities and are subject to alternative income opportunities, representing greater constraints in terms of their mobility, thus increasing their exposure to natural disasters and other unexpected CC events (Costello et al. 2009).

The relationship between being exposed to natural disasters and being a victim of self-directed violence or interpersonal has been established. Several works, e.g., the ones from Clayton (2019), Miles-Novelo and Anderson (2019), Molyneaux et al. (2020), Whittenbury (2013), and UN Women (2016), have shown that in addition to being a major threat to the planet's ecosystems, CC may be linked with the increase in interpersonal violence. The impact of fast global warming in increasing the incidence of violent behaviour (Anderson and DeLisi 2011; Miles-Novelo and Anderson 2019) has also been demonstrated. The relationship between CC and violent behaviour has been established both by the direct effects of global warming in increasing irritability, aggression and violence; or, indirectly, by increasing the risk of children and adolescents becoming adults subjected to violence, with a special focus also on women; as well as how CC could impact the economic and political stability of countries with livelihoods at risk, promoting migration and violent conflict between groups (Anderson and DeLisi 2011). Experiencing natural disasters can increase the rate of violence both in the short- and long-term, triggering mental distress and anger (Parkinson 2017; Rezaeian 2013), as well posing significant and increasing challenges for the public health (Johnson 2020).

The occurrence of extreme weather events has been associated with an increase in GBV, namely the occurrence of domestic violence, intimate partner violence and other forms of abuse against women and girls (e.g., Molyneaux et al. 2020; Parkinson 2017; Sety 2012). Such evidences have been found both in developing countries as Chad and Somalia, but also in developed countries, examples being Australia and the United States of America (Anderson and Delisi 2011; Whittenbury 2013). The connection between domestic violence and extreme weather disasters was proven in the study by Parkinson (2017), which sought to analyse and debate three important explanatory hypotheses: (i) the fact that the occurrence of the disaster makes it possible to reveal the situations of abuse that already exist; (ii) the fact that disasters can accentuate female vulnerability to the experience of male aggression; and (iii) the "culture of denial", perceptible by the existence of numerous factors, e.g., risk to confidentiality and fear of hurting family members and communities, fear of retaliation from partners, that inhibit women from reporting situations of abuse suffered. The impact of natural disasters and their relationship to GBV, namely domestic violence, on women's mental health has been demonstrated (Caridade et al. 2022). PTSD, depression, and other significant mental health problems have been identified as the most prevalent in post-disaster by women, and other minorities, i.e., poor, elderly, or primary caregivers of children (Bell and Folkerth 2016). Others authors, e.g., Memon (2020), Molyneaux et al. (2020), Rahman (2013) or Whittenbury (2013), confirmed that women experience greater inequalities during any climate-stressed situation and also react differently, developing more symptoms of PTSD and other mental health issues. Social and gender inequalities tend to be exacerbated with CC, resulting in an increase in its impact on the global well-being of minority groups and women victims in a more particular way.

Responding to the Impact of CC: Opportunities Addressing Mental Health and Well-Being

The occurrence of climate events tends to attract the attention of political agendas, in terms of responses to be defined immediately, but also in terms of recovery and preparation/prevention strategies for future events. This has been considered fundamental to enhance the increase of political changes and to promote the learning/preparation of organizations (Crow and Albright 2021). In this sense, it has been argued that CC post-recovery should not be based only on the reconstruction of physical structures, often damaged by climatic events, but should seek broader social objectives that include the well-being and resilience (Orr et al. 2020). This is particularly important in the case of marginalized populations, which present more vulnerabilities in terms of material, social, economic and mental conditions for the recovery required after exposure to extreme events. It is essential to consider different trajectories and typologies of resilience of populations, addressing resilience as a skill but also as a process, to better guide the intervention after the events. Prior to climatic disasters, it is then essential to assess and to know, those that constitute the main recovery skills to better guide the development of procedures to be implemented in disaster preparedness programs. In turn, understanding resilience as a process will make it possible to better understand what stage of resilience the individual exposed to the event is in, i.e., whether faced with preventive resilience (presilience) or progressive resilience (proresilience) and thus better guide the intervention to follow (Bogardi and Fekete 2018).

Although the negative impact of CC on mental health is known, the research shows that CC may also inspire individuals to engage in more positive and adaptive behaviours, such as compassion, optimism, altruism, promoting a life meaning and personal development, motivated by CC post-traumatic growth. For instance, some particular emotional responses are associated with several mental health results connected with greater well-being (Hayes et al. 2018; Hicks et al. 2020; Ramsay and Manderson 2011).

In order to analyse and problematize this positive impact of CC on mental health and well-being it is proposed in this chapter to distinguish hedonic and eudaimonic well-being, inspired in the ancient philosophy of Aristippus (Keyes and Annas 2009). According to this perspective, hedonic well-being suggests positive emotional sates, such as satisfaction, whereas eudaimonic well-being means to feeling meaningful, pursuing the right ends and using personal strengths and competences. In this eudaimonic well-being, the focus is pursuing virtues and excellences, for example bravery, generosity, wisdom and being fair; dimensions that are fundamentally worth being pursued and do not derive their worth from the external benefits, or materialistic objects (Huppert et al. 2006; Keyes 2013; Ryan and Deci 2001).

In this sense, some authors, such as Tiatia-Seath et al. (2020) and Venhoeven et al. (2013), have argued that behaving in a pro-environmental way may reduce hedonic well-being, while increasing eudaimonic well-being. The same group of authors suggest that defending the environment can be unachievable, thus making determined for this purpose an arduous load. Progressing in a pro-environmental direction will necessarily involve a less consuming behaviour. Consequently, pro-environmental actions can enhance personal eudaimonic well-being, and not hedonic well-being. As a result, changing the focus from commodities and money, to the achievement of constructive relationships, individual developmental and life meaning, contributes to eudaimonic well-being and is also a more pro-environmental behaviour. In this sense, it is important to inform individuals that their pro-environmental behaviour can increase their own well-being (Venhoeven et al. 2013).

Another interesting analysis of the fact that CC is able to activate positive changes in individuals experience is reflected in the construct of post-traumatic growth.

Tedeschi and Calhoun (2004a, p. 1) define post-traumatic growth as "the experience of positive change that occurs as a result of the struggle with highly challenging life crises", produced by specific events. These events are those that severely challenge and shake individual's way of understanding the world; how they make sense of their daily live and came from circumstances that are uncontrollable, potentially irreversible and unusual (Calhoun and Tedeschi 2014). These events, that are traumatic, force individuals to reconsider the basic assumptions about "who they are, what people around them are like, what kind of world they live in, or what the future e may hold" (Tedeschi and Calhoun 2004b, p. 6). Considering the CC mental health impacts summarized in this chapter, the authors (Calhoun and Tedeschi 2014) consider that CC effects on humanity can trigger traumatic experiences and that paradoxical changes can occur as a result of post-traumatic growth, because people have been often surprised by finding things of value that they never anticipated before, as they are just trying to survive and get through whatever trauma they have experienced.

Tedeschi and Calhoun (2004b) proposed that post-traumatic growth can be revealed in a diversity of behaviours, that the authors categorized in five domains. The first one—personal strength—where people discovered that they are stronger than they thought they were, that they endured things that before the traumatic events, they did not think they could handle out. The second domain—appreciation of life—where people have more gratitude than they had before, and recognized that life has a lot to offer that they may overlooked and taken granted. So, they appreciate often simpler things of life, that are free, rather than the achievements. In the third domain—new possibilities—aftermath of trauma, individuals develop a sense that there are opportunities that they have now that might not have seen before, because so many thinks have been taken from them or lost that, what they have been doing before, they cannot continue to do. The fourth domain—changes in their relationships with other people—involve a deeper emotional connection with people more compassion and empathy for people. The fifth and last domain—spiritual and existential growth—is where people have to confront with the kind of big questions about life that may not have taken on before paint much attention, because of their traumatic experiences. So, they might ask questions of why they been have been suffering or why they have to suffer in life or what's the point of living through difficulties, what does it all mean. It is also suggested that post-traumatic growth can be faced as a continuing process, so it is not like one gets there, but learn and develop new ways of living, promoting individual's strength and the ability to change.

The direct and indirect effects of CC can be considered traumatic events that may produce distressing responses in persons experiencing those events. The authors of this chapter would like to highlight that trauma is not being presented as something good, since post-traumatic growth occurs in the context of suffering and significant struggle. What is important to be clarified in the context of this chapter is that, despite the several behavioural responses that are hostile, some persons may report positive transformations in the context of CC events, which cannot be certainly considered a universal positive response. Although pointing out this positive outcome, individuals who have been exposed to trauma, in particular CC related, may need professional help, being necessary that countries are able to provide adequate mental health support through community and institutional health services. Ultimately, post-traumatic growth is a pathway to more resilience, because people do become better, being able to manage traumas in the future, developing a deeper understanding of how to live life well.

Building the resilience that is necessary in response to being exposed to natural risks and CC collides, however, with social and gender inequalities that promote vulnerability to violence, as known, clarified through this chapter. This demands a transformative approach that is able to address the underlying structural causes of discriminatory inequalities and the social role of gender, and also implementing measures that can challenge these same inequalities, providing support services to female victims and community members, as well as raising awareness to the problem of violence and how it can be exacerbated in the CC post-disaster period (Masson et al. 2019).

Conclusions

CC may be considered one of the greatest hazards to mental health in the contemporary societies. Understanding the impacts of CC on mental health is, then, a subject of significance and deserves considerable attention.

Acknowledgements related to CC and mental health can be difficult because there is a chance of pathologizing 'normal' reactions to CC and not paying attention to the really mental health impacts of CC. Also, a further important consideration to be concerned about this topic is that previous mental health circumstances or problems can be activated by CC (Swim et al. 2011). As mentioned before in this chapter, mental health assessment must include well-being and psychopatology. In this sense, the researchers have a general tendency to theorize mental health exclusively as mental illness, using instruments that measure only mental health problems. Few empirical studies have focused on instruments evaluating the potentialities for mental health of extreme weather events, assessing dimensions as life meaning or post-traumatic growth (Hayes et al. 2018). Further understanding on this could make clearer the diversity of mental health effects from deriving from CC and also facilitate to recognize the several dimensions that can influence the positive mental health results and contribute to construct a psychosocial resilience.

In terms of practical implications, it is proposed in this chapter that psychological interventions in the context of CC should be considered, assisting the empowerment of individuals and the development of emotional resiliency. In order to achieve this goal, the provision of resources, such as more ecological literacy of population, promoting greater individual awareness, and training of mental health professionals on post-traumatic growth and resilience interventions, must be considered.

Finally, considering the direct and indirect effects/impacts of CC in exacerbating social inequalities, in increasing the vulnerability of certain social groups such as women, children, elderly, people with disadvantaged socio-economic status or people with any disability to experiencing interpersonal violence and aggression, with implications for their mental health and global well-being, further challenges must be considered, as defended by Caridade et al. (2022): (i) more research is needed in this field to gain a more solid understanding of the diverse gender issues related to CC in both developing and developed countries; (ii) it is necessary to identify and discuss future theoretical and practical perspectives to address the direct and indirect effects that the CC has on GBV, focusing on the joint energies of researchers from diverse disciplines, collaborating and cooperating with the policymakers; and (iii) it is important to adopt mainstream gender and intersectional approaches to deal with violence and to better target effective interventions that consider the cumulative impacts of race, gender and economics. Additionally, further research is required to allow to apply theories of post-traumatic growth to CC, as well as additional studies on life meaning and resilience are necessary.

Despite the important contribution given in this chapter to the systematization of the information on CC and mental health and well-being, the present chapter is based on an expert-driven literature review, which is a limitation. Systematic reviews and meta-analysis on the subject will need to be developed in future studies. Accordingly, the authors of this chapter would like to use this opportunity to encourage the researchers from different scientific domains to further increase the research on the topic, exploring the connection between CC and mental health, so that the acquired knowledge can be translated into effective worldwide policy actions, investing in the promotion of pro-environmental behaviour in populations through formal and informal education in schools and organizations, better mapping the social responses to CC and developing finner policies at social level, thus contributing to combat this reality. It is also important to focus on the development of better and more adequate resilience measures, which can be triggered by exposure to certain types of hazard impact levels.

Funding This study was conducted at the Psychology Research Centre (PSI/01662), School of Psychology, University of Minho, supported by the Foundation for Science and Technology (FCT) through the Portuguese State Budget (Ref.: UIDB/PSI/01662/2020).

References

Albrecht G, Sartore GM, Connor L, Higginbotham N, Freeman S, Kelly B, Stain H, Tonna A, Pollard G (2007) Solastalgia: the distress caused by environmental change. Australian Psychiatry 15(Suppl 1):8

American Psychiatric Association (APA) (2015) Understanding mental disorders: your guide to DSM5. DC Psychiatric Publishing, Washington

Anderson CA, DeLisi M (2011) Implications of global climate change for violence in developed and developing countries. In: J. Forgas A, Kruglanski, K Williams (eds) The psychology of social conflict and aggression. Psychology Press, New York, pp 249–265

Azuma K, Ikeda K, Kagi N, Yanagi U, Hasegawa K, Osawa H (2014) Effects of water-damaged homes after flooding: health status of the residents and the environmental risk factors. Int J Environ Health Res 24(2):158–175
Barkin JL, Buoli M, Curry CL, von Esenwein SA, Upadhyay S, Kearney MB et al (2021) Effects of extreme weather events on child mood and behavior. Dev Med Child Neurol 63(7):785–790
Basha C, Srinu G, Rao L (2017) Evolution of concept of sustainable development, its deterrence by emerging climate change and a way forward. Eco Env Cons 23(3):1615–1622
Bell SA, Folkerth LA (2016) Women's mental health and intimate partner violence following natural disaster: a scoping review. Prehosp Disaster Med 31(6):648–657
Bogardi J, Fekete A (2018) Disaster-related resilience as ability and process: a concept guiding the analysis of response behavior before, during and after extreme events. Am J Clim Chang 7:54–78
Calhoun LG, Tedeschi RG (eds) (2014) Handbook of posttraumatic growth: research and practice. Routledge
Caridade S, Vidal D, Dinis MA (2022) Climate change and gender-based violence: outcomes, challenges and future perspectives. In: Vidal DG, Pimenta Dinis MA, Dias RC, Filho W (eds) Sustainable policies and practices in energy, environment and health research: addressing cross-cutting issues. Springer, pp 167–176
Carvalho M, Matos MG (2019) Saúde versus doença mental [Health versus mental disease]. In: Nunes L, Fonte C, Alves S, Sani A, Estrada R, Caridade S (eds) Comportamento e Saúde Mental Dicionário Enciclopédico. Pactor, pp 190–191
Charlson F, Ali S, Benmarhnia T, Pearl M, Massazza A, Augustinavicius J, Scott JG (2021) Climate change and mental health: a scoping review. Int J Environ Res Public Health 18(9):4486
Christenson E, Elliot M, Banerjee O, Hamrick L, Bartram, J (2014) Climate-related hazards: a method for global assessment of urban and rural population exposure to cyclones, droughts and floods. Int J Environ Res Public Health 11(2):2169–2192
Cianconi P, Betrò S, Janiri L (2020) The impact of climate change on mental health: a systematic descriptive review. Front Psychiatry 11
Clayton S (2019) Psychology and climate change. Psychologist 40(3):167–173
Costello A, Abbas M, Allen A et al (2009) Managing the health effects of climate change. Lancet 373:1693–1733
Crow D, Albright E (2021) Community disaster recovery: moving from vulnerability to resilience. Cambridge University Press
Cunsolo WA, Stephenson E, Allen J, Bourque F, Drossos A, Elgarøy S, Kral MJ, Mauro I, Moses J, Pearce T, MacDonald JP, Wexler L (2014) Examining relationships between climate change and mental health in the Circumpolar North. Reg Environ Change 15(1):169–182
Doherty TJ, Clayton S (2011) The psychological impacts of global climate change. Am Psychol 66(4):265–276
Dumont C, Haase E, Dolber T, Lewis J, Coverdale J (2020) Climate change and risk of completed suicide. J Nerv Ment Dis 208(7):559–565
European Union Agency for Fundamental Rights (FRA) (2014) Violence against women: an EU-wide survey. Main results report [Cited 2021 Feb 5]. Available from: https://fra.europa.eu/en/publication/2014/violence-against-women-eu-wide-survey-main-results-report
Fernandez A, Black J, Jones M, Wilson L, Salvador-Carulla L, Astell-Burt T, Black D (2015) Flooding and mental health: a systematic mapping review. PLoS ONE 10(4):e0119929
Fonte C (2019) Bem-Estar [Well-Being]. In: Nunes L, Fonte C, Alves S, Sani A, Estrada R, Caridade S (eds) Dicionário Enciclopédico Comportamento e Saúde Mental. Pactor, pp 234–235
Fonte C, Silva I, Vilhena E, Keyes C (2020) The Portuguese adaptation of the mental health continuum-short form for adult population. Community Ment Health J 56(2):368–375
Filho WL, Dinis MAP, Sivapalan S, Begum HNg TF, Al-Amin AQ, Alam GM, Sharifi A, Salvia AL, Kalsoom Q, Saroar M, Neiva S (2021a) Sustainability practices at higher education institutions in Asia. Int J Sustain High Educ

Filho WL, Sima M, Sharifi A, Luetz JM, Salvia AL, Mifsud M, Olooto FM, Djekic I, Anholon R, Rampasso I, Donkor F, Dinis MAP, Klavins M, Finnveden G, Chari MM, Molthan-Hill P, Mifsud A, Sen SK, Lokupitiya E (2021b) Handling climate change education at universities: an overview. Environ Sci Eur 33: 1–19. Article 109

Filho WL, Eustachio JHPP, Dinis MAP, Sharifi A, Venkatesan M, Donkor FK, Doni F, Abubakar IR, Cichos K, Vargas-Hernández J (2022a) Transient poverty in a sustainable development context. Int J Sustain Dev World Ecol

Filho WL, Wall T, Barbir J, Alverio GN, Dinis MAP, Ramirez J (2022b) Relevance of international partnerships in the implementation of the UN sustainable development goals. Nat Comm 13(1). Article 613

Hanna R, Oliva P (2016) Implications of climate change for children in developing countries. Future Child 26(1):115–132

Hayes K, Blashki G, Wiseman J, Burke S, Reifels L (2018) Climate change and mental health: risks, impacts and priority actions. Int J Ment Heal Syst 12:28

Hicks et al., 2020Hicks TA, Bountress KE, Resnick HS, Ruggiero KJ, Amstadter AB (2020) Caregiver support buffers posttraumatic stress disorder symptoms following a natural disaster in relation to binge drinking. Psychol Trauma Theory Res Prac Policy. Advance online publication

Huppert F, Baylis N, Keverne B (eds) (2006) The science of well-being. Oxford University Press, Oxford

Islam N, Winkel J (2017) Climate change and social inequalities. Department of Economics and Social Affairs United Nations, New York

Johnson W (2020) Using precision public health to manage climate change: opportunities, challenges, and health justice. J Law Med Ethics 48:681–693

Keyes CLM (2002) The mental health continuum: from languishing to flourishing in life. J Health Soc Behav 43:207–222

Keyes CLM (2005) Mental illness and/or mental health? Investigating axioms of the complete state model of health. J Consult Clin Psychol 73:539–548

Keyes CLM (2013) Mental well-being: international contributions to the study of positive mental health. Springer, Dordrecht

Keyes CLM, Annas J (2009) Feeling good and functioning well: distinctive concepts in ancient philosophy and contemporary science. J Posit Psychol 4(3):197–201

Massazza A, Ardino V, Fioravanzo R (2022) Climate change, trauma and mental health in Italy: a scoping review. Eur J Psychotraumatol 13(1):1–16

Masson V, Benoudji C, Reyes S, Bernard G (2019) How violence against women and girls undermines resilience to climate risks in Chad. Disasters 43(S3):S245–S270

Memon FS (2020) Climate change and violence against women: study of a flood-affected population in the rural area of Sindh, Pakistan. Pak J Women's Stud Alam-e-Niswan 27(1):65–85

Miles-Novelo A, Anderson CA (2019) Climate change and psychology: effects of rapid global warming on violence and aggression. Curr Clim Chang Rep 5:36–46

Molyneaux R, Gibbs L, Bryant R, Humphreys C, Hegarty K, Kellett C, Gallagher H et al (2020) Interpersonal violence and mental health outcomes following disaster. Bjpsych Open 6(1):E1

North CS, Pfefferbaum B (2013) Mental health response to community disasters: a systematic review. JAMA 310(5):507–518

Obrien LV, Berry HL, Coleman C, Hanigan IC (2014) Drought as a mental health exposure. Environ Res 131:181–187

Orr P, Morse-Jones S, Jones R, White C, Manmiz N (2020) Measuring recovery from extreme weather events. Available from https://www.climatexchange.org.u

Parkinson D (2017) Investigating the increase in domestic violence post disaster: an Australian case study. J Interpers Violence 34(11):2333–2362

Rahman S (2013) Climate change, disaster and gender vulnerability: a study on two divisions of Bangladesh. Am J Hum Ecol 2(2):72–82

Ramsay T, Manderson L (2011) Resilience, spirituality and posttraumatic growth: reshaping the effects of climate change. In: Weissbecker (ed.) Climate change and human well-being). Springer, New York, p 165–84

Rezaeian M (2013) The association between natural disasters and violence: a systematic review of the literature and a call for more epidemiological studies. J Res Med Sci 18(12):1103–1107

Ryan R, Deci E (2001) On happiness and human potentials: a review of research on hedonic and eudaimonic well-being. Annu Rev Psychol 52:141–166

Samantha KS, Teaghan L, Hogg ZL, Iain W (2021) From anger to action: differential impacts of eco-anxiety, eco-depression, and eco-anger on climate action and wellbeing. J Clim Chang Health 1:100003

Sanson AV, Van Hoorn J, Burke SEL (2019) Responding to the impacts of the climate crisis on children and youth. Child Development Perspectives 13:201–207

Sety M (2012) Domestic violence and natural disasters. Australian domestic and family violence clearinghouse thematic review 3:1–10

Swim J, Clayton S, Doherty T, Gifford R, Howard G, Reser J, Stern P, Weber E (2010) Psychology and global climate change: addressing a multifaceted phenomenon and set of challenges. A report of the American Psychological Association task force on the interface between psychology and global climate change

Swim JK, Stern PC, Doherty TJ, Clayton S, Reser JP, Weber EU, Gifford R, Howard GS (2011) Psychology's contributions to understanding and addressing global climate change. Am Psychol 66(4):241–250

Tedeschi RG, Calhoun LG (2004a) A clinical approach to posttraumatic growth. In: Linley A, Joseph S (eds) Positive psychology in practice. Wiley, pp 405–419

Tedeschi RG, Calhoun LG (2004b) Posttraumatic growth: conceptual foundations and empirical evidence. Psychol Inq 15(1):1–18

Tiatia-Seath J, Tupou T, Fookes I (2020) Climate change, mental health, and well-being for pacific peoples: a literature review. Contemp Pac 32(2):399–430

To P, Eboreime E, Agyapong V (2021) The impact of wildfires on mental health: a scoping review. Behav Sci 11(9):126

United Nations (2015) Transforming our world: the 2030 agenda for sustainable development. Resolution adopted by the general assembly on 25 September 2015, A/RES/70/1 Geneva

United Nations (2021) sustainable development goals progress chart 2021. Statistics Division, Department of Economic and Social Affairs, United Nations, New York

UN Women (2016) Annual report 2016–2017 [cited 2021, Feb 3]. Available from https://www.unwomen.org/en/digital-library/publications/2017/6/annual-report-2016-2017

van Nieuwenhuizen A, Hudson K, Chen X et al (2021) The effects of climate change on child and adolescent mental health: clinical considerations. Curr Psychiatry Rep 23:88

Venhoeven LA, Bolderdijk JW, Steg L (2013) Explaining the paradox: how pro-environmental behaviour can both thwart and foster well-being. Sustainability 5(4):1372–1386

Watts N, Amann M, Ayeb-Karlsson S, Belesova K, Bouley T, Boykoff M, Byass P, Cai W, Campbell-Lendrum D, Chambers J, Cox PM, Daly M, Dasandi N, Davies M, Depledge M, Depoux A, Dominguez-Salas P, Drummond P, Ekins P, Flahault A, Frumkin H, Georgeson L, Ghanei M, Grace D, Graham H, Grojsman R, Haines A, Hamilton I, Hartinger S, Johnson A, Kelman I, Kiesewetter G, Kniveton D, Liang L, Lott M, Lowe R, Mace G, Odhiambo Sewe M, Maslin M, Mikhaylov S, Milner J, Latifi AM, Moradi-Lakeh M, Morrissey K, Murray K, Neville T, Nilsson M, Oreszczyn T, Owfi F, Pencheon D, Pye S, Rabbaniha M, Robinson E, Rocklöv J, Schütte S, Shumake-Guillemot J, Steinbach R, Tabatabaei M, Wheeler N, Wilkinson P, Gong P, Montgomery H, Costello A (2017) The Lancet countdown: tracking progress on health and climate change. Lancet 389(10074):1151–1164

Westerhof G, Keyes C (2010) Mental illness and mental health: the two continua model across the lifespan. J Adult Dev 2:110–119

Whittenbury K (2013) Climate change, women's health, wellbeing and experiences of gender based violence in Australia. In: Alston M, Whittenbury K (eds) Research, action and policy: addressing the gendered impacts of climate change. Springer, Australia, pp 207–220

World Health Organization (2008) The global burden of disease: 2004 update. WHO Press, Geneva

World Health Organization (2015a) WHO calls for urgent action to protect health from climate change—sign the call. http://www.who.int/globalchange/global-campaign/cop21/en/

World Health Organization, 2015bWorld Health Organization (2015b) Weekly epidemiological record. Cholera 90: 517–544

Younan D, Li L, Tuvblad C, Wu J, Lurmann F, Franklin M, Berhane K, Chen J (2018) Long-term ambient temperature and externalizing behaviors in adolescents. Am J Epidemiol 187(9):1931–1941

Carla Alexandra Martins Fonte Associate Professor at Faculty of Human and Social Sciences /University Fernando Pessoa, Porto, Portugal. Has a Ph.D. in Psychology (University of Minho), Master in Psychology in University of Porto and degree in Psychology, also in University of Minho. Specialist Psychologist in Clinical and Health Psychology and Psychotherapy. Psychologist/Psychotherapist at the Hospital School of the University Fernando Pessoa and researcher in the area of mental health in clinical and non clinical populations, well-being, happiness and flow in different life contexts through the lens of Positive Psychology.

Sónia Maria Martins Caridade Ph.D. in Psychology of Justice, is an assistant professor at the School of Psychology at the University of Minho and an integrated member of the Center for Research in Psychology. She is coordinator for academic mobility at the UM School of Psychology and coordinator of internships in the Master's in Psychology of Justice at UM. Her main line of research is within the scope of violence in intimate relationships (adults and juveniles), domestic violence and juvenile delinquency. Her main line of research is within the scope of violence in intimate relationships (adults and juveniles), domestic violence, and juvenile delinquency. She is a reviewer in several international scientifc journals.

Maria Alzira Pimenta Dinis Ph.D., M.Sc. is an Associate Professor at the Faculty of Science and Technology (FCT), University Fernando Pessoa (UFP), Porto, Portugal. Graduated in Metallurgical Engineering and Master Degree in Materials Engineering at the Faculty of Engineering, University of Porto, Portugal. PhD degree in Earth Sciences at FCT, UFP. Researcher at FP-ENAS, UFP Energy, Environment and Health Research Unit in environment-related topics. Experienced author, editor and reviewer, with scientific outputs translated by a number of publications. Sustainability expertise.

Chapter 4
Climate Change Impact on Mental Health: Is Nature Fighting Us Back?

Isabel Maria Sousa Lopes Silva, Gloria Svietlana Jólluskin García, Paulo José Ribeiro Cardoso, and Allison Margaret Byrne

Abstract The relationship between sense of place, environment, and wellbeing, as well as the implications of climate change on human health have long been recognised. However, distress and loss of comfort caused by degradation of the environment, home and sense of belonging do not seem to have received equal attention, despite the severity of the consequences for mental health. We present a narrative review which aims to reflect on the pathways through which climate change can affect individuals' mental health. Psychosocial reactions that have received more attention are those that occur in situations of environmental disaster, namely acute stress disorder and post-traumatic stress disorder. However, psychosocial reactions to climate change are not always of this magnitude, making them invisible to governments and communities. And when these reactions do become visible, usually through the adoption of high-risk coping behaviour, the association with climate change is not always recognised. In this chapter, we also discuss the role of the media in these reactions, as indirect exposure to climate change messages can also have significant effects on mental health, altering individuals' risk perception and consequence predictions, and affecting their beliefs and confidence to move into more climate protective behaviours.

The relationship between sense of place, environment, and well-being, as well as the implications of climate change on human health, have long been recognised by researchers, although not always by the general population.

I. M. S. L. Silva (✉) · G. S. J. García · P. J. R. Cardoso · A. M. Byrne
Universidade Fernando Pessoa, Praça 9 de Abril, 349, 4249-004 Porto, Portugal
e-mail: isabels@ufp.edu.pt

G. S. J. García
e-mail: gloria@ufp.edu.pt

A. M. Byrne
e-mail: abyrne@ufp.edu.pt

P. J. R. Cardoso
Universidade Lusíada do Porto, Praça 9 de Abril, 349, 4249-004 Porto, Portugal

W. Leal Filho et al. (eds.), *Climate Change and Health Hazards*, Climate Change Management, https://doi.org/10.1007/978-3-031-26592-1_4

Research on the effects of nature on health has focused mainly on disease or on reparative effects (reduction of stress, anxiety, anger, depression, substance abuse and attention deficits, recovery after surgical treatments, etc.). However, scrutiny of existing research has also allowed us to confirm that a connection with nature is a significant factor in well-being, having deep emotional effects on individuals. A narrative literature review conducted by Silva et al. (2022) underlines the fact that contact and feelings of connection with nature are associated with a more meaningful life in general, with higher levels of happiness, life satisfaction, health perception, psychological, social and emotional well-being, positive affect, mindfulness, vitality, autonomy, personal growth, and pro-sociality; it fosters interest in acting for the common good, improves cognitive functioning, creative problem solving, and generosity and has a positive effect on aspirations.

Psychosocial reactions that occur in situations of environmental disasters have also been an area of concern for researchers. But distress and loss of comfort caused by degradation of the environment, home and sense of belonging do not seem to have received equal research attention, despite the severity of the consequences for mental health.

In this chapter, we present a narrative review which aims to reflect on the ways in which climate change can affect individuals' mental health, from the reactions to major events to more discrete reactions, often invisible to governments. Consequently, this review has no predetermined research question, specified search strategy or protocol, as its intention is to furnish a wider and comprehensive overview on this topic, providing state-of-the-art information and putting it into perspective. The literature search was conducted in the Online Knowledge Library (B-on), as it aggregates numerous databases from different scientific fields, allows for the selection of peer reviewed papers and facilitates access to papers' full texts. The papers were selected with the intention of demonstrating the diversity of psychosocial effects associated with climate change at the present time. From the selected papers, we also identified and analysed the sources consulted, in order to delve deeper into the concepts.

Psychosocial Reactions to Environmental Disasters

Environmental disasters are events, frequently related to climate change, that can be designated as catastrophic due to the massive destruction of property or threat to life involved. Catastrophic situations are sudden events, almost always unpredictable, of natural or technological origin, capable of creating victims and massive damage, seriously affecting people's safety, the living conditions of populations and the socio-economic conditions of the country (Baum 1987), and which may lead to the development of individual crisis and/or trauma.

According to the United Nations *Atlas of Mortality and Economic Losses from Weather, Climate and Water Extremes* (United Nations 2021), from 1970 to 2019, natural disasters were responsible for 50% of all disasters, 45% of all reported deaths and 74% of all reported economic losses. In that period, more than 11,000 disasters attributed to these hazards were reported, amounting to more than two million deaths and $3.64 trillion in losses. And despite more than 91% of those deaths occurred in developing countries, this is not just a distant problem. See, for example, what happened in Europe during this period, in which 1,672 disasters occurred and 159,438 citizens died due events such as floods, storms and extreme temperatures.

In parallel to these disasters, we should not forget a series of other environmental events and situations that, on another scale, generate an imbalance in human beings when they perceive that their usual skills and strategies to deal with challenging circumstances are ineffective to cope with the new situation (Silva 2009). If the problem cannot be solved, avoided, or redefined, the tension can reach a peak and the person goes into imbalance, becoming vulnerable and feeling that they cannot control their own life (Gilliland and James 1988; Roberts 1990).

Both of these occurrences, environmental disasters and crises, can lead to emergency situations and result in psychological trauma. The NHS Department of Health Emergency Preparedness Division (2009) stresses that major events and disasters can challenge the beliefs we hold about ourselves, our families and friends, and the world, namely the beliefs that the world is essentially a good place, that life has meaning, purpose, value and is worthwhile. Psychological trauma occurs precisely when such events challenge these assumptions and push the person beyond their tolerance limit, when they can be so damaging that they make the person question and modify their fundamental views of the world (NHS Department of Health Emergency Preparedness Division 2009). Psychological reactions can be very different depending on whether it is a singular event causing psychological trauma, or long-term events that can result in complex psychological trauma, where reactions such as anticipation, mass denial, dissociation, self-anaesthesia, and anger may be present (North Atlantic Treaty Organisation [NATO] 2008).

The confrontation with extreme situations may make us question our view of the world, our values, and what we consider to be safe, fair and predictable, and may generate a feeling of powerlessness and insecurity (Maia 2007), but it may also offer an opportunity for personal growth and development of new skills and resources to deal with new situations that may occur in the future (Roberts 1990; Gilliland and James 1988).

Most people do not remain permanently in crisis, restoring their balance in a few weeks or months after the event and reducing the discomfort experienced. For others this homeostatic mechanism is less effective. Some may experience an increased sense of efficacy and the belief in their ability to manage future challenges, with what we can designate as post-traumatic growth occurring. However, others will also experience adverse mental health effects from ecological disasters. Crisis situations may, therefore, have serious psychological consequences that very often are underestimated, even though these may frequently be more serious than the physical effects (Silva 2009).

Crisis and catastrophic situations are potentially traumatic events that may jeopardize normal functioning mechanisms and may lead to psychopathology. However, research shows that, although exposure to threatening or adverse situations may indeed increase the probability of developing psychopathology, it does not inevitably cause psychological distress, and there are individuals who show no signs of considerable distress after such an experience (Maia 2007). There is, therefore, great variability in individual reactions to these situations.

Exposure to situations in which the person feels their life is threatened, or real death or serious injury occur, or in which direct or indirect extreme exposure to aversive details of traumatic events take place (e.g., what often happens to field professionals; the almost real time exposure to tragic events through the media) can generate significant psychological disturbance (American Psychiatric Association [APA] 2014).

In most cases where there is emotional distress and disturbance after the event, this effect tends to be very short-lived, however, in some people it becomes chronic. Moreover, when the disaster results from human action, the effects of the event tend to last longer than when it results from a natural phenomenon (Baum 1987).

Hofmeister and Navarro (2017, p. 36), reflecting on the various definitions of "victim" presented in the literature, conclude that they all have two aspects in common: the image of someone "who has suffered harm and injury from forces beyond their control" and "the idea of the victim as being in a weak and defenceless state, in need of protection."

Shepherd and Hodgkinson (1990) also draw attention to the to "hidden victims" of disasters, presenting a classification of victims that specifically recognises their existence: (1) primary victims, those on the front lines who received the most exposure to the event (visible victims); (2) secondary victims, grieving family members and friends; (3) third-level victims, which include rescue personnel; (4) fourth-level victims, into which falls the community involved in the disaster, including those who went to the scene to offer help; (5) fifth-level victims, in which are included those people who present distress despite not having been directly involved; and (6) and sixth-level victims, which refers to those people who might have happened to be primary victims (often seen by others as someone "blessed by luck") or who are indirectly or vicariously involved.

It is worth highlighting two groups of "hidden victims" who are often neglected (Silva 2009):

a. Professionals acting in disaster situations and volunteers who are exposed to potentially traumatic stressful situations, having to help victims and to deal with their own feelings and exposure to danger.
b. Victims who result from exposure through the media. There is a strong relationship between the level of exposure given by the media to a disaster and the degree of anxiety experienced by people who were not directly exposed and for whom the disaster does not pose an immediate threat.

The psychological reactions of victims should be framed considering the phase in which they are: threat or impact phase; response phase (short term); and recovery phase (long term). In each of these stages, the reactions (and their meanings) may be different in terms of behaviours, thoughts and emotions presented by the victims (Silva 2009).

In the impact phase, it is usual for individuals in the affected area to mobilize immediately in terms of rescues, securing shelter, safety, and food/water for citizens. In that first moment, they essentially organize to address priorities (attending to the injured, identifying the most fragile members of the community), using their knowledge and waiting for resources from outside the community. Only after these priority needs are met, does attention begin to be paid to the mental health-related needs of the individuals affected by the event.

NATO (2008) draws attention to the enormous spectrum of psychosocial reactions to disaster situations and major events, pointing out that in such situations: (a) distress following these situations is very common; (b) in most cases this distress is transient and not associated with any dysfunction; (c) in some people this distress may last longer and be more disabling; (d) most people do not need mental health care provided by a professional in this field, although a minority may need such care; (e) screening, surveillance, and clinical assessment may be necessary for a sizeable group of people who may be at risk; and (f) a small percentage of those affected may require long-term mental health care to meet their needs.

In fact, most stress reactions to disasters are mild and transient (Laurendeau et al. 2007), although some people experience more intense reactions that generally do not become chronic problems. Among the first reactions to the event are behaviours that can be described as passivity versus rebellion. It is possible to find victims with passive behaviour who quietly obey the instructions given by rescue team professionals, as well as to find victims who present oppositional behaviour to retain some control over the situation, trying to maintain their routines, even when they are in shelters, and who may present a demanding and complaining attitude (Cohen 1990).

Parson (2008) gives a comprehensive description of traumatic stress responses in disasters:

- Neurophysiological responses: body tension; fatigue; nightmares; flashbacks; anxiety; bodily injury, malaise and pain; exhaustion due to not sleeping; cardiovascular reactivity and feeling of racing heart; gastrointestinal changes; decreased sex drive; lower back pain; irritability; feeling easily frightened; insomnia; hypervigilance; headache; changes in appetite; breathing problems.
- Responses related to personal experience: dissociation (feeling of unreality, "blankness," and being "disconnected"); low self-esteem due to feelings of failure and low self-efficacy; absence of joy and pleasure; persistent search for security; loss of hope, faith, will, motivation, or purpose in life; feeling of deep emptiness; feeling of injustice; feeling of being "in pieces" or fragmented; difficulty making decisions; internal feeling of exhaustion and chaos; emotional numbness; active expectation of future catastrophe; mental disorientation/disorganization; feeling overwhelmed by unwanted memories and affections; narcissism and resentment against authorities; self-blaming; apathy; feeling deeply distant from the rest of society; lack of trust; fear of intimacy; grief resulting from loss; shame; survivor's guilt; guilt for what you did to stay alive; feeling neglected and abandoned.
- Emotional responses: fear; shock; anxiety; anger; terror; dramatic mood swings; sadness; depression; sense of loss; shame; resentment against persons who caused

the event; desire for revenge; phobias; obsession with death, loss, fire, smoke, and/or debris.

- Cognitive responses: mental confusion; concentration problems; attention problems; memory difficulties; amnesia (problems remembering aspects of the traumatic event); denial; vivid and repetitive memories of the trauma; decreased effectiveness in decision making; unwanted memories; impoverished attention span; excessive worry.
- Behavioural responses: social isolation; increased interpersonal conflict; overprotection of significant others; avoidance of intimacy; avoidance of situations; instrumental substance abuse.

After exposure to a traumatic event, it is possible for the victim to present with an acute stress disorder. The APA. (2014, p. 337) defines acute stress disorder as "the development of characteristic symptoms lasting from 3 days to 1 month after exposure to one or more traumatic events." These symptoms may begin soon after the traumatic event and can be very diverse (APA 2014):

1. Intrusive symptoms, including intrusive, involuntary, recurrent memories of the traumatic event that cause discomfort; recurrent dreams related to the event, which also generate discomfort; dissociative reactions, such as flashback experiences, in which the individual feels or acts as if the traumatic event is occurring again and in which there may be a complete loss of awareness of the surrounding environment; intense or prolonged psychological distress or intense physiological reactions when the individual is exposed to stimuli (internal or external) that trigger memories of the traumatic event.
2. Negative mood, manifested by a persistent inability to experience positive emotions such as happiness, satisfaction, or feelings of love. Feelings of sadness and depression are frequently presented by victims and are often associated with difficulty in focusing attention and making decisions, feelings of personal ineffectiveness, guilt, hopelessness, social isolation, and lack of hope.
3. Dissociative symptoms, including the existence of an altered sensation of the reality of the surrounding environment and the inability to remember aspects of the traumatic event. The subjective sensation of feeling disconnected, of dullness (emotional anaesthesia), or of absence of emotional reaction, often described by individuals as feeling like the "living dead" (sometimes these latter symptoms are interpreted as positive by those around the victim, but may be nothing more than a way of avoiding pain that is felt to be unbearable) and depersonalization (the individual may experience persistent or recurrent feelings of being separated from their own body and mental processes, feeling as if they are an observer of their own life, or experiencing life and the world as unreal, as if they were a dream).
4. Avoidance symptoms. The individual strives to avoid memories, thoughts, or emotions associated with the traumatic event that might generate uneasiness and may even avoid external memories (e.g., people, places, conversations, activities, objects, and situations) that might awaken the trauma-related memories, thoughts, or emotions.

5. Activation symptoms. Individuals may present sleep disturbance; impulsive behaviour; irritable behaviour and anger outbursts, which may manifest through physical or verbal aggression towards people and/or objects; hostility; self-destructive behaviours and adoption of risky behaviours (e.g., substance abuse); hypervigilance, often associated with feelings of constant threat, fear, and anxiety; concentration difficulties; and startle response.

Based on literature reviews, Pereira (2015) and NATO (2008) systematize the symptoms of acute stress reaction into four major categories:

1. Emotional reactions, such as emotional shock, depression, anxiety and panic, guilt, anger, fear, despair, irritability, affective blunting, numbness, feelings of grief and vulnerability, helplessness/abandonment; fear that the situation will recur, and anhedonia;
2. Cognitive reactions, such as scattered attention, difficulty concentrating, difficulty making decisions, low self-efficacy and low self-esteem, disbelief, denial, memory changes, confusion, distortion, intrusive thoughts and concern dissociation, hypervigilance;
3. Physical reactions, such as hypertension, tachycardia, hyperventilation, fatigue, insomnia, hyper-alertness, somatic complaints (e.g., headache), nausea, thirst, appetite changes (usually decreased appetite), chills and sweats, decreased energy;
4. Behavioural reactions, such as "fight or flight" behaviours, immobilization, automatic obedience, alienation, withdrawal from activities, distrust, problems at work, conflict and agitation, regressive behaviours, and lack of energy; and
5. Social reactions, such as regressive behaviours, withdrawal, irritability, interpersonal conflicts, and avoidance are considered.

These symptoms can cause malaise and deficits in functioning in several domains relevant to the individual's life (such as interpersonal and work relationships). Regarding interpersonal relationships, withdrawal and angry outbursts are not uncommon, especially when the partner is not also a survivor of the event (Hodgkinson and Stewart 1991). The survivor may feel that the other person cannot understand because they were not present, and the sense of comfort is sometimes only found in the company of other survivors.

Also deserving of attention is "survivor's guilt," which Hodgkinson and Stewart (1991) describe an existential guilt, in which the victim wonders why they survived when others did not—"Why me?", "Why did God choose me?", "Why did I survive? I'm old and so many children died…"—and guilt focused on actions or on the lack of them—"Did I do enough?", "Could I have saved more people?"—, which may be especially intense when parents survive their children or when there is competition for survival. These authors further distinguish between "static guilt" and "animated guilt", considering that "static guilt" is characterized by keeping the victim stuck in the experience, unable to move forward—a situation in which the role of the professional will be to assist the person in progressing from static guilt to an animated guilt and from there to responsibility and behaviour that relieves the guilt—and identifying "animated guilt" as a guilt that enables the person to perform self-analysis and evolve towards responsibility.

Hodgkinson and Stewart (1991) warn that in some circumstances guilt assumes a symbolic function: by continuing to suffer, the person serves as a living testimony, paying homage to family members, friends, or other people who did not survive. This guilt will be a way of proving one's loyalty to these people, so recovering from the event can be interpreted as betrayal, as forgetting the deaths and suffering, or even as the individual's forgiveness of those who were considered responsible. These authors also consider that guilt presupposes the presence of choice and the power to exercise it, which, in fact, may not have been possible at the time of impact. Guilt, according to this perspective, can be perceived as an attempt to deny or annihilate the feeling of hopelessness and helplessness, that is, the feeling that nothing was or could have been done to prevent it from happening (Hodgkinson and Stewart 1991).

Defences such as repression, denial, and isolation are natural, and may even be necessary in the short term, and for some people it is the only way to allow them to continue to perform painful tasks, such as identifying family members and significant others in mortuary spaces (Hodgkinson and Stewart 1991). These defences may persist or give way to another period of open expression of emotions.

Sometimes survivors develop a strong awareness of victimization, which in extreme cases may reach the paranoid level, leading them to be suspicious of those who offer help and to be suspicious and fearful that these offers of help are false (Hodgkinson and Stewart 1991). According to Hodgkinson and Stewart (1991), this seems to have a few reasons behind it:

- Victims may perceive their experience as so precious that it should not be shared with those who will not be able to truly understand it. Victims may select groups from which "those who cannot understand" are excluded. But at the same time, it can cause the victim to become "frozen" in time and isolated in their victim identity;
- Sometimes victims perceive help as making them feel weak and diminished;
- It is possible that disaster victims have never before experienced such intense emotions in their lives and may be ashamed of their feelings or afraid that they are going "crazy". Even for those who have knowledge about emotional help and the professionals who provide it, receiving such help may not be easy.

According to NATO (2008), major events and disasters can challenge the beliefs we have about ourselves, especially the beliefs that the world is essentially a good place, that life and events have meaning and purpose, and that life is valuable and precious. Often survivors are unable to get through the experience of a normal grieving process because they cannot find a spiritual or ideological explanation for what happened (Silva 2009).

Stress reactions after a traumatic event are normal reactions to abnormal situations, and they can even be adaptive. However, in some cases, they may evolve into psychopathology (Pereira 2015). The diagnosis of post-traumatic stress disorder can be made even when these symptoms have a late onset, 6 or more months after the event. In this condition, as we described before, the intrusive symptoms that occur immediately after the traumatic event may be present (APA 2014; Hodgkinson and Stewart 1991).

In the clinical pictures of post-traumatic stress disorder, we can also find the avoidance of disturbing or intrusive memories, thoughts or emotions linked to the event (internal stimuli), as well as of people, places, conversations, activities, objects and situations (external stimuli); negative changes in cognitions (inability to remember important aspects of the event; excessively negative and persistent beliefs or expectations about oneself, others, and/or the world, e.g., "I'm bad," "I can't trust anyone," "The world is too dangerous," "My nervous system has been ruined forever") and mood—on the one hand, fear, feelings of horror, anger, feelings of guilt and/or shame, which are often reflected in disengagement from meaningful activities, and on the other hand, persistent inability to experience positive emotions such as happiness, satisfaction, and feelings of love.

Similarly to what happens with acute stress disorder, in post-traumatic stress disorder significant changes in activation and reactivity associated with the event may occur, such as (APA 2014; Crocq 2002): irritable behaviours and angry outbursts; reckless and self-destructive behaviours; hypervigilance; exaggerated startle response; concentration difficulties; sleep disturbance. These symptoms cause significant discomfort or difficulties in social, occupational, or other areas relevant to the individual (APA 2014). Dissociative symptoms, such as depersonalization and derealization, should also not be overlooked (APA 2014).

Prevalence of post-traumatic stress disorder (5.8%–90%), anxiety disorders (0.04%–84%) and depressive disorders (5.9%–81%) resulting from natural disasters and extreme weather events varies considerably, most likely because of differences in terms of sociodemographic variables, type of events and their severity, time of assessment and methodology (Rataj et al. 2020).

Post-traumatic stress disorder is associated with other mental health problems, such as higher incidence of suicide and suicidal ideation, depression, anxiety, insomnia, violence, aggression, interpersonal difficulties, and job-related difficulties, alcohol, or other substances abuse, as well as higher probability of engaging in risk behaviours such as smoking and unhealthy eating habits (Clayton et al. 2021).

"Invisible" Psychosocial Reactions to Climate Change

Psychosocial reactions to climate change are not always visible and their association with climate change is not always recognised. However, research shows that climate change and its effects can generate distress, anxiety, worry, and/or fear and it can result in what has been called "psychoterratic syndromes", a set of responses to a negatively felt and negatively perceived environmental change (Albrecht 2011, p. 48). Although it is very difficult to establish a causal link between climate and environmental crisis and mental ill-health, Clayton (2020) also identifies terms such as "ecoanxiety", "climate anxiety", "climate grief" and "eco distress" (p. 2) to describe these negative responses, and Galway et al. (2019) employ the term "earth-related states" (p. 6/14). But other concepts can be found in scientific literature, such as nature-deficit disorder (Louv 2005), a term that refers to the human costs of alienation from nature (diminished use of the senses, attention difficulties, and higher rates of physical and emotional illness), eco-paralysis (Albrecht, 2011) a term used to designate people who become so distressed by environmental issues that they are unable to act, a situation that is sometimes misinterpreted as apathy, and "solastalgia", which is "the distress that is produced by environmental change impacting on people while they are directly connected to their home environment" (Albrecht et al. 2007, p. 44). To address mental-ill health in response to climate change and ecological adversity, Thoma et al. (2021) introduced the concept of Clinical Ecopsychology.

Research has shown that even gradual and slow environmental changes, and chronic and subacute climate events (such as heat waves, air pollution, flooding, drought, chronic degradation of places by sea level rise, etc.) have significant effects on people's mental health, including on children and adolescents (van Nieuwenhuizen et al. 2021). Children and young adults aged between 16 and 25 years have been revealed to be very or extremely worried about climate change (59%); more than half of them exhibit feelings of sadness, anxiety, guilt, annoyance, powerlessness and helplessness; more than 45% report that those feelings are affecting their daily life and functioning; 75% think that their future is terrifying; and 83% believe that people are failing in taking care of the planet (Hickman et al. 2021).

Increasing Temperatures and Extreme Heat Events

Increased average temperatures and decreased air quality have consequences in the type of activities that people engage in, principally in terms of outdoor activities and recreation (Clayton et al. 2021). It is possible that people try to adjust themselves to the temperature change by exercising in indoor environments, but reducing access to the restorative potential of outdoor environments may also result in an increase in stress and in the loss of the long-term emotional benefits of outdoor physical activity (Clayton et al. 2021).

Increased temperatures are related to negative mental health outcomes, namely to mood and anxiety disorders, schizophrenia, vascular dementia and suicide levels, as well as to higher use of emergency mental health services (Clayton et al. 2021). Higher temperatures represent an additional source of stress for people who are already psychologically at risk, overwhelming their coping skills, but they can also have less obvious negative effects on mental health in the general population, generating a "low-level stress burden which may not manifest as a clinically diagnosable mental health problem" (Clayton et al. 2021, p. 33). Furthermore, high temperatures are associated with higher levels of negative emotions, feelings of fatigue and increased suicide levels, but also with lower positive emotions (Clayton et al. 2021).

Qiu et al. (2022) demonstrated that, for each 5 °C increase in short-term exposure to cold season temperature, the relative risk of acute hospital admission increased by 3.66% for depression, by 3.03% for schizophrenia, and by 3.52% for bipolar disorder.

A review of studies on heat and interpersonal aggression also reveals a causal and positive relationship between these variables. Different hypotheses have been presented to try and explain this association between heat and violence (Clayton et al. 2021), such as: (a) it is possible that heat has an impact on arousal and irritability, with consequent decreases in attention and self-regulation; (b) heat can increase negative and hostile thoughts; (c) higher temperatures are associated with loss of sleep, which can in turn increase irritability; and (d) heat can have a negative effect on cognitive function, reducing the capability to resolve a conflict without violence.

Droughts

Drought is also associated with psychological distress (generalized anxiety and depression), particularly for farmers and farming communities, who see themselves facing increased workload with reduced "likelihood of a good crop, raising the risk of financial hardship, and causing uncertainty about the future". Droughts may also create a sense of shame and humiliation, degrading the land and the farmer's connection to it and it has been associated with suicide (Clayton et al. 2021, p. 34).

Moreover, droughts may force migration from communities with depressed economies, changing family and community structures, for example increasing symptoms of anxiety and depression, and engendering tension within family and social networks, often because of the poor reception of immigrants by receiving communities, particularly for those who have a strong sense of identity linked to the land, as "drought can lead to environmental degradation of one's home, affecting people's ability to interact with their environment and causing additional issues related to rupture of place bonds, culture change and loss, and altered community and family dynamics" (Vins et al. 2015, p. 13259).

Air Pollution

Air pollution increases the risk of anxiety and other mental disorders, impairs cognitive function, predicts lower ratings for happiness and life satisfaction, degrades social behaviour (it is associated with higher irritability and higher rates of criminal behaviour), and also increases recourse to mental health services (Clayton et al. 2021).

Newbury et al. (2019) demonstrated that there are significant associations between outdoor exposure to nitrogen dioxide, nitrogen oxides, and particulate matter and reports of psychotic experiences during adolescence. Shin et al. (2018) found that high concentrations of PM_{10}, NO_2 and CO are positively associated with the prevalence of high stress, poor quality of life, depressiveness, diagnosis of depression, and suicide ideation after adjusting for confounding factors. In the cold season, an increased short-term exposure to $PM_{2.5}$ and NO_2 is related with an increase in the risk of acute hospital admissions for psychiatric disorders (Qui et al. 2022)—exposure to $PM_{2.5}$ is associated with hospital admission rates for depression, schizophrenia, and bipolar disorder; and increase in NO_2 is linked to hospital admission rates for depression and schizophrenia.

Moreover, the effects of air pollution exposure on mental health are not just momentary, they have been shown to have long lasting consequences—for example, early-life exposure to nitrogen oxides is significantly associated with general psychopathology at 18 years of age (Reuben et al. 2021). Shin et al. (2018) underline that long-term exposure to ambient air pollution can be an independent risk factor for mental health disorders (from subjective stress to suicide ideation).

Many explanatory hypotheses have been presented for the association between air pollution and adverse psychiatric health outcomes (Qiu et al. 2022): increased inflammatory responses triggered by air pollutants can exacerbate depressive or manic symptoms in people with psychiatric disorders; particulate matter can lead to adverse neuro-inflammatory and autoimmune responses, which can trigger/worsen psychiatric conditions; air pollutants may increase glucocorticoid activity and stress hormone cortisol concentrations, triggering psychiatric episodes; traffic pollutants (such polycyclic aromatic hydrocarbons and ultrafine particles) have the potential to lead to brain atrophy and cognitive impairment, etc.

Although scientific literature is not consensual with regard to the association of air pollution and mental health impairment (see, for e.g., Zijlema et al. 2016 study), the evidence collected makes clear the need to deepen the scientific knowledge on this issue.

Deforestation

Forestry area at country level has been shown to be associated with a lower prevalence of mental health disorders (Bolton et al. 2022), but deforestation has been occurring either as a consequence of climatic events, or because of direct human action. Populations believe that forests are a source of health and protection from various types of stress, and news concerning deforestation has a significant psychosocial impact, increasing their ecological awareness and, consequently, increasing their level of stress (Cianconi et al. 2020).

In addition to the loss of the positive effects of contact with nature in promoting health, optimizing individuals' functioning and preventing diseases and dysfunctions, deforestation can have an adverse impact on well-being. In fact, "people believe that an important world heritage has been damaged and lost", a feeling that is known as biospheric concern (Cianconi et al. 2020, p. 8). Landscape modification can make people develop a profound sense of loss of connection and detachment from the environment they know (solastalgia) and can cause eco-anxiety.

Climate Change and Media

Climate change is frequently perceived as a temporally and spatially more distant danger that does not directly affect us. The media have been assuming an important role increasing awareness of the present danger. University students strongly agree on the importance of the media in the awareness of climate change (Rahman et al. 2014). Actually, the main sources of information about climate global warming are television (82.8%), the Internet (56.2%) and social networks (49.4%) (Morote and Hernández 2022).

Information about ecological crises and public debates on the need for a sustainable lifestyle have been promoting feelings of shame, guilt, anger, fear, grief and anxiety in communities, but also of hope (Neckel and Hasenfratz 2021). In this context, those feelings assume enormous relevance and can have ambivalent repercussions. On one hand, they can increase motivation to adopt more environmentally responsible behaviours. This favours a conscious attention to the problems of global warming and ecological crises, with acceptance of the idea that climate change is a major threat, combat of which is an absolute necessity. On the other hand, they can also generate emotional resistance, often manifested as criticism of moralizations and prohibitionist policies (Neckel and Hasenfratz 2021).

The way we see climate change and behave (or fail to do so) to protect the environment does not result purely from rational-cognitive processes. Affective-emotional processes also assume an important role in the way we understand climate change and behave towards it—emotions are probably the biggest promoter of change, the biggest barrier to change, and often the biggest reward of change.

Public discussion has been valuing what can be called "persuasive communication"—a communication based on the belief that active change through human action first implies an awareness of the need for change, and this awareness is demonstrated by feelings such as anxiety and fear (Neckel and Hasenfratz 2021). Climate anxiety has been identified as the consequence of people's perception of climate change (Clayton 2020). But, should there be a limit to this? What should be considered "reasonable anxiety"?

Conclusion

Narrative reviews, such as this, have the potential for bias in the paper selection process and in the literature analysis, often being accused of lack of transparency, as they allow researchers, even unintentionally, to give a high value to some papers while entirely ignoring others. But it is also true that, if well conducted, these reviews allow us to present a broader view of the topic, a view that is not limited by indexers and strict criteria which, in spite of intending to ensure scientific rigour in the review, often force us to ignore relevant scientific publications.

This literature review allows us to conclude that we are far from having a complete comprehension of the connections between climate change and its consequences on mental health. The study of these complex connections is a challenge—there is considerable heterogeneity in the climate change aspects assessed and a huge diversity in the measures adopted; there are several direct and immediate effects, but also indirect short-term effects as well as long term ones; different populations present distinct levels of vulnerability for mental health problems; aetiology of mental health disorders is very complex, etc.

Nevertheless, research conducted in recent years alerts us to the multiplicity of risks that climate change poses to mental health, and not only to the risks associated with environmental disasters, but also risks associated with climate changes of smaller magnitude, previously "invisible" to governments, communities and to mental health professionals. Research also points out the need to educate the media, as its approach to environmental issues can have powerful effects on risk perception and consequence predictions, as well as in individuals' beliefs and their confidence to adopt climate protective behaviours.

It is time to stop and think: is Nature fighting us back? For how long will we ignore the consequences of our actions on our own mental health before starting to address this problem? What are our primary world beliefs? Do we feel that the world is harmless or threatening, stable or fragile, just or unjust, meaningful or meaningless, improvable or too hard to improve, beautiful or ugly (Clifton and Kim 2020)? How do those beliefs affect our expectations and responsibilities toward future actions in order to protect Nature and, consequently, ourselves?

References

Albrecht G (2011) Chronic environmental change: emerging 'psychoterratic' syndromes. In: Weissbecker I (ed) Climate change and human well-being. Int and Cult Psychol. Springer, New York, pp 43–56. https://doi.org/10.1007/978-1-4419-9742-5_3

Albrecht G et al (2007) Solastalgia: the distress caused by environmental change. Australas Psychiatry 15(1):S95–S98. https://doi.org/10.1080/10398560701701288

American Psychiatric Association (2014) DSM-V: manual de diagnóstico e estatística das perturbações emocionais. Climepsi Editores, Lisbon

Baum A (1987) Toxins, technology, and natural disasters. In: VandenBos GR, Bryant, BK (eds) Master lectures series. Cataclysms, crises, and catastrophes: psychology in action. American Psychological Association, Washington DC, pp 5–49. https://doi.org/10.1037/11106-001

Bolton AV, Montag D, Gallo V (2022) Global forestry areas, deforestation and mental health: a worldwide ecological study. J Clim Chang Health 6:100–109. https://doi.org/10.1016/j.joclim.2021.100109

Cianconi P, Betrò S, Janiri L (2020) The impact of climate change on mental health: a systematic descriptive review. Front Psychiatry 11:74. https://doi.org/10.3389/fpsyt.2020.00074

Clayton S (2020) Climate anxiety: psychological responses to climate change. J Anxiety Disord 74:102263. https://doi.org/10.1016/j.janxdis.2020.102263
Clayton S, Manning CM, Speiser M, Hill AN (2021) Mental health and our changing climate: impacts, inequities, responses. American Psychological Association and ecoAmerica, Washington DC. https://www.apa.org/news/press/releases/mental-health-climate-change.pdf
Clifton J, Kim ES (2020) Healthy in a crummy world: implications of primal world beliefs for health psychology. Med Hypotheses 135:109463. https://doi.org/10.1016/j.mehy.2019.109463
Cohen RE (1990) Post-disaster mobilization and crisis counselling: guidelines and techniques for developing crisis-oriented services for disaster victims. In: Yeager KR, Roberts AR (eds) Crisis intervention handbook—assessment, treatment and research. Wadsworth Publishing Company, New York, pp 279–299
Crocq L (2002) Special teams for medical/psychological intervention in disaster victims. World Psychiatry 1(3):154–155
Galway LP, Beery T, Jones-Casey K, Tasala K (2019) Mapping the solastalgia literature: a scoping review study. Int J Environ Res Public Health 16:2662
Gilliland BE, James RK (1988) Crisis intervention strategies. Brooks/Cole Publishing Company, California
Hickman C et al (2021) Climate anxiety in children and young people and their beliefs about government responses to climate change: a global survey. Lancet Planet Health 5(12):e863–e873. https://doi.org/10.1016/S2542-5196(21)00278-3
Hodgkinson PE, Stewart M (1991) Coping with catastrophe. A handbook of disaster management. Routledge, London
Hofmeister U, Navarro S (2017) A psychosocial approach in humanitarian forensic action. The Latin American perspective. Forensic Sci Int 280:35–43. https://doi.org/10.1016/j.forsciint.2017.08.027
Laurendeau M-C, Labarre L, Senécal G (2007) The psychosocial dimension of health and social service interventions in emergency situations. Open Med J 1(2):e102–e106
Louv R (2005) Last child in the woods: saving our children from nature-deficit disorder. Algonquin Books, USA
Maia AC (2007) Factores preditores de PTSD e critérios de selecção em profissionais de actuação na crise [PTSD predictors and selection criteria in crisis professionals]. In: Sales L (ed) Psiquiatria de catástrofe [Disaster psychiatry]. Almedina, Coimbra, pp 263–276
Morote A-F, Hernández M (2022) What do school children know about climate change? A social sciences approach. Soc Sci 11:179. https://doi.org/10.3390/socsci11040179
Neckel S, Hasenfratz M (2021) Climate emotions and emotional climates: the emotional map of ecological crises and the blind spots on our sociological landscapes. Soc Sci Inf 60(2):253–271. https://doi.org/10.1177/0539018421996264
Newbury JB et al (2019) Association of air pollution exposure with psychotic experiences during adolescence. JAMA Psychiat 76(6):614–623. https://doi.org/10.1001/jamapsychiatry.2019.0056
NHS Department of Health Emergency Preparedness Division (2009) NSH emergency planning guidance. Planning for the psychosocial and mental health care of people affected by major incidents and disasters: interim national strategic guidance. NHS, London. https://www.the-eps.org/wp-content/uploads/dept-of-health-guidance.pdf
North Atlantic Treaty Organisation (2008). Psychosocial care for people affected by disasters and major incidents. A model for designing, delivering and managing psychosocial services for people involved in major incidents, conflict, disasters and terrorism. NATO. https://www.coe.int/t/dg4/majorhazards/ressources/virtuallibrary/materials/Others/NATO_Guidance_Psychosocial_Care_for_People_Affected_by_Disasters_and_Major_Incidents.pdf
Parson ER (2008) Victims of disasters: helping people recover—from acute distress to healing and integration. http://www.giftfromwithin.org/pdf/victims.pdf
Pereira M (2015) Intervenção psicológica em crise e catástrofe [Psychological intervention in crisis and catastrophe]. Ordem dos Psicólogos Portugueses, Lisbon

Qiu X et al (2022) Associations of short-term exposure to air pollution and increased ambient temperature with psychiatric hospital admissions in older adults in the USA: a case–crossover study. Lancet Planet Health 6(4):e331–e341. https://doi.org/10.1016/S2542-5196(22)00017-1

Rahman MS, Mohamad OB, Zarim ZB (2014) Climate change: a review of its health impact and perceived awareness by the young citizens. Glob J Health Sci 6(4):196–204. https://doi.org/10.5539/gjhs.v6n4p196

Rataj E, Kunzweiler K, Garthus-Niegel S (2020) Extreme weather events in developing countries and related injuries and mental health disorders—a systematic review. BMC Public Health 16:1020. https://doi.org/10.1186/s12889-016-3692-7

Reuben A et al (2021) Association of air pollution exposure in childhood and adolescence with psychopathology at the transition to adulthood. JAMA Netw Open 4(4):e217508. https://doi.org/10.1001/jamanetworkopen.2021.7508

Roberts AR (1990) Crisis intervention handbook—assessment, treatment and research. Wadsworth Publishing Company, New York

Shepherd M, Hodgkinson PE (1990) The hidden victims of disaster: helper stress. Stress Medicine 6:29–35

Shin J, Park JY, Choi J (2018) Long-term exposure to ambient air pollutants and mental health status: a nationwide population-based cross-sectional study. PLoS One 13(4):e0195607. https://doi.org/10.1371/journal.pone.0195607

Silva I (2009) O papel da Psicologia no acompanhamento de vítimas, familiares e profissionais intervenientes em catástrofes [The role of Psychology in accompanying victims, relatives and professionals involved in disasters]. In: Pereira MFTP (ed) CSI Catástrofes [CSI Catastrophes]. Edições Universidade Fernando Pessoa, Porto, pp 177–194

Silva I, Jólluskin G, Cardoso P (2022) Positive ecological psychology: contributes to the promotion of healthy lives and well-being. In: Filho WL, Vidal DG, Dinis MAP, Dias RC (eds) Sustainable policies and practices in energy, environment and health research. World Sustainability Series. Springer, Cham, pp 261–275 https://doi.org/10.1007/978-3-030-86304-3_16

Thoma MV, Rohleder N, Rohner SL (2021) Clinical ecopsychology: the mental health impacts and underlying pathways of the climate and environmental crisis. Front Psychiatry 12:675936. https://doi.org/10.3389/fpsyt.2021.675936

United Nations (2021) Atlas of mortality and economic losses from weather, climate and water extremes. UN. https://news.un.org/en/story/2021/09/1098662

van Nieuwenhuizen A, Hudson K, Chen X, Hwong AR (2021) The effects of climate change on child and adolescent mental health: clinical considerations. Curr Psychiatry Rep 23(12):88. https://doi.org/10.1007/s11920-021-01296-y

Vins H, Bell J, Saha S, Hess JJ (2015) The mental health outcomes of drought: a systematic review and causal process diagram. Int J Environ Res Public Health 12(10):13251–13275. https://doi.org/10.3390/ijerph121013251

Zijlema WL et al (2016) The association of air pollution and depressed mood in 70,928 individuals from four European cohorts. Int J Hyg Environ Health 219(2):212–219. https://doi.org/10.1016/j.ijheh.2015.11.006

Isabel Maria Sousa Lopes Silva Ph.D. in Psychology from University of Porto. Associate Professor at Fernando Pessoa University (UFP), where she is responsible for the Behavioural and Social Sciences Research Centre (FP-B2S) and where she coordinates the Psychometry Laboratory. She is a clinical psychologist in practice at the University Fernando Pessoa's Teaching Hospital. She holds a specialist title in Clinical and Health Psychology and in Community Psychology awarded by the Portuguese Order of Psychologists.

Gloria Svietlana Jólluskin García Ph.D. in Psychology from University of Santiago de Compostela. Associate Professor at University Fernando Pessoa (UFP). Research member of the

Behavioural and Social Sciences Research Centre (FP-B2S). Member of the Psychometrics Laboratory at UFP. Holds a specialist title in Social Psychology and in Psychology of Justice awarded by the Portuguese Order of Psychologists.

Paulo José Ribeiro Cardoso Ph.D. in Advertising and Public Relations from Autonomous University of Barcelona. Professor at Lusíada University of Porto and at University Fernando Pessoa, in the fields of Marketing and Communication. As a researcher, he has published and disseminated his work in scientific journals and at national and international conferences. As a consultant he has worked in the field of consumer behaviour, brand identity and communication planning.

Allison Margaret Byrne Master in Computer Science, University of the West of England. B.A. in Portuguese, University of London, King's College. Member of the Behavioural and Social Sciences Research Centre (FP-B2S). Member of LACI Neuroscience Group at UFP. Her research interests include language development, language learning and production, as well as translation.

Part II
Climate Change Impacts on Human and Environmental Health

Chapter 5
Industrial Waste Management in Ghana: Environmental Challenges and Climate Change Impacts on Human Health

Justice Kofi Debrah, Godfred Kwesi Teye, Zujaja Wahaj, and Maria Alzira Pimenta Dinis

Abstract Waste having a significant and relevant material presence in our everyday life was not considered worthy of human attention up until the 1990s. Most probably, this delayed recognition of waste is due to the fact that it has been mistakenly understood as something that is supposed to remain invisible to the human eye in a sense that it is assumed to be occupying those spaces that are not inhabited by human beings. Contemporary times, however, are marked by giving due importance to waste. It is seen as a crucial environmental and health challenge for species' survival on earth and is therefore placed center stage in the United Nations Sustainable Development Goals (SDGs) that aim to protect the planet. More recently, SDG 12, ensuring sustainable consumption and production patterns, has alerted to fast-growing waste management (WM) environmental issues. With the earth being viewed as a limited natural resource and having the inadequate capacity to accommodate wastes leading to adverse environmental and health consequences, caution must be taken to understand the associations between global consumption, total waste, and wasting patterns. Conceptualizing waste as a globally circulating material, particularly in terms of greenhouse gas emissions (GHG) such as carbon dioxide (CO_2), methane (CH_4),

J. K. Debrah (✉)
Faculty of Science and Technology, University Fernando Pessoa (UFP), Praça 9 de Abril 349, 4249-004 Porto, Portugal
e-mail: 40511@ufp.edu.pt

G. K. Teye
Key Laboratory of Integrated Regulation and Resource Development on Shallow Lakes of Ministry of Education, College of Environment, Hohai University, No.1 Xikang Road, Nanjing 210098, China
e-mail: teyegodfred@hhu.edu.cn

Z. Wahaj
Faculty of International Business and Management, National University of Sciences and Technology, Islamabad, Pakistan
e-mail: zujaja.wahaj@nbs.nust.edu.pk

M. A. P. Dinis
UFP Energy, Environment and Health Research Unit (FP-ENAS), University Fernando Pessoa (UFP), Praça 9 de Abril 349, 4249-004 Porto, Portugal
e-mail: madinis@ufp.edu.pt

W. Leal Filho et al. (eds.), *Climate Change and Health Hazards*, Climate Change Management, https://doi.org/10.1007/978-3-031-26592-1_5

and sulphur oxide (SO_x), the governance of waste has become a global concern. This chapter focuses on WM in Ghana, a middle-income economy country in the West of Africa. Due to rapid industrialization, Ghana is witnessing an ever-escalation of the country's contribution to global environmental waste issues, especially GHG emissions. This study aims to identify and address the challenges associated with managing waste in Ghana. Additionally, it proposes specific measures to mitigate climate change and its effects on the environment and human health resulting from WM. A descriptive cross-sectional study was conducted with 101 responses received from WM experts based on the aim. The data were analysed using the statistical package for social sciences (SPSS) version 27. 72% of respondents were male, with most completing first and second degrees and have been working between 11 and 15 years. Although most of the respondents' report waste being poorly managed, 93% consider that the associated causes of improper WM include inadequate infrastructure, lack of landfills, lack of tools and logistics, and equipment. All respondents believed that the challenges affect climate change through the emission of GHGs and particulate matter, resulting in global warming and impacting human health.

Introduction

Waste management (WM) is a significant issue in developing countries (Jude et al. 2019; Debrah et al. 2021a; Debrah et al. 2021). It is an escalating challenge in terms of environmental degradation, economic sustainability (Vitorino de Souza Melaré et al. 2017; Ferronato and Torretta 2019) and associated social issues such as poverty and climate change (Debrah et al. 2022c). WM in developing countries is intensifying due to spontaneous population growth (Schneider et al. 2017), rapid urbanization (Awasthi et al. 2019), weak sanitation infrastructure (Teye et al. 2019), and poor policy implementation. All these factors adversely affect the environment and human health (Vidal et al. 2019; Debrah et al. 2021b, 2022b). According to Kaza et al. (2018), 2.01 billion tons of waste are globally generated in a year, with approximately 1.33 billion tons of solid waste (66%) coming from developing countries (Bundhoo 2018; Odonkor and Sallar 2021).

In Sub-Saharan Africa (SSA), 174 million tons of waste were generated in 2016 (Kaza et al. 2018), increasing to 180 million tons in 2019 (Ayeleru et al. 2020). Waste generation is expected to triple by 2050 in SSA (Kaza et al. 2018) due to the fast industrialization agenda of most governments within the sub-region. However, 44% - 55% of waste is collected (AUDA-NEPAD 2019; Yeo et al. 2020), including medical, industrial, commercial, and residential waste. It is also estimated that about 4% of the collected waste is recycled, with the remaining sent to uncontrolled dumpsites and landfills, mostly non-engineered (Idowu et al. 2019).

In Ghana, about 23,300 tons of waste are generated daily (Debrah et al. 2022b). This presents a challenge since the WM infrastructural gap, which is caused by insufficient funding for WM activities, lack of enforcement of sanitation-related laws, and change in the pattern of consumption by the citizenry (Sharma et al. 2020), contributes towards inefficient WM services. Moreover, the infrastructural gap in managing waste generated by industry, households, and commercial entities i.e., markets and shops, results in the release of gases that pollute the environment as well as the transmission of diseases that affect human health (Debrah et al. 2021b). The WM situation is considered as a challenge by stakeholders due to the various environmental and health impacts ranging from air pollution, climate change and growth of micro-organisms (Nikiema and Asiedu 2022; Chen et al. 2020). Additionally, most residential, commercial, institutional, and industrial wastes generated in SSA are either uncontrollably burnt or incinerated, consequently leading to the emission of air pollutants such as dioxins, particulate matter, and heavy metals that may be toxic to public health (EPA 2020). These hazardous waste pollutants are mainly generated by entities such as hospitals, higher education institutions (HEI), and industries are harmful when not adequately managed. Furthermore, these wastes could cause chronic diseases and contamination of soil and water bodies (Manzoor and Sharma 2019), resulting in bio-accumulation and bio-magnification (Zenker et al. 2014), negatively impacting human health.

With Ghana being no exception, improper WM in the developing countries contributes to the factors such as GHG and particulate matter, affecting climate change. According to Wei et al. (2016), developing countries contribute 39–47% of methane (CH_4), carbon dioxide (CO_2), sulphur oxide (SO_x), and nitrous oxide (NO_x) emissions adding up to the effects of climate change. Studies conducted by Nathaniel et al. (2020), Bhat et al. (2021) and Maja and Ayano (2021) have shown that one of the best ways to manage climate change effects is through proper WM which includes pollution control and environmental protection. Most of the industrial waste in SSA is disposed at non-engineered landfills, where it anaerobically decomposes to emit some of the GHG (Debrah et al. 2022b), such as CH_4, CO_2, SO_x and NO_x, as well as particulate matters, which contribute to climate change and its adverse effects on the environment and human health. Some of these effects may range from extreme weather conditions such as bushfires, heatwaves, and floods, rising sea levels (Leigh et al. 2015; Mora et al. 2018; Aghakouchak et al. 2020), poor air quality, the spread of infectious diseases (Zscheischler et al. 2018), food insecurity and threats to water bodies (Youn et al. 2014). Given the anticipated problems associated with the effects of climate change on the environment and health, proper WM is thus essential. Yet, it is a challenge since it is contingent upon education, attitudinal change, provision of infrastructure and awareness (Debrah et al. 2021a), and stakeholder participation (Oduro-Appiah et al. 2019; Teye et al. 2019; Kotei et al. 2020; Pandey et al. 2022) within the SSA, and also in Ghana. The main aim of this study is to present the challenges associated with WM in Ghana and how stakeholders, which include WM companies, academic, civil society organizations (CSOs), non-governmental organizations (NOGs), and community-based organizations (CBOs), are helping in the management. Additionally, the study identifies the effects of WM on climate change and human health. Finally, it proposes specific measures to mitigate the climate change impacts arising from improper WM and its effects on the health of the environment.

Material and Methods

Study Area

Field research was carried out in five regions of Ghana. These included Greater Accra, Ashanti, Central, Western and Eastern, as shown on the map of Ghana in Fig. 5.1. The selection of regions was based on their high level of waste generation resulting from high population size and several institutions, for instance, educational, religious, hospitals, and industries established in Ghana. The chosen regions generated an average waste of 0.73 kg/person/day (Miezah et al. 2015). Per a recent population census from Ghana Statistical Service (GSS), the selected regions have a total population of 18,742,214 out of the 30,832,019 total population in Ghana, representing approximately 60.8% of the population residing in the selected regions (GSS 2021). Given the problems, namely non-engineered landfills, poor WM infrastructure, inadequate policy implementation, and lack of enforcement of environmental sanitation laws associated with WM in Ghana within these five regions, it is essential to assess how wastes, especially the industrial, hospital, and HEI are managed, the effects of climate change on the environment and humans, and the challenges therein. Furthermore, to implement strategies that can be adapted to mitigate climate change caused by improper WM of industrial, institutional, and commercial waste.

Study Design

A multi methodology approach with a descriptive cross-sectional design for this study was used. According to Estellita Lins et al. (2019) and Mingers and Brocklesby (1997), a multi-methodology requires the combination of different methods in data collection. This includes employing qualitative and quantitative methods such as observations, surveys, interviews, and others (Creswell 2014) to prevent biases (Driessnack et al. 2007). Therefore, this study was structured into site visits for observation (qualitative) and structured questionnaires (quantitative) used to conduct the survey.

Survey

A structured survey questionnaire was developed and hosted in a Google form link. The questionnaires were categorized into sociodemographic sections, i.e., gender, age, education level, organization, and how to deal with the WM situation in Ghana.

A preliminary pilot test was conducted, receiving seven responses from Meridian WM Services in Tema, Ghana, within three (3) days. The responses helped the authors to improve the content and remove ambiguity from the survey questionnaires, ensuring the clarity of the feasibility of the study.

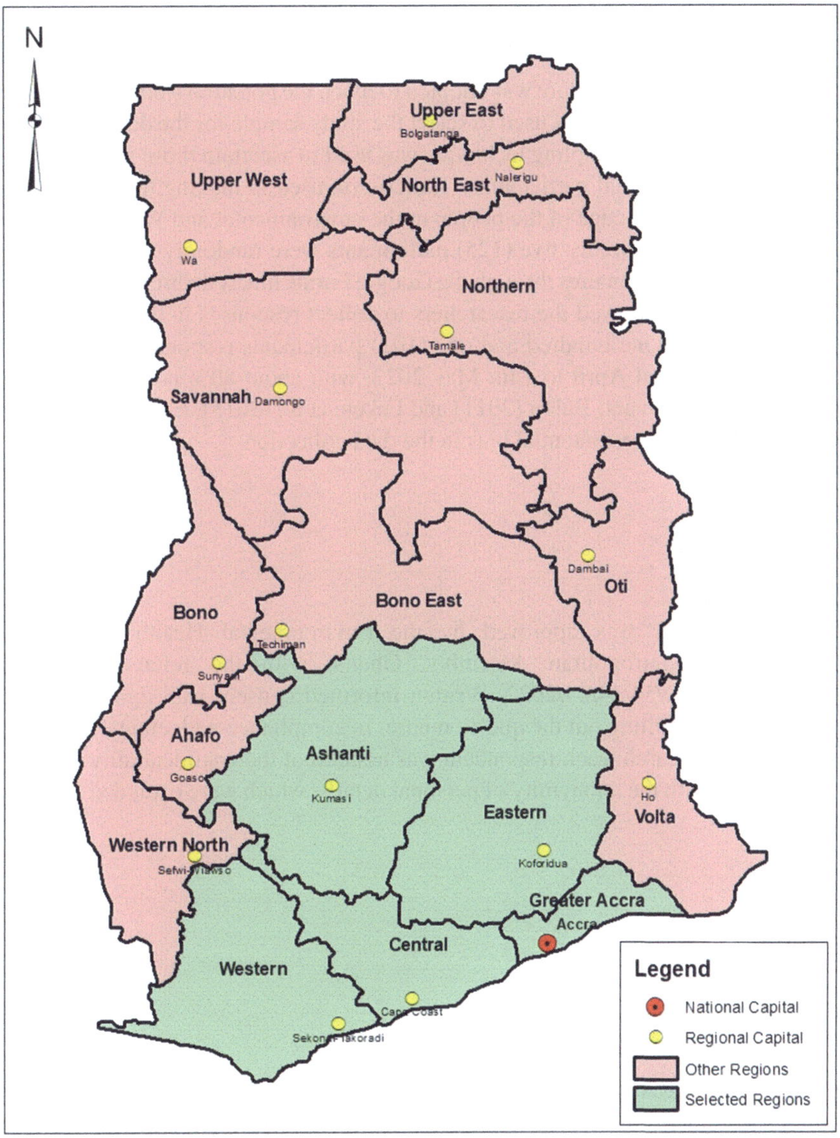

Fig. 5.1 Map of Ghana indicating Greater Accra, Ashanti, Central, Western and Eastern, the regions used to carry out this study

Sampling Technique

A two-step sampling procedure was adopted to reach the potential study participants. Firstly, random sampling was used to select the study sample for the data collection. Secondly, a purposive sampling technique was used to ascertain those who qualified to be chosen as potential participants. The criteria used in making this choice was the background knowledge of the people in the environmental and WM field.

One hundred and twenty-five (125) participants were randomly selected for the distribution of questionnaires through the Google Forms link sent through WhatsApp and/or email. This allowed the researchers to collect responses in the selected five regions in Ghana. One hundred and one (101) participants responded to the questionnaires from 22nd April to 14th May 2022, with about 80% participation rate. According to Arfken and Balon (2011) and Fekete et al. (2015), a participation rate within 60–90% reduces potential bias in the data collection.

Ethical Issues

The present study was approved by the Environmental Health Department of the Accra Metropolitan Assembly, Ghana, with the reference number AMA/OKS/EHSD/54/14/04/2022. Written informed consent was signed by each participant before filling out the questionnaire. In compliance with ethical standards in conducting research, each respondent was assured of the confidentiality of information, along with the anonymity of personal details, which were excluded from the questionnaires.

Results

Demographic Characteristics

This study engages WM experts to seek their views on the WM situation in Ghana, its effect on climate change, and human health. From the demographic data shown in Fig. 5.2, 54% of the participants work in a private institution, with the remaining 46% in the government sector. The gender male/female ratio is 72:28, with a mean age of 38.3 ± 13.2. All the participants have completed some level of education from secondary to doctorate. The majority of the respondents (47%) have completed a university first degree, 6% have a doctorate with only 2% have completed secondary education. However, 54% occupy senior positions in their respective organizations, with 14% being high management members. Most of the participants (32%) of this study have worked between 11 and 15 years, with none having a total working experience from 26 years upwards.

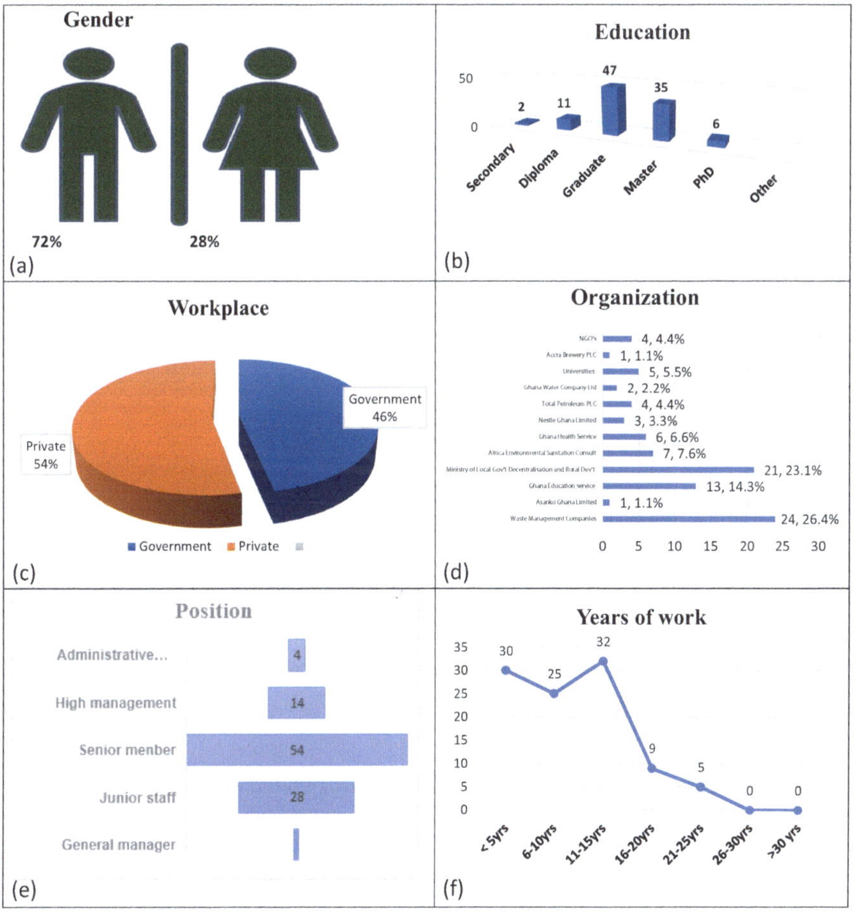

Fig. 5.2 Demography data of WM experts (n = 101). **a** gender, **b** educational level, **c** workplace, **d** organization, **e** position, and **f** year of work of the participant

Waste Management Experts' Open-Ended Responses

Table 5.1 shows "yes" and "no" closed-ended responses while Fig. 5.3 shows the responses to multiple-choice closed-ended responses from the WM experts who participated in the survey. Among the respondents, seventy-eight (78) participants, representing 77.2%, believe that waste is not correctly disposed of in Ghana, with many associated challenges (93.1%) such as waste financing, WM infrastructure, logistics, and equipment. Virtually all the WM experts (101 participants) acknowledged that improper WM is a threat to human health, with only 5.9% believing poor WM has no effect on climate change.

Table 5.1 Closed-ended responses by WM experts on the effects of waste on climate change and human health ($n = 101$)

Variable	n (%)
Are wastes properly disposed in Ghana?	
Yes	23 (22.8)
No	74 (77.2)
Do you know of any challenges associated with waste management in Ghana?	
Yes	94 (93.1)
No	7 (6.9)
Is improper waste management a threat to human health?	
Yes	101 (100)
No	0
Does improper waste management have an effect on climate change?	
Yes	95 (94.1)
No	6 (5.9)
Total	101 (100)

Note n = number respondents, %, percentage

Waste Management Professionals' Responses to the Opened-Ended Questions

Tables 5.2, 5.3 and 5.4 show the responses to opened-ended questions from the WM experts. In Table 5.2, a few WM experts gave reasons why they believe waste is properly managed in Ghana. Tables 5.3 and 5.4 on the other hand indicate opinions expressed by WM practitioners on challenges associated with WM, the effect of poor WM on climate change, and mitigation measures against climate change.

Field Trip Observation

WM disposal is a significant problem in Ghana since most waste landfills are non-engineered with no space to dispose of the waste. This creates a heap of waste in communities due to the inability of trucks to haul waste to the landfills.

Figure 5.4 indicates waste disposal sites of a community without a communal container and non-engineered landfills.

For example, Fig. 5.4**a** shows a community site where waste dumped is hauled to the landfill site (see Fig. 5.4**b**) for disposal. However, at the time this study was carried out, the site had no communal 12 m^3 waste container and refused is dumped on the ground with the fire smouldering. Figure 5.4**b** gives evidence of non-engineered landfills where most of the waste generated in Greater Accra, Ghana, is dumped.

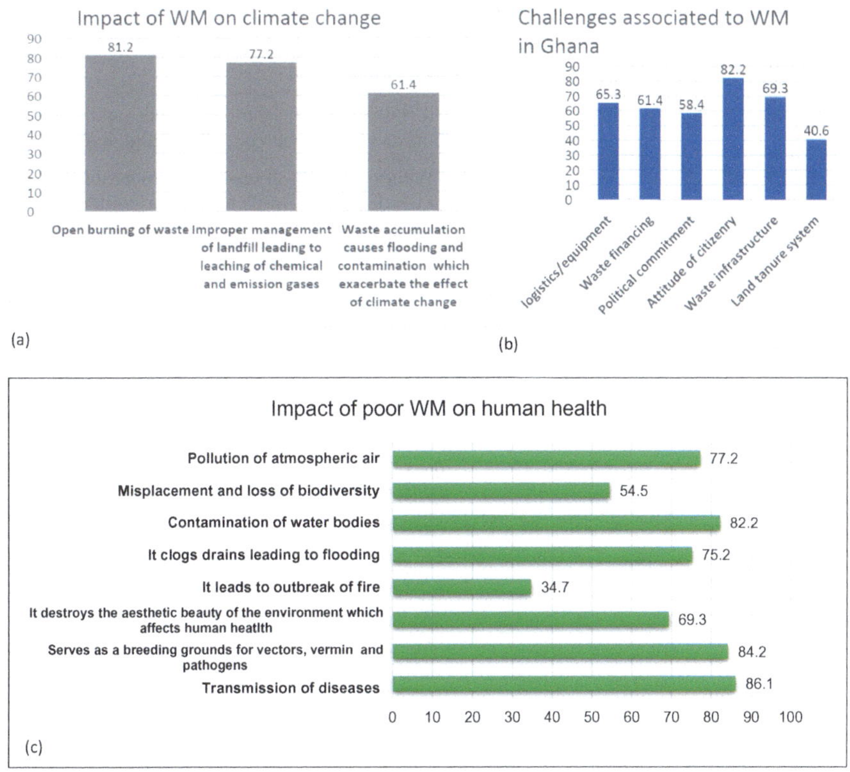

Fig. 5.3 Impact assessment of WM on human health and climate change; a, b, and c ($n = 101$). **a** impact of WM on climate change, **b** challenges associated with WM in Ghana, and **c** impact of poor WM on human health

Discussion

Improper WM coupled with the rapid pace of industrialization in developing countries (Mohanty et al. 2020), including SSA (Debrah et al. 2021), could contribute to the recent global warming temperature rise of 0.85 °C (NASA 2021), adversely impacting climate change and human health. Considering the above information, the stakeholders, for example, governments, private WM companies, NGOs, CSOs, academia, CBOs, and individuals, are tasked to adequately protect and manage waste that will lead to environmental problems.

The study data indicates that 72% of WM experts are males with a mean age and standard deviation of 38.3 ± 13.2 and have their first or master's degree certificates. The high percentage difference between the men and women in positions in this study is not restricted to WM and other professions in the developed world (Sociovision and Wippermann 2010). Furthermore, 54% of the WM experts work with private organizations, including WM companies, mining companies, food processing companies,

Table 5.2 Opened-ended statements by WM experts on the effect of waste on climate change and human health

Question	Coded statements
If yes, briefly explain your answer	"It is deposited in central containers and waste bins and finally hauled to the landfill site"
Question being: Are waste properly disposed of in Ghana?	"Within our areas of operations, more than 50% of residents have waste receptacles with which waste is stored. There is a certified waste contractor who collects this stored waste and safely disposes of them"
	"Attitude of the citizens and inadequate logistics and resources"
	"Trucks are designed to pick the waste from domestic homes"
	"Most of our municipal waste is disposed of at the Landfill site and material recovery facilities (MRF) in Ghana"
	"Waste management in Ghana are in two folds, i.e., public waste and domestic waste. The waste is sent to the transfer station within each district and finally to the landfill site"
	"About 60–70% of municipal waste is collected and disposed of at the final disposal site. Almost 10% of waste is recycled. Most industrial wastes are managed properly due to the enforcement of the laws of Environmental Protection Agency (EPA)"
	'Waste is disposed of through engineered landfills and composting plants or MRFs"
	'There are waste bins and communal refuse containers in most regions in Ghana. Furthermore, there are final disposal sites and recycling plants"
	'Indiscriminate dumping of refuse onto the earth's surface without any treatment"
	"Waste is sent to the final disposal site"
	"Waste is disposed of in storage receptacles"
	"Waste hauled is either sent to a recycling center or a landfill site"
	"Most waste is not properly disposed of, but Zoomlion Ghana limited is helping clean up the country and has advised people to dispose of waste"
If necessary, please indicate any additional comments on the previous questions you consider important but were not addressed?	"Inclusion of waste management in our academic curricula right from the kindergarten"
	"Citizens should segregate waste at the point of generation"

Table 5.3 Coding of opened-ended responses by WM experts' responses on the effects of poor WM on climate change and human health

Question	Coded statements	*n*	%
If yes, briefly explain your answer (*n* = 88)/Question being: Do you know of any challenges associated with waste management in Ghana?	Lack of landfill facilities	16	12.6
	Inadequate logistics for waste companies	12	9.5
	Lack of law enforcement	7	5.5
	Indiscriminate waste disposal	13	10.2
	Lack of adequate waste management infrastructure	5	3.9
	Poor road infrastructure	2	1.7
	Lack of education	13	10.2
	Lack of government involvement	3	2.4
	Most wastes are uncollected by waste companies on time	3	2.4
	Lack of waste bins	6	4.7
	Lack of trained personnel	4	3.1
	Lack of segregation	12	9.5
	Inability to recycle/treatment centers	12	9.5
	Lack of payment waste services	4	3.1
	Land acquisition for landfills	4	3.1
	Financial	3	2.4
	Poor city's planning	2	1.5
If yes, briefly explain your answer (*n* = 94)/Question being: Is improper waste management a threat to human health?	Outbreak of communicable diseases	88	66.2
	Impact air and soil quality	21	15.7
	Destroy land and aquatic system	5	3.8
	Lead to flooding/chocked drains	6	4.5
	Breeding site for insects, vermin and diseases vectors	8	6.0
	Contaminate water bodies (industrial waste)	5	3.8
If yes, briefly explain your answer (*n* = 85)/Question being: Does improper waste management have an effect on climate change?	Emissions of Green House Gases (GHGs) affect climate change	53	57.0
	Waste burning contributes to depleting the ozone layer	15	16.1
	Burning waste affects the respiratory system	8	8.6
	Waste affects water bodies	3	3.2
	Excessive heat/heat radiation	2	2.2
	Burning waste led to global warming	7	7.5
	Irregular rainfall pattern	4	4.3
	Flooding for climate change	1	1.1

(continued)

Table 5.3 (continued)

Question	Coded statements	*n*	%
What are the other effects of poor waste management on climate change?	Irregular rainfall patterns	3	27.3
	Destruction of soil structure	1	9.1
	Famine	1	9.1
	Acid rain	1	9.1
	Destruction of aquatic life	4	36.3
What are some of the measures that could be adapted to mitigate the effects of climate change?	Education on proper waste management for individuals	21	21.2
	Eliminate waste burning	8	8.1
	Engaging in a greener economy (tree planting)	15	15.2
	Engaging in proper waste management practices (sorting)	27	27.3
	Enforcement of sanitation laws	11	11.1
	Adapting policies on climate change	4	4.0
	Adopting renewable energy policies	5	5.1
	Adopting sustainable means of transport	1	1.0
	Irrigation methods to combat rain patterns	1	1.0
	Motivation for adherence to laws	1	1.0
	Resourcing waste management institutions adequately	1	1.0

and oil companies, with most working between 11 and 15 years with vast knowledge and experience.

Findings of this study echo comments similar to other developing countries. For instance, about 77% of the WM professionals' respondents expressed that waste is inappropriately managed. This is similar to existing research findings, see, for example, Bundhoo (2018), Ferronato and Torretta (2019), Debrah et al. (2021a), for SSA and Niyongabo et al. (2019), Debrah et al. (2021a, b), Odonkor and Sallar (2021) and Debrah et al. (2022a), for Ghana. For the majority of the respondents (93%), the challenges associated with poor management of waste include lack of finances, lack of logistics and equipment, lack of adequate WM infrastructure, lack of policy and related regulations, and inadequate law enforcement. Similar findings were also enumerated by Leal Filho et al. (2016) study on benchmarking approaches and methods in the field of urban WM.

Other factors include segregation of waste before disposal due to lack of education, poor city planning, lack of landfill facilities, inability to recycle waste generated, and inadequate government support to the WM service providers. These WM challenges are similar to other studies conducted in most countries within the SSA (Sibanda et al. 2017; Dawodu et al. 2021; Lissah et al. 2021).

WM, primarily industrial, medical, and commercial waste can be detrimental to public health and affects climate change when improperly managed. The participants of this study indicated that improper WM could lead to transmission or outbreak and

Table 5.4 Opened-ended statement by WM experts before coding

Question	Coded statements
If yes, briefly explain your answer (n = 88)/Question being: Do you know of any challenges associated with waste management in Ghana?	"Lack of landfills to dispose waste generated"
	"Waste management companies do not have the tools, logistics, and equipment to perform their task"
	"Various metropolitan, municipal and district assemblies cannot enforce their sanitation by-laws to the latter"
	"Both government and waste management companies lack the infrastructure to manage waste in Ghana"
	"Lack of good attitude towards managing waste by the citizenry due to the absence of education by the experts"
	"Private waste companies collect waste due to lack of government involvement"
	"Most towns and individuals lack waste bins and containers to depose the waste generated"
	"Most waste companies lack trained personnel to manage waste generated"
	"Most of the waste companies in Ghana lack equipment in managing waste generated"
	"Lack of recycling plants in the treatment of waste generated"
	"Most cities do not have lands to construct landfills"
	"Poor city planning and road network in Ghana makes most waste uncollected in cities"
If yes, briefly explain your answer (n = 94)/Question being: Is improper waste management a threat to human health?	"Waste can cause an outbreak of diseases and other health issues"
	"Waste pollutes the air and soil, making it unhealthy"
	"Waste from the mining destroys the aquatic ecosystem"
	"Waste block drains and cause flooding when it rains"
	"Waste serves as breeding sites for insects and animals"
	"Waste is thrown into water bodies which contaminate it"
If yes, briefly explain your answer (n = 85)/Question being: Does improper waste management have an effect on climate change?	"Emissions of greenhouse gases (GHGs) like methane pollute the atmosphere and affect climate change"

(continued)

Table 5.4 (continued)

Question	Coded statements
	"Frequent burning of waste can deplete the ozone layer, leading to climate change"
	"Choosing a burning as a form of waste management can free chlorofluorocarbons (CFCs) into the atmosphere faster and even cause respiratory problems"
	"Burning waste leads to global warming"
	"Improper waste management could affect climate change because too much waste could cause flooding which would bring about changes in the climate"
What are other effects of poor waste management on climate change, if any?	"Affects rainfall patterns"
	"Destruction of soil structure and soil minerals necessary for plant growth"
	"Climate change brings about famine"
	"Soil contamination, water contamination, extreme weather caused by climate change, air contamination, harm towards animal and marine life, human damage, warmer air, creating a disastrous greenhouse effect"
What are some of the measures that could be adapted to mitigate the effects of climate change?	"Education to the citizenry for proper management of waste in their household"
	"Punish offenders who burn waste"
	"Beautification by planting trees and greening"
	"Enforcement of laws on sanitation"
	"Sorting of waste before collecting"
	"Growing of more trees, proper legal mining, proper waste disposal, enforcement of sanitation by-laws by assemblies"
	"Stakeholder engagement in waste management practices"
	"Education should be intensified on proper waste management"
	"Streamline policies direction regarding waste management at the governmental level"

spread of diseases like diarrhoea, typhoid, worm infestation, and cholera. The poorly managed waste also serves as breeding sites for vectors, vermin, and insects (Debrah et al. 2021b). Few of these vermin, such as cockroaches and mosquitoes, transmit pathogens that are causal agents for diseases, as shown in Fig. 5.4. Additionally, the poorly managed waste blocks drain to cause flooding when it rains, which threatens

(a) (b)

Fig. 5.4 **a** Waste disposal at communal dumping sites and **b** non-engineered landfill sites (Date of photograph: 26/04/2022 @ 11:47 a.m.)

human health, as confirmed by other studies (Sano et al. 2016; Suleman 2016; Ayilara et al. 2020; Yu et al. 2020; Debrah et al. 2021b).

In relation to climate change, 95% of the respondents indicated that improper WM contributes to climate change. The experts attributed the causative agents of climate change to indiscriminate burning of the waste, which releases particulate matter and other gases to pollute the air. Also, as shown in Fig. 5.4, the landfills leach toxic chemicals into the soil and water bodies and emit GHGs such as CH_4, CO_2, SO_x, and NO_x. GHG exacerbates climate change, which results in extreme global temperatures warming the earth (Mikhaylov et al. 2020; Shen et al. 2020), leading to morbidity and death of humans (Khatana et al. 2022) as well as injuries by heat stress (Occupational Safety and Health Administration (OSHA 2005). Furthermore, the improper management of waste results in the emission of GHGs and depletes the ozone layer, thereby affecting a change in climatic conditions. Moreover, these activities destroyed soil structure and soil mineral content. In addition, they changed the rainfall pattern, causing feminine in most developing countries that depend on only rainfall patterns for agricultural production (Tarafdar 2022).

Although, education is identified as a key resource that has the potential to ensure proper WM practices by creating awareness and change in attitudes and behaviour (Zsóka et al. 2013; Niekerk 2014; Kumar and Padmaja 2017; Debrah et al. 2021a). The WM experts i.e., the respondents from this study, believe that some Ghanaians do not see the essence of segregating waste products in their homes and industries. Hence, this results in burning and dumping in any available space, which could contribute to the emission of GHG and particulate matter (Gautam and Agrawal 2021). Therefore, it is crucial to impart proper education on how to segregate waste created by the citizens in developing countries. In addition, this effort should be backed with effective enforcement of sanitation laws to curtail the usual ways of burning waste as a means of disposal, which adds particulate matter and gases to the atmospheric air, affecting a change in the climate.

Conclusion

Improper WM contributes to climate change, threatening human health. In Ghana, WM is a challenge regarding global environmental sustainability, especially because of the GHG emissions. Given this, a multi-method approach with a descriptive cross-sectional study was used to seek information from WM experts in five selected regions in Ghana within the period of 22nd April 2022 to 14th May 2022. The respondents of this study identified some of the challenges associated with WM in Ghana as lack of waste financing, infrastructure needs, scarcity of tools and logistics, and absence of engineered landfills. Other reported factors are the lack of enforcement of sanitation and related laws, lack of adequately trained personnel with the capacity to handle WM, and lack of education to sort waste.

The poor WM due to lack of equipment, waste infrastructure, and logistics results in the accumulation of waste in cities and landfills, which contributes to the emission of GHGs, including CH_4, CO_2, SO_x, NO_x, and particulate matters. Therefore, to reduce GHGs emissions associated with poor WM, the government and the major stakeholders in the industry must ensure the enforcement of environmental sanitation and related laws. This should include the provision of waste bins to individual homes. Also, there must be education on the segregation of waste at all levels and the essence of not burning waste indiscriminately. Non-engineered landfills could include planting of shrubs and trees, thus helping to reduce the emission of GHGs. Finally, WM establishments at the institutional level must partner with other stakeholders to move from the normal disposal of waste at the non-engineered landfills to sanitary landfills.

To address climate change and its health implications in Ghana and SSA, governments must streamline policy directions on WM. In addition, governments must partner with private investors and ensure that WM companies are equipped with new technologies, benefiting from financial investment. This will change WM practices from the usual collection and disposal at the non-engineered landfill to segregation, recycling, and sanitary landfills to reduce GHGs emissions, assisting climate change and its effect on human health.

Limitations and Future Research

The findings reported in this paper reflect the challenges associated with WM in Ghana and its impacts on the environment and human health. However, it is essential to note that despite these findings, some limitations must be considered:

Firstly, the data collected was limited to five regions in Ghana with a high concentration of industrial, commercial, and institutional waste generation. Therefore, any generalization must be cautious. Hence, future research should be designed to compare the selected studied regions with the remaining areas in the country. Additionally, a study to assess the WM challenges faced by stakeholders in Ghana is recommended. The insights obtained would allow the policy makers to formulate a sustainable framework to address WM challenges.

References

Aghakouchak A, Chiang F, Huning LS, et al (2020) The annual review of earth and planetary sciences is online at earth. Annu Rev Earth Planet Sci 48:20.1–20.30. https://doi.org/10.1146/annurev-earth-071719

Arfken CL, Balon R (2011) Declining participation in research studies. Psychother Psychosom 80:325–328. https://doi.org/10.1159/000324795

AUDA-NEPAD (2019) What a waste: innovations in Africa's waste material management. https://www.nepad.org/blog/what-waste-innovations-africas-waste-material-management

Awasthi MK, Zhao J, Soundari PG, et al (2019) Sustainable management of solid waste. Amsterdam, Elsevier

Ayeleru O, Dlova S, Akinribide OJ, Ntuli F et al (2020) Challenges of plastic waste generation and management in sub-Saharan Africa: a review. Waste Manage 24–42. https://doi.org/10.1016/j.wasman.2020.04.017

Ayilara MS, Olanrewaju OS, Babalola OO, Odeyemi O (2020) Waste management through composting: challenges and potentials. Sustain 12:1–23. https://doi.org/10.3390/su12114456

Bhat SA, Bashir O, Bilal M, Ishaq A, Mehraj UDD, Rohitashw K, Rouf AB, Farooq S (2021) Impact of COVID-related lockdowns on environmental and climate change scenarios. Environ Res. https://doi.org/10.1016/j.envres.2021.110839

Bundhoo ZMA (2018) Solid waste management in least developed countries: current status and challenges faced. J Mater Cycles Waste Management 20:1867–1877. https://doi.org/10.1007/s10163-018-0728-3

Chen D, Faibil D, Agyemang M (2020) Evaluating critical barriers and pathways to implementation of e-waste formalization management systems in Ghana: a hybrid BWM and fuzzy TOPSIS approach. Environ Sci Pollut Res. https://doi.org/10.1007/s11356-020-10360-8

Creswell JW (2014) A concise introduction to mixed methods research. SAGE Publications, London

Dawodu A, Oladejo J, Tsiga Z, et al (2021) Underutilization of waste as a resource: bottom-up approach to waste management and its energy implications in Lagos, Nigeria. Intell. Build. Int.

Debrah JK, Carlotto IN, Vidal DG, Dinis MAP (2021) Managing medical waste in Ghana-the reality. Int J Environ Stud. https://doi.org/10.1080/00207233.2021.1994752

Debrah JK, Vidal DG, Dinis MAP (2021a) Raising awareness on solid waste management through formal education for sustainability: a developing countries evidence review. Recycling 6:1–21. https://doi.org/10.3390/recycling6010006

Debrah JK, Vidal DG, Dinis MAP (2021b) innovative use of plastic for a clean and sustainable environmental management: learning cases from Ghana. Africa. Urban Sci 5:12. https://doi.org/10.3390/urbansci5010012

Debrah JK, Vidal DG, Dinis MA (2022a) Sustinable pharmaceutical waste management: pharmacist and patients perception in ghanaian hospitals. In: handbook of sustainability science in the future. Springer, Cham

Debrah JK, Vidal DG, Dinis MAP (2022b) Environmental waste sustainability: organic valorisation and socioeconomic benefits towards sustainable development in Ghana. In Leal Filho W, Vidal DG, Dinis MAP (eds) Sustainable policies and practices in energy, environment and health research. world sustainability series. Springer, Cham, pp 425–437

Debrah JK, Vidal DG, Dinis MAP (2022c) Recovering from COVID–19 environment and social impacts in Sub-Saharan Africa: the role of social engagement. In Handbook of sustainability science in the future, Filho, Wal. Springer, Cham

Driessnack M, Sousa VD, Mendes IAC (2007) An overview of research designs relevant to nursing: part 3: Mixed and multiplemethods. Rev Lat Am Enfermagem 15:1046–1049. https://doi.org/10.1590/s0104-11692007000500025

Environmental Protection Agency (2020) Best practices for solid waste management: best practices for solid waste management: a guide for decision-makers in developing countries. a guide decision makers dev countries. EPA, pp. 1–166

Estellita Lins MP, Netto SOA, de Castro Lobo MS (2019) Multimethodology applied to the evaluation of Healthcare in Brazilian municipalities. Health Care Management Sci 22:197–214. https://doi.org/10.1007/s10729-018-9432-z

Fekete C, Segerer W, Gemperli A, Brinkhof MW (2015) Participation rates, response bias and response behaviours in the community survey of the Swiss Spinal Cord Injury Cohort Study (SwiSCI). BMC Med Res Methodol 15:1–14. https://doi.org/10.1186/s12874-015-0076-0

Ferronato N, Torretta V (2019) Waste mismanagement in developing countries: a review of global issues. Int J Environ Res Public Health 16

Gautam M, Agrawal M (2021) Greenhouse gas emissions from municipal solid waste management: a review of global scenario. In: Muthu SS (ed) Carbon footprint case studies: municipal solid waste management, sustainable road transport and carbon sequestration. Springer Singapore, Singapore, pp 123–160

Ghana Statistical Service (2021) Ghana 2021 population and housing census (PHC). Ghana Statistical Service, Accra

Idowu IA, Atherton W, Hashim K et al (2019) An analysis of the status of landfill classification systems in developing countries: Sub Saharan Africa landfill experiences. Waste Manag 87:761–771. https://doi.org/10.1016/j.wasman.2019.03.011

Jude SAA, Prabhu S, Veerapandian JP, et al (2019) Automatic waste segregation and monitoring system of municipal solid waste

Kaza S, Yao L, Bhada-Tata P, Woerden V (2018) What is waste 2. A global snapshot of solid waste management to 2050. World Bank Publications, The World Bank Group, Washington, DC 20433

Khatana SAM, Werner RM, Groeneveld PW (2022) Association of extreme heat with all-cause mortality in the contiguous US, 2008–2017. 5:2008–2017. https://doi.org/10.1001/jamanetworkopen.2022.12957

Kotei P, Annang T, Yirenya-Tawiah D (2020) Stakeholder participation for sustainable solid waste management in Ga west municipality, Accra–Ghana. Am J 3:44–60

Kumar PVS, Padmaja P (2017) Knowledge, attitude, practices of biomedical waste management among nursing students and staff in a tertiary care hospital. Ann Int Med Dent Res 3:248–251. https://doi.org/10.21276/aimdr.2017.3.4.cm1

Leal Filho W, Brandli L, Moora H et al (2016) Benchmarking approaches and methods in the field of urban waste management. J Clean Prod 112:4377–4386. https://doi.org/10.1016/j.jclepro.2015.09.065

Leigh C, Bush A, Harrison ET et al (2015) Ecological effects of extreme climatic events on riverine ecosystems: Insights from Australia. Freshw Biol 60:2620–2638. https://doi.org/10.1111/fwb.12515

Lissah SY, Ayanore MA, Krugu JK, et al (2021) Managing urban solid waste in Ghana: Perspectives and experiences of municipal waste company managers and supervisors in an urban municipality. PLoS One 16

Maja MM, Ayano SF (2021) The impact of population growth on natural resources and farmers' capacity to adapt to climate change in low-income countries. Earth Syst Environ 5:271–283. https://doi.org/10.1007/s41748-021-00209-6

Manzoor J, Sharma M (2019) Impact of biomedical waste on environment and human health. Environ Claims J 31:311–334. https://doi.org/10.1080/10406026.2019.1619265

Miezah K, Obiri-Danso K, Kádár Z et al (2015) Municipal solid waste characterization and quantification as a measure towards effective waste management in Ghana. Waste Manage 46:15–27. https://doi.org/10.1016/j.wasman.2015.09.009

Mikhaylov A, Moiseev N, Aleshin K, Burkhardt T (2020) Global climate change and greenhouse effect. Entrep Sustain Issues 7:2897–2913. https://doi.org/10.9770/jesi.2020.7.4(21)

Mingers J, Brocklesby J (1997) Multimethodology: Towards a framework for mixing methodologies. Omega 25:489–509. https://doi.org/10.1016/S0305-0483(97)00018-2

Mohanty S, Saha S, Santra GH, Kumari A (2020) Future perspective of solid waste management strategy in India. In: Baskar C, Ramakrishna S, Baskar S, et al. (eds) Handbook of solid waste management: sustainability through circular economy. Springer Singapore, Singapore, pp 1–36

Mora C, Spirandelli D, Franklin EC et al (2018) Broad threat to humanity from cumulative climate hazards intensified by greenhouse gas emissions. National Climate Change 8:1062–1071

NASA (2021) Global climate change (global temperature anomalies from 1880 to 2021). In NASA's goddard inst. Sp. Stud. New York. https://svs.gsfc.nasa.gov/4964

Nathaniel SP, Nwulu N, Bekun F (2020) Natural resource, globalization, urbanization, human capital, and environmental degradation in Latin American and Caribbean countries. Environ Sci Pollut Res. https://doi.org/10.1007/s11356-020-10850-9

Nikiema J, Asiedu Z (2022) A review of the cost and effectiveness of solutions to address plastic pollution. Environ Science and Pollution Research 29:24547–24573. https://doi.org/10.1007/s11356-021-18038-5

Niyongabo E, Jang YC, Kang D, Sung K (2019) Generation, management practices and rapid risk assessment of solid medical wastes: a case study in Burundi. J Mater Cycles Waste Management 21:950–961. https://doi.org/10.1007/s10163-019-00854-0

Occupational Safety and Health administration (OSHA) (2005) Protecting workers from the effects of heat 2004–2005. OSHA

Odonkor ST, Sallar AM (2021) Correlates of household waste management in Ghana: implications for public health. Heliyon 7:e08227. https://doi.org/10.1016/j.heliyon.2021.e08227

Oduro-Appiah K, Afful A, Kotey VN, de Vries N (2019) Working with the informal service chain as a locally appropriate strategy for sustainable modernization of municipal solid waste management systems in lower-middle income cities: Lessons from Accra, Ghana. Resources 8:1–16. https://doi.org/10.3390/resources8010012

Pandey J, Kaushal RK, Shukla SP (2022) Behavioural and structural analysis of stakeholders in E-waste management. Mater Today Proc 51:416–421. https://doi.org/10.1016/j.matpr.2021.05.566

Sano D, Amarasiri M, Hata A et al (2016) Risk management of viral infectious diseases in wastewater reclamation and reuse: Review. Environ Int 91:220–229. https://doi.org/10.1016/j.envint.2016.03.001

Schneider P, Anh LH, Wagner J et al (2017) Solid waste management in Ho Chi Minh City, Vietnam: Moving towards a circular economy? Sustain 9:1–20. https://doi.org/10.3390/su9020286

Sharma HB, Vanapalli KR, Cheela VRS et al (2020) Challenges, opportunities, and innovations for effective solid waste management during and post COVID-19 pandemic. Resour Conserv Recycl 162:105052. https://doi.org/10.1016/j.resconrec.2020.105052

Shen M, Huang W, Chen M et al (2020) (Micro)plastic crisis: Un-ignorable contribution to global greenhouse gas emissions and climate change. J Clean Prod 254:120138. https://doi.org/10.1016/j.jclepro.2020.120138

Sibanda LK, Obange N, Awuor FO (2017) Challenges of Solid Waste Management in Kisumu, Kenya. Urban Forum 28:387–402. https://doi.org/10.1007/s12132-017-9316-1

Sociovision S, Wippermann C (2010) Women in executive positions barriers and bridges. In Fed. Minist. Fam. Aff. Sr. Citizens, Women Youth—BMFSFJ 11018 Berlin. http://www.bmfsfj.de/RedaktionBMFSFJ/Broschuerenstelle/Pdf-Anlagen/frauen-in-fuehrungspositionen-englisch,property=pdf,bereich=bmfsfj,sprache=de,rwb=true.pdf

Suleman Y (2016) Solid waste disposal and community health implications in Ghana: Evidence from Sawaba, Asokore Mampong municipal assembly. J Civil Environ Eng 5. https://doi.org/10.4172/2165-784x.1000202

Tarafdar JC (2022) Role of soil biology on soil health for sustainable agricultural production. In Giri B, Kapoor R, Wu Q-S, Varma A (eds) Structure and functions of pedosphere. Springer Nature Singapore, Singapore, pp 67–81

Teye GK, Kumi JA, Jingyu H, Buabeng T (2019) Private sector participation in solid waste management: the case of Ledzokuku-Krowor Municipal Assembly, Accra -Ghana. Int J Sci Res 75:45–56. https://doi.org/10.21506/j.ponte.2019.01.27

van Niekerk I (2014) Waste management behaviour: a case study of school children in Mpumalanga. Potchefstroom Campus of the North-West University, South Africa

Vidal DG, Maia RL, Oliveira GM, et al (2019) Cities challenges in the contemporary societies: Urban sustainability and environmental issues. Sociol Line 119–138. https://doi.org/10.30553/sociologiaonline.2019.20.6

Vitorino de Souza Melaré VA, Montenegro González S, Faceli K, Casadei V (2017) Technologies and decision support systems to aid solid-waste management: a systematic review. Waste Manag 59:567–584. https://doi.org/10.1016/j.wasman.2016.10.045

Wei T, Dong W, Yan Q et al (2016) Developed and developing world contributions to climate system change based on carbon dioxide, methane and nitrous oxide emissions. Adv Atmos Sci 33:632–643. https://doi.org/10.1007/s00376-015-5141-4

Yeo D, Dongo K, Mertenat A et al (2020) Material flows and greenhouse gas emissions reduction potential of decentralized composting in sub-Saharan Africa: A case study in Tiassalé, Côte D'Ivoire. Int J Environ Res Public Health 17:1–15. https://doi.org/10.3390/ijerph17197229

Youn SJ, Taylor WW, Lynch AJ et al (2014) Inland capture fishery contributions to global food security and threats to their future. Glob Food Sec 3:142–148

Yu H, Sun X, Solvang WD, Zhao X (2020) Reverse Logistics Network Design for Effective Management of Medical Waste in Epidemic Outbreak: Insights from the Coronavirus Disease 2019 (COVID-19) in Wuhan. SSRN Electron J. https://doi.org/10.2139/ssrn.3538063

Zenker A, Cicero MR, Prestinaci F et al (2014) Bioaccumulation and biomagnification potential of pharmaceuticals with a focus to the aquatic environment. J Environ Manage 133:378–387. https://doi.org/10.1016/j.jenvman.2013.12.017

Zscheischler J, Westra S, Van Den Hurk BJJM et al (2018) Future climate risk from compound events. Nat Clim Chang 8:469–477

Zsóka Á, Szerényi ZM, Széchy A, Kocsis T (2013) Greening due to environmental education? Environmental knowledge, attitudes, consumer behavior and everyday pro-environmental activities of Hungarian high school and university students. J Clean Prod 48:126–138. https://doi.org/10.1016/j.jclepro.2012.11.030

Justice Kofi Debrah is Ph.D. candidate in Ecology and Environmental Health at Faculty of Science and Technology, University Fernando Pessoa (UFP). His area of research interest is in Environmental waste management, medical waste management, environmental education, environmental sustainability and sustainable development.

Godfred Kwesi Teye is Ph.D. Candidate at the College of Environment, Hohai University, China. His research areas are transition metal-modified magnetite nanostructured photocatalysis for degradation of pharmaceutical residues in water, environmental sustainability, Wastewater management, Environment health and Solid waste management.

Zujaja Wahaj is an Assistant Professor of Sociology at NUST Business School, National University of Sciences and Technology, Islamabad, Pakistan. She is a grant winner of several funded research projects on the ethnography of health organizations, particularly HIV and AIDS clinics functioning in South Asia. Her current research explores collaborative challenges of our time such as the gender/health dimensions of climate change. In convergence with the research on sociology of vulnerable populations, her recent scholarship is focused on aspects of gender/class nexus of digital platforms, COVID-19, and employees well-being with the aim of developing inclusive understanding towards the sensitive emotional struggles of marginalized people.

Maria Alzira Pimenta Dinis Habilitation, Ph.D., M.Sc., and Associate Professor with Habilitation at University Fernando Pessoa (UFP), Porto, Portugal. Researcher at FP-ENAS, UFP Energy, Environment and Health Research Unit, Porto, Portugal. Experienced author, editor and reviewer. Sustainability expertise.

Chapter 6
Climate Change and Transmissible Diseases

Maria Pia de Melo Alvim Ferraz Tavares and Gisela Marta Teixeira de Sousa Oliveira

Abstract Infectious diseases may be the result of either pathogen increasing frequency of transmission within a given population or expansion of pathogen host range alongside with the increasing of its geographical distribution. Vector transmitted new pathogens may emerge driven by environmental factors and hazards such as droughts, floods, wildfires, or heatwaves, which are foreseen to become more frequent due to climate change. Environmental factors and hazards may be the cause in the change of vectors' geographical distribution or its increase within a particular geographical area. Infectious agents that may cross species and environmental barriers are climate sensitive therefore, its study is important.

Human health is tightly connected to the health of animals and the environment. Animal feed, human nutrition, animal and human health are closely linked with water contamination, air pollution and the loss of biodiversity. The way climate change, as well as other health determinants, affects host defences and viral pathogenesis deserves scientific research and governments and stakeholders' attention to support preparedness plans and capacity of health systems to face emergence infection diseases.

This chapter presents a state-of-the-art of the interactions of climate change effects on transmissible diseases based on (1) authors previous experience and knowledge on the subjects hereby addressed; (2) information from several institutional sources

M. P. de M. A. Ferraz Tavares (✉)
Departamento de Engenharia Metalúrgica e de Materiais, Faculdade de Engenharia da Universidade do Porto, Universidade do Porto, Rua Roberto Frias, 4200-465 Porto, Portugal
e-mail: mpferraz@fe.up.pt; mpferraz@ineb.up.pt

Instituto de Engenharia Biomédica, Universidade do Porto, Rua Alfredo Allen 208, 4200-135 Porto, Portugal

G. M. T. de S. Oliveira
UFP Energy, Environment and Health Research Unit (FP-ENAS), University Fernando Pessoa, Praça Nove de Abril, 349, 4249-004 Porto, Portugal
e-mail: gisela@ufp.edu.pt

M. P. de M. A. Ferraz Tavares
i3S, Instituto de Investigação e Inovação em Saúde da Universidade do Porto, R. Alfredo Allen 208, 4200-135 Porto, Portugal

W. Leal Filho et al. (eds.), *Climate Change and Health Hazards*, Climate Change Management, https://doi.org/10.1007/978-3-031-26592-1_6

such as the World Health Organization, the United Nations Framework Convention for Climate Change, the Intergovernmental Panel on Climate Change, the United Nations Economic and Social Council, the European Commission of the European Union and The Lancet Countdown; and (3) an oriented search for scientific publications using the criteria described below. Authors analysed publications in Web of Science and PubMed databases choosing title/abstract/keywords screening (and full-text when necessary) and using the keywords: "Climate changes"; "Infectious diseases", "SARS-CoV-2", "Transmissible diseases". Detailed information was extracted from the selected papers and the institutional sources based on full-text screening.

Climate Change Influence on Life

The detrimental influence of human activity on the atmosphere have been explained by scientists and addressed by politicians for decades. The depletion of the protective ozone layer at the stratospheric zone is one of the well-known examples and a success case resulting from the singular concerted action of several countries adopting less harmful technologies. Nevertheless, reducing the impacts of climate change driven by human activity has not yet received the same global support as the ozone depletion: gaseous (pollutants and GHG–greenhouse gases) emissions to the atmosphere are continuously increasing year after year, despite scientific evidence of the negative consequences of those emissions (Gulev et al. 2021; Dhakal et al. 2022). The most important forces driving climate change arise from the pressure of exponential population growth, and the demand to fulfil the increasing need of essential goods and services; and the on-going intensive resources exploitation and use, still focused on economic benefit purposes rather than the improvement of human development and well-being. Some collective agreements have been put in force, such as the Kyoto Protocol or the Paris Agreement, however, the voluntary national measures and commitments foreseen in these international arrangements, often fail because they do not translate into the needed transformative actions that would effectively mitigate climate change (UNFCCC 2021).

Climate change represents a threat to humanity in every dimension and it is already recognized that, globally, climate change is the biggest public health danger in the twenty-first century, putting at risk the right to health foreseen in the Paris Agreement. According to reports (WHO 2018, 2021) from the World Health Organization (WHO), there is evidence that climate change is linked to a growing number of infectious health outcomes that are spreading to new regions.

A growing number of infectious diseases are associated with climate change effects that are unbalanced. Weather parameters contribute in a nonlinear way to infectious disease transmission (Metcalf et al. 2017) and have been shown to be one of the key drivers of the emergence, re-emergence, and spread of infectious disease (Hess et al. 2020; Semenza et al. 2016). Climate influences the survival, reproduction and distribution of disease pathogens and vectors, as well as their transmission and geographical distribution (Semenza and Suk 2018). Climate changes can decrease suitability for transmission in certain areas and for other infectious diseases it may increase (Watts et al. 2021).

Climate change impacts result from the likelihood of a threat, the level of exposure to that threat and the vulnerability of a population or an ecosystem to be affected by the threat. These three elements that compose climate change risks—a danger, the magnitude of the identified danger and the resilience capability of ecosystems and populations—, are extremely complex in their interaction which are recognized to impact not only the environment but to affect the social and economic levels (European Commission 2021a). In addition, it is likely that, in the midterm future, climate change will pose some threats to Earth that are yet to be scientifically identified (European Commission 2021b).

The Importance of Reducing GHG Emissions

The greenhouse effect of some gases in the atmosphere represents an accumulation of chemical energy with a consequential temperature elevation of the air masses containing these GHG which are mainly in the troposphere. The troposphere is the atmosphere layer adjacent to the lithosphere (the Earth surface), both being part of the biosphere where life on Earth exists. Besides water vapour, carbon dioxide (CO_2) is the most abundant GHG with a mean residence time in the atmosphere estimated to be about 100 years. This implicates that the present atmospheric CO_2 concentration reflects the accumulation of CO_2 emissions for the last 100 years. This fact is extremely important because it reveals that if at this moment, the world was capable of totally stop anthropogenic CO_2 emissions, although the atmospheric CO_2 concentration would drop, it would take nearly 100 years to achieve the natural balance of the carbon cycle. However, the positive impacts on Earth climate that would come from the hypothetic annulation of anthropogenic CO_2 emissions would be observed after a lag time sufficient for achieving balance through the dissipation of the additional energy accumulated in the atmosphere due to all GHG (IPCC 2021). The troposphere is in permanent interaction with the lithosphere especially by mass and heat transfer processes through what consists of the natural cycles of water, carbon, or nitrogen. Oceans play a very important role in all the natural cycles because they represent the majority area of Earth, thus the major surface for interaction with the troposphere. Therefore, over 90% of the Earth excess heat is accumulated in oceans, causing oceans global temperature raise and the huge volume of water in oceans strongly influences Earth temperature (Cheng et al. 2020).

Emissions of greenhouse gases and pollutants to the atmosphere have been changing the natural pattern behaviour of climate at a concerning rhythm causing disruption on other Earth systems, particularly in the biosphere (Gulev et al. 2021). However, global awareness of humankind frailty before nature's own course has recently become more evident due to the COVID-19 pandemic. In fact, until 2020, not even the enormous destruction and misery caused by two sequential world wars have paused services, commercial and industrial activities or mobility at a global scale as COVID-19 confinement measures did. If additional proofs were needed to show the direct consequence of human activity on gaseous emissions to the atmosphere, COVID-19 restrictions to many activities, adopted all-over the world in the first semester of 2020, caused a significant reduction on those emissions, especially those from fossil fuels, clearly demonstrating the human-induced effect. This never experienced before, world-wide and collective stand-by human behaviour provided the time-window opportunity and the momentum to reflect and raise awareness on climate change impacts and to call for global and effective commitment to prevent, or at least minimize, an irreversible climate change situation on Earth. However, the economic slowdown has not translated into an improvement of climate change mitigation because the urge to reach a fast economic recovery is undermining several previous achievements on sustainable development. Therefore, despite the verified reduction on GHG emissions in 2020 and 2021, the goals established under the Paris Agreement are not on track: atmospheric concentrations of GHG continued to increase, the present global average Earth temperature has raised to 1.2 °C above pre-industrial levels and the accumulated loss of over 50 million hectares of forests in 2015 to 2020 continues to degrade biodiversity (United Nations Economic and Social Council 2021). These, alongside with other drivers for climate change, will have impacts on human health.

The impacts of climate change on Earth, considering it as a living system, will unfold both expectable and unpredictable consequences on human health. According to the Intergovernmental Panel on Climate Change (IPCC) observations and projections on climate change and due to the increase in GHG concentrations in the atmosphere and of oceans temperature, weather patterns are likely to continue to change and extreme weather events will be more frequent (IPCC WGII 2022). In Europe, it is predicted that atmospheric temperature will continue to rise and heat waves will increase in duration, frequency and intensity (Martinez et al. 2019). Differential temperature increases are expected in different regions of Europe (Karwat and Franzke 2021). Dry summers, with low precipitation alongside with increasing global temperatures will leed to frequent and intense droughts, rising pressure on water resources. Additionally, heavy rainstorms and flash floods are projected to become also more frequent (European Commission 2021a).

Therefore, it is probable that these changes will affect disease transmission. Water-related hazards such as floods or droughts, heat waves, hurricanes and blizzards occurrence will be geographic dependent and different in intensity, occurrence and gravity, but whatever the level of destruction caused, they will directly affect the balance in ecosystems. The pace of living species adaptation to climate change impacts is rather slow when compared to the rate that climate change is happening. Alongside with other climate change consequences like desertification, this imbalance is leading to the loss of biodiversity, favouring the dominance of some species over others, the introduction of new pests and disease agents in established ecosystems and inducing the shift of endemic diseases to new geographic zones.

Salt-marsh mosquitoes breeding habitats may become decreased or be eliminated due to the projected rise in sea level. Endemic viruses of these salt-marsh habitats would also be eliminated due to bird and mammalian hosts extinction. Alternatively, inland intrusion of salty water may turn former fresh water habitats into salt-marsh areas which could support vector and host species displaced from former salt-marsh habitats (Hales et al. 2002).

These harmful effects of climate change on transmissible diseases occurrence, severity and dissemination are channelled into the biosphere with multiplication of effects like a cascade process through different routes: air, water, land use and food chains and their synergetic relations. Knowledge on the complexity of Earth climate system is evolving quickly and the body of scientific evidence is enlarging. In some regions, the combined effect of the increasing CO_2 concentration and the atmospheric temperature is also likely to alter the ripening seasons for many crops that may occur earlier than the usual. In addition to the crops maturation periods shift in time, the changes in rainfall patterns caused by droughts or intensive precipitation events may favour the occurrence of pests or fungi infestations on crops leading to food contamination with mycotoxins (Medina et al. 2017). Human ingestion of mycotoxins by rotten food cause mycotoxicosis, a type of fungi infections with serious acute and chronic effects on human health (Awuchi et al. 2022). This example of climate change chain effects represents risks for the security and safety of food with expectable consequences for both human and livestock health.

Transmission and Infection Routes Influenced by Climate and Environmental Changes

Based on transmission mode, infectious diseases may be classified into direct and indirect. Direct infections spread through direct contact from person to person or through droplet/aerosol exposure. On the other hand, indirect transmission needs an intermediate vector organism (mainly arthropods) or a physical vehicle like food, soil or water. Another criterium to classify infectious diseases is their natural reservoirs. In the case of human reservoirs, the infections are classified as anthroponoses, being zoonoses those who have animal reservoirs. Anthroponoses and zoonoses are directly and indirectly transmitted, while vector-borne diseases are indirectly transmitted.

Infectious agents can be viruses, bacteria, fungi, protozoa and multicellular parasites which may be very different in size, type and mode of transmission. Microorganisms that cause anthroponoses have adapted to the humans as their dominant, exclusive, host. Oppositely, natural reservoirs of zoonoses infectious agents are non-human species.

Climate Sensitivities of Infectious Diseases

Climate variability affects the transmission lifecycle of pathogens therefore, influencing infectious diseases occurrence. Transmission cycles that require a vector or non-human host are more susceptible to external environmental influences than those diseases which include only the pathogen and the human (Paz et al. 2021). Climate change may have direct impacts on pathogens survival, reproduction, and life cycle, and indirect influences on pathogens habitat, environment, and competitors. Therefore, pathogens distribution in number, season and geographic location may also change.

Vector survival and reproduction, vector's biting rate, and pathogen's incubation rate within the vector organism are the main vector-borne disease transmission determinants. The infective incubation time of a vector-borne agent in its vector organism is quite sensitive to temperature changes. Besides temperature, other important determinants for the agent, vector and host include precipitation patterns, sea level elevation, wind and daylight period.

Infectious agent (viruses, bacteria, fungi, protozoa and multicellular parasites) and their associated vector organism do not have mechanisms to regulate their internal temperature and fluid levels, being those directly determined by the climate of their habitats. Consequently, pathogenic agents or vector species can survive and reproduce in a limited range of climatic conditions.

Geographical distribution is the result of disease carrying vectors natural adaptation to climate conditions because vectors undergo evolutionary responses to adapt to increasing temperatures (Gubler et al. 2001). As an example, malaria mosquito vector is known to be temperature sensitive, only surviving in hot climate habitats, so, due to increasing temperatures in former cooler climates, such as the African highlands, it is expected that this disease may emerge in other geographic regions that are becoming warmer.

Due to fast environmental changes, survival for many species depends on moving to find adequate climates (Alfano et al. 2019). Many species' geographic ranges are expected to shift a hundred kilometres or more in the next century (Burrows et al. 2014; Chen et al. 2011). Therefore, parasites and pathogens of these animals will go into these new environments posing real threats to global health namely by spillover (a process that allow animal pathogens to make the leap to human hosts) (Jones et al. 2008; Plowright et al. 2017). The magnitude of spillover events depends on opportunity and compatibility.

Adaptation to climate change may also cause the migration of mammals' species to territories with at least one unfamiliar species, forcing the share of habitats. This global re-organization of mammal populations is foreseen to have a great impact on mammalian microbiome in general and in human microbiome in particular (Carlson et al. 2022).

Most human and domestic animal pathogens causing the most significant diseases, based on morbidity and mortality, are climate sensitive, therefore both indicators are affected by climate change (Brugueras et al. 2020; Semenza and Suk 2017). Vector-borne diseases (VBDs) are transmitted by arthropod vectors that, being ectothermic, are particularly sensitive to changes in climatic conditions (Rocklov and Dubrow 2020). Habitat availability influence arthropods density, distribution and abundance. Temperature has an impact in pathogen maturation and replication in mosquitoes being able to increase or decrease the probability of infection. For those diseases transmitted by vectors that have aquatic developmental stages in their lifecycles, precipitation has a strong influence. Changes in precipitation patterns may have direct consequences on infectious disease outbreaks.

Precipitation increases food supplies favouring the growth of vertebrate reservoirs. Heavy rainfalls may cause flooding and decrease vector populations through the elimination of larval habitats and the creation inadequate environments for vertebrate reservoirs. On the other hand, floods may have an indirect impact increasing the likelihood of vector-human contact by forcing insect or rodent vectors to seek refuge in houses. Air humidity influence transmission of vector-borne diseases, particularly for insect vectors, like mosquitoes and ticks, which dehydrate under dry conditions so, decreasing their survival. Relative humidity has been found to be one of the most critical determinants in climate-disease models, such as in Dengue fever (Hales et al. 2002) and Lyme disease models (Mount et al. 1997; Cecílio et al. 2022).

Waterborne infections occur by contact of humans with contaminated drinking or recreational water, or food. Contaminated water results from extreme weather events, such as floods and droughts, or inappropriate water treatment and sewage disposal. Rainfall influences the transport and dissemination of waterborne infectious agents, while temperature affects their growth and survival, and may change the seasonality and geographical distribution of waterborne diseases.

Concerning marine environments, warm temperatures create favourable conditions for red tides, caused by microalgae blooms, favouring the incidence of shellfish poisoning (Epstein 1993). Increasing sea surface temperatures can indirectly influence the viability of enteric pathogens such as *Vibrio cholerae* by increasing their reservoir's food supply (Colwell 1996). Air temperatures also have been linked to hospital admissions of Peruvian children with diarrhoea disease (Checkley et al. 2000).

Storms can contaminate watersheds by transportation of human and animal faecal products and other wastes into groundwater. Water contamination following heavy rains has been documented for *Cryptosporidium*, *Giardia*, and *E. coli* (Paz et al. 2021) being this phenomena increased in locations with high soil saturation conditions due to microbial transport easiness (Rose et al. 2001). On the opposite hand, in developing countries, water scarcity and the use of contaminated water have been associated with diarrheal disease outbreaks.

Climate change alters the hydrological cycle on Earth in unexpected manners. Severe weather events, flooding, storm surges, and droughts disrupt the hydrological cycle (Semenza et al. 2012). The consequences of abrupt climate changes are more difficult to deal in terms of public health then gradual ones. Extreme precipitation may trigger a chain of secondary events. Cascading risks depend on existing vulnerabilities in society that get exacerbated by climate change (Semenza 2021).

The transmission of food-borne diseases (FBD) pathogens is complex because many of these pathogens thrive in the environment, are resistant to heat stress, and have low infective doses. Concerning FBD, in developed countries, the most common bacterial cause of diarrhoeal disease is *Campylobacter*. These bacteria present very strong association with seasonality and climate variability.

An abridged list of major climate change impacts on human health due to infections is presented in Table 6.1. The information presented in the table does not intend to be fully comprehensive but rather indicative of some infections that are influenced by climate changes.

The process by which virus are able to overcome interspecies barriers and infect other animal species is called zoonotic spillover. Climate conditions can affect some of these barriers therefore facilitating or buffering spillover showing the relevance of understanding how climate changes affects the cross-species pathogen transmissions. Directly, climate can promote a pathogen's survival, development and dissemination enhancing spillover. Indirect climate influence is not only more important but also much more complex. Climate change leads to losses in biodiversity, shifts in species and shrinking of ecosystems which can enhance spillovers. Climate can also have an impact in transmission chains because it affects inter-human transmission. It is accepted that climate stability enhances continuous transmission. Oppositely, extreme climate events force strain selection due to the exposure of new niches, also favouring spillover.

Concerning the host, environmental factors have increased the incidence of lifestyle diseases such as diabetes, and consequently have affected host immunity and inflammatory response (Fattorini and Regoli 2020). As far as SARS-CoV-2 virus is concerned, diabetes affects host viral ACE-2 receptor expression levels, therefore, altering host response to virus infection (Comunian et al. 2020).

Air, water and aerosols' transmission of SARS-CoV-2 may also be influenced by global climate changes. The effect of climatic factors on the development of herd immunity to SARS-CoV-2 and to other viruses, still require further investigation being a matter of discussion once this relationship is not entirely clear. The development of a clinical response to a virus infection depends on several factors. Virus exposure dose affects not only transmission capability but also infection outcomes. Climate change effects may eventually lead to changes in infection doses and infection formulations resulting into more severe outcomes. The efficiency of host innate and adaptive immune responses are host factors affecting disease development upon virus infection.

Malnutrition leads to impaired immunity and recent studies show that Vitamin D deficiency increase the risk of viral infections (Zhou et al. 2019). Moreover, air pollutants and GHG may affect antiviral host response. It is thought that this mechanism may explain, for example, the occurrence of initial SARS-CoV-2 outbreak in Italy, initially in Lombardy, which is industrial area, thus with considerable atmospheric pollution. Lombardy population is known to have chronic vitamin D deficiency, striking more seriously during cloudy winters (Ferrari et al. 2019; Conticini et al. 2020). This could also be associated with high levels of greenhouse gases and anthropogenic aerosols, which absorb part of the natural UV-B radiation, making it

Table 6.1 Health risks driven by climate change effects caused by human activities

Human activities	Climate and environmental changes		Health risks
Gaseous emissions (GHG and pollutants) to the atmosphere derived from: • Urbanization • Industrialization • Agriculture intensification	• CO_2 and other GHGs atmospheric concentration increase • Increase of particulate matter and pollutants concentration in the air • Atmospheric temperature increase • Ocean warming	Change in climate patterns with increased frequency and severity of extreme weather events: • Heat waves • Blizzards and hurricanes • Storms and heavy rain	• Increase of transmissible infections through air and water channels due to the occurrence of plagues, pests and infestations • Populations displacemet due to floods
		Ocean acidification and surface temperature increase	• Toxic algal blooms
		Storms and intense precipitation increase pools for arthropod hosts proliferation and breeding	• Rift valley fever • West Nile virus fever • Increase *Aedes aegypti* mosquito breeding sites: Dengue • Oropouche virus fever • Malaria • Heart disease caused by *Coxsackie* B virus
		Longer and warmer summers Changes in rainfall patterns	• Mycotoxicosis
Land use change by: • Agricultural intensification • Intensive livestock production • Construction of dams, canals, irrigation systems • Mineral and energy resources exploitation • Urban sprawl • Industrialization • Large built infrastructures • Deforestation, reforestation and new habitation	• Land contamination • Water contamination • Deforestation • Reforestation • Desertification • Ecosystems and habitats disruption • Loss of biodiversity • Decrease in rainfall amounts, drought processes	Crop insecticides lead to vector resistance	• Malaria
		Increase in rodent abundance	• Venezuelan haemorrhagic fever • Leptospirosis • Hantavirus pulmonary syndrome
		Increase larval contact due to moist soil	• Helminthiasies
		Antibiotic and pesticides residues in water and soil from intensive cultivation to increase food production	• Food-borne infections (like diarrhoea or gastroenteritis)
		Increase in snail host habitat	• Schistosomiasis
		Increase contact with sand fly vectors	• Visceral leishmaniasis • Cutaneous leishmaniasis

(continued)

Table 6.1 (continued)

Human activities	Climate and environmental changes		Health risks
		Increase tick hosts, outdoor exposure	• Lyme disease
Intensive use of water resources and poor water management due to: • Urbanization and urban crowding • Population density and poor housing and sanitary conditions • Industrialization • Soil imperviousness by the building environment	• Surface waters scarcity • Aquifers depletion • Increase water contamination • Decrease on the availability of safe water for human consumption	Decrease sanitation and hygiene	• Cholera
		Water contamination by: • Faecal and wastewater sources • Toxic chemicals • Microbiological contaminants • Metabolic disruption chemicals	• Liver failure by *Hepatitis* A and E virus • Wound infection by *V. alginolyticus* and *V. parahaemolyticus*
		Increase *Aedes aegypti* mosquito breeding sites	• Dengue
		Increase proximity of sand fly vectors	• Cutaneous leishmaniasis
		Increase larval contact due to moist soil	• Helminthiasies

less available for cutaneous synthesis of vitamin D, leading to a growing deficiency (Wacker and Holick 2013).

Several studies have connected climate changes with gut microbiome composition changes, which may explain impaired immune defences against virus infections (Sepulveda and Moeller 2020; Phillips 2009). Human microbiome includes the genomes of microbiota inhabiting humans, composed by protozoa, archaea, eukaryotes, viruses and mainly bacteria that live symbiotically with humans. These microorganisms are responsible for vitamin production, food metabolization, and the regulation of immune responses against pathogens. The nature of human microbiome, which affects immunity, is expected to change in time, not only due to accelerated lifestyles, but also as a consequence of climate change effects on weather patterns, soil microbiota, food chains, nutrition and living conditions.

Recommendations and Conclusions

Health systems have to be prepared for changes in health outcomes, namely threats from infectious disease transmission and pathways. Several food-, vector- and water-borne diseases are climate-sensitive. The consequences of climate change on diseases incidence and prevalence are difficult to evaluate due to the complexity of disease transmission mechanisms and its underlying drivers and determinants. However, it is important to be aware of the importance of climate-sensitiveness in transmissible diseases.

Some patterns and mechanisms of transmissible diseases are not yet fully understood due to the complex nature of the phenomena involved and, therefore, require an integrated and holistic analysis of several factors and determinants focusing biological, ecological environmental, climatic and demographic data. This analysis would build upon the current knowledge to accelerate the construction and validation of transmissible diseases models, which, in turn, would make possible the formulation of predictive models to address the expected impacts of climate change in different evolutionary and socioeconomic scenarios. These predictive models will support risks assessment, alert systems for disease outbreaks and preparedness of governments and other decision makers to respond to the emergence and spread of transmissible diseases.

Additionally, it is important to establish fast, efficient and accurate communication channels among, not only the health community and policy makers, but the civil society as well to raise awareness and to prevent misinformation and panic. The academia and health workers should be involved in these communication systems and in educational programs to strength preparedness capacity and public awareness by delivering clear and simple climate-health messages.

During epidemiologic transition most of infectious disease control measures have been neglected. However, due to COVID-19 pandemic these measures have received renewed attention; therefore, several developments in the field of infectious diseases control have emerged. The multiple effects of the latest pandemic on health, mental-health, society mobility and socio-economic conditions clearly demonstrated countries unpreparedness and the urgent need for a global, centralised communication, report and management of emergences to provide jointed responses. It is mandatory to develop global policies for transmissible diseases control programs, to articulate national health action plans at a global scale, to establish strategies and resilience measures to prevent and reduce the climate changes impacts in transmissible diseases worldwide. These concerted actions and efforts would improve the global management of climate-sensitive infectious diseases as was the case of the recent processes' acceleration in SARS-CoV-2 vaccine development.

References

Alfano N, Tagliapietra V, Rosso F, Manica M, Arnoldi D, Pindo M, Rizzoli A (2019) Changes in microbiota across developmental stages of *aedes koreicus*, an invasive mosquito vector in Europe: indications for microbiota-based control strategies. Front in Microbiol 10. https://doi.org/10.3389/fmicb.2019.02832

Awuchi CG, Ondari EN, Nwozo S, Odongo GA, Eseoghene IJ, Twinomuhwezi H, Ogbonna CU, Upadhyay AK, Adeleye AO, Okpala COR (2022) Mycotoxins' toxicological mechanisms involving humans, livestock and their associated health concerns: a review. Toxins 14(3):167. https://doi.org/10.3390/toxins14030167

Brugueras S, Fernández-Martínez B, Martínez-de la Puente J, Figuerola J, Porro TM, Rius C, Larrauri A, Gómez-Barroso D (2020) Environmental drivers, climate change and emergent diseases transmitted by mosquitoes and their vectors in southern Europe: a systematic review. Environ Res 191:110038. https://doi.org/10.1016/j.envres.2020.110038

Burrows MT, Schoeman DS, Richardson AJ, Molinos JG, Hoffmann A, Buckley LB, Moore PJ, Brown CJ, Bruno JF, Duarte CM, Halpern BS, Hoegh-Guldberg O, Kappel CV, Kiessling W, O'Connor MI, Pandolfi JM, Parmesan C, Sydeman WJ, Ferrier S, Williams KJ, Poloczanska ES (2014) Geographical limits to species-range shifts are suggested by climate velocity. Nature 507(7493):492–495. https://doi.org/10.1038/nature12976

Carlson CJ, Albery GF, Merow C, Trisos CH, Zipfel CM, Eskew EA, Olival KJ, Ross N, Bansal S (2022) Climate change increases cross-species viral transmission risk. Nature. https://doi.org/10.1038/s41586-022-04788-w

Cecílio P, Cordeiro-da-Silva A, Oliveira F (2022) Sand flies: basic information on the vectors of leishmaniasis and their interactions with *Leishmania* parasites. Commun Biol 5(1):305. https://doi.org/10.1038/s42003-022-03240-z

Checkley W, Epstein LD, Gilman RH, Figueroa D, Cama RI, Patz JA, Black RE (2000) Effect of El Nino and ambient temperature on hospital admissions for diarrhoeal diseases in Peruvian children. Lancet 355(9202):442–450. https://doi.org/10.1016/s0140-6736(00)82010-3

Chen IC, Hill JK, Ohlemuller R, Roy DB, Thomas CD (2011) Rapid range shifts of species associated with high levels of climate warming. Science 333(6045):1024–1026. https://doi.org/10.1126/science.1206432

Cheng L, Abraham J, Zhu J, Trenberth KE, Fasullo J, Boyer T, Locarnini R, Zhang B, Yu F, Wan L, Chen X, Song X, Liu Y, Mann ME (2020) Record-setting ocean warmth continued in 2019. Adv Atmos Sci 37(2):137–142. https://doi.org/10.1007/s00376-020-9283-7

Colwell RR (1996) Global climate and infectious disease: the cholera paradigm. Science 274(5295):2025–2031. https://doi.org/10.1126/science.274.5295.2025

Comunian S, Dongo D, Milani C, Palestini P (2020) Air pollution and Covid-19: the role of particulate matter in the spread and increase of covid-19's morbidity and mortality. Int J Environ Res Public Health 17(12). https://doi.org/10.3390/ijerph17124487

Conticini E, Frediani B, Caro D (2020) Can atmospheric pollution be considered a co-factor in extremely high level of SARS-CoV-2 lethality in northern Italy? Environ Pollut 261:114465. https://doi.org/10.1016/j.envpol.2020.114465

Dhakal S, Minx JC, Toth FL, Abdel-Aziz A, Meza MJF, Hubacek K, Jonckheere IGC, Kim Y-G, Nemet GF, Pachauri S, Tan XC, Wiedmann T (2022) Emissions trends and drivers. In: Shukla PR, Skea J, Slade R et al. (eds) Climate change 2022: mitigation of climate change. Contribution of working group III to the sixth assessment report of the intergovernmental panel on climate change. Cambridge University Press, Cambridge, p 131. https://doi.org/10.1017/9781009157926.004

Epstein PR (1993) Algal blooms in the spread and persistence of cholera. Biosystems 31(2–3):209–221. https://doi.org/10.1016/0303-2647(93)90050-m

European Commission (2021a) Commission Staff working document: impact assessment report. Accompanying the document: communication from the commission to the European parliament, the council, the economic and social committee and the committee of the regions: forging a climate-resilient Europe—the new Eu strategy on adaptation to climate change. SWD(2021) 25 final. European Union, Brussels

European Commission (2021b) Communication from the commission to the European parliament, the council, the economic and social committee and the committee of the regions: forging a climate-resilient Europe—the new Eu strategy on adaptation to climate change. COM(2021) 82 final. European Union, Brussels

Fattorini D, Regoli F (2020) Role of the chronic air pollution levels in the Covid-19 outbreak risk in Italy. Environ Pollut 264:114732. https://doi.org/10.1016/j.envpol.2020.114732

Ferrari D, Lombardi G, Strollo M, Pontillo M, Motta A, Locatelli M (2019) Association between solar ultraviolet doses and vitamin D clinical routine data in European mid-latitude population between 2006 and 2018. Photochem Photobiol Sci 18(11):2696–2706. https://doi.org/10.1039/C9PP00372J

Gubler DJ, Reiter P, Ebi KL, Yap W, Nasci R, Patz JA (2001) Climate variability and change in the United States: Potential impacts on vector- and rodent-borne diseases. Environ Health Persp 109:223–233. https://doi.org/10.1289/ehp.109-1240669

Gulev SK, Thorne PW, Ahn J, Dentener FJ, Domingues CM, Gerland S, Gong D, Kaufman DS, Nnamchi HC, Quaas J, Rivera JA, Sathyendranath S, Smith SL, Trewin B, Shuckmann KV, Vose RS (2021) Changing state of the climate system. In: Masson-Delmotte V, Zhai P, Pirani A et al. (eds) Climate change 2021: the physical science basis. Contribution of working group I to the sixth assessment report of the intergovernmental panel on climate change. Cambridge University Press, Cambridge, p 213

Hales S, de Wet N, Maindonald J, Woodward A (2002) Potential effect of population and climate changes on global distribution of dengue fever: an empirical model. Lancet 360(9336):830–834. https://doi.org/10.1016/S0140-6736(02)09964-6

Hess J, Boodram LG, Paz S, Stewart Ibarra AM, Wasserheit JN, Lowe R (2020) Strengthening the global response to climate change and infectious disease threats. BMJ 371:m3081. https://doi.org/10.1136/bmj.m3081

IPCC (2021) Climate change 2021: the physical science basis. Contribution of working group I to the sixth assessment report of the intergovernmental panel on climate change. IPCC sixth assessment report (AR6). United Nations, Cambridge

IPCC WGII (2022) Summary for policymakers. In: Pörtner H-O, Roberts DC, Tignor M et al. (eds) Climate change 2022: impacts, adaptation and vulnerability. The working group II contribution to the IPCC sixth assessment report Cambridge University Press, Switzerland, p 40

Jones KE, Patel NG, Levy MA, Storeygard A, Balk D, Gittleman JL, Daszak P (2008) Global trends in emerging infectious diseases. Nature 451(7181):990–993. https://doi.org/10.1038/nature06536

Karwat A, Franzke CLE (2021) Future projections of heat mortality risk for major European cities. Weather Clim Soc 13(4):913–931. https://doi.org/10.1175/wcas-d-20-0142.1

Martinez GS, Linares C, Ayuso A, Kendrovski V, Boeckmann M, Diaz J (2019) Heat-health action plans in Europe: challenges ahead and how to tackle them. Environ Res 176:108548. https://doi.org/10.1016/j.envres.2019.108548

Medina A, Akbar A, Baazeem A, Rodriguez A, Magan N (2017) Climate change, food security and mycotoxins: do we know enough? Fungal Biol Rev 31(3):143–154. https://doi.org/10.1016/j.fbr.2017.04.002

Metcalf CJE, Walter KS, Wesolowski A, Buckee CO, Shevliakova E, Tatem AJ, Boos WR, Weinberger DM, Pitzer VE (2017) Identifying climate drivers of infectious disease dynamics: recent advances and challenges ahead. Proc Biol Sci 284(1860). https://doi.org/10.1098/rspb.2017.0901

Mount GA, Haile DG, Daniels E (1997) Simulation of management strategies for the blacklegged tick (Acari: Ixodidae) and the Lyme disease spirochete, Borrelia burgdorferi. J Med Entomol 34(6):672–683. https://doi.org/10.1093/jmedent/34.6.672

Paz S, Majeed A, Christophides GK (2021) Climate change impacts on infectious diseases in the Eastern Mediterranean and the Middle East (EMME)-risks and recommendations. Clim Change 169(3–4):40. https://doi.org/10.1007/s10584-021-03300-z

Phillips ML (2009) Gut reaction: environmental effects on the human microbiota. Environ Health Perspect 117(5):A198–205. https://doi.org/10.1289/ehp.117-a198

Plowright RK, Parrish CR, McCallum H, Hudson PJ, Ko AI, Graham AL, Lloyd-Smith JO (2017) Pathways to zoonotic spillover. Nat Rev Microbiol 15(8):502–510. https://doi.org/10.1038/nrmicro.2017.45

Rocklov J, Dubrow R (2020) Climate change: an enduring challenge for vector-borne disease prevention and control. Nat Immunol 21(5):479–483. https://doi.org/10.1038/s41590-020-0648-y

Rose JB, Epstein PR, Lipp EK, Sherman BH, Bernard SM, Patz JA (2001) Climate variability and change in the United States: potential impacts on water- and foodborne diseases caused by microbiologic agents. Environ Health Perspect 109(Suppl 2):211–221. https://doi.org/10.1289/ehp.01109s2211

Semenza JC (2021) Lateral public health: advancing systemic resilience to climate change. Lancet Reg Health Eur 9:100231. https://doi.org/10.1016/j.lanepe.2021.100231
Semenza JC, Herbst S, Rechenburg A, Suk JE, Höser C, Schreiber C, Kistemann T (2012) Climate change impact assessment of food—and waterborne diseases. Crit Rev Environ Sci Technol 42(8):857–890. https://doi.org/10.1080/10643389.2010.534706
Semenza JC, Lindgren E, Balkanyi L, Espinosa L, Almqvist MS, Penttinen P, Rocklov J (2016) Determinants and drivers of infectious disease threat events in Europe. Emerg Infect Dis 22(4):581–589. https://doi.org/10.3201/eid2204
Semenza JC, Suk JE (2017) Vector-borne diseases and climate change: a European perspective. FEMS Microbiology Lett 365(2). https://doi.org/10.1093/femsle/fnx244
Semenza JC, Suk JE (2018) Vector-borne diseases and climate change: a European perspective. FEMS Microbiol Lett 365(2). https://doi.org/10.1093/femsle/fnx244
Sepulveda J, Moeller AH (2020) The effects of temperature on animal gut microbiomes. Front Microbiol 11:384. https://doi.org/10.3389/fmicb.2020.00384
UNFCCC (2021) Nationally determined contributions under the Paris agreement—revised synthesis report by the secretariat of the conference of the parties serving as the meeting of the parties to the Paris agreement in its third session at Glasgow, 31 October to 12 November 2021. United Nations
United Nations Economic and Social Council (2021) Progress towards the sustainable development goals—report of the secretary-general. United Nations, New York
Wacker M, Holick MF (2013) Sunlight and vitamin D: a global perspective for health. Dermatoendocrinol 5(1):51–108. https://doi.org/10.4161/derm.24494
Watts N, Amann M, Arnell N, Ayeb-Karlsson S, Beagley J, Belesova K, Boykoff M, Byass P, Cai W, Campbell-Lendrum D, Capstick S, Chambers J, Coleman S, Dalin C, Daly M, Dasandi N, Dasgupta S, Davies M, Di Napoli C, Dominguez-Salas P, Drummond P, Dubrow R, Ebi KL, Eckelman M, Ekins P, Escobar LE, Georgeson L, Golder S, Grace D, Graham H, Haggar P, Hamilton I, Hartinger S, Hess J, Hsu SC, Hughes N, Jankin Mikhaylov S, Jimenez MP, Kelman I, Kennard H, Kiesewetter G, Kinney PL, Kjellstrom T, Kniveton D, Lampard P, Lemke B, Liu Y, Liu Z, Lott M, Lowe R, Martinez-Urtaza J, Maslin M, McAllister L, McGushin A, McMichael C, Milner J, Moradi-Lakeh M, Morrissey K, Munzert S, Murray KA, Neville T, Nilsson M, Sewe MO, Oreszczyn T, Otto M, Owfi F, Pearman O, Pencheon D, Quinn R, Rabbaniha M, Robinson E, Rocklov J, Romanello M, Semenza JC, Sherman J, Shi L, Springmann M, Tabatabaei M, Taylor J, Trinanes J, Shumake-Guillemot J, Vu B, Wilkinson P, Winning M, Gong P, Montgomery H, Costello A (2021) The 2020 report of The Lancet Countdown on health and climate change: responding to converging crises. Lancet 397(10269):129–170. https://doi.org/10.1016/S0140-6736(20)32290-X
WHO (2018) COP24 special report: health & climate change. World Health Organization, Geneva
WHO (2021) COP26 special report on climate change and health: the health argument for climate action. World Health Organization, Geneva
Zhou YF, Luo BA, Qin LL (2019) The association between vitamin D deficiency and community-acquired pneumonia: a meta-analysis of observational studies. Medicine 98(38):e17252. https://doi.org/10.1097/md.0000000000017252

Maria Pia de Melo Alvim Ferraz Tavares is a Professor at the Faculdade de Engenharia da Universidade do Porto (FEUP) and a Researcher in Instituto de Investigação e Inovação em Saúde in Universidade do Porto (i3S/INEB). She worked in multidisciplinary subjects, e.g. infectious diseases, biomaterials with antimicrobial properties, antimicrobial resistance mechanisms (ARM) and alternatives to ARM, microorganisms' multispecies biofilms. She has published 61 documents, with 1860 citations, with h-index 24 (SCOPUS).

Gisela Marta Teixeira de Sousa Oliveira is Science Manager and Researcher in Environmental Engineering fields at Universidade Fernando Pessoa. She has over 2000 hours of experience as teacher and as instructor in food technology and environmental engineering subjects. Her scientific research interests include: air pollution; GHG emissions; environmental engineering; circular economy; sustainable development and cross-cutting subjects. Published 17 articles, 13 book sections and 2 books. Participated in 8 projects. Participates in scientific evaluation committees and as journal guest editor and reviewer.

Chapter 7
What Will Be the Most Critical Driver of Population Dynamics: Climate Change, the COVID-19 Pandemic, or Both?

Khandaker Jafor Ahmed, Nishat Tasneem, Bijoya Saha, and Shah Md Atiqul Haq

Abstract The world has already experienced many extreme weather events and the consequences of these events are severe and will become even more severe in scope and intensity. In addition to the risks and crises associated with climate change, the world has also had to deal with the COVID-19 pandemic, resulting in many deaths and socio-economic and psychological consequences. Since both (climate change and pandemics) affect human life in different ways, including population dynamics in terms of fertility, mortality and migration, this study aims to critically review the literature and discuss whether one or both issues will be the most influential driving force. The study also aims to look at the effects on population dynamics by visiting the arguments, results and implications of different studies to understand the diversified effects on society. The examination of the relevant studies shows that the frequency and magnitude of extreme climate events and COVID-19 are spatially different and the impacts of both events are geographically uneven and impose a double burden on humans and their well-being. For example, due to their vulnerability to the adverse effects of climate change and the recent pandemic, poorer countries have a lower adaptive capacity and generally experience difficulties rebuilding and recovering from losses and damages. In addition, the results of the studies included in the assessment indicate that climate change and the COVID-19 pandemic are likely to affect population dynamics—mortality, migration and fertility. Extreme weather events and the recent pandemic may contribute to mortality differences, migration decisions, and fertility intentions, given the experience of weather events and the recent pandemic. Moreover, the effects of climate change and the pandemic are mixed, with either upward or downward trends, or sometimes both, in demographic components. This study argues for serious consideration of these issues as they will

---

K. J. Ahmed (✉)
Walsh School of Foreign Service, Georgetown University, 37th and O St NW, Washington, District of Columbia 20057, United States
e-mail: ka906@georgetown.edu; k.m.jafor1991sust@gmail.com

N. Tasneem · B. Saha · S. M. A. Haq
Department of Sociology, Shahjalal University of Science and Technology, Sylhet 3114, Bangladesh

W. Leal Filho et al. (eds.), *Climate Change and Health Hazards*, Climate Change Management, https://doi.org/10.1007/978-3-031-26592-1_7

affect population dynamics closely linked to socio-economic, cultural, and political issues. There is an urgent need for detailed studies of the inevitable consequences of climate change and the recent pandemic by examining their impact on population dynamics.

Introduction

Climate change and related extreme climate (or weather) events are becoming more severe in their frequency and intensity (IPCC 2022) and are affecting the population in a variety of ways (Jiang and Hardee 2011; Muttarak 2021). On top of these climate events, our world has witnessed severe diseases and pandemics, some of which have been triggered by climate change. COVID-19 is the worst pandemic that has affected every country in the world, causing many deaths and affecting every aspect of human life (Mofijur et al. 2021; Walker et al. 2020). The impact of extreme climate events or the pandemic on the population is not concentrated in one region, it has spread to every part of the world, and both are a threat to human well-being. However, poorer countries with limited resources and opportunities suffer the most from their influence.

Changes in population size, density, and distribution are induced by changes in three core population dynamics: mortality, migration, and fertility. Climate change and the global pandemic are likely to impact population dynamics negatively. This paper reviews the literature to understand in what ways climate change and the COVID-19 pandemic influence mortality, migration, and fertility. This review involves knowing how mortality, migration, and fertility respond to climate change and the COVID-19 pandemic and which one is the most influential driver of change in population dynamics.

To find relevant articles, multiple databases such as Web of Science, Scopus, CINAHL, and PubMed were chosen and accessed through the online library of the University of Adelaide. These four databases were selected for search queries since these were widely used in the systematic literature reviews in the research of climate change, COVID-19 pandemic and population dynamics (Booth et al. 2021; Fatemi et al. 2017; Harville et al. 2010; Obokata et al. 2014; Zotti et al. 2013). The search terms included population dynamics (e.g., mortality, migration, and fertility), climate change terms (e.g., climate, temperature, precipitation), disaster terms (earthquake, hurricane, storm, tsunami, famine, flood, drought, war, bombing, pandemic), and COVID-19 pandemic. Full texts of published journal articles were critically reviewed to identify trends and determinants of mortality, migration, and fertility in the context of climate change and the COVID-19 pandemic. In this review, only peer-reviewed journal publications written in English were included in the analysis when context and timeframe do not bind these.

Climate Change and COVID-19 Pandemic: A Global Issue

Climate change is not concentrated in a particular region of the world, it is a global issue (Wake 2015), and the effects of climate change are already being felt in a significant portion of the world (IPCC 2022). Due to global warming, weather patterns are shifting in ways that can negatively affect human well-being (IPCC 2022; Lyon et al. 2022). These effects include heat waves, droughts, hurricanes, floods, glacier melt, altered ecosystems, and rising sea levels (IPCC 2022). These changes are likely to impact the dynamics of the population, including fertility, mortality, and migration rates (Muttarak 2021). Acknowledging the impacts of climate change in different parts of the world vary (Arnell et al. 2019), counting on their mitigation strategies and adaptive capacities, these effects are irreversible, and research and action to tackle these must look beyond the near future (Lyon et al. 2022).

On the other hand, the novel coronavirus disease 2019 pandemic, named COVID-19 on 11 February 2020, has spread rapidly worldwide (Fauci et al. 2020; Gates 2020). Since the inception of the first case of COVID-19 found in Wuhan, Hubei province, China and reported to the World Health Organization (WHO) on 31 December 2019, it takes only a month to declare a Public Health Emergency of international concern before declaring a global pandemic on 11 March 2020 (WHO 2020). According to the World Health Organization situation report (99) data of 3 July 2022, the globally reported confirmed cases are 54, 63, 63, 985 with 63, 36, 802 deaths. The distribution of confirmed cases in different regions is presented in Fig. 7.1. The European region, followed by the American region, is in the top place in confirmed cases and reported deaths (WHO 2022a b). The cases and deaths of COVID-19 show exponential growth worldwide.

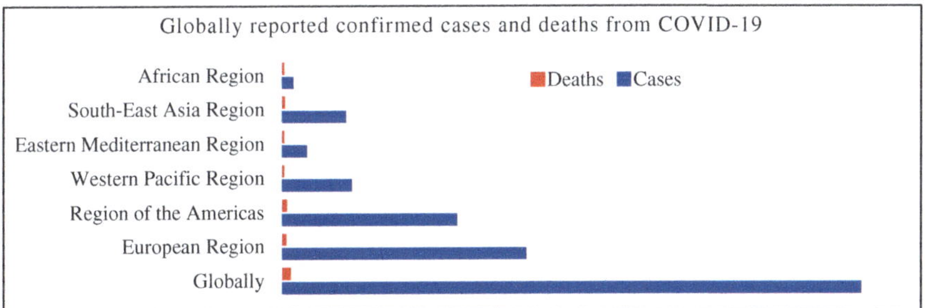

	Globally	European Region	Region of the Americas	Western Pacific Region	Eastern Mediterranean Region	South-East Asia Region	African Region
Deaths	6336802	2027968	2762527	238904	343596	790178	173616
Cases	546363985	228917538	163205242	64433670	22044303	58628247	9134221

Fig. 7.1 Globally reported confirmed cases and deaths from COVID-19, by WHO Region, as of 3 July 2022 (*Data source* WHO COVID-19 Weekly Epidemiological Update, Edition 99 [6 July 2022])

The world has already experienced several outbreaks, epidemics, and pandemics in the last century. Influenza, HIV/AIDS, Cholera, Smallpox, Plague, SARS, Swine flu, Ebola, Zika, Dengue, etc. Beyond the effects of pandemics on morbidity and mortality, they also bring impacts on physical and mental health, society, and the economy (Almond 2006; Bloom and Cadarette 2019; Galea et al. 2020; Kwong et al. 2012; Qiu et al. 2018; Reid 2005). The COVID-19 pandemic has ravaged human societies worldwide from its early appearance in Wuhan, China, throughout the winter of 2019 (Lupton and Willis 2021). The pandemic has an uneven distribution of its impact on many nations and demographic factors such as age and sex, as well as its impact on population health, years of life lost, and lifespan (Aburto et al. 2021; Dowd et al. 2020; Pifarré I Arolas et al. 2021).

Climate Change and Population Dynamics

This section aims to review recent research on these topics and lay out what we know about them so far. Rapid climate change has the potential to affect all three of these aspects of the trajectory of human life: mortality, fertility, and migration.

Impact on Mortality

Mounting evidence points to a connection between human mortality and climate change. Climate change can bring short-term and long-term effects on mortality rates. An estimation shows that there will be 74 deaths per 100,000 people worldwide in 2100 because of climate change (Carleton et al. 2022). The exact number of people die each year from infectious diseases like tuberculosis and HIV/AIDS, is shown in Fig. 7.2. According to the study, an additional 85 deaths per 100,000 could be expected by 2100. If current emission levels are not reduced, there is a real possibility that climate change will cause more people to die. It is estimated that 125 people will lose their lives to cancer for every 100,000 people on the planet. This mortality rate is comparable to the current death rate caused by cardiovascular disease (Carleton et al. 2022).

Extreme temperatures directly influence mortality rates, regardless of age or gender. According to Faye et al. (2021), those over 55 years and females were more susceptible to heat wave-related mortality in Senegal. The study also revealed that young children aged 0 to 5 years are at extreme risk of passing away. Chen et al. (2019) found that the elderly, those with respiratory ailments, and those who were illiterate were most at risk during cold spells in China aged 0 to 5 years are at extreme risk of passing away.

According to Lu et al. (2019), exposure to temperatures that are too hot or too cold has a negative impact on one's likelihood of dying. Between 2002 and 2011, they studied the impact of temperature on Hong Kong's cause-specific mortality.

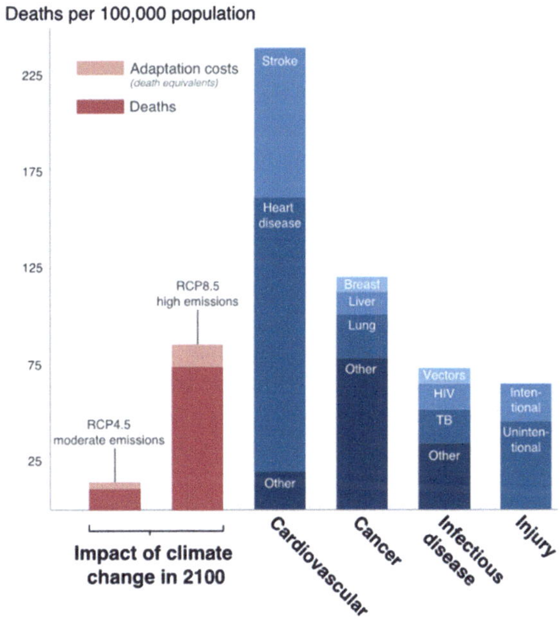

Fig. 7.2 Climate change is comparable to death causes in 2100 (*Source* Modified from Carleton et al. 2022)

They found that their estimates of the heat and cold exposures and the various causes of death differ from one another in a statistically significant manner. Extreme hot and cold temperatures are linked to an increased risk of death for any reason, including cardiovascular and respiratory diseases. According to their findings, the Chinese population is more susceptible to the adverse effects of cold weather than the favourable impact of hot weather (Lu et al. 2019). In China's subtropical regions, both high and low temperatures were linked to an increased mortality rate (Yi and Chan 2015).

This literature review demonstrates that climate change, directly and indirectly, correlates with mortality risk. However, this pattern of relationships does not appear to be consistent across the study areas. Depending on the region, gender, and age group, the pattern of relationships emerges and differs. This section highlights that patients with coronary heart disease, respiratory disease, women, the elderly, and people living in temperate continental and monsoon zones are vulnerable to climate change mortality. Extreme heat or cold makes humans susceptible to many diseases. Non-accidental deaths from cardiovascular, respiratory, and stroke diseases are all up during extreme heat and cold waves.

Impact on Migration

Environmental changes in human–environment interactions have been shown to be a significant driver of human migration. Short-distance internal migration has been found to benefit significantly from temperature variations and episodic disasters in general. When it comes to rainfall and migration, there are conflicting reports. Climate change and migration dynamics are discussed in this section of this paper.

Climate change adversely impacts agricultural production, a significant channel through which migration volume occurs in many parts of the world. According to Bohra-Mishra et al. (2017), climate variables influenced rural-to-urban migration in the Philippines. They found that increased temperatures and typhoon activity increased out-migration, while precipitation had no consistent or significant effect. Temperature and typhoons negatively affected rice yields, a significant predictor of agricultural productivity, and caused more out-migration from agriculturally dependent and rural provinces.

A large body of research on human movement in developing countries has revealed significant social and economic barriers to migration (Bohra-Mishra et al. 2017; Carrico and Donato 2019; Gray and Mueller 2012; Gray and Wise 2016; Thiede et al. 2016). Bangladesh's environmental vulnerability makes it an excellent place to study migration and the environment (Carrico and Donato 2019). In southwestern Bangladesh, Carrico and Donato (2019) found that migration increased most during dry spells, though first trips increased after warm spells and average rainfall. Some evidence links community-level migrant networks to migration after environmental shocks, but there is no conclusive evidence that migrant relatives play a role. Gray and Mueller (2012) found that moderate floods have little impact on mobility in rural Bangladesh, and crop failures affect mobility because families in severely affected areas are more likely to move.

Gray and Wise (2016) found that climate variability affects migration in Kenya, Uganda, Nigeria, Burkina Faso, and Senegal. Migration increases as Uganda's temperature disparity widen. Kenya and Burkina Faso have widening temperature gaps, and there's no evidence that Nigeria and Senegal are linked. Nawrotzki and DeWaard (2016) examined the seven-year risks of Mexican immigration to the U.S. after a climate shock. Migration is less likely three years after climatic stress because families have adjusted to that shock. According to Thiede and Gray (2017), climate variability and climate impacts on migration are more important for short-distance migration.

According to the results and findings of these pieces of research, migration is a direct result of climate change's influence on the environment. It is anticipated that climate change will lead to an increase in the number of people moving and being displaced. Multi-component events like migration influence household decisions and individual choices about where to relocate. The destruction of crops and homes caused by a severe flood, typhoon, or drought can sometimes result in the displacement of an entire community. Temperatures that are too high can affect people's future income, economic systems, and strategies. This, among other things, can force people to relocate.

Impact on Fertility

Human reproductive processes and their outcomes will likely be altered due to climate change, among other things. At the community and individual levels, we can anticipate a diverse array of responses to the effects of climate change. Barreca et al. (2018) measured that temperature and weather shock significantly impacted American women's fertility rate between 1931 and 2010. They found a significant drop in birth rates for eight to ten months following an extremely hot day, followed by a partial comeback in birth rates at 11, 12, and 13 months. In a study based on household panel data spanning over two decades, Sellers and Gray (2019) found that current delays in the monsoon's arrival are associated with a greater desire to have another baby and a decrease in the use of contraceptive measures and family planning, but not with an increase in births. Long periods of above-average temperatures reduce fertility goals and increase contraceptive use among poor and rural women.

Climate change increases seasonal variation in conception rates and lower annual conception rates. Hajdu and Hajdu (2021) found that hot temperatures reduce conception rates in the first few weeks after exposure but rebound. Simon (2017) found that households in humid areas of rural Mexico are more likely to have a child after above-average rainfall than in dry areas. In Sub-Saharan Africa, Thiede et al. (2022) found the opposite.

Climate change-induced fluctuations in food production could affect fertility. Fertility is affected by food insecurity in numerous ways. When it comes to fertility, high temperatures and precipitation have a direct and indirect impact on crop production. However, when it comes to infant mortality, high temperatures and precipitation have an indirect and direct impact on crop production, as Chen et al. (2021) found in their study in Bangladesh.

This section reviews the literature to understand the relationship between climate change and fertility rates. This relationship summarises direct and indirect climate change effects on fertility. It demonstrates the adverse impact on individuals. However, this pattern of relationships does not appear to be consistent across all regions and by the level of climate vulnerabilities and impacts. In terms of geography, sociodemographic and cultural factors, the patterns of relationships differ. In this review, it has been found that climate change harms human fertility and health. It affects fertility short- and long-term. These changes affect infant mortality. These changes affect women, pregnant women, newborns, children, and the elderly. Temperature is a risk factor for abortion and stillbirth. Climate change will affect families' well-being, such as their ability to maintain a standard of living, agricultural production, livelihood sources, and access to staple food, health access, and educational facilities and access.

COVID-19 Pandemic and Population Dynamics

This section reviews the published literature to understand in which ways and to what extent the COVID-19 pandemic influenced the change in mortality, migration, and fertility rates worldwide.

Impact on Mortality

Most nations' life expectancy increased steadily in the twentieth and twenty-first centuries before the COVID-19 pandemic (United Nations 2022). In developed countries, advancements in old age have been the driving force behind recent increases in life expectancy (Aburto et al. 2020). However, this COVID-19 pandemic sparked a worldwide mortality crisis, creating new difficulties for population health in a setting where life expectancy progression paths have grown increasingly diverse. Males often have greater COVID-19 mortality rates than females, with older age groups seeing the highest case-fatality rates (Ahrenfeldt et al. 2021; Levin et al. 2020). Mortality from other sources of death was also indirectly impacted by the pandemic. In contrast to lockdowns, which may have decreased the number of accident fatalities, new research has emphasised the effects of postponing treatments or putting off seeking care for malignancies or cardiovascular problems (Calderon-Anyosa et al. 2021; Hanna et al. 2020; Wu et al. 2021).

Infection and death patterns of COVID-19 seem to vary by country. Surprisingly, richer countries with better healthcare resources have higher death rates than India and other Southeast Asian nations (Jain et al. 2020). There has been evidence that deaths due to other co-morbidities have increased during the pandemic, for instance, a significant rise in the number of diabetes-related fatalities in most American states or cities (Ran et al. 2021). It has been stated that COVID-19 has substantially influenced the healthcare system, which might lead to an elevated death toll. Hesse is a state in central Germany with a population of over 6 million, which saw a rise in the overall mortality rate during the COVID-19-related lockdown compared to almost the same timeframe in 2019 (Nef et al. 2021). In addition, the fatalities attributable to cardiovascular disease rose by 7.6% (Nef et al. 2021).

Estimates provided by the World Health Organization (WHO) indicate that the complete death toll related explicitly or implicitly to COVID-19, which is referred to as "excess mortality", was roughly 14.9 million between 1 January 2020 and 31 December 2021 (range of 13.3 million to 16.6 million) (WHO 2022a b). Excess mortality comprises fatalities linked to COVID-19 either directly due to the illness or indirectly due to the pandemic's effects on healthcare systems and society.

Impact on Migration

COVID-19 has been seen as the greatest danger to human life and the greatest worldwide health crisis of the twentieth century (Chakraborty and Maity 2020). COVID-19's breakout, according to Chakraborty and Maity (2020), has the potential to considerably slow down global economic growth while also having a detrimental influence on human life. COVID-19 is said to act as a push factor for internal and international migration (Kanu 2020). The economic repercussions of measures to stop the virus' spread are particularly dangerous for migrant workers (Newland 2020). According to Abdullah et al. (2021), in 2021, the number of employed individuals in Malaysia rose much higher than the number of unemployed. This indicates that the labour market is seeing an increase in demand. This may be due to the COVID-19 pandemic's positive impacts on employment and economic growth, resulting in new jobs. Additionally, it is also prevalent that since the emergence of COVID-19, people living abroad frequently return to their countries or ask the government to take them back to their own countries (Kanu 2020).

According to OECD (2020), the COVID-19 issue has long-term impacts on migratory movements. Before the pandemic, 5.3 million people moved permanently to the OECD in 2019, comparable numbers in 2017 and 2018. Permanent labour migration increased by more than 13%. Almost all OECD nations banned foreigners' access to their countries when the pandemic started. New visas and permits granted in OECD nations declined by nearly 46% during the first half of 2020 due to a drop in resident permits. The most significant drop ever recorded is at this point. The decrease was 72% in the second quarter. The OECD region expected international migration to reach a record low in 2020. According to OECD (2020), there are compelling indications that mobility won't quickly return to where it was. This is brought on by a decline in the need for labour, enduringly strong travel restrictions, extensive teleworking among highly trained professionals, and remote learning among students.

The pandemic scenario will probably be dreadful in nations with weak governance and those already coping with crises, such as wars and climatic and natural disasters. The food and agriculture industry is a significant pathway via which COVID-19 impacts humans (Smith and Wesselbaum 2020). FAO (2020) reports that since February 2020, food prices have climbed by more than 10% in Bolivia, Belarus, Myanmar, and Ghana and more than 20% in Sudan, Guyana and Zambia. In this situation, people may travel more often to diversify their household income and stabilise their spending to survive external shocks to food availability (Barret 2010). International migration, as well as rural-to-urban migration, are projected to rise as a result of food insecurity. The COVID-19 problem may cause a structural shift in migration patterns if individuals avoid COVID-19 destination nations that are severely affected and relocate to other countries, such as the United States, Italy, or Spain (Smith and Wesselbaum 2020).

The pandemic might have terrible impacts on the target countries because of economic downturns, remittance restrictions, and the postponement of planned migration, all of which have taken place within these countries (IOM 2020). In fact, following the outbreak in some nations, remittance inflows to underdeveloped countries began to fall (Takenaka 2020). Haslag and Weagley (2021) found that COVID-19 considerably affected more than 12% of moves between April 2020 and December 2021, with a noticeable shift toward smaller cities and areas with lower costs of living, cheaper taxes, and regions with less pandemic-related restrictions.

Impact on Fertility

When expressing a wish for children, fertility intention needs to consider different factors, for instance, the optimum number of kids, gender, duration, and spacing between pregnancies (Shreffler et al. 2016). The economy, fertility regulations, education level, the atmosphere, and public services are a few of the numerous variables that impact fertility goals as a systemic issue (Lau et al. 2018). The COVID-19 crisis caused financial and employment instability, reduced access to healthcare and affected various other facets of life (Aassve et al. 2020; Lindberg et al. 2020). Specifically resulting from the COVID-19 pandemic, a new normal reality has evolved. It has an unparalleled influence on the everyday lives of millions of people, and we find it impossible to predict what could happen in the long run. In contrast, decisions concerning parenting are connected to future thinking, and people's imagination and future narratives may be at least as essential for this decision as past or current circumstances (Vignoli et al. 2020a, 2020b).

Economic uncertainty caused by the pandemic seems to have had a comparable influence on men's and women's reproductive intentions; nevertheless, the pandemic's uneven economic implications also created and will have heterogeneous and immediate impacts on fertility motivations (Arpino et al. 2021). Individuals with more susceptible employment had a lower chance of definitely planning to have a(nother) child in the near term and a greater risk of abandoning their pre-COVID-19 fertility plan, although modifications in fertility plans were unaffected by occupation (Arpino et al. 2021).

The COVID-19 pandemic has posed additional physical and mental health difficulties. Fertility decisions are influenced by the psychological stress of being at home alone and worrying about the future (Lebel et al. 2020). The likelihood of a couple's relationship breaking down or having fewer children increases when anxiety and psychological stress play a more significant role in the dynamic (Vieira et al. 2020). However, on the contrary, staying at home and being alone may allow people to spend more time with their families and spouses, improving the quality of their relationships and inspiring more people to have kids (Ahmed et al. 2020; Szabo et al. 2020). After the COVID-19 pandemic, three out of every ten couples of reproductive age who had shown an interest in getting pregnant withdrew their plans (Zhu et al. 2020), which is supported by other evidence suggesting that when the COVID-19 pandemic began, there were fewer women who wanted to become pregnant, and these findings might have long-term consequences for fertility rates (Kahn et al. 2021).

The reproductive consequences of COVID-19 are not likely to be uniformly spread across nations. The pandemic has already had varying effects on various countries, such as regarding time. In addition, economic and demographic backdrop features may enhance or limit people's pre-crisis reproduction intentions, decreasing (or otherwise) the uncertainty resulting from the ongoing health crises and the poor economic pattern (Caltabiano et al. 2017; Vignoli et al. 2020c). Constraints on physical intergenerational support have been imposed as a result of the physical separation that is considered necessary by COVID-19 containment strategies. A shortage of childcare resources may have a higher impact on reproductive intentions in countries where grandparental caregiving is more intense (Price et al. 2018). Fertility programs have been revised downward in every country, although not in the same manner. Fertility plans have shifted little in Germany and France, but many people are still planning or delaying starting a family. In Italy, however, the abandonment rate is substantially greater than in other nations, while the proportion of those who decide to delay their intentions is smaller (Luppi et al. 2020).

Women's reproductive health is one of the many areas that has been affected by COVID-19; thus, the availability of effective contraceptives that are also inexpensive and acceptable has never been more critical (Dasgupta et al. 2020). Many women and couples are at risk of being unable to receive vital therapy for their reproductive health because of the COVID-19 outbreak. Breakdowns in the supply chain have made it difficult for many women to get contraceptive supplies, which has decreased manufacturing, distribution, and availability (UNFPA 2020). If women's and couples' demands for family planning are not met, the number of unintended births will increase. In low- and middle-income countries, the usage of long- and short-acting reversible contraception dropped by 10% due to COVID-19, according to Riley et al. (2020). An additional 49 million women could not access modern contraception, resulting in an additional 15 million unwanted births per year in these nations. One scenario presented by Micelli et al. (2020) is the situation in which 37% of respondents who had intended to have a child before the pandemic altered their views, mostly owing to financial concerns brought on by the recession. Guetto et al. (2020), who polled Italians of reproductive age and performed online experiments on typical narratives of the future, found that pre-pandemic fertility plans decrease as the expected length of the health emergency grows.

Discussion

With climate change, biodiversity loss, environmental degradation, and the escalating COVID-19, we are in a critical period of human history. Climate change has largely influenced the progress in human development, such as reducing income and increasing hunger in low-income countries (UNDP 2020), and it has been making it more difficult to achieve the Sustainable Development Goals (SDGs) of the United Nations even before the COVID-19 pandemic (Stern and Valero 2021).

The increasing consequences of climate change have become very visible worldwide with severe heat waves, floods, wildfires and tropical cyclones (IPCC 2022).

The world has seen a lot of dramatic changes after WHO declared the novel coronavirus (COVID-19) a global pandemic on 11 March 2020 (WHO 2020). Before the pandemic, climate change and associated disasters (including extreme weather or climate events) were on top of discussions. While the concept of climate change is old but a pandemic and its outrageous consequences are pretty new to our generation, COVID-19 seems to act as a means of letting us know what emergencies are. COVID-19 is now considered the most significant biological disaster in human history and affects every aspect of human well-being. Like climate change, its devastating impacts are seen all over and in every field, including the world economy, healthcare sectors, and social and political (Walker et al. 2020). The COVID-19 pandemic, in conjunction with climate change, will undoubtedly influence the evolution of population dynamics. Our world has already realised the importance of climate change in population forecasts. This COVID-19 pandemic should be considered in climate change modelling and population forecasts.

It has been found that climate change (e.g., Barreca et al. 2018; Hajdu and Hajdu 2021) and the pandemic (Arpino et al. 2021; Riley et al. 2020) have adverse impacts on birth outcomes. If we consider the mortality factor, climate change (Carleton et al. 2022) and COVID-19 (Ahrenfeldt et al. 2021; Levin et al. 2020; WHO 2022a b) increase the global mortality rate. Climate change has a definite and significant impact on population movement (Carrico and Donato 2019; Gray and Wise 2016; Thiede et al. 2016), while the migration tendency during COVID-19 evolved as a complex pattern (OECD 2020). Both climate change and the pandemic are likely to impact the determinants of population change directly and indirectly in many ways. As climate change does, the impact of COVID-19 is unevenly distributed across the regional and global levels, and adaptation and mitigation strategies should be context-specific. To some extent, COVID-19 is more likely to worsen the ongoing impact of climate change in many parts of the world. Undoubtedly, it could be implied that both climate change and COVID-19 are critical drivers of population changes, and directly or indirectly, both influence population dynamics overall. There is a need for a sustainable, inclusive, and resilient global recovery addressing the impact of both climate change and COVID-19.

The impact of climate change and the COVID-19 outbreak has created dire conditions for human life in every sector, including the global economy, healthcare, social and political, as well as population dynamics in terms of mortality, fertility, and migration. Our study had limitations as it represents an exploration of the most critical driving force of population dynamics between climate change and the COVID-19 outbreak. Most of the literature mentioned in this study was issued recently. Due to time constraints, we could not include all recent literature. Growing worldwide attention to climate change has been overlooked by the pandemic in recent times. As the effects of climate change and the COVID-19 outbreak are becoming severe these days, it is necessary to study the connection between the COVID-19 outbreak and climate change in recent times. This article suggests that COVID-19 and how climate change affects population dynamics should be studied more thoroughly.

As previous studies are limited in several ways, future studies should explore the following issues to provide policymakers with more persuasive and rigorous guidance. As increasing the impacts of climate change and the COVID-19 outbreak may destabilise existing adaptation strategies, future studies that can contribute to policy-making should assess the effectiveness of adaptation strategies in different scenarios. Furthermore, in this study, the consequences of climate change and COVID-19 should be better estimated using a combination of adaptive capacity and predicted trends under current conditions. The main focus of such research should be on developing or emerging countries as they are most vulnerable to climate change and the COVID-19 pandemic. As climate and health data are not available in most developing countries, access to health data and its linkage to climate changes and susceptibility to COVID-19 must be easily achieved. Furthermore, that study requires a more profound knowledge of population dynamics' determinants. The authors suggest that local and state authorities adopt rapid short-term and long-term initiatives to protect the health of vulnerable and economically disadvantaged people, particularly those affected by the pandemic and the climate crisis. Policymakers and other people in charge should devise and include practical ways to deal with population dynamics caused by climate change and COVID-19.

Conclusion

With the increasing impacts of climate change, the recent COVID-19 pandemic has placed a multidimensional burden on human society in terms of health, economic, social, psychological, and population dynamics issues. In this study, we discuss how climate change and the COVID-19 pandemic affect components of population dynamics such as fertility, mortality, and migration. We also present some figures showing scenarios, not only at the global level but also at the regional level, for changes in fertility, mortality, and migration trends due to climate change and the recent pandemic. Results based on the different studies included in this study show diversification and spatial variation in terms of impacts, trends, and directions. In addition, significant changes in the effects of climate change and the recent pandemic COVID-19 in different forms such as trends in mortality, fertility, and migration compared to periods without a pandemic and without extreme weather events. Therefore, differences in trends depend to a large extent on the socio-economic, cultural, political, health, and demographic policies of the affected country, as well as how the nation responds and encourages its population to increase adaptive capacity. This study reveals that both issues can exacerbate loss and damage, impede normal development, and cause changes in elements of population structure by affecting fertility, mortality, and migration. The study urges the integration of the unavoidable impacts of climate change and COVID-19 into future research, exploration, and understanding of population dynamics so that policies and preparations can be made in advance to maintain human well-being.

To date, climate change can be considered the most critical driver of population dynamics than the COVID-19 pandemic, but it is important to investigate the question in future. Both will make population projections difficult. Research endeavour will be increasingly necessary for the short and long term to identify distinguishable trends in mortality, migration, and fertility and navigate the significant determinants influencing the trends. It can be a debatable issue in future to distinguish which one of climate change and COVID-19 is the most influential driver of population change.

References

Aassve A, Cavalli N, Mencarini L, Plach S, Livi Bacci M (2020) The COVID-19 pandemic and human fertility. Science 369(6502):370–371. https://doi.org/10.1126/science.abc9520

Abdullah B, Sareya R, Kele ATA, Boroh RP, Naim S (2021) An overview on COVID-19 and its impact on migration in Malaysia. J Int Bus Econ Entrepr 6(2):66–73

Aburto JM, Kashyap R, Schöley J, Angus C, Ermisch J, Mills MC, Dowd JB (2021) Estimating the burden of the COVID-19 pandemic on mortality, life expectancy and lifespan inequality in England and Wales: a population-level analysis. J Epidemiol Community Health 75(8):735–740

Aburto JM, Villavicencio F, Basellini U, Kjærgaard S, Vaupel JW (2020) Dynamics of life expectancy and life span equality. Proc Natl Acad Sci 117(10):5250–5259

Ahmed D, Buheji M, Fardan SM (2020) Re-emphasising the future family role in "care economy" as a result of covid-19 pandemic spillovers. Am J Econ 10(6):332–338

Ahrenfeldt LJ, Otavova M, Christensen K, Lindahl-Jacobsen R (2021) Sex and age differences in COVID-19 mortality in Europe. Wien Klin Wochenschr 133(7):393–398

Almond, D. (2006). Is the 1918 influenza pandemic over? Long-term effects of in utero influenza exposure in the post-1940 U.S. population. J Polit Econ 114(4), 672–712. https://doi.org/10.1086/50715

Arnell NW, Lowe JA, Challinor AJ et al (2019) Global and regional impacts of climate change at different levels of global temperature increase. Clim Change 155:377–391. https://doi.org/10.1007/s10584-019-02464-z

Arpino B, Luppi F, Rosina A (2021) Changes in fertility plans during the COVID-19 pandemic in Italy: the role of occupation and income vulnerability.

Barreca A, Deschenes O, Guldi M (2018) Maybe next month? Temperature shocks and dynamic adjustments in birth rates. Demography 55(4):1269–1293. https://doi.org/10.1007/s13524-018-0690-7

Barrett CB (2010) Measuring food insecurity. Science 327(5967):825–828

Bloom DE, Cadarette D (2019) Infectious disease threats in the twenty-first century: strengthening the global response. Front Immunol 10. https://doi.org/10.3389/fimmu.2019.00549

Bohra-Mishra P, Oppenheimer M, Cai R, Feng S, Licker R (2017) Climate variability and migration in the Philippines. Popul Environ 38(3):286–308. https://doi.org/10.1007/s11111-016-0263-x

Booth A, Reed AB, Ponzo S, Yassaee A, Aral M et al (2021) Population risk factors for severe disease and mortality in COVID-19: A global systematic review and meta-analysis. PLoS ONE 16(3):e0247461. https://doi.org/10.1371/journal.pone.0247461

Calderon-Anyosa RJ, Kaufman JS (2021) Impact of COVID-19 lockdown policy on homicide, suicide, and motor vehicle deaths in Peru. Prev Med 143:106331

Caltabiano M, Comolli CL, Rosina A (2017) The effect of the great recession on permanent childlessness in Italy. Demogr Res 37:635–668

Carleton T, Jina A, Delgado M, Greenstone, M, Houser T, Hsiang S, Hultgren A, Kopp RE, McCusker KE, Nath I, Rising J, Rode A, Seo HK, Viaene A, Yuan J, Zhang AT (2022) Valuing

the global mortality consequences of climate change accounting for adaptation costs and benefits. Quart J Econ. https://doi.org/10.1093/qje/qjac020
Carrico AR, Donato K (2019) Extreme weather and migration: evidence from Bangladesh. Popul Environ 41(1):1–31. https://doi.org/10.1007/s11111-019-00322-9
Chakraborty I, Maity P (2020) COVID-19 outbreak: Migration, effects on society, global environment and prevention. Sci Total Environ 728:138882
Chen J, Yang J, Zhou M, Yin P, Wang B, Liu J, Chen Z, Song X, Ou CQ, Liu Q (2019) Cold spell and mortality in 31 Chinese capital cities: Definitions, vulnerability and implications. Environ Int 128:271–278. https://doi.org/10.1016/j.envint.2019.04.049
Chen M, Haq SMA, Ahmed KJ, Hussain AHMB, Ahmed MNQ (2021) The link between climate change, food security and fertility: the case of Bangladesh. PLoS ONE 16. https://doi.org/10.1371/journal.pone.025819
Dasgupta A, Kantorová V Ueffing P (2020) The impact of the COVID-19 crisis on meeting needs for family planning: a global scenario by contraceptive methods used. Gates Open Res 3:1661
Dowd JB, Andriano L, Brazel DM, Rotondi V, Block P, Ding X, Mills MC (2020) Demographic science aids in understanding the spread and fatality rates of COVID-19. Proc Natl Acad Sci 117(18):9696–9698
FAO (2020) Impact of COVID-19 on informal workers. https://www.fao.org/family-farming/detail/en/c/1275270/
Fatemi F, Ardalan A, Aguirre B, Mansouri N, Mohammadfam I (2017) Social vulnerability indicators in disasters: Findings from a systematic review. International Journal of Disaster Risk Reduction 22:219–227. https://doi.org/10.1016/j.ijdrr.2016.09.006
Fauci AS, Lane HC, Redfield RR (2020) Covid-19—navigating the uncharted. N Engl J Med 382(13):1268–1269. https://doi.org/10.1056/NEJMe2002387
Faye M, Dème A, Diongue AK, Diouf I (2021) Impact of different heat wave definitions on daily mortality in Bandafassi. Senegal. Plos ONE 16. https://doi.org/10.1371/journal.pone.0249199
Galea S, Merchant RM, Lurie N (2020) The mental health consequences of COVID-19 and physical distancing: The need for prevention and early intervention. JAMA Intern Med. https://doi.org/10.1001/jamainternmed.2020.1562
Gates B (2020) Responding to Covid-19—a once-in-a-century pandemic? N Engl J Med 382(18):1677–1679. https://doi.org/10.1056/NEJMp2003762
Gray CL, Mueller V (2012) Natural disasters and population mobility in Bangladesh. Proc Natl Acad Sci USA 109(16):6000–6005. https://doi.org/10.1073/pnas.1115944109
Gray C, Wise E (2016) Country-specific effects of climate variability on human migration. Clim Change 135(3–4):555–568. https://doi.org/10.1007/s10584-015-1592-y
Guetto R, Bazzani G, Vignoli D (2020). Narratives of the future shape fertility in uncertain times: evidence from the COVID-19 pandemic (No. 2020_11). Universita'degli Studi di Firenze, Dipartimento di Statistica, Informatica, Applicazioni "G. Parenti"
Hajdu T, Hajdu G (2021) Temperature, climate change, and human conception rates: evidence from Hungary. J Popul Econ. https://doi.org/10.1007/s00148-020-00814-1
Hanna, TP, King WD, Thibodeau S, Jalink M, Paulin GA, Harvey-Jones E, Aggarwal A. (2020) Mortality due to cancer treatment delay: systematic review and meta-analysis. BMJ 371:m4087.
Harville EW, Xiong X, Buekens P (2010) Disasters and perinatal health: a systematic review. Obstet Gynecol Surv 65(11):713–728. https://doi.org/10.1097/OGX.0b013e31820eddbe
Haslag, PH, Weagley D (2021) From LA to boise: how migration has changed during the COVID-19 pandemic.
International Organization for Migration (IOM) (2020) COVID-19 analytical snapshot #16: international remittances. https://www.iom.int/sites/default/files/documents/covid-19_analytical_snapshot_16_-_international_remittances.pdf
IPCC (2022) Climate change 2022: impacts, adaptation and vulnerability. Working group II contribution to the IPCC sixth assessment report. Retrieved 26 July, 2022, from https://www.ipcc.ch/report/sixth-assessment-report-working-group-ii/

Jain VK, Iyengar K, Vaish A, Vaishya R (2020) Differential mortality in COVID-19 patients from India and western countries. Diabetes Metab Syndr 14(5):1037–1041

Jiang L, Hardee K (2011) How do recent population trends matter to climate change? Popul Res Policy Rev 30(2):287–312

Kahn LG, Trasande L, Liu M, Mehta-Lee SS, Brubaker SG, Jacobson MH (2021) Factors associated with changes in pregnancy intention among women who were mothers of young children in New York City following the COVID-19 outbreak. JAMA Netw Open 4(9):e2124273–e2124273

Kanu IA (2020) Covid-19 pandemic and the health of (African) migrants. AMAMIHE J Appl Phil 18:2.

Kwong JC, Ratnasingham S, Campitelli MA, Daneman N, Deeks SL, Manuel DG, Crowcroft NS (2012) The Impact of Infection on Population Health: Results of the Ontario Burden of Infectious Diseases Study (burden of Infectious Diseases in Ontario). 7(9):e44103. https://doi.org/10.1371/journal.pone.0044103

Lau BHP, Huo R, Wang K, Shi L, Li R, Mu S, Chan CHY (2018) Intention of having a second child among infertile and fertile women attending outpatient gynecology clinics in three major cities in China: a cross-sectional study. Hum Reprod Open 4:hoy014

Lebel C, MacKinnon A, Bagshawe M, Tomfohr-Madsen L, Giesbrecht G (2020) Elevated depression and anxiety symptoms among pregnant individuals during the COVID-19 pandemic. J Affect Disord 277:5–13

Levin AT, Hanage WP, Owusu-Boaitey N, Cochran KB, Walsh SP, Meyerowitz-Katz G (2020) Assessing the age specificity of infection fatality rates for COVID-19: systematic review, meta-analysis, and public policy implications. Eur J Epidemiol 35(12):1123–1138

Lindberg LD, VandeVusse A, Mueller J, Kirstein M (2020) Early Impacts of the COVID-19 pandemic: findings from the 2020 Guttmacher survey of reproductive health experiences. Guttmacher Institute, New York. https://www.guttmacher.org/report/early-impacts-covid-19-pandemic-findings-2020-guttmacher-survey-reproductive-health.

Lu M, Zhao X, Wei D, Zhang J, Li Y, He J, Xia H, Qiu X (2019) Ambient temperature and reproductive health outcomes. In: Ambient temperature and health in China. Springer, Singapore, pp 95–103. https://doi.org/10.1007/978-981-13-2583-0_6

Luppi F, Arpino B, Rosina A (2020) The impact of COVID-19 on fertility plans in Italy, Germany, France, Spain, and the United Kingdom. Demogr Res 43:1399–1412

Lupton D, Willis K (2021). The COVID-19 crisis. Soc Perspect. Routledge, London.

Lyon C, Saupe EE, Smith CJ, Hill DJ, Beckerman AP, Stringer LC, Marchant R, McKay J, Burke A, O'Higgins P, Dunhill AM, Allen BJ, Riel-Salvatore J, Aze T (2022) Climate change research and action must look beyond 2100. Glob Change Biol 28:349–361. https://doi.org/10.1111/gcb.15871

Micelli E, Cito G, Cocci A, Polloni G, Russo GI, Minervini A, Coccia ME (2020) Desire for parenthood at the time of COVID-19 pandemic: an insight into the Italian situation. J Psychosom Obstet Gynecol 41(3):183–190

Mofijur M, Fattah I, Alam MA, Islam A, Ong HC, Rahman S, Najafi G, Ahmed SF, Uddin MA, Mahlia T (2021) Impact of COVID-19 on the social, economic, environmental and energy domains: Lessons learnt from a global pandemic. Sustain Produ Consum 26:343–359. https://doi.org/10.1016/j.spc.2020.10.016

Muttarak, R (2021) Demographic perspectives in research on global environmental change. International Institute for Applied Systems Analysis. https://arxiv.org/pdf/2102.00757

Nawrotzki RJ, DeWaard J (2016) Climate shocks and the timing of migration from Mexico. Popul Environ 38(1):72–100. https://doi.org/10.1007/s11111-016-0255-x

Nef HM, Elsässer A, Möllmann H, Abdel-Hadi M, Bauer T, Brück M, Dörr O (2021) Impact of the COVID-19 pandemic on cardiovascular mortality and catherization activity during the lockdown in central Germany: an observational study. Clin Res Cardiol 110(2):292–301

Newland, K (2020). Will international migration governance survive the COVID-19 pandemic. Migration Policy Institute

Obokata R, Veronis L, McLeman R (2014) Empirical research on international environmental migration: a systematic review. Popul Environ 36:111–135. https://doi.org/10.1007/s11111-014-0210-7

OECD (2020) COVID-19 crisis puts migration and progress on integration at risk. Retrieved 2 July 2022, from https://www.oecd.org/migration/covid-19-crisis-puts-migration-and-progress-on-integration-at-risk.htm

Pifarré i Arolas H, Acosta E, López-Casasnovas G, Lo A, Nicodemo C, Riffe T, Myrskylä M (2021) Years of life lost to COVID-19 in 81 countries. Sci Rep 11(1):1–6

Price, D, Ribe, E, Gessa, G, Glaser, K (2018) Grandparental childcare: grandparenting practices around the world, pp 43–62. https://doi.org/10.2307/j.ctv7h0tzm.8

Qiu W, Chu C, Mao A, Wu J (2018) The impacts on health, society, and economy of SARS and H7N9 outbreaks in China: A case comparison study. J Environ Public Health. https://doi.org/10.1155/2018/2710185

Ran J, Zhao S, Han L, Ge Y, Chong MK, Cao W, Sun S. (2021) Increase in diabetes mortality associated with COVID-19 pandemic in the U.S. Diabet Care 44(7):e146–e147

Reid A (2005) The effects of the 1918–1919 influenza pandemic on infant and child health in Derbyshire. Med Hist 49(1):29–54. https://doi.org/10.1017/S0025727300008279

Riley T, Sully E, Ahmed Z, Biddlecom A (2020) Estimates of the potential impact of the COVID-19 pandemic on sexual and reproductive health in low-and middle-income countries. Int Perspect Sex Reprod Health 46:73–76

Sellers S, Gray C (2019) Climate shocks constrain human fertility in Indonesia. World Dev 117:357–369. https://doi.org/10.1016/j.worlddev.2019.02.003

Simon DH (2017) Exploring the influence of precipitation on fertility timing in rural Mexico. Popul Environ 38(4):407–423. https://doi.org/10.1007/s11111-017-0281-3

Shreffler KM, Tiemeyer S, Dorius C, Spierling T, Greil AL, McQuillan J (2016) Infertility and fertility intentions, desires, and outcomes among U.S. women. Demogr Res 35:1149–1168

Smith MD, Wesselbaum D (2020) COVID-19, food insecurity, and migration. J Nutr 150(11):2855–2858

Stern N, Valero A (2021) Innovation, growth and the transition to net-zero emissions. Res Policy 50(9):104293

Szabo TG, Richling S, Embry DD, Biglan A, Wilson KG (2020) From helpless to hero: Promoting values-based behavior and positive family interaction in the midst of COVID-19. Behav Anal Pract 13(3):568–576

Takenaka AK, Gaspar R, Villafuerte J, Narayanan B (2020) COVID-19 impact on International migration, remittances, and recipient households in developing Asia

Thiede BC, Gray CL (2017) Heterogeneous climate effects on human migration in Indonesia. Popul Environ 39(2):147–172. https://doi.org/10.1007/s11111-016-0265-8

Thiede BC, Ronnkvist S, Armao A, Burka K (2022) Climate anomalies and birth rates in sub-Saharan Africa. Climat Change 171(1–2). https://doi.org/10.1007/s10584-021-03273-z

Thiede B, Gray C, Mueller V (2016) Climate variability and inter-provincial migration in South America, 1970–2011. Glob Environ Chang 41:228–240. https://doi.org/10.1016/j.gloenvcha.2016.10.005

UNFPA (2020) Impact of the COVID-19 pandemic on family planning and ending gender-based violence, female genital mutilation and child marriage. https://www.unfpa.org/sites/default/files/resource-pdf/COVID-19_impact_brief_for_UNFPA_24_April_2020_1.pdf

United Nations (2022) World population prospects—population division—united nations. Retrieved 28 June 2022, from http://population.un.org/wpp/

United Nations Development Programme. (2020). The next frontier: human development and the anthropocene human development report 2020. https://hdr.undp.org/content/human-development-report-2020

Vieira CM, Franco OH, Restrepo CG, Abel T (2020) COVID-19: The forgotten priorities of the pandemic. Maturitas 136:38–41

Vignoli D, Bazzani G, Guetto R, Minello A, Pirani E (2020a) Uncertainty and narratives of the future: a theoretical framework for contemporary fertility. Analysing Contemporary fertility 25–47. https://doi.org/10.1007/978-3-030-48519-1_3

Vignoli D, Guetto R, Bazzani, G, Pirani, E, and Minello A (2020b) A reflection on economic uncertainty and fertility in Europe: The narrative framework. Genus 76(1). https://doi.org/10.1186/s41118-020-00094-3

Vignoli D, Tocchioni V, Mattei A (2020c) The impact of job uncertainty on first-birth postponement. Adv Life Course Res 45:100308

Wake B (2015) Global versus local. Nature. Clim Change 5:974. https://doi.org/10.1038/nclimate2858

Walker P, Whittaker C, Watson OJ, Baguelin M, Winskill P, Hamlet A, Djafaara BA, Cucunubá Z, Olivera Mesa D, Green W, Thompson H, Nayagam S, Ainslie K, Bhatia S, Bhatt S, Boonyasiri A, Boyd O, Brazeau NF, Cattarino L, Cuomo-Dannenburg G, Ghani AC (2020) The impact of COVID-19 and strategies for mitigation and suppression in low- and middle-income countries. Science (New York, NY) 369(6502):413–422. https://doi.org/10.1126/science.abc0035

WHO (2020) Novel coronavirus (2019-nCoV) situation report 51. Retrieved 26 April 2020, from https://www.who.int/docs/default-source/coronaviruse/situation-reports/20200311-sitrep-51-covid-19.pdf?sfvrsn=1ba62e57_10

WHO (2022a) The true death toll of COVID-19: estimating global excess mortality. Retrieved 28 June 2022a, from https://www.who.int/data/stories/the-true-death-toll-of-covid-19-estimating-global-excess-mortality

WHO (2022b) COVID-19 weekly epidemiological update. Edition 99. Retrieved 13 July 2022b, from https://www.who.int/publications/m/item/weekly-epidemiological-update-on-covid-19---6-july-2022b

Wu J, Mamas MA, Mohamed MO, Kwok CS, Roebuck C, Humberstone B, Gale CP (2021) Place and causes of acute cardiovascular mortality during the COVID-19 pandemic. Heart 107(2):113–119

Yi W, Chan APC (2015) Effects of temperature on mortality in Hong Kong: a time series analysis. Int J Biometeorol 59(7):927–936. https://doi.org/10.1007/s00484-014-0895-4

Zhu C, Wu J, Liang Y, Yan L, He C, Chen L, Zhang J (2020) Fertility intentions among couples in Shanghai under COVID-19: A cross-sectional study. Int J Gynecol Obstet 151(3):399–406

Zotti M, Williams A, Robertson M, Horney J, Hsia J (2013) Post-disaster reproductive health outcomes. Matern Child Health J 17(5):783–796. https://doi.org/10.1007/s10995-012-1068-x

Khandaker Jafor Ahmed is a Postdoctoral Fellow in the Walsh School of Foreign Service at Georgetown University, Washington DC. In this position, he is working on a project focused on the population and environment in Bangladesh with a multidisciplinary team of demographers and environmental scientists who have been conducting research in Bangladesh for more than a decade. He received his doctoral degree in Environmental Studies from the Department of Geography, Environment and Population at the University of Adelaide (UoA), Australia. His research interest falls within the broader field of population and environment, with a special focus on human fertility. Given his academic qualification (Bachelor's and Master's degrees in Sociology, and PhD in Environmental Studies) and research experience (as a Research Associate in the Centre for Climate Change and Environmental Research at BRAC University, Bangladesh), his research seeks to explain the impact of climate change and extreme climate events on population dynamics (mortality, migration, and fertility).

Nishat Tasneem is a post-graduate student in the Department of Sociology, Shahjalal University of Science and Technology (SUST), Bangladesh. She completed her Bachelor's degree in the same department, and her research interests lie in the area of climate change, pandemics, and health.

Bijoya Saha is an undergraduate student in the Department of Sociology, Shahjalal University of Science and Technology (SUST), Bangladesh. Her research interests fall within the broader field of environmental degradation and demographic change, with a special focus on fertility change.

Shah Md Atiqul Haq Ph.D. is currently a Professor in the Department of Sociology at Shahjalalal University of Science and Technology (SUST), Bangladesh. He recently completed a three-year post-doctoral fellowship (FNRS) at the Demographic Research Centre, Université Catholique de Louvain (UCL), Belgium. In 2013, he received PhD from the Department of Asian and International Studies, City University of Hong Kong. Before his doctoral studies, he completed a master's thesis in human ecology at Vrije Universiteit Brussel (VUB). His studies at VUB and his education in sociology at SUST prompted him to conduct multidisciplinary research on the environment and society, demographic dynamics and climate change.

Chapter 8
The Impacts of Climate Change on Human Wellbeing in the Municipality of Porto—An Analysis Based on Remote Sensing

Hélder Lopes, Paula Remoaldo, Vítor Ribeiro, and Javier Martín-Vide

Abstract Climate Change and Global Warming are affecting the environment and consequently people's lives and property due to an increase in the frequency and magnitude of extreme phenomena. This is particularly problematic in cities and is becoming an eco-environmental issue and a public health concern. In addition to global and regional warming there are the increasingly frequent and intense heat-waves and, in cities, a surplus heat input produced by the Urban Heat Island Effect (UHI). Land Surface Temperature (LST) is widely used to assess and analyse the UHI based on remote sensing procedures. This chapter uses the coastal municipality of Porto, located in the northwest of mainland Portugal, as a case study to evaluate the relationship between the spatial variation of the LST in summer and the main influencing factors in several dimensions. Nine scenes from Landsat 8–OLI [band 10 (10.60–11.19 μm)] were acquired between 2013 and 2018, with a maximum cloud cover of 10.0% and distributed along the four seasons of the year. The processed scenes were obtained directly from the Global Estimate of LST in the LANDSAT repository, and from the emissivity based on the Normalized Difference Vegetation Index (NDVI). The gaps produced by the clouds were filled using a 9 × 9 matrix moving average. Non-advective synoptic situations were predominant, considering that they usually reduce the intensity and magnitude of the UHI and homogenize the Air Temperature (AT). The meteorological data regarding the days of scene acquisition (namely, air temperature, wind speed and atmospheric pressure) resulted from the analysis of IPMA's (Portuguese Institute for Sea and Atmosphere) data. In a

H. Lopes (✉) · P. Remoaldo · V. Ribeiro
Lab2PT e Departamento de Geografia da Universidade Do Minho, Braga, Portugal
e-mail: d12906@ics.uminho.pt

H. Lopes · J. Martín-Vide
Grupo de Climatología, Departamento de Geografía de la IdRA, Universidad de Barcelona, Barcelona, Spain
e-mail: jmartinvide@ub.edu

V. Ribeiro
Escola Superior de Educação Paula Frassinetti, ESEPF, Porto, Portugal

W. Leal Filho et al. (eds.), *Climate Change and Health Hazards*, Climate Change Management, https://doi.org/10.1007/978-3-031-26592-1_8

first stage, different factors were combined using Analytic Hierarchy Process (AHP) based on expert's opinion. The main goal of this research is to identify the potential impact on the health of the most vulnerable groups, in a context of climate change by applying the risk hazard-exposure-vulnerability method (designated as Heat Risk Index—HRI). Based on this methodological proposal, it is possible to identify areas where urban growth has resulted in greater UHI intensity and risk increase. In these areas there is also an increase in the amount of available energy that will be radiated into the atmosphere, and which will be transferred by convection as sensible heat, thus becoming responsible for air heating. In view of the thermal stress problems caused during heatwaves, in dense or consolidated urban areas, it is not possible to make major changes regarding morphology and the inclusion of large green areas. It is necessary to consider measures capable of mitigating potential health problems arising from the effects of climate change. It should be noted that the Portuguese city of Porto presents the greatest climatic risk at European level. This is happens mainly for two reasons: (i) extreme weather events will be increasingly frequent and intense, as climate projections indicate; and (ii) adaptation is not yet on the political agenda. The adaptation of urban areas to climate change scenarios should start as soon as possible, prioritizing solutions based on a systemic perspective. Local urban decisions have global consequences.

Introduction

The past three decades have been consecutively the warmest since 1850 (IPCC 2021). Scientific evidence suggests that in the near future air temperature will be 1.5 °C above pre-industrial levels if greenhouse gas emissions continue to rise to current levels (Schaeffer et al. 2012; Allen et al. 2019; Mondal et al. 2022).

Emissions contribute to the occurrence of episodes of extreme heat, namely associated with the greater frequency of heatwaves and droughts (Carvalho et al. 2021; Geirinhas et al. 2021). They also affect agricultural production and food distribution (Bogale and Tolossa 2021), ecosystems (Talukder et al. 2022; Leal Filho et al. 2021) endangering too many human lives (Simpson et al. 2021; Urry 2015). There is also growing concern about how the risks associated with extreme heat may affect residents in urban areas (who are already over 55% of the world's population and are projected to reach 70% by 2050) (Cohen 2001; UNWTO 2019).

Temperatures in urban areas are higher than those in surrounding rural areas, which contributes to a phenomenon called Urban Heat Island (UHI) (Lemus-Canovas et al. 2020; Lopes 2008; Lopes et al. 2022a, b; Martin-Vide et al. 2015; Martin-Vide and Moreno-Garcia 2020). Its main consequences are increased mortality and morbidity, human bioclimatic discomfort, energy consumption, and depleted air quality (Jay et al. 2021). In particular, the effects on human health can be very serious, eventually reducing productivity levels or affecting urban tourists' experience at certain times of the day (Lopes et al. 2021b).

This being so, measures capable of mitigating the negative impacts of extreme heat events are needed (Laukkonen et al. 2009; Leal Filho et al. 2018). Likewise, it is crucial to assess the physical and socio-economic characteristics of urban areas. Effective reduction of potential heat-related impacts requires localized adaptation, considering the identification of risk areas at a fine scale (such as a census tract).

Climate change is happening at a faster rate than humans can respond. Mitigation and adaptation to the increased risk are essential to face heat-related disasters. Hence the relevance of the holistic spatial representation of risk (hazard, exposure, vulnerability, adaptive capacity, and resilience—Allen et al. 2019; IPCC 2018; Leal Filho et al. 2018; Martín and Paneque 2022). The methodological practice in this thematic area is still limited due to the absence of major studies, particularly at local/regional geographical scales. Studies also accommodate the interests of politicians and are often not consistent with more robust scientific methodologies capable of meeting the needs of end users (e.g., local communities, tourists, and other stakeholders). In such a context, adapting to climate risks is a new challenge for many cities. The municipality of Porto, located in Portugal (Southern Europe), was selected as a case study.

This investigation aims to evaluate the relationship between the spatial variation of the LST in summer and the main influencing factors in several risk dimensions, namely with the creation of a new index—Heat Risk Index (HRI). The main objectives are to establish those areas most susceptible to extreme heat, namely when it comes to the most vulnerable groups (e.g., the elderly) and support any measures to be defined in the medium and long term needed to prepare both decision-makers and technicians for the new societal challenges in this territory. Advances in remotely sensed thermal data have been useful, not only in studying the UHI phenomenon, but also in assessing health risk and vulnerability associated with climate extremes. Based on the concept of risk as a function of hazard, exposure, and vulnerability (sensitivity and adaptive capacity), we evaluated the current health risk from extreme heat in Porto during the summer season between 2013 and 2018, using nine remote sensing scenes from LANDSAT 8-OLI and socioecological indicators. After selecting indicators and extracting risk domains, data were calibrated by five experts who were responsible for assigning weights, based on an Analytic Hierarchy Process (AHP). We also discussed the implications of our findings for adaptation-based planning. We would like to point out that our study is the first to use this type of approach in the study area.

This research proves to be very useful, as it allows the creation of an index that reflects the vulnerability, especially of the most susceptible groups, to the risk of extreme heat in a context of climate change.

After the introduction, the next section is utterly dedicated to how climate extremes can be dealt with in the context of climate change. Section 'Methods and Data' presents the methodology underlying the study, followed by its major findings in section 'Results'. Finally, there is the study's discussion and its main conclusions, and some recommendations.

Addressing the Risk of Extreme Heat Alongside Mitigation and Adaptation to Climate Change in Cities

Although the definition and approach of risk has been debated for a long time, the concept has multiple terminologies and consensus is far from being reached, namely in its application to urban areas.

The scientific literature on climate risks shows that this concept has been widely used within the various assessment reports of the Intergovernmental Panel on Climate Change (IPCC) (IPCC 2007, 2014, 2021). The risk of climate extremes is usually expressed through Crichton's risk triangle (Crichton 1999), which includes the dimensions of hazard, exposure, and vulnerability. However, the Special Report on Risk Management from Extreme Events and Disasters and the Sixth Assessment Report of the IPCC (AR6) both reinforce the importance of integrated risk assessment, with the objective of reducing the occurrence of disasters and managing susceptibility and vulnerability (IPCC 2021).

Also, our thermal health risk assessment approach was based on the risk principles designated in the IPCC AR5 (IPCC 2014: 5), namely:

1. Hazard—"the potential occurrence of a natural or human-induced physical event or trend or physical impact that may cause loss of life, injury, or other health impacts, as well as damage and loss to property, infrastructure, livelihoods, service provision, ecosystems, and environmental resources";
2. Exposure—"the presence of people, livelihoods, species or ecosystems, environmental functions, services, and resources, infrastructure, or economic, social, or cultural assets in places and settings that could be adversely affected";
3. Vulnerability—"the propensity or predisposition to be adversely affected [and] encompasses a variety of concepts and elements, including sensitivity or susceptibility to harm and lack of capacity to cope and adapt".

Despite its inclusion in several international reports, it was not always possible to integrate all risk dimensions, namely in regards to the relationship between causes and effects on population (Field et al. 2012; Martín and Paneque 2022; Wolf and McGregor 2013). Over time, efforts have been made to contribute to this physical-natural-anthropic integration. Three periods of evolution of risk theory can be identified (Martín and Paneque 2022; Table 8.1):

1. 1st phase—*risk-hazard paradigm*—based on the frequency and magnitude of hazard in the face of extreme episodes, it is related to determinism in the late nineteenth century (Wisner 2016) and lasted until the 2000s;
2. 2nd phase—*vulnerability paradigm*—the integration of people in the uncertainty of natural systems was considered relevant, namely when determined through the awareness and perception of risk (Kates 1962; Mitchell 2008; Martín and Paneque 2022). This paradigm asserted itself during the Kobe World Conference on Disaster Reduction (2005), where the integration of internal capabilities (knowledge and resources) to deal with hazards was valued (Ben Wisner and Walker 2005; O'Brien et al. 2006);

Table 8.1 Evolution of risk theory paradigms

Risk paradigm	Period of greatest influence	Significance	Used formulas	Major risk-related events
Risk-Hazard	Until 2000	Intensity and number of repetitions of an extreme phenomenon in a given community	Hazard (Frequency × magnitude)	First World Conference on Natural Disasters in Yokohama, Japan from May 23 to 27, 1994
Vulnerability	2000–2015	Susceptibility of a given community exposed to suffering damage from an extreme phenomenon	Hazard (Frequency × magnitude) × Vulnerability (Exposure × susceptibility)	World Conference on Disaster Reduction, in Kobe, Hyogo, Japan, from 18 to 22 January 2005 Kobe, the Hyogo Framework for Action 2005–2015

(continued)

Table 8.1 (continued)

Risk paradigm	Period of greatest influence	Significance	Used formulas	Major risk-related events
Resilience	2015–now	Ability of a given community to absorb the effects of an extreme phenomenon and recover from its consequences and to prepare and adapt to reduce the effects of future events	Hazard (Frequency × magnitude) + Exposure + Vulnerability $\text{Vulnerability} = \frac{\text{susceptibility}}{\text{Resilience (coping capacity} \times \text{adaptative capacity)}}$	Third UN World Conference on Disaster Reduction in Sendai, Japan, on March 18, 2015 Sendai Framework for Disaster Risk Reduction 2015–2030

Source Authors' own elaboration

3. 3rd phase—*resilience paradigm*—recognition of the adaptation mechanisms of different groups in the community and the way they act proactively to protect themselves. This paradigm asserted itself during the Sendai conference in 2015 and implied a change from a negative to a positive view on people's adaptive and self-protective capacity (Lo et al. 2017; Wisner 2016). An important element in the risk equation is uncertainty about tomorrow (the future—Aitsi-Selmi et al. 2015). This paradigm was further reinforced during COVID-19, with the inclusion of the individual's subjective perception of various risks (Lopes et al. 2021a).

Despite this, different areas of knowledge and authors tend to establish their own definitions according to the objectives of their studies (Kruse et al. 2017). We share the opinion of several authors about the processes that can condition the susceptibility to risk and the adaptive capacity of populations in different territories (Bohle et al. 1994; Schmidtlein et al. 2008; Cutter et al. 2010; Lin et al. 2017), namely: (i) the economic capacity of the groups involved; (ii) the political or social power established; and (iii) the patterns of social reproduction of practices that are perpetuated in the history of the territories. In addition to the physical-territorial dimensions (e.g., climate, exposure, topography, and physiography), one should not forget about social (e.g., building typology and land use), demographic (e.g., number of inhabitants and distribution of the population by functional groups) and economic (e.g., proportion of unemployed population, prevalence of disadvantaged classes—Cutter et al. 2012; de Loyola Hummell et al. 2016; Estoque et al. 2020; Willis and Fitton 2016) aspects. All these elements must be considered.

We feel confident in the research structure we chose to follow, as well as our focus on the determining elements (magnitude and frequency of hazard, exposure, susceptibility, ability to cope and adaptive capacity), as opposed to any jargon (sometimes disputed) exclusively intrinsic to the concepts of vulnerability and resilience (Martín and Paneque 2022).

Methods and Data

Study Area

This research is applied to the municipality of Porto, a Portuguese coastal area positioned in the northwest of mainland Portugal and part of the Iberian Peninsula. It is the second largest Portuguese municipality and, by 2021, it had 231,962 inhabitants (2.2% of the population in the country) (INE 2021). The municipality has an area of 41.42 km^2 and a population density of 5,600 inhabitants per km^2, according to the latest available census data.

It has a privileged geographical location along the Atlantic Ocean coast from west to east, being bordered in the south by the Douro River. The municipality has a hot summer Mediterranean climate influenced by the Atlantic Ocean (Lopes et al. 2021b).

Geographic location, socio-economic and environmental scenario make this municipality a living laboratory for research on the effects of climate risks and on the vulnerability and adaptability of the local community. The profound social and urban transformation that the municipality has been experiencing during the past century is fundamental for the integration of the positive and negative impacts associated with changes in urban-landscape planning (Lopes et al. 2021a; Madureira et al. 2011; Rio Fernandes et al. 2018; Vidal et al. 2022).

Several policies have also been created in the last 30 years, contributing to the affirmation of the city of Porto on an international scale (Table 8.2). Some of the actions and policies sought to adapt the city to the urgency of adaptation to climate change. More recently, COVID-19 has brought about some additional measures and more sustainable strategies to the municipality in the post-pandemic period (Lopes et al. 2022b).

The global warming trend makes it essential to change urban strategies, which implies planning the future based on the identification of risks and the vulnerability of the population. The meteorological records of the Porto—Serra do Pilar station allow us to identify a temperature increase trend (Lopes et al. 2021b), which follows the estimates released for this latitude in the various climate forecast models (Fig. 8.1). This increase has become evident in the observed average and maximum temperature peaks since the end of the XX century. Regarding precipitation, it was not possible to find any type of trend due to its very intense intra and interannual variability.

Adding to this problem, the intense artificialization of the territory introduces profound changes in temperature and wind direction. These conditions have repercussions on the risk associated with extreme heat in dense urban areas during the summer, caused by the Urban Heat Island effect and aggravated in some cases by situations of extreme heatwaves (Alcoforado and Andrade 2006; Alcoforado et al. 2009; Lopes et al. 2021b, 2022a; Madureira et al. 2021; Lopes 2022; Teixeira et al. 2022).

Developing Risk Indicators in an Adaptation Context to Climate Change

The methodology of this study is anchored to assumptions used in other studies conducted in the past ten years and based on remote sensing and socio-ecological data regarding the cities of Milan, Padua, Turin, Bologna, Genoa, Florence, Rome, Bari, Naples, Palermo and Catania (Italy—Morabito et al. 2015), Birmingham (United Kingdom—Tomlinson et al. 2011), London (United Kingdom—Wolf and McGregor, 2013) and Chongqing (China—Zhang et al. 2019). Our research uses the variables from these studies, adding other equally relevant ones, namely the existence of shelters and health facilities nearby. Among the most used socioecological indicators are,

Table 8.2 Most relevant initiatives of transformation related to tourism in Porto

Period	Main relevant policies and programs	Most relevant transformation initiatives in Porto	Main implications
1990's	1991—Incentive system for the modernization of trade (SIMC) 1992—Support program for traditional arts and crafts 1992—European single market 1993—Approval of the Municipal Master Plan of Porto 1993—Approval of the regulation for the attribution of the 'Tradition stores' insignia 1994—Program to support the modernization of trade (PROCOM)	1996—Historic center as UNESCO World Heritage Site 1998—Ibero-American Summit 1991—Opening of Bom Sucesso Shopping Center 1995—McDonalds in downtown Porto	First measures to regulate land use and occupation and attempt to implement measures to support the modernization of economic activity. Relevance of UNESCO's attribution for the integration of the city in major tourist destinations
2000's	2000—Incentive system for commercial urbanism projects (URBCOM) 2001—Urban Rehabilitation Program within the scope of Porto 2001 2005—Incentive systems for trade modernization projects (MODCOM)	2001—European Capital of Culture 2001—Restoration of public space (e.g., Jardim da Cordoaria, Praça da Batalha, Praça de D. João I) 2002—Opening of the Water Pavilion 2002—Metro system opening 2004—SRU Porto Vivo 2004—Renovation and expansion of the Bessa Stadium 2004—Construction of the Estádio do Dragão 2004—Euro 2004 2005—SRU—Porto Vivo Masterplan 2005—Construction of Casa da Música 2006—Renovation of Avenida dos Aliados and Praça da Liberdade 2006—Inauguration of the transparent building 2007—Shopping Gran Plaza (La Vie) 2008—Shopping Trindade 2009—Opening of the 'A vida portuguesa' store in Porto 2009—Opening of SeaLife 2009—First AirBnB accommodation in Porto	Competitiveness of the city through betting on culture and diversifying the tourist offer. Renovation of vast areas of the urban fabric

(continued)

Table 8.2 (continued)

Period	Main relevant policies and programs	Most relevant transformation initiatives in Porto	Main implications
2010's–	2012—Review of the Municipal Master Plan of Porto 2013—Trade Measure Investe 2014—Pedestrianization of Rua das Flores 2015–Trade Measure Investe (phase 2) 2016—'Porto de Tradição' Program 2019—Municipal fund to support establishments and entities ('Porto de Tradição') 2020—2nd Review of the Porto Municipal Master Plan	2010—Lonely Planet travel guide editor considered Porto among the top 10 tourist destinations in Europe 2012—European best destination 2014—European best destination 2015—Livraria Lello & Irmão recognized internationally for its association with the Harry Potter saga, by J.K. Rowling and collection of an entrance fee 2017—Porto and its gastronomic treasures were the scene of the CNN television program "Unknown Pieces", by Anthony Bourdain 2017—Starbucks arrive in Porto 2017—European best destination 2017—'Porto tradition' program 2020—Europe's Leading City Break Destination 2020—First pedestrianization measures for areas on certain days of the week associated with the COVID-19 period 2021—Metro line expansion (Pink line) 2021—New City Park in East side (Rio Tinto—Campanhã)	Recognition of the merit of the Porto brand Reformulation of measures to refunctionalize the use of public space and adapt to COVID-19

Source Own elaboration

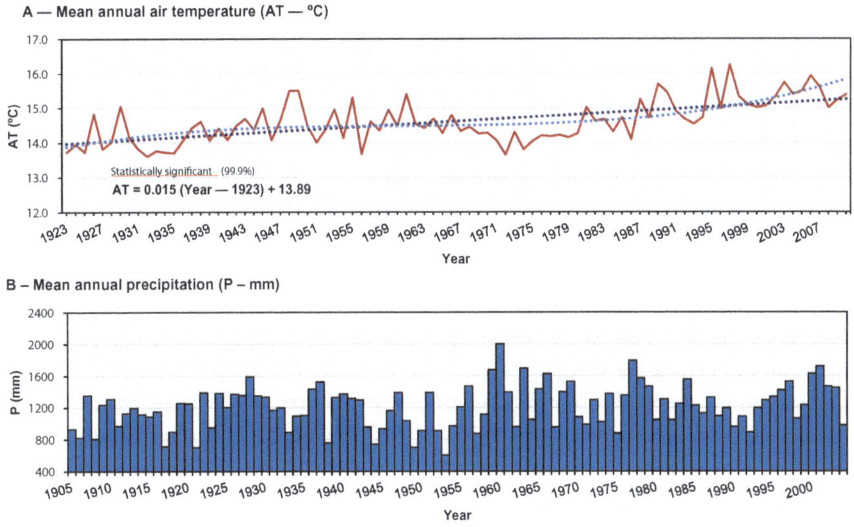

Fig. 8.1 Mean annual air temperature and precipitation registered at Porto (Serra do Pilar) station between 1905 and 2010 (*Source* Authors' own elaboration, based on IPMA's—Portuguese Institute for Sea and Atmosphere—data)

for example, population without high school diplomas (de Loyola Hummell et al. 2016; Swart et al. 2012; Noi and Nitivattananon 2015; Parikh et al. 2017), unemployed and retired (Cutter et al. 2012; Estoque et al. 2020; Fekete et al. 2010; Swart et al. 2012) and aged over 65 (Estoque et al. 2020; Borchers Arriagada et al. 2020; Loughnan et al. 2012; Hua et al. 2021).

For this purpose, several indicators were combined to allow the analysis of hazard-exposure-vulnerability. Our research follows the current paradigm of risk theory (based on resilience).

Indicators for Hazard Analysis

Land Surface Temperatures (LST) are a useful dataset, as shown in earlier studies of health risk and vulnerability associated with heat stress (Tomlinson et al. 2011; Estoque et al. 2020; Morabito et al. 2015; Wolf and McGregor 2013). Based on that and in response to the lack of measured Air Temperature (AT) data needed to support an infra-city analysis, we used the 2013–2018 summer (June, July and August) LST data from LANDSAT 8-OLI, with maximum cloud cover of 10.0%. Using the inverse of Planck's Law, it was possible to determine the LST, identified by Weng et al. (2014) as the "skin temperature" of the earth's surface.

The processed scenes were obtained directly from the Global Estimate of Surface Temperature in the LANDSAT repository (Parastatidis et al. 2017), and from the emissivity based on the NDVI (Normalized Difference Vegetation Index). The gaps produced by the clouds were filled using a 9 × 9 matrix moving average. Likewise, non-advective synoptic situations were predominant, considering the fact that these reduce the UHI intensity and magnitude and homogenize the AT in urban areas (Heaviside et al. 2015).

Meteorological data obtained from the LANDSAT scenes (namely, AT, wind speed and atmospheric pressure), and concerning a 9-day period, resulted from the analysis of IPMA's data for the Porto reference station (Pedras Rubras) and the metadata of the archive. Table 8.3 summarizes the meteorological conditions associated with the extraction of each of the scenes.

Based on the thermal infrared band 10 (10.30–11.30 μm) LSTs were derived on LANDSAT 8-OLI at a spatial resolution of 100 m and resampled to 30 m through the cubic convolution resampling method. For the surface emissivity in the LST calculation, the methodology proposed by Sobrino et al. (2008) e Lemus-Canovas et al. (2020) was used.

The identification of the limits of the areas exposed to the UHI was defined based on the method proposed by Ma et al. (2010) (Eq. 8.1):

$$LST > \mu\text{LST} + (0,5 \times \sigma\text{LST}) \tag{8.1}$$

where LST is the Land Surface Temperature, μLST is the mean of the Land Surface Temperature, and σLST is the standard deviation of the Land Surface Temperature.

Some authors refer to the effectiveness of using spectral indices to overcome the problem of mixed pixels in areas with complex land uses (Byomkesh et al. 2012; Mushore et al. 2018). NDVI is used to determine the density of vegetation cover. It is obtained by using the following Eq. (8.2):

$$NDVI = \frac{[d(band5) - d(band4)]}{[d(band5) + d(band4)]} \tag{8.2}$$

where *d* corresponds to the digital number (DN) of bands 4 and 5 of the Landsat 8-OLI image.

The Environmental Criticality Index (ECI) was calculated through the combination of LST and NDVI (Ranagalage et al. 2017; Senanayake et al. 2013a). LST and NDVI values were extended with the histogram equalization method for pixel values between 1 and 255, in order to enhance the contrast and clarity that results from the initial layer and in order to avoid infinite values of the ECI, resulting from the presence of NDVI values equal to 0 (Senanayake 2013). The ECI was obtained through the following Eq. (8.3):

$$ECI_{(LST-Veg)} = \frac{LST(stretched1 - 255)}{NDVI(stretched1 - 255)} \tag{8.3}$$

Table 8.3 Summary of weather conditions referring to the LANDSAT 8-OLI satellite scenes

Date	Hour (a.m. – GMT + 00)	Solar Azimute	Solar Elevation	Air Temperature (AT) (°C)	Wind Speed (WS) (km.h^{-1})	Wind Direction	Pressure (hPa)	Relative Humidity (RH_ %)
04/06/2013	11:15:55	132.33	65.99	26	13	ESE	1 004.26	34
20/06/2013	11:15:47	129.42	66.27	16	19	NNW	1 013.18	68
09/07/2014	11:13:42	128.69	64.57	23	15	NNE	1 010.21	47
29/08/2015	11:13:39	144.37	54.23	23	9	NNE	1 007.23	78
14/07/2016	11:13:24	132.33	63.17	26	19	ENE	1 009.22	32
01/07/2017	11:13:38	128.09	65.33	19	19	NNE	1 018.14	52
02/08/2017	11:13:49	134.53	60.72	21	11	WNW	1 012.19	56
18/06/2018	11:12:23	131.42	65.17	29	11	NW	1 011.20	40
21/08/2018	11:12:56	143.09	55.34	26	9	W	1 006.24	57

Source Own elaboration

where *LST* (stretched 1–255) refers to the extended values of the *LST*, while *NDVI* (stretched 1–255) corresponds to the widening of the values of the *NDVI* pixels. Five classes were defined, using the quantile classification method (very low, low, moderate, high and very high). Although the water bodies produce the lowest NDVI values, they do not represent environmentally critical areas and therefore had to be excluded from the analysis (Senanayake et al. 2013a).

Indicators for the Exposure Analysis

Exposure was defined in the context of this study as an index referring to the presence of people who could be negatively affected by extreme heat. We used population density as an indicator of heat exposure. This indicator has been used in other studies of wellbeing risk and vulnerability associated with extreme heat (Tomlinson et al. 2011; Estoque et al. 2020; Buscail et al. 2012; Morabito et al. 2015). First, we calculated a spatially heat exposure indicator, determining the overall heat exposure index for each area to be consistent with how the hazard index was produced. Population density was calculated considering the population and area of each census tract.

Indicators for the Vulnerability Analysis

Vulnerability is a multivariate problem that captures all dimensions of a society, either directly using variables or through their proxies (Maier et al. 2014). Vulnerability was assessed as a sensitivity function and adaptive capacity. Considering the data availability, the final sensitivity factors unfolded in two domains:

(i) population sensitivity, including age [indicated by proportion of young population (<15 years) and proportion of elderly population (≥65 years)] and socioeconomic structure (indicated by incidence of poverty, namely retired and unemployed population); and
(ii) sensitivity associated with built structures, using the variables of old buildings (before 1960), buildings without masonry plates and buildings with 5 or more floors.

All data were extracted from the 2011 census tract. There is still no data for these variables with the same level of disaggregation for 2021.

The adaptive capacity was based on the use of variables related to the presence of green spaces, shelter areas and health services and the literacy of the resident population, namely the percentage of the population with a higher education level.

Data on green areas were obtained through the analysis of satellite images and the respective NDVI of the reference period indicated in Table 8.3. For this purpose, the Fractional Vegetation Cover (FVC) was calculated. The FVC was calculated based on the formulation proposed by Carlson et al. (1995) (Eq. 8.4):

$$FVC = (NDVI*)^2 \quad (8.4)$$

where FVC is the vegetation fraction and *NDVI** corresponds to the normalized *NDVI*. *FVC* translates the percentage of vegetation at ground level in each area of the census tract based on its vertical projection (Yang et al. 2017). $FVC = 0$ has no vegetation and $FVC = 100$ has dense vegetation.

Shelters consist of all equipment or services that allow individuals to protect themselves in situations of intense heat. Coffee-shops, churches, train and metro stations, and shopping centers were considered. Five minutes on foot ranges were evaluated, according to the methodology of accessibility network analysis proposed by several authors (e.g., Lopes et al. 2019; Ribeiro et al. 2014; Ribeiro 2014). Using the same methodology, proximity to health units was identified as 15 min in an emergency vehicle and 5 min on foot. After performing the analysis, the number of service areas for each census tract was counted.

Comparative weights were determined based on an AHP process of pairwise comparison (Estoque et al. 2020). Five academic and professional experts in the thematic were asked about the weights to be assigned. The assigned values are within the variety considered in other studies (e.g., Estoque et al. 2020). The aggregation of indicators and the use of weights was in line with the IPCC recommendations on the derivation of vulnerability indices, namely that "a climate vulnerability index is typically derived by combining, with or without weighting, several indicators assumed to represent vulnerability" (IPCC 2014: 1775).

The Composite Risk Index—Heat Risk Index (HRI)

The composite risk index arises from the need for integrative approaches that expand the calculation of probabilities associated with a hazard without excluding its unintended consequences (Cutter et al. 2012). For example, areas with greater exposure do not always overlap areas of greater socio-economic vulnerability (Chow et al. 2012; Schmidtlein et al. 2008; Cutter and Finch 2008).

It should be noted that before any aggregation, the data values of the indicators were first normalized to a common It should be noted that before any aggregation, the data values of the indicators were first normalized to a common value range to make them comparable to each other. We used the min–max normalization method between 0 and 1 (e.g., Estoque et al. 2020; Estoque and Murayama 2014; OECD 2008). The indicators were on an interval scale or measurement ratio, which made it possible to use the same normalization method for all of them. Once again, we used the AHP method to assign the underlying weights to the three risk domains. Thus, it was decided to give greater weight to hazard (55%), followed by exposure (35%) and, finally, vulnerability (15%). Above all, these considerations reveal that the hazard component still gets most of the focus, with the remaining components being increasingly relevant for adaptation to climate change. The sum of the different weights of all domains allowed to derive the Heat Risk Index (HRI). Classification of the data in all variables and indices was based on the quantile method according to 5 categories (very low, low, moderate, high, and very high).

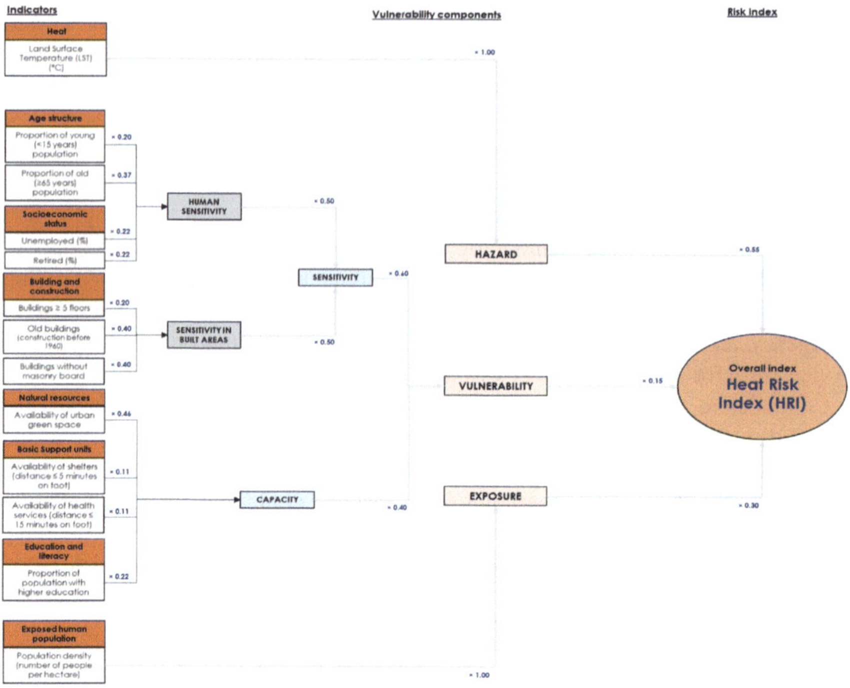

Fig. 8.2 Summary of variables considered for the composite index—Heat Risk Index (HRI) (*Source* Authors' own elaboration, based on AHP method)

The variables used and the weights assigned to each category are summarized in Fig. 8.2 and in Table 8.4.

Results

LST distribution patterns influence susceptibility to climate hazards from heat extremes in urban areas. The radiative and thermal properties of materials influence the amount of solar radiation that is reflected and stored in the lower urban atmosphere (Lopes 2008; Lopes et al. 2022a). The analysis of these patterns allows us to identify the places where there is a greater accumulation and production of heat. Typically, these areas are associated with urban surfaces, where there is a high thermal load (resulting in higher temperatures associated with the UHI). The places with the lowest heat production and storage correspond to the Urban Cool Island (UCI).

Table 8.4 Indicators selected for the analysis of the Heat Risk Index (HRI)

Risk domain	Indicator	Unit of measurement	Relevance	Bibliographic reference	Data sources
Hazard	Land Surface Temperature (LST)	ºC	Thermal load related to surface materials (natural and artificial). Overheated areas (with higher LST) have adverse effects on thermal comfort and public health	de Sherbinin and Bardy (2015), Estoque et al. (2020), Heltberg and Bonch-Osmolovskiy (2011), Lopes et al. (2022a, b), Pramanik et al. (2022)	Landsat 8 OLI/TIRS; Earth Explorer/ /US Geological Survey
Exposure	Population Density	Inhabitants/km^2	It indicates greater human activity and heat production. The higher the population density in the built environment, the higher the energy consumption and the release of anthropic heat, contributing to the (potential) urban heat island	Buscail et al. (2012), Dong et al. (2020), Estoque et al. (2020), Hua et al. (2021), Morabito et al. (2015), Pramanik et al. (2022), Tomlinson et al. (2011)	Census Tract 2011, BGRI, INE

(continued)

Table 8.4 (continued)

Risk domain	Indicator	Unit of measurement	Relevance	Bibliographic reference	Data sources
Vulnerability Sensitivity (Buildings)	Buildings with 5 or more floors	%	The verticalization of the building conditions the ventilation and renewal of hot air. It also contributes to overheating. Thus, the greater the height of buildings, the least air circulation to the surface. Among the main causes are aerodynamic roughness and lower solar and terrestrial radiation loss, sensible heat and pollutants in the lower atmosphere	Samuelson et al. (2020)	Census Tract 2011, BGRI, INE
	Old buildings (constructions pre-1960)	%	Expresses the aging of the existing building. It is assumed that pre-1960 buildings are more susceptible to climate risks, when compared to the rest of the housing stock	Murphy (2009), Nahlik et al. (2017)	Census Tract 2011, BGRI, INE

(continued)

Table 8.4 (continued)

Risk domain	Indicator	Unit of measurement	Relevance	Bibliographic reference	Data sources
	Buildings without masonry plates	%	Indicates the structural fragility of the building. The bigger the proportion of buildings with less stable structures in an urban area, the greater its economic sensitivity to climatic extremes	White-Newsome et al. (2012)	Census Tract 2011, BGRI, INE
Vulnerability Sensitivity (economic status)	Young (≤15 years) and elderly people (≥65 years)	%	It reveals the importance of these two groups when it comes to fragility in situations of extreme weather. The greater the proportion of elderly people, as well as very young children, in relation to the total resident population, the higher the respective degree of sensitivity, in terms of public health	Borchers Arriagada et al. (2020), Estoque et al. (2020), Hua et al. (2021), Loughnan et al. (2012)	Census Tract 2011, BGRI, INE

(continued)

Table 8.4 (continued)

Risk domain	Indicator	Unit of measurement	Relevance	Bibliographic reference	Data sources
	Retired	%	It refers to the reduced economic and human potential of pensioner and retired population to respond and adapt to climatic extremes. The greater the proportion of the population dependent on the State or on the family, the lower the capacity to adaptation	Estoque et al. (2020), Swart et al. (2012)	Census Tract 2011, BGRI, INE
	Unemployed	%	It indicates the existence of economic constraints in a certain population. Areas with high unemployment generally reflect situations of low economic vitality and, consequently, a weak capacity to adaptation	Cutter et al. (2012), Fekete et al. (2010)	Census Tract 2011, BGRI, INE

(continued)

Table 8.4 (continued)

Risk domain	Indicator	Unit of measurement	Relevance	Bibliographic reference	Data sources
Vulnerability Coping and Adaptive capacity	Population with higher education	%	Refers to the general level of information and awareness of a given population about the impact of climate change, as well as the appropriate individual responses (measures to be taken)	de Loyola Hummell et al. (2016), Noi and Nitivattananon (2015), Parikh et al. (2017), Swart et al. (2012)	Census Tract 2011, BGRI, INE
	Shelters (cafes, metro and train stations, shopping)	% (service area of at least one shelter less than 5 min away)	Relative distance (walking time) of the population of a given area to the nearest shelter. The faster you access places of shelter, the greater the ability to respond to extreme weather events and emergency situations	Aminipouri et al. (2016), Liu et al. (2020)	Survey conducted by the research team and data provided by the Porto City Council
	Health services (HS)	% (service area with at least one HS less than 5 min on foot and 15 min by emergency vehicle)	Refers to relative distance (time to travel on foot or in an emergency vehicle) of the population to a hospital or other health unit. The faster you access health services, the greater the ability to respond, in case of extreme weather events that trigger emergency situations	Hansen et al. (2013), Uejio et al. (2011)	Survey conducted by the research team and data provided by the Porto City Council

(continued)

Table 8.4 (continued)

Risk domain	Indicator	Unit of measurement	Relevance	Bibliographic reference	Data sources
	Fractional Vegetation Cover (FVC)	%	Refers to the percentage of green spaces (Fractional Vegetation Cover—FVC) calculated based on satellite images of the summer period for the LANDSAT 8-OLI scenes selected in the study	Buscail et al. (2012), Carlson and Ripley (1997), Estoque et al. (2020), Gandini et al. (2021), Pramanik et al. (2022)	Landsat 8-OLI/TIRS; Earth Explorer/ /US Geological Survey

Source Authors' own elaboration

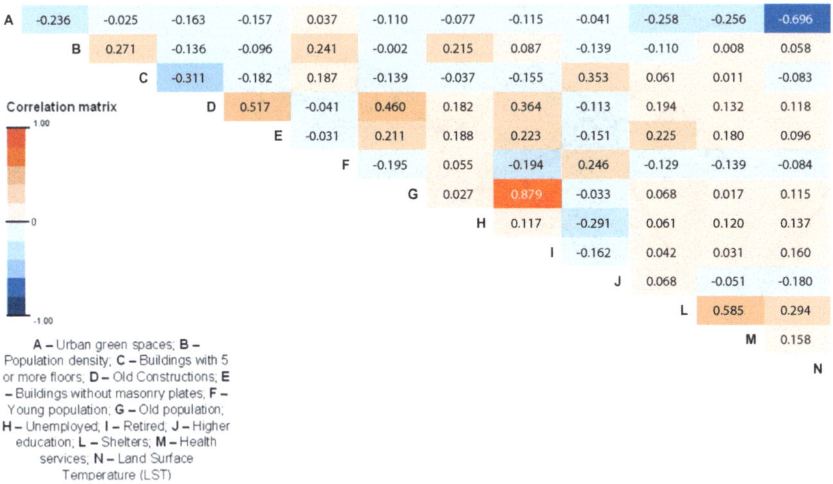

	B	C	D	E	F	G	H	I	J	L	M	N
A	-0.236	-0.025	-0.163	-0.157	0.037	-0.110	-0.077	-0.115	-0.041	-0.258	-0.256	-0.696
B		0.271	-0.136	-0.096	0.241	-0.002	0.215	0.087	-0.139	-0.110	0.008	0.058
C			-0.311	-0.182	0.187	-0.139	-0.037	-0.155	0.353	0.061	0.011	-0.083
D				0.517	-0.041	0.460	0.182	0.364	-0.113	0.194	0.132	0.118
E					-0.031	0.211	0.188	0.223	-0.151	0.225	0.180	0.096
F						-0.195	0.055	-0.194	0.246	-0.129	-0.139	-0.084
G							0.027	0.879	-0.033	0.068	0.017	0.115
H								0.117	-0.291	0.061	0.120	0.137
I									-0.162	0.042	0.031	0.160
J										0.068	-0.051	-0.180
L											0.585	0.294
M												0.158
N												

Fig. 8.3 Correlation between the variables considered for the extraction of the Heat Risk Index (HRI) in the municipality of Porto (*Source* Authors' own elaboration, based on INE [2012] and Landsat 8-OLI scenes between 2013 and 2018)

The spatial distribution of LST makes it possible to determine the places of production and accumulation of heat caused by the emission of long-wavelength radiation and the release of sensible heat. The areas with the highest LST are formed in the industrial area of Ramalde, in the central urban core of Porto—Aliados and in the Via de Cintura Interna (VCI)—extension between Estádio do Dragão and Mercado Abastecedor (to the northeast).

The areas with the lowest heat retention and production and lowest LST are identified in the Parque da Cidade do Porto, in the university campus of Asprela, and in the sector between Parque Oriental and Bairro do Largateiro. Few areas with lower LST are found in the center of Porto, being reduced to very small green areas, whose effects are not aggregating to the reduction of the UHI intensity (Fig. 8.3).

The LST standard deviation and amplitude observed are higher in summer than in winter and in the case of Porto it is estimated that in summer it reaches around 2.3 °C. There is a clear inverse correlation between the LST and NDVI variables ($p = -0.696$—moderate correlation) (Fig. 8.4). In any case, Porto has undergone a process of demographic growth related to an intense urbanization process in the last decades. This has resulted in land use changes and a significant rise in greenhouse gas emission (GGE) (Madureira et al. 2021; Rafael et al. 2016). The high rate of urbanization in Porto led to a colossal decrease in green spaces, namely in the central areas (Madureira et al. 2011). Vegetation cover plays an important role in minimizing environmental problems. The removal of vegetation intensifies environmental criticality, while acting as one of the main contributions to the formation of the UHI, namely

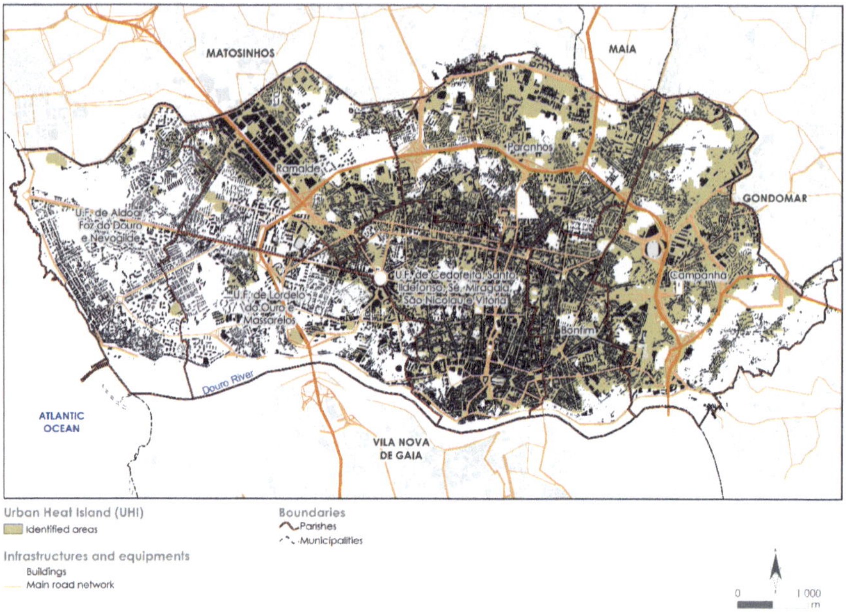

Fig. 8.4 Urban Heat Island in summer in the municipality of Porto (*Source* Authors' own elaboration, based on Landsat 8-OLI scenes between 2013 and 2018)

during the summer (Foley et al. 2005; Wong and Yu 2005; Lemus-Canovas et al. 2021). It is recognized that these factors make the city extremely vulnerable to the effects of climate change, despite the influence of its proximity to Douro river and the Atlantic Ocean (Monteiro et al. 2012; Monteiro and Madureira 2009).

Another problem is the predominance of an older population in these more central areas, usually residing in older dwellings ($p = 0.460$). These more vulnerable individuals live in less equipped housing and, in some situations, on higher floors (5th floor or higher) and are more susceptible to environmental problems caused by heat extremes.

The highest ECI is located in industrial areas (e.g., Ramalde industrial area). This is caused by the higher albedo that is associated with earth-flax or zinc sheets (0.50–0.70). Earth-flax or zinc sheets absorb less energy, but this energy is not eliminated or used due to the nature of the material (Forman 2014) (Fig. 8.5). The main paved roads, with heavy traffic (e.g., VCI), are also critical environmental areas. These have low albedo and are responsible for storing solar radiation during the day and converting it into thermal energy at night (Bowler et al. 2010; Senanayake et al. 2013a) through the process of terrestrial irradiation.

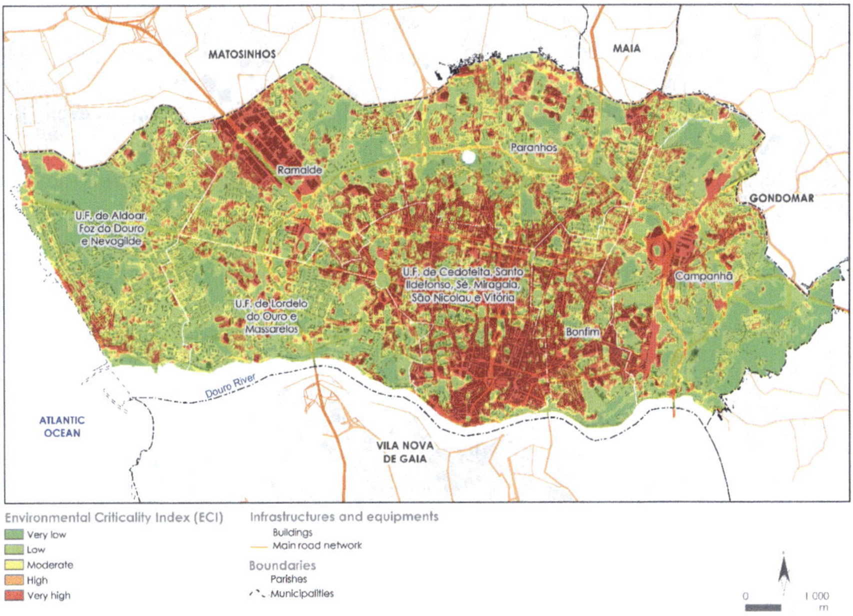

Fig. 8.5 Environmental Criticality Index (ECI) in summer in the municipality of Porto (*Source* Authors' own elaboration, based on Landsat 8-OLI scenes between 2013 and 2018)

These problems associated with danger in situations of heat stress are often related to a greater number of people exposed (exposure) and to groups with greater vulnerability (e.g., the elderly, with more degraded housing, with lower academic qualifications). Integrated risk management makes it possible to assess the ability to create mitigation and adaptation options to deal with climate change.

Areas with higher heat anomalies also have greater socioeconomic vulnerability and less adaptive capacity. These are the areas where the implementation of adaptation and mitigation measures is most important. The main causes that can justify these results are the greater waterproofing of the soil and the smaller number of green spaces in the city (Hami et al. 2019).

Heat stress showed a decreasing spatial trend from the main urban area to the bordering areas along the coast and to the East. The danger associated with extreme heat was particularly intense in the parishes affected by the UHI, namely the U.F. de Cedofeita, Santo Ildefonso, Sé, Miragaia, São Nicolau e Vitória (79.4%), Bonfim (76.8%) and Paranhos (71.5%).

The social vulnerability of the city edge blocks was less intense, and the statistical subsections of the city center showed a high to very high vulnerability index. The parish of Ramalde, the U.F. de Aldoar, Foz do Douro e Nevogilde and the U.F. de Lordelo de Ouro e Massarelos, located to the west of the municipality, presented a very low vulnerability index (Fig. 8.8).

The lowest vulnerability indices are related to the concentration of younger, wealthier people, new built-up areas, more green areas and better natural ventilation conditions in this section of the city.

The spatial pattern of human exposure showed that the range of population density amplitude was relatively small in the city of Porto. This highlights the pattern of strong urban pressure. Population density showed a decreasing spatial trend from the city center to the surrounding areas. The intensity of human exposure had a gradient difference between the most densely occupied and bordering areas.

The highest exposure values were found in four distinct areas in terms of sociodemographic occupation:

(i) In the historic center of Porto—the parishes of Bonfim and U.F. de Cedofeita, Santo Ildefonso, Sé, Miragaia, São Nicolau e Vitória correspond to the most densely inhabited areas, where the population is aging and more expensive rents due to real estate speculation are associated with tourism growth.
(ii) The parish of Campanhã has a higher incidence of unemployment (triples the values recorded in the western parishes of the city), with 43% of its population living in social housing, which highlights existing socio-economic and environmental challenges (Monteiro et al. 2013; Vidal et al. 2022).
(iii) The parish of Ramalde has a high number of exposed middle-class population, according to the analysis carried out by Graça et al. (2018) and Graça et al. (2017) on the provision of ecosystem services related to urban green spaces.
(iv) The coast (namely, Foz do Douro) has environmentally strengthened areas and a population with high social expectations, related to a higher average monthly income, the existence of a young adult population and high levels of academic training.

32.6% of the territory of Porto is located in an area of high HRI. The parish of Bonfim has the largest territorial area susceptible to risk. 41.4% of the area of this parish has a very high HRI (Fig. 8.6). The territories that deserve particular attention are located in areas of high to very high HRI with a more vulnerable elderly population. Census tract with $\geq$35% and <50% and >50% aged 65 and over were considered. Note the existence of some areas with very high risk and more than 50% of the elderly population, as in Ramalde or Paranhos (the latter being associated with the existence of a nursing home—Lar de Burgos) (Fig. 8.7).

These risks may exceed the physiological adaptive capacity of vulnerable population groups and limit the well-being. Individuals over 65 years of age are consistently the most vulnerable and may exacerbate associated health issues.

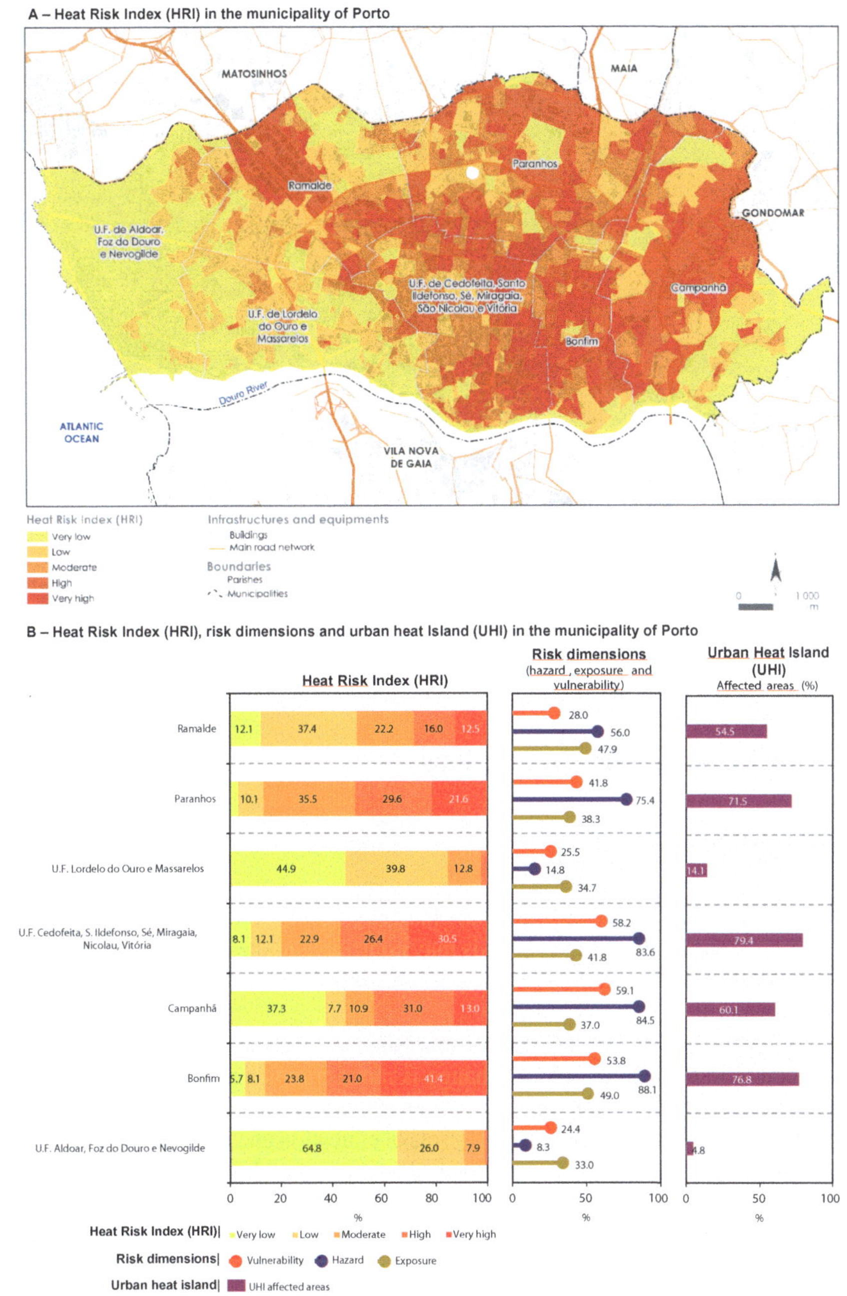

Fig. 8.6 Spatial representation of Heat Risk Index (HRI) in the municipality of Porto (A) and summary of Heat Risk Index (HRI), risk dimensions and Urban Heat Island (UHI) in the municipality of Porto (*Source* Authors' own elaboration)

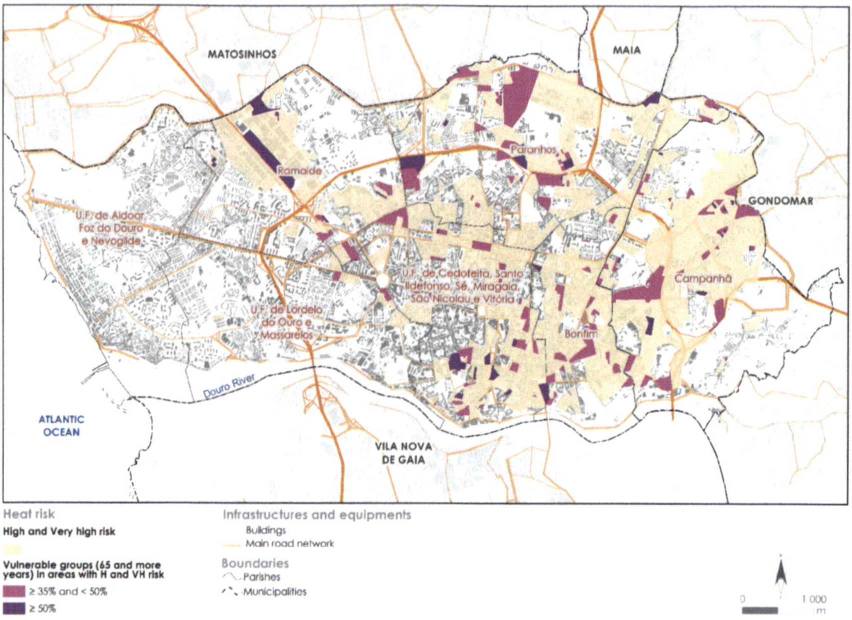

Fig. 8.7 High and very high risk of heat and vulnerable groups in the municipality of Porto (*Source* Authors' own elaboration)

Lessons Learned and Conclusions

Main Conclusions

The results of the analysis indicate that changes in LST could contribute to appreciable reductions in HRI. Such analysis may contribute to the discussion of the type of adaptation actions likely to alleviate possible impacts of heat stress under climate change scenarios.

Empirical results show that a relationship between levels of socioeconomic segregation and vulnerability to climate change is more complex than evidenced in this investigation. Reducing vulnerabilities is very difficult, even in the medium and long term. It is essential to enhance benefits, levels of education and improve health conditions, each one of them long-term changes in society, and therefore understandably hard to implement.

The improvement of the HRI should be based on regulatory options that can help minimize these problems. A few alternatives could be implemented in the short term, such as more directive and communicational measures (Fain et al. 2018; Lapola et al. 2019; Lopes et al. 2022a), namely risk communication about heatwaves and educational literacy (Lapola et al. 2019). This improvement in communication should be aimed mainly at the most vulnerable population in the municipality in the face of extremes of heat (the elderly), who tend to be less connected to digital media (social networks, websites or video streaming) and more to their caregivers and to television and radio (Lane et al. 2014; Grineski et al. 2015). The effects of heat on people's well-being are very diverse and can range from heat discomfort due to heat syncope (fainting) to heat exhaustion or clinical heat stroke.

The HRI mapping also serves to validate some of the most or least attractive areas considering their susceptibility to risk, namely on very hot days during the summer. In 2021, Madureira et al. (2021) proposed a very simple methodological approach to guide citizens around the city when extreme temperatures occur based on data captured at the census tract scale through a remote sensing analysis of LANDSAT 8-OLI scenes. Lopes et al. (2022a) proposed an analysis of areas of greater and lesser thermal comfort through transects of AT measurement obtained from mobile campaigns. The results of our study show some similarities. Taking into account the results obtained from the HRI, there are a series of recommendations that should be provided to the population in order for them to avoid certain visit times to places of greater risk. Some of the city's green spaces (namey the City Park, the Palácio de Cristal Gardens and the Serralves Foundation Gardens) and water points (drinking and regular fountains) are great places to head to on hot days.

In the medium and long term, some options can be considered when it comes to reducing areas susceptible to HRI. Cooling city roofs is one of the options, turning them into "green" roofs. Evapotranspiration from small vegetation contributes to lowering the temperature inside houses during the summer (Georgi and Dimitriou 2010; Akbari 2002; Vidal et al. 2022; Teixeira et al. 2022). An extensive range of academic literature focusing on adapting cities to climate change describes the range of co-benefits of green solutions as part of urban governance, including important contributions to the health and well-being of residents, notably by reducing air pollution (Amato-Lourenço et al. 2016; Bernatzky 1982; Rutt and Gulsrud 2016; Senanayake et al. 2013a, b).

In the city of Porto, there were several initiatives aimed at tackling the lack of green areas, namely increasing the number of trees in the streets and parks with trees in an urban context. Like the western side of the city of Porto (Parque da Cidade), the eastern side was provided, as of 2021, with a large park along the banks of the river Tinto (the most eastern section of the municipality of Porto). There were also several initiatives by the Municipality of Porto to systematize the city's tree inventory and efforts to encourage the planting of new trees (through collective or individual initiatives) in various parts of the urban space. However, in addition to environmental concerns, there are some urban requalification and construction interventions which can contribute to the emergence of new risk areas.

Other problems may arise associated with urban rehabilitation and requalification processes within the scope of extended operations financed by community funds. These may have consequences related to soil compaction and waterproofing and increase of the thermal load. The following operations ought to be mentioned:

1. the Pink Metro Line (Station S. Bento—Hospital S. António—Galicia—Casa da Música connection)—with implications for the urban structure and, in some areas, for the increase in the subsurface urban heat island (UHI_{sub});
2. the refurbishment project of the former slaughterhouse into an anchor facility for the rehabilitation of the eastern sector of the city, with a view to increasing social, economic and cultural cohesion;
3. the occasional renovation works in the territory—the new Mercado do Bolhão and the new bridge over the Douro River are perhaps the most emblematic works, but other interventions and constructions follow, namely in the Foz do Douro area (hotel and gated communities).

The diversification of infrastructure and equipment may have positive and negative impacts on the various dimensions of risk. The reduction of the HRI must prioritize ecosystems, through solutions tailored to the different types of land uses within the urban core and the need to live with uncertainty. It will be increasingly necessary to carry out monitoring, in order to reduce risks and expand the spectrum of solutions to be implemented.

The Use of HRI in Land Policy and Planning

The systematization of geographic information on our maps should not be confined to academic halls but rather be conveyed to decision makers. It is widely known that the mapping of risk areas has sought to integrate policy and practice in public services, even though the various embryonic studies did not have this objective in mind (Preston et al. 2011). This makes it difficult to reach this target audience nowadays. We believe that bringing this mapping exercise to the municipal and intra-urban geographic scale will make it possible to support decision-makers and the most vulnerable people themselves. They may feel more entitled to at least discuss these data, based on active citizenship initiatives.

Most professionals are relatively up to date with scientific information on climate extremes, climate change and their impacts. There is also recognition of the pro-activity of the municipality of Porto towards adaptation in a context of climate change. However, there are evidently some shortcomings, namely when it comes to the consideration of thermal risks in a holistic way. This is mainly due to the lack of assessment and monitoring of vulnerabilities and risks at the local level, which may contribute to the adoption of stricter adaptation and mitigation measures. However, the municipality's ability to prepare for tensions and changes or to respond to the effects of climate change depends on several factors (cultural, political, economic and development context), data availability and the quality of assessments.

Limitations and Future Directions

This research offers a starting point for testing, an essential stage towards building essential information to deal with upcoming extreme events and climate change scenarios. Some limitations need to be addressed in future investigations. First, we can point out the unavailability and/or non-existence of certain data that would be relevant and interesting to include in this methodology. For example, the lack of data on morbidity and mortality at such a fine scale (such as the census tract) makes it impossible to compare the collected data and carry out a cause-effect analysis at specific times of the year (e.g., during heatwaves). The availability of these mortality data in open access databases could facilitate comparative analyses and contribute to the refinement of the analysis on the risk of health issues associated with the increase in intense heat phenomena. Furthermore, the fact that this study is purely numerical prevents us from evaluating other subjective dimensions that are likely to be found in this study area and using other qualitative analysis tools (questionnaire surveys, focus groups or workshops).

In terms of vulnerability components, adaptive capacity proved to be the most difficult to determine, because of the variety of possible indicators and the existence of some of a subjective nature. Furthermore, data referring to population and distribution of buildings in the census tract date back to 2011, and it is likely that there have been changes in both since then. The last Census (2021) has not yet published the more recent data. Lastly, another limitation refers to the time spent in the field, as it is necessary to evaluate the data collected to ensure that it is accurate.

This methodological approach, combining multiple dimensions related to socio-environmental characteristics, proved to be a useful tool in mitigating the effects of potential risks in the municipality of Porto. This research allows for the continuation of research studies with the inclusion of additional indicators and definition of horizons for the present and future in relation to climatic extremes.

To complement this analysis, we consider that it would be technically appropriate to find adaptation and regulation strategies for areas in higher risk just as it would be interesting to apply different comparative analysis models.

Appendix

See Fig. 8.8.

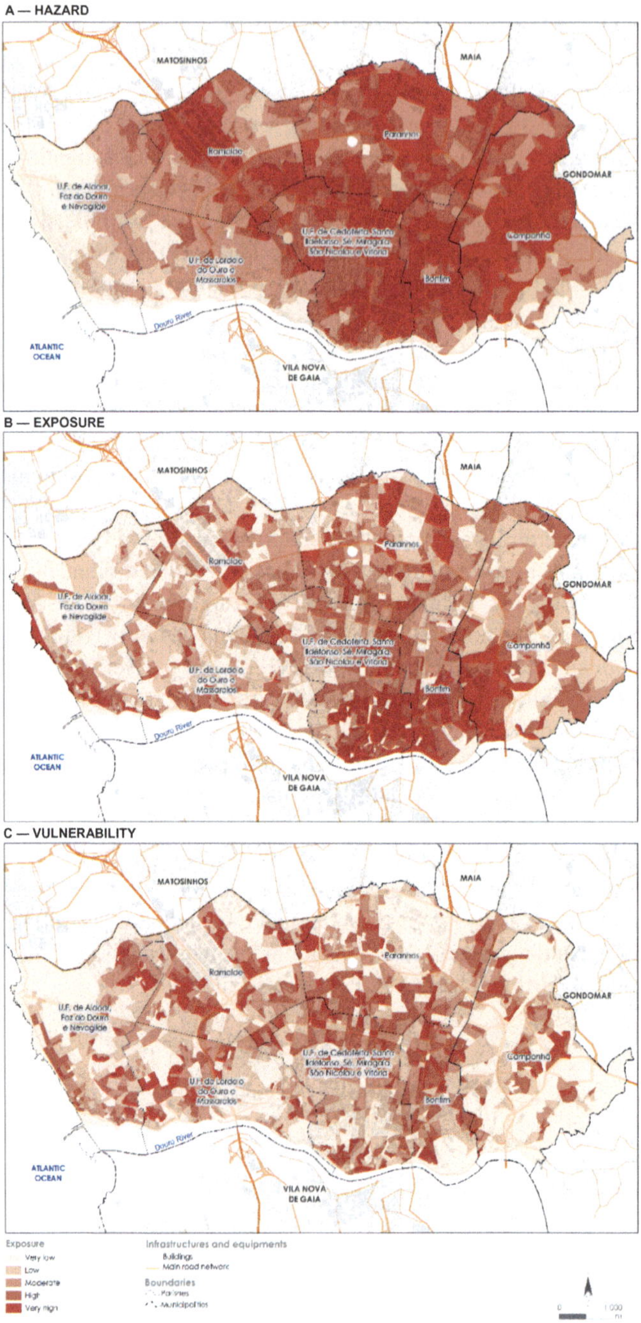

Fig. 8.8 Risk dimensions in the municipality of Porto. **A** Hazard, **B** Exposure and **C** Vulnerability (*Source* Authors' own elaboration)

References

Aitsi-Selmi A, Egawa S, Sasaki H, Wannous C, Murray V (2015) The Sendai framework for disaster risk reduction: Renewing the global commitment to people's resilience, health, and well-being. Int J Disaster Risk Sci 6:164–176

Akbari H (2002) Shade trees reduce building energy use and CO2 emissions from power plants. Environ Pollut 116:S119–S126

Alcoforado MJ, Andrade H (2006) Nocturnal urban heat island in Lisbon (Portugal): main features and modelling attempts. Theor Appl Climatol 84:151–159

Alcoforado MJ, Andrade H, Lopes A, Vasconcelos J (2009) Application of climatic guidelines to urban planning: the example of Lisbon (Portugal). Landsc Urban Plan 90:56–65

Allen M, Antwi-Agyei P, Aragon-Durand F, Babiker M, Bertoldi P, Bind M, Brown S, Buckeridge M, Camilloni I, Cartwright A (2019) Technical summary: Global warming of 1.5 °C. An IPCC Special Report on the impacts of global warming of 1.5 °C above pre-industrial levels and related global greenhouse gas emission pathways, in the context of strengthening the global response to the three. IPCC, Geneve

Amato-Lourenço LF, Moreira TCL, Arantes BL, Silva Filho DF, Mauad T (2016) Metrópoles, cobertura vegetal, áreas verdes e saúde. Estud Avançados 30:113–130

Bernatzky A (1982) The contribution of tress and green spaces to a town climate. Energy Build 5:1–10

Bogale GA, Tolossa TT (2021) Climate change intensification impacts and challenges of invasive species and adaptation measures in Eastern Ethiopia. Sustain Environ 7:1875555

Bohle HG, Downing TE, Watts MJ (1994) Climate change and social vulnerability: toward a sociology and geography of food insecurity. Glob Environ Chang 4:37–48

Borchers Arriagada N, Bowman DMJS, Palmer AJ, Johnston FH (2020) Climate change, wildfires, heatwaves and health impacts in Australia. In: Extreme Weather Events and Human Health. Springer, pp 99–116

Bowler DE, Buyung-Ali L, Knight TM, Pullin AS (2010) Urban greening to cool towns and cities: a systematic review of the empirical evidence. Landsc Urban Plan 97:147–155

Buscail C, Upegui E, Viel J-F (2012) Mapping heatwave health risk at the community level for public health action. Int J Health Geogr 11:1–9

Byomkesh T, Nakagoshi N, Dewan AM (2012) Urbanization and green space dynamics in Greater Dhaka Bangladesh. Landsc Ecol Eng 8:45–58

Carlson TN, Gillies RR, Schmugge TJ (1995) An interpretation of methodologies for indirect measurement of soil water content. Agric For Meteorol 77:191–205

Carvalho D, Cardoso Pereira S, Rocha A (2021) Future surface temperature changes for the Iberian Peninsula according to EURO-CORDEX climate projections. Clim Dyn 56:123–138

Chow WTL, Chuang W-C, Gober P (2012) Vulnerability to extreme heat in metropolitan Phoenix: spatial, temporal, and demographic dimensions. Prof Geogr 64:286–302

Cohen JE (2001) World population in 2050: assessing the projections, in: Conference Series-Federal Reserve Bank of Boston. Federal Reserve Bank of Boston; 1998, pp 83–113

Crichton D (1999) The risk triangle. In: Ingleton, J. (ed) Natural disaster management. Tudor Rose, London

Cutter SL, Finch C (2008) Temporal and spatial changes in social vulnerability to natural hazards. Proc Natl Acad Sci 105:2301–2306

Cutter SL, Burton CG, Emrich CT (2010) Disaster resilience indicators for benchmarking baseline conditions. J Homel Secur Emerg Manag 7

Cutter SL, Boruff BJ, Shirley WL (2012) Social vulnerability to environmental hazards. In: Hazards vulnerability and environmental justice. Routledge, pp 143–160

de Loyola Hummell BM, Cutter SL, Emrich CT (2016) Social vulnerability to natural hazards in Brazil. Int J Disaster Risk Sci 7:111–122

Estoque RC, Murayama Y (2014) Social–ecological status index: a preliminary study of its structural composition and application. Ecol Indic 43:183–194

Estoque RC, Ooba M, Seposo XT, Togawa T, Hijioka Y, Takahashi K, Nakamura S (2020) Heat health risk assessment in Philippine cities using remotely sensed data and social-ecological indicators. Nat Commun 11:1–12

Fain SJ, Quiñones M, Álvarez-Berríos NL, Parés-Ramos IK, Gould WA (2018) Climate change and coffee: assessing vulnerability by modeling future climate suitability in the Caribbean island of Puerto Rico. Clim Change 146:175–186

Fekete A, Damm M, Birkmann J (2010) Scales as a challenge for vulnerability assessment. Nat Hazards 55:729–747

Field CB, Barros V, Stocker TF, Dahe Q (2012) Managing the risks of extreme events and disasters to advance climate change adaptation: special report of the intergovernmental panel on climate change. Cambridge University Press

Foley JA, DeFries R, Asner GP, Barford C, Bonan G, Carpenter SR, Chapin FS, Coe MT, Daily GC, Gibbs HK (2005) Global consequences of land use. Science 309:570–574

Forman RTT (2014) Urban ecology: science of cities. Cambridge University Press

Geirinhas JL, Russo A, Libonati R, Sousa PM, Miralles DG, Trigo RM (2021) Recent increasing frequency of compound summer drought and heatwaves in Southeast Brazil. Environ Res Lett 16:34036

Georgi JN, Dimitriou D (2010) The contribution of urban green spaces to the improvement of environment in cities: case study of Chania, Greece. Build Environ 45:1401–1414

Graça MS, Gonçalves JF, Alves PJM, Nowak DJ, Hoehn R, Ellis A, Farinha-Marques P, Cunha M (2017) Assessing mismatches in ecosystem services proficiency across the urban fabric of Porto (Portugal): the influence of structural and socioeconomic variables. Ecosyst Serv 23:82–93

Graça M, Alves P, Gonçalves J, Nowak DJ, Hoehn R, Farinha-Marques P, Cunha M (2018) Assessing how green space types affect ecosystem services delivery in Porto, Portugal. Landsc Urban Plan 170:195–208

Grineski SE, Collins TW, McDonald YJ, Aldouri R, Aboargob F, Eldeb A, Romo Aguilar ML, Velázquez-Angulo G (2015) Double exposure and the climate gap: changing demographics and extreme heat in Ciudad Juárez, Mexico. Local Environ 20:180–201

Hami A, Abdi B, Zarehaghi D, Maulan SB (2019) Assessing the thermal comfort effects of green spaces: a systematic review of methods, parameters, and plants' attributes. Sustain Cities Soc 49:101634

Heaviside C, Cai X, Vardoulakis S (2015) The effects of horizontal advection on the urban heat island in Birmingham and the West Midlands, United Kingdom during a heatwave. Q J R Meteorol Soc 141:1429–1441

Hua J, Zhang X, Ren C, Shi Y, Lee TC (2021) Spatiotemporal assessment of extreme heat risk for high-density cities: a case study of Hong Kong from 2006 to 2016. Sustain Cities Soc 64:102507

INE (2021) População residente (N.o) por Local de residência (resultados preliminares Censos2021) e Sexo; Decenal—INE, Recenseamento da população e habitação—Censos 2021 [WWW Document]. Result. Prelim. do Recens. da Popul. e habitação—Censos 2021. https://ine.pt/xportal/xmain?xpid=INE&xpgid=ine_indicadores&indOcorrCod=0010745&contexto=pi&selTab=tab0

IPCC (2014a) Climate change 2013—the physical science basis. Clim Chang Phys Sci Basis. Géneve, Switzerland

IPCC (2014b) Climate change 2014: mitigation of climate change, working group III contribution to the fifth assessment report of the intergovernmental panel on climate change. Géneve, Switzerland

IPCC (2007) Synthesis report. An assessment of the intergovernmental panel on climate change, IPCC Plenary XXVII: Valencia, Spain

IPCC (2018) Global warming of 1.5 C, An IPCC Special Report on the impacts of global warming. Geneva

IPCC (2021) AR6 climate change 2021: the physical science basis—working group I contribution to the sixth assessment report of the intergovernmental panel on climate change. Géneve, Switzerland

Jay O, Capon A, Berry P, Broderick C, de Dear R, Havenith G, Honda Y, Kovats RS, Ma W, Malik A (2021) Reducing the health effects of hot weather and heat extremes: from personal cooling strategies to green cities. Lancet 398:709–724
Kates RW (1962) Hazard and choice perception in flood plain management. University of Chicago, Chicago
Kruse S, Abeling T, Deeming H, Fordham M, Forrester J, Jülich S, Karanci AN, Kuhlicke C, Pelling M, Pedoth L (2017) Conceptualizing community resilience to natural hazards–the emBRACE framework. Nat Hazards Earth Syst Sci 17:2321–2333
Lane K, Wheeler K, Charles-Guzman K, Ahmed M, Blum M, Gregory K, Graber N, Clark N, Matte T (2014) Extreme heat awareness and protective behaviors in New York City. J Urban Heal 91:403–414
Lapola DM, Braga DR, Di Giulio GM, Torres RR, Vasconcellos MP (2019) Heat stress vulnerability and risk at the (super) local scale in six Brazilian capitals. Clim Change 154:477–492
Laukkonen J, Blanco PK, Lenhart J, Keiner M, Cavric B, Kinuthia-Njenga C (2009) Combining climate change adaptation and mitigation measures at the local level. Habitat Int 33:287–292
Leal Filho W, Icaza LE, Neht A, Klavins M, Morgan EA (2018) Coping with the impacts of urban heat islands. A literature based study on understanding urban heat vulnerability and the need for resilience in cities in a global climate change context. J Clean Prod 171:1140–1149
Leal Filho W, Azeiteiro UM, Balogun AL, Setti AF, Mucova SR, Ayal D, Totin E, Lydia AM, Kalaba FK, Oguge NO (2021) The influence of ecosystems services depletion to climate change adaptation efforts in Africa. Sci Total Environ 779:146414
Lemus-Canovas M, Martin-Vide J, Moreno-Garcia MC, Lopez-Bustins JA (2020) Estimating Barcelona's metropolitan daytime hot and cold poles using Landsat-8 Land Surface Temperature. Sci Total Environ 699:134307
Lin KHE, Lee HC, Lin TH (2017) How does resilience matter? An empirical verification of the relationships between resilience and vulnerability. Nat Hazards 88:1229–1250
Lo ST, Chan EY, Chan GW, Murray V, Abrahams J, Ardalan A, Kayano R, Yau JW (2017) Health emergency and disaster risk management (Health-EDRM): developing the research field within the Sendai framework paradigm. Int J Disaster Risk Sci 8:145–149
Lopes A (2008) O sobreaquecimento das cidades. Causas e medidas para a mitigação da ilha de calor de Lisboa. Territorium, pp 39–52
Lopes H (2022) A influência do clima e da morfologia urbana na atividade turística da Área Metropolitana do Porto: perceções e simulação do conforto bioclimático dos turistas. Doctoral Thesis, University of Minho and University of Barcelona
Lopes H, Ribeiro V, Remoaldo PC (2019) Spatial accessibility and social inclusion: the impact of Portugal's last health reform. GeoHealth 3(11):356–368. https://doi.org/10.1029/2018GH000165
Lopes H, Remoaldo PC, Ribeiro V, Martín-Vide J (2021a) Perceptions of human thermal comfort in an urban tourism destination–a case study of porto (Portugal). Build Environ 108246
Lopes H, Remoaldo PC, Ribeiro V, Martín-Vide J (2021b) Effects of the COVID-19 pandemic on tourist risk perceptions—the case study of Porto. Sustainability 13:6399
Lopes H, Remoaldo P, Ribeiro V, Martín-Vide J (2022a) Pathways for adapting tourism to climate change in an urban destination–evidences based on thermal conditions for the Porto Metropolitan Area (Portugal). J Environ Manage 315
Lopes H, Remoaldo P, Ribeiro V, Martín-Vide J (2022b) Análise do Ambiente Térmico Urbano e Áreas Potencialmente Expostas ao Calor Extremo no município do Porto (Portugal). Cuad Geogr Rev Colomb Geogr 31:1–17
Loughnan M, Nicholls N, Tapper NJ (2012) Mapping heat health risks in urban areas. Int J Popul Res 2012
Ma Y, Kuang Y, Huang N (2010) Coupling urbanization analyses for studying urban thermal environment and its interplay with biophysical parameters based on TM/ETM+ imagery. Int J Appl Earth Obs Geoinf 12:110–118
Madureira H, Andresen T, Monteiro A (2011) Green structure and planning evolution in Porto. Urban For Urban Green 10:141–149

Madureira H, Monteiro A, Cruz S (2021) Where to go or where not to go—a method for advising communities during extreme temperatures. Climate 9:49
Maier G, Grundstein A, Jang W, Li C, Naeher LP, Shepherd M (2014) Assessing the performance of a vulnerability index during oppressive heat across Georgia, United States. Weather Clim Soc 6:253–263
Martin-Vide J, Moreno-Garcia MC (2020) Probability values for the intensity of Barcelona's urban heat island (Spain). Atmos Res 240:104877
Martin-Vide J, Sarricolea P, Moreno-García MC (2015) On the definition of urban heat island intensity: the "rural" reference. Front Earth Sci 3:1–3. https://doi.org/10.3389/feart.2015.00024
Martín Y, Paneque P (2022) Moving from adaptation capacities to implementing adaptation to extreme heat events in urban areas of the European Union: introducing the U-ADAPT! research approach. J Environ Manage 310:114773
Mitchell JK (2008) Perspectives on alternatives: differentiation and integration in pursuit of a better fit between society and nature. Prog Hum Geogr 32:451–458
Mondal SK, Huang J, Wang Y, Su B, Kundzewicz ZW, Jiang S, Zhai J, Chen Z, Jing C, Jiang T (2022) Changes in extreme precipitation across South Asia for each 0.5 °C of warming from 1.5 °C to 3.0 °C above pre-industrial levels. Atmos. Res 266: 105961
Monteiro A, Madureira H (2009) Porto-an urban area on the way to happiness, in: Proceedings of CITTA 2nd annual conference on planning research: planning in times of uncertainty
Monteiro A, Fonseca L, Almeida M, Sousa C, Velho S, Carvalho V (2012) Atlas da saúde e da doença–vulnerabilidades climáticas e socioeconómicas na Grande Área Metropolitana do Porto e Concelho do Porto. Proj. ONDAS (PTDC/SAUESA/73016/2006), FCT–COMPETE I II
Monteiro A, Carvalho V, Oliveira T, Sousa C (2013) Excess mortality and morbidity during the July 2006 heat wave in Porto, Portugal. Int J Biometeorol 57:155–167. https://doi.org/10.1007/s00484-012-0543-9
Morabito M, Crisci A, Gioli B, Gualtieri G, Toscano P, Di Stefano V, Orlandini S, Gensini GF (2015) Urban-hazard risk analysis: mapping of heat-related risks in the elderly in major Italian cities. PLoS One 10:e0127277
Mushore TD, Mutanga O, Odindi J, Dube T (2018) Determining extreme heat vulnerability of Harare Metropolitan City using multispectral remote sensing and socio-economic data. J Spat Sci 63:173–191
Noi LVT, Nitivattananon V (2015) Assessment of vulnerabilities to climate change for urban water and wastewater infrastructure management: case study in Dong Nai river basin, Vietnam. Environ Dev 16:119–137
O'Brien G, O'keefe P, Rose J, Wisner B (2006) Climate change and disaster management. Disasters 30: 64–80
OECD (2008) Handbook on constructing composite indicators: methodology and user guide. OECD Publishing
Parastatidis D, Mitraka Z, Chrysoulakis N, Abrams M (2017) Online global land surface temperature estimation from Landsat. Remote Sens 9:1208
Parikh K, Parikh J, Kumar M (2017) Vulnerability of Surat, Gujarat to flooding from Tapi River: a climate change impact assessment. Vayu Mandal 43:120–129
Preston BL, Yuen EJ, Westaway RM (2011) Putting vulnerability to climate change on the map: a review of approaches, benefits, and risks. Sustain Sci 6:177–202
Rafael S, Martins H, Sá E, Carvalho D, Borrego C, Lopes M (2016) Influence of urban resilience measures in the magnitude and behaviour of energy fluxes in the city of Porto (Portugal) under a climate change scenario. Sci Total Environ 566:1500–1510
Ranagalage M, Estoque RC, Murayama Y (2017) An urban heat island study of the Colombo metropolitan area, Sri Lanka, based on Landsat data (1997–2017). ISPRS Int J Geo-Information 6:189
Ribeiro V (2014) Exclusão social: um fenómeno invisível. Sílabas & Desafios

Ribeiro V, Remoaldo P, Gutiérrez J (2014) Mapping transport disadvantages of elderly people in relation to access to bus stops: contribution of geographic information systems. In: Accessibility and spatial interaction. Edward Elgar Publishing

Rio Fernandes J, Carvalho L, Chamusca P, Mendes T (2018) O Porto e a Airbnb. Porto: Book Cover

Rutt RL, Gulsrud NM (2016) Green justice in the city: a new agenda for urban green space research in Europe. Urban For Urban Green 19:123–127

Schaeffer M, Hare W, Rahmstorf S, Vermeer M (2012) Long-term sea-level rise implied by 1.5 C and 2 C warming levels. Nat Clim Chang 2:867–870

Schmidtlein MC, Deutsch RC, Piegorsch WW, Cutter SL (2008) A sensitivity analysis of the social vulnerability index. Risk Anal An Int J 28:1099–1114

Senanayake I, Welivitiya W, Nadeeka PM (2013a) Remote sensing based analysis of urban heat islands with vegetation cover in Colombo city, Sri Lanka using Landsat-7 ETM+ data. Urban Clim 5:19–35

Senanayake I, Welivitiya W, Nadeeka PM (2013b) Urban green spaces analysis for development planning in Colombo, Sri Lanka, utilizing THEOS satellite imagery–a remote sensing and GIS approach. Urban For Urban Green 12:307–314

Simpson NP, Mach KJ, Constable A, Hess J, Hogarth R, Howden M, Lawrence J, Lempert RJ, Muccione V, Mackey B (2021) A framework for complex climate change risk assessment. One Earth 4:489–501

Sobrino JA, Jiménez-Muñoz JC, Sòria G, Romaguera M, Guanter L, Moreno J, Plaza A, Martínez P (2008) Land surface emissivity retrieval from different VNIR and TIR sensors. IEEE Trans Geosci Remote Sens 46:316–327

Swart R, Fons J, Geertsema W, van Hove B, Gregor M, Havranek M, Jacobs C, Kazmierczak A, Krellenberg K, Kuhlicke C (2012) Urban vulnerability indicators. A joint report of ETC-CCA and ETC-SIA. Copenhagen ETC-CCA ETC-SIA Tech. Rep. 1, 2012

Talukder B, Ganguli N, Matthew R, Hipel KW, Orbinski J (2022) Climate change-accelerated ocean biodiversity loss & associated planetary health impacts. J Clim Chang Heal: 100114

Teixeira CP, Fernandes CO, Ahern J (2022) Adaptive planting design and management framework for urban climate change adaptation and mitigation. Urban For Urban Green: 127548

Tomlinson CJ, Chapman L, Thornes JE, Baker CJ (2011) Including the urban heat island in spatial heat health risk assessment strategies: a case study for Birmingham, UK. Int J Health Geogr 10:1–14

UNWTO (2019) World urbanization prospects 2018: highlights. New York

Urry J (2015) Climate change and society. In: Why the social sciences matter. Springer, pp 45–59

Vidal DG, Dias RC, Teixeira CP, Fernandes CO, Filho WL, Barros N, Maia RL (2022) Clustering public urban green spaces through ecosystem services potential: a typology proposal for place-based interventions. Environ Sci Policy 132: 262–272. https://doi.org/10.1016/j.envsci.2022.03.002

Weng Q, Fu P, Gao F (2014) Generating daily land surface temperature at Landsat resolution by fusing Landsat and MODIS data. Remote Sens Environ 145:55–67

Willis I, Fitton J (2016) A review of multivariate social vulnerability methodologies: a case study of the River Parrett catchment, UK. Nat Hazards Earth Syst Sci 16:1387–1399

Wisner B (2016) Vulnerability as concept, model, metric, and tool. In: Oxford Research Encyclopedia of natural hazard science

Wisner B, Walker P (2005) Beyond Kobe. In: A proactive look at the world conference on disaster reduction, pp 18–22

Wolf T, McGregor G (2013) The development of a heat wave vulnerability index for London, United Kingdom. Weather Clim Extrem 1:59–68

Wong NH, Yu C (2005) Study of green areas and urban heat island in a tropical city. Habitat Int 29:547–558

Yang L, Jia K, Liang S, Wei X, Yao Y, Zhang X (2017) A robust algorithm for estimating surface fractional vegetation cover from landsat data. Remote Sens 9:857

Hélder Silva Lopes (WoS: https://www.webofscience.com/wos/author/record/ADP-8422-2022) is invited assistant professor of the Department of Geography of the University of Minho, where he teaches in areas of Geography, Climate Change and Civil Protection, Biostatistics, Landscape, Cartography, Statistics for Geography and Human Geography of Portugal. He is a researcher at the Laboratory of Landscape, Heritage and Territory (Lab2PT) and a collaborator member of the Climatology Group of the Instituto de l'Aigua (IdRA, Climatology Group—Barcelona). Main research interests are Environmental Perceptions; Heritage and Territory; Landscape Assessment; Urban Tourism; Environmental Management; Citizen Science; Climate Change; and Regional and Local Development.

Paula Remoaldo (WoS: https://www.webofscience.com/wos/author/record/29377575), a full professor at the University do Minho, in the Department of Geography, has extensive experience in researching tourism themes (Scopus Index h12). She has a PhD in Human Geography with co-direction of the University of Louvain-la-Neuve (Belgium). Her research interests are Regional and Local Development, Cultural Tourism, Creative Tourism and Health Geography. Between 2017 and 2020 she was the Head of the Department of Geography and in the period 2017–2020 the Director of Lab2PT (Laboratory of Landscape, Heritage and Territory—Excellent Evaluation). From 2021 to 2023 is Co-Coordinator of SpaceR (Space and Representation of Lab2PT). She is the author of 155 papers dealing with Cultural Tourism, Creative Tourism, Regional and Local Development, and Climate Change. She has been involved in several territorial and tourism planning projects.

Vitor Ribeiro (WoS: https://www.webofscience.com/wos/author/record/14961662), is a Geography Teacher at Geography Department, at Minho University, and at Basic Education Department of ESEPF. He owns a PhD in Geography and a Post-Graduation in education technologies and is an integrated member at Lab2PT—Landscape, Heritage, and Territory Laboratory. Research interests are GIS, Tourism, Transport, Rural and Urban Development, Climate Change and Geographic Education. As a result, he participated in several national and international projects, and lead several publications on those research subjects.

Javier Martin-Vide (WoS: https://www.webofscience.com/wos/author/record/29149109), has a Degree in Mathematics (1977) and doctorate in Geography and History (1982). Full professor on Climatology in the University of Barcelona (Spain). Numerary academician of the Royal Academy of Sciences and Arts of Barcelona and correspondant of the Royal Academy for Overseas Sciences (Belgium) and the Academy of Sciences of Malaga. Expert on statistical analysis of precipitation, weather types, urban climatology and climate change. He has participated in several European projects (ADVICE, IMPROVE, etc.). Some of his scientific contributions are the discovery of the Western Mediterranean Oscillation and the proposal of the Concentration Index and the Disparity Index. He has supervised 34 doctoral thesis. He has been the first president of the Spanish Association of Climatology, president of the Spanish Association of Geography, and director of the Water Research Institute of the University of Barcelona. He has published 30 books and about 400 articles.

Chapter 9
The Impact of Climate Change on Water Resources and Human Health—Examples from Croatia and Bosnia and Herzegovina

Darija Vukić Lušić, Daniel Maestro, Arijana Cenov, Maja Radišić, Dražen Lušić, Marin Glad, Željko Linšak, Nerma Maestro, Dijana Tomić Linšak, and Josip Rubinić

Abstract Due to the increase in the frequency and duration of extreme weather conditions, the main impacts of climate changes will have multiple effects on water resources. Longer periods with elevated air temperatures and more frequent occurrence of extreme precipitation are expected. It will lead to lower availability of water resources and more frequent catastrophic flooding. Greater spring utilization will further aggravate the occurrence of karst springs turbidity as well as coastal aquifers salinization. The negative impact on surface and recreational waters is also to be expected, especially in the case of inadequately built rainwater and urban wastewater collection systems. This paper presents the results of analyses of four pilot areas, three in Croatia: (a) Monitoring the water level and degrees of salinization of drinking water sources (Vrana Lake, Cres Island); (b) Water turbidity (Rijeka Municipality water resources); (c) Rainwater and wastewater impact on the bathing seawater quality (Šibenik); and one in Bosnia and Herzegovina: (d) The water supply in flood conditions (Maglaj Municipality). Descriptive statistical methods were used to analyse the observed data in the pilot areas studied. Two types of models were used in the implementation of modelling—a hydrological model based on balance estimates and machine learning models. These phenomena, along with other environmental factors (air, soil, food quality deterioration), can represent a negative impact on human health. Their monitoring requires a multidisciplinary approach. The aim of this paper was to assess the negative impacts of climate change on human health in the case studies described and to suggest possible ways to mitigate these negative impacts.

D. V. Lušić (✉) · A. Cenov · D. Lušić · M. Glad · Ž. Linšak · D. T. Linšak
Department of Environmental Health, Faculty of Medicine, University of Rijeka, Braće Branchetta 20, 51000 Rijeka, Croatia
e-mail: darija.vukic.lusic@medri.uniri.hr

A. Cenov
e-mail: arijana.cenov@zzjzphz.hr

D. Lušić
e-mail: drazen.lusic@medri.uniri.hr

W. Leal Filho et al. (eds.), *Climate Change and Health Hazards*, Climate Change Management, https://doi.org/10.1007/978-3-031-26592-1_9

Introduction

Climate change directly or indirectly affects many sectors, including health. In the sectors identified as most vulnerable to climate change, the negative impacts of climate change occur not only in isolated or in particular sectors. They occur usually as combined impacts with manifestations and interactions in multiple areas, which in turn affect the health sector. Rising air and water temperatures, the frequency and magnitude of extreme events, summer heat waves, changing precipitation patterns including heavy rains, storm surges, hail, floods (especially in urban areas), and droughts are all events that threaten the availability and quality of these resources and increase the risk of the spread of waterborne and foodborne diseases (Semenza et al. 2016).

The main impacts of climate change, particularly due to the increased frequency and duration of extreme weather conditions and other key climate indicators, are

M. Glad
e-mail: marin.glad@zzjzpgz.hr

Ž. Linšak
e-mail: zeljko.linsak@uniri.hr

D. T. Linšak
e-mail: dijanatl@uniri.hr

D. V. Lušić · A. Cenov · M. Glad · Ž. Linšak · D. T. Linšak
Department of Environmental Health, Teaching Institute of Public Health of Primorje-Gorski Kotar County, Krešimirova 52a, 51000 Rijeka, Croatia

D. V. Lušić · D. Lušić
Center for Advanced Computing and Modelling, University of Rijeka, 51000 Rijeka, Croatia

D. Maestro
Department of Environmental Health, Institute for Public Health of the Federation of Bosnia and Herzegovina, Tahtali Sokak 17, 71000 Sarajevo, Bosnia and Herzegovina
e-mail: d.maestro@zzjzfbih.ba

Department of Environmental Health, Faculty of Health Studies, University of Sarajevo, Stjepana Tomića 1, 71000 Sarajevo, Bosnia and Herzegovina

M. Radišić · J. Rubinić
Faculty of Civil Engineering Rijeka, Radmile Matejčić 3, 51000 Rijeka, Croatia
e-mail: maja.radisic@gradri.uniri.hr

J. Rubinić
e-mail: jrubinic@gradri.uniri.hr

D. Lušić
Faculty of Health Studies, University of Rijeka, Viktora Cara Emina 5, 51000 Rijeka, Croatia

N. Maestro
Department of Communal Hygiene, Institute for Public Health of Canton Sarajevo, Bosnia and Herzegovina, Dr. Mustafe Pintola 1, Ilidža, Sarajevo 71000, Bosnia and Herzegovina
e-mail: nerma.maestro@zzjzks.ba

as follows: increased mortality, changes in the epidemiology of chronic noncommunicable diseases and acute infectious diseases, deterioration in the quality of all components of the environment, air, soil, water, threats to food and drinking water safety, and the emergence of a range of pollutants in the environment. Deterioration of water safety due to microbiological and chemical contamination, as well as reduced availability of water supplies due to changes of macro- and micro-climatic conditions, pose a significant vulnerability and future burden to the health care system (EEA 2017).

The paper presents the results of analyses of the situation related to the described phenomena in four pilot areas: three in the Dinaric Karst in Croatia: (1) Water sources for water supply from crypto-depression Vrana Lake on Cres, (2) Water sources Rječina and Zvir near Rijeka, and (3) Coastal water quality in Šibenik, as well as one in the territory of Bosnia and Herzegovina: (4) Water supply under flood conditions—Municipality of Maglaj. In the aforementioned pilot areas, the impact of climate change on water resources, and thus on health risks, was analysed from different aspects of occurrence, and promising measures to reduce these potential negative impacts were studied.

Vrana Lake on the island of Cres is a natural phenomenon with about 220 million m^3 of fresh water on the otherwise waterless karst island. It is the only source of water supply for about 10,000 inhabitants of the islands of Cres and Lošinj and many times more tourists during the summer seasons (Croatian Bureau of Statistics 2022). The trends of water level fluctuations in the lake, the impact that climate change could have on inflows into the lake system by the end of this century, the risk of seawater intrusion into the lake system, and protective measures to avoid endangering the water supply and thus worsening health, were analysed. In the pilot area of the water supply sources of the City of Rijeka, the problem of turbidity of the drinking water of the water supply sources was analysed, as well as the associated health risks and possible protective measures related to drought (Vukić Lušić et al. 2011). The consequences of climate change on water resources were also analysed in the pilot area of Maglaj (Institute for Public Health of the FBiH; Hydro-Engineering Institute Sarajevo 2017). In this case, they were analysed in the context of flood events and floods caused by them, as well as associated health risks and necessary preventive protection measures. The fourth pilot area deals with various aspects of climate change on water resources. In this case, the seawater in the Šibenik region and its quality on the beaches (Ministry of Economy and Sustainable Development 2022).

These phenomena, together with other environmental factors (reduced food safety, deterioration of air and soil quality), have adverse effects on human health, and their monitoring requires a multidisciplinary approach. A comprehensive analysis of test data from different components of the environment (soil, water, air, food) will allow a more accurate assessment of the magnitude of the impacts. The document also provides proposals for measures to adapt to climate change for the health sector.

Descriptive statistics were used to describe the results: relative frequencies, the arithmetic mean (AVER) as a measure of the mean, the range of the data (MIN–MAX) as a measure of the dispersion of the data, and graphically. A regression analysis was used to determine the trends of the observed parameters. Normality of the data distribution is tested using the Kolmogorov-Smirinov test. Of the nonparametric tests, the Mann Whitney test was used, at a significance level of $p < 0.05$. Two types of models were used to implement the modelling of climate change impacts on water systems. The first is a hydrological model based on the spatial distribution of relevant climatological parameters. Average annual surface runoff was estimated using the Langbein method (1962) adapted to a GIS environment (Horvat and Rubinic 2006). The analyses used historical data on annual precipitation and mean annual air temperatures (1951–2008), as well as data based on climate models 1951–2100. Two climate models were used for this purpose: ALADIN (Bubnová et al. 1995) and RegCM3 (Pal et al. 2007). In addition, models from the field of machine learning were used (Hall et al. 2009; Witten and Frank 2000)—neural networks, using a multilayer perceptron with the error-backpropagation algorithm, with the model—learning from the historical data—giving a prediction of possible changes in modified climate conditions.

Case Studies

Vrana Lake, Cres Island, HR—Reduced Lake Levels and Increasing Salinity of Drinking Water

Vrana Lake on the island of Cres is an exceptional natural phenomenon of cryptodepression (bottom at a height of 61.3 m below sea level) with up to 220 million m^3 of drinking water in the highly deficient karst area of the islands of Cres and Lošinj. Although the lake is located only 3–5 km from the sea, it is a freshwater system (chloride content varies within narrow limits of about 25 mg/L with an average of 58 mg/L), from which about 2.2 million m^3 of excellent quality drinking water is provided annually for water supply. It shows a daily peak production of about 0.2 m^3s^{-1}. Figure 9.1 shows the location of the lake in relation to the surrounding island area. Figure 9.2 shows a cross-section of the lake and the island with a schematic model of the maintenance of the equilibrium of the salt and fresh water in the island aquifer. It has been elaborated according to the principles of the hydrostatic Ghyben-Herzberger law (Verruijt 1968), but with additional elements that depend on groundwater flow dynamics, conditions associated with dynamic elements (De Weist 1965).

Monitoring of the state of the lake system, whose selected indicators are included in Fig. 9.3a–d, dates back to 1929, when complete data on water level fluctuations in the lake and fallen precipitation were available.

Annual precipitation varies from 698 (2011) to 1743 mm (1960) with an average of 1075 mm during the analysed period 1929–2016. The middling water level of the lake is 12.64 m above sea level, and the highest ever recorded was 16.86 m above sea level (1938), while the lowest was 9.05 m above sea level (2012). The abstraction of water from the lake for water supply had begun in 1952, with minimal quantities

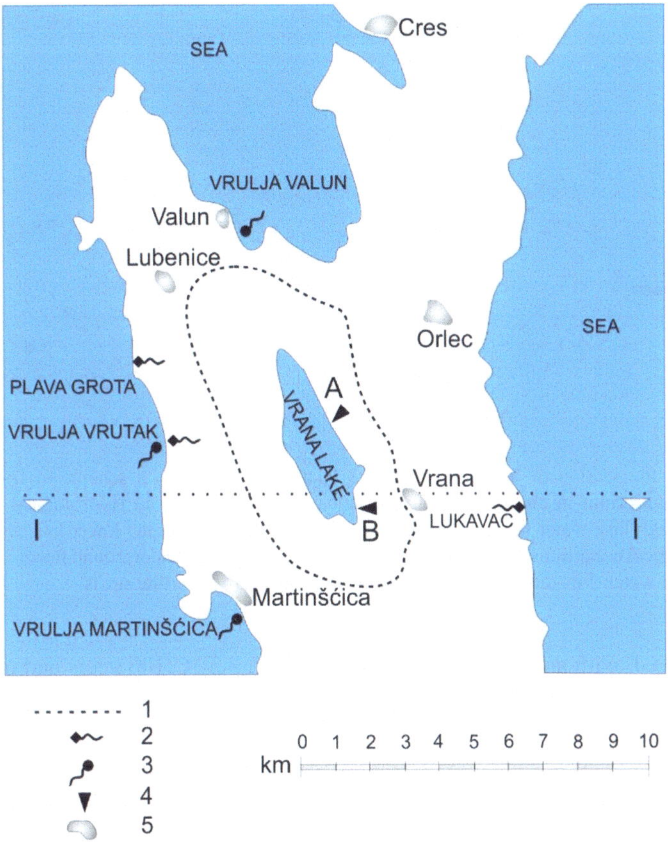

Fig. 9.1 Position of Vrana Lake and its catchment on the island of Cres (Katalinić et al. 2007). Legend: (1) orographic catchment boundary, (2) coastal brackish spring, (3) coastal spring, (4) hydrological station, (5) settlement

pumped in the early years of the use. The registered data have been preserved since 1967. At that time, the annual average was about 4 L/s, while during the last ten years the average annual abstraction has stabilized at about 70 L/s. It has an average annual maximum of 74 L/s (2010) and a maximum daily abstraction of about 200 L/s. In total, over 95 million m^3 of water has been pumped out of the lake to this date, which is over 40% of the average volume of water in the lake before the abstraction began. These abstractions influenced the effect of lowering the water level in the lake and, together with the influence of current changes of climatic conditions (increasing air temperatures and thus an increase in evaporation from the surface of the lake and evapotranspiration in the catchment area), have resulted in a significant downward trend of the minimum water level in the lake (3.77 per 100 years). For the period after 1979, a very pronounced trend of increasing water temperatures in the lake

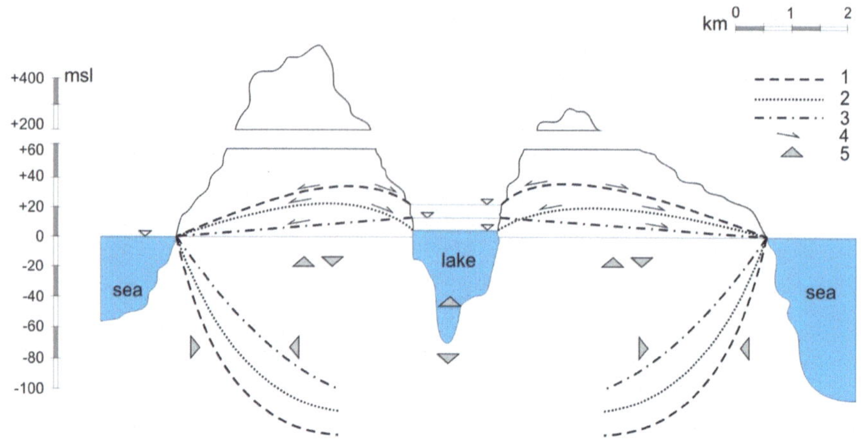

Fig. 9.2 Profile section of Vrana Lake on the Island of Cres, with a scheme of freshwater—saltwater equilibrium in characteristic hydrological states. Legend: (1) Hydrological conditions during the high lake water levels, (2) Hydrological conditions during the lower lake water levels, (3) Hydrological conditions during the dry periods, (4) Gradients of underground flows, (5) General underground water flow directions and freshwater—saltwater borderline shifts

was observed, with annual average values of about 3.3 °C/100 years and maximum values of up to 5.3 °C/100 years (Rubinić et al. 2017).

The possibility of using water from the lake is based on the fact that lowering the water level in the lake also reduces water losses from the lake system. It is possible to pump only so much water from the lake that does not further disturb the equilibrium of the lake system with the sea, thus reducing the losses due to the sinking from the lake system. The lake is a very inert system, therefore the amplitude of water level fluctuations in the lake should be taken into account. It depends not only on water abstraction and loss, but also on fallen precipitation and inflows, as well as evaporation from the lake surface. According to Petrik (1961), it is assumed that the minimum water level in the lake, which ensures the freshwater stability of the lake, is 3 m above sea level. However, the results of the analysis of the possible effects of climate change on the lake system (Rubinić et al. 2011) indicate that in the future there may be a further reduction of inflows into the lake system and thus a greater range of fluctuation of water levels in the lake. Different climate scenarios were analysed, of which the scenario modelled by the RegCM3 model (Pal et al. 2007) is an optimistic scenario and the Aladdin model (Bubnová et al. 1995) is a pessimistic one (Fig. 9.4a and b).

A study related to the assessment of climate change in relation to Vrana Lake was the subject of two projects, the CCWaterS project completed ten years ago (Rubinić et al. 2011) and the current project Management of Karst Coastal Aquifers Endangered by Climate Change (UKV) (KK.05.1.1.02.0022, 2022). The goal was to ensure adequate protection of the lake system from salinization, and to ensure water supply in the future without increased health risks and desalination needs. This is because

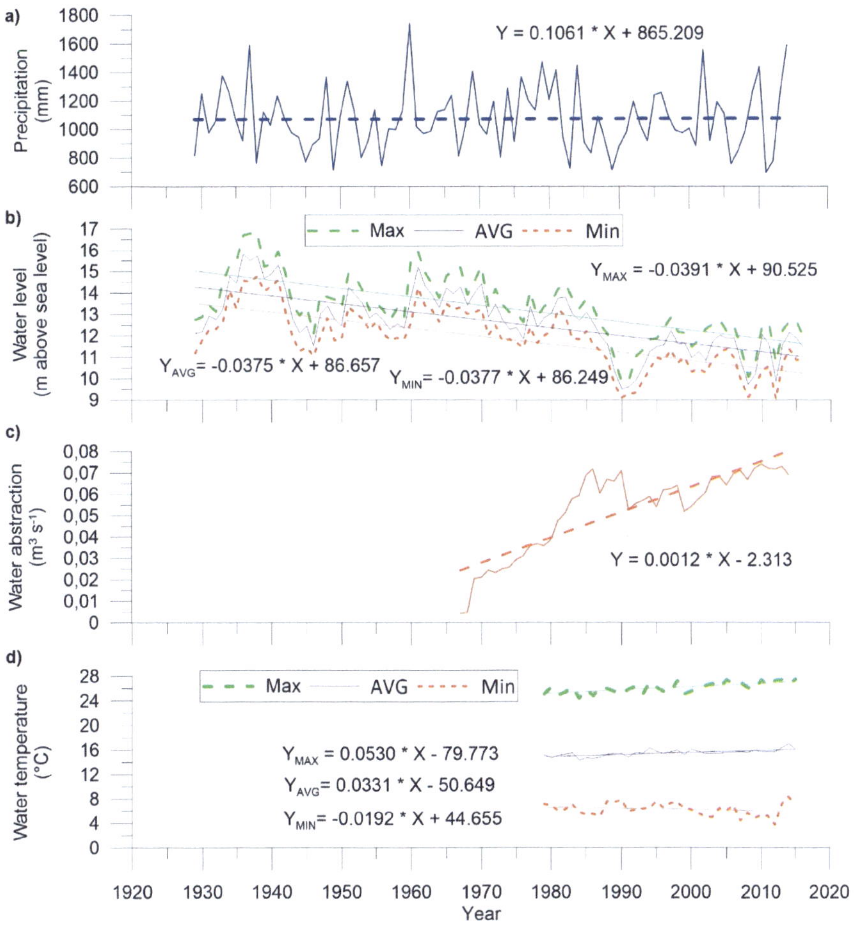

Fig. 9.3 Trends of particular hydrological indicators of Vrana Lake on the island (1929–2016): **a** precipitation, **b** water level, **c** water abstraction, and **d** water temperature (Rubinić et al. 2017)

the lake system is inert and has an extremely large volume of water relative to inflow (water exchange time over 30 years). Therefore, once elevated concentrations of seawater in the lake system would be unlikely to normalize in the foreseeable future. Seawater intrusion could endanger the water supply of more than 10,000 inhabitants of the islands of Cres and Lošinj, and many times more tourists during the summer (Croatian Bureau of Statistics 2022). It may also lead to the increase of the bromide content, a precursor of the disinfection by-product trihalomethanes (THM), which, in the long term, increases the risk of cancer in humans (Chowdhury 2022). The correlation analysis of the results of drinking water samples taken from the distribution network in the area of the islands of Cres and Lošinj during the period 2013–2021 ($N = 62$) has showed a significant correlation between the electrical conductivity of the

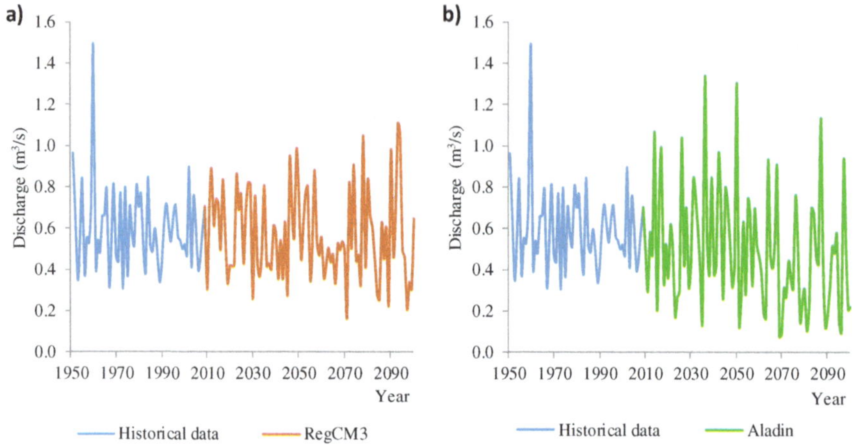

Fig. 9.4 Presentation of time changes of annual recharge into Vrana Lake generated for the period 2010–2100 evaluated by climatological models: **a** RegCM3 and **b** Aladin (Rubinić et al. 2011)

water and the concentration of THMs, mainly $CHClBr_2$ and $CHBr_3$ ($r_s = 0.50$ and $r_s = 0.36$, respectively, $p < 0.05$). In addition, frequent use of water with higher sodium chloride content may cause negative effects on the cardiovascular system and the prevalence of hypertension (Hoque et al. 2016). Without the planned construction of desalination and conditioning facilities for water abstracted from the lake, even more seawater intrusion into the lake system could jeopardize water supplies and increase health risks because water might not be suitable for consumption.

Currently, the lake is not at increased risk of losing its stability, but it is necessary to actively monitor the state of the lake system and the impact of abstraction from the system. This is especially true for future conditions under anticipated climate change and its consequences for the lake system. It mostly refers to reduced inflows and increased water temperatures as well as the evaporation from the lake surface. As possible measures to reduce the risk of saline seawater entering the lake system, deep observation wells will need to be installed along the margins of the lake system to monitor changes in the dynamic balance of fresh and saline water in the Island Lake aquifer. The first two of these 200-m wells, reaching deep below sea level, were constructed in 2021/2022. In addition, it is necessary to influence water consumption so that it does not increase in the future, both by reducing losses and by use of secondary resources for certain categories of use. This refers to giving preference to the use of rainwater and treated wastewater for irrigation in agriculture, horticultural zones and a planned golf course.

Sources of Drinking Water of the City of Rijeka, Croatia—Water Turbidity

The area of Primorje-Gorski Kotar County is rich in groundwater in the karst water system, which is the main source of drinking water in the wider area (~90%). Groundwater is generally a desirable source of drinking water, as the negative impact on groundwater quality is less pronounced when compared to surface waters. The water supply of the City of Rijeka and its surroundings includes six typical karst springs: Rječina, Zvir I, Martinšćica, Perilo, Dobra, and Dobrica, of which the most important ones are Rječina and Zvir I. The rate of connection of the population to the public water supply system in this area is over 99%, which is significantly higher than the Croatian national average (87.7%) (Croatian Institute of Public Health 2021).

The Rječina spring, with an average annual yield of around 7.4 m^3s^{-1} and a maximum of 63 m^3s^{-1}, dries up regularly during the dry season, when the water needs are the greatest. The number of days without overflow of the Rječina spring depends on hydrological conditions in the current year and is on average 2–3 months. The water springs at an altitude of 325 m above sea level and the spring itself is fed from the mountainous hinterland of Rijeka. This is one of the most water-rich areas in Croatia in terms of total precipitation (up to 4000 mm) (Rubinić and Sarić 2005). In the catchment area of the spring (the mountain massif and the Grobničko polje plateau) there are no major urban settlements, traffic areas and significant activities developed. Therefore, the water possesses good characteristics and the bacteriological feature deteriorates only during the heavy rains after a long dry period. Due to the general favourable altitude, the City of Rijeka's water supply system is almost completely gravitationally connected to the Rječina source when its yield allows it, and is used up to a maximum of 2.0 m^3s^{-1}.

In dry seasons, when the Rječina spring is less productive or dries up, the springs on the coast are activated, and the most important source of water in this area is the spring Zvir I. It is located in a canyon on the right bank of the Rječina river, almost in the centre of the City of Rijeka. It has been fed by waters from the area of the Grobničko polje and Gumanac plateaus. It is the largest permanent karst spring in Kvarner Bay with the maximum yield of 2000 L/s. It has never run dry and is therefore considered as one of the most important water resources of the water supply system of Rijeka. The disadvantage when compared to the Rječina spring is that the water for water supply must be pumped to higher altitudes and at several altitude zones, which significantly increases the cost of the water. Additionally, the spring is located in the city centre at the sea level, which makes its protection much more difficult. Therefore, the water of the Zvir I spring is constantly being contaminated with bacteria, and the intensity of pollution is more pronounced during the periods of heavy rainfall. This is due to the washout of the basin surface, the contaminants accumulated in the karst aquifers during dry periods, as well as the reactivation of previously deposited sediments in the karst aquifers. As a result, increased turbidity and bacteriological contamination occur more frequently, especially at higher water flow rates, following extended dry periods. Figure 9.5a–c shows that the leaching effects of bacteriological contamination are the highest after the first occurrence of higher flows at the Rječina spring and later decrease, regardless of the increase in flow.

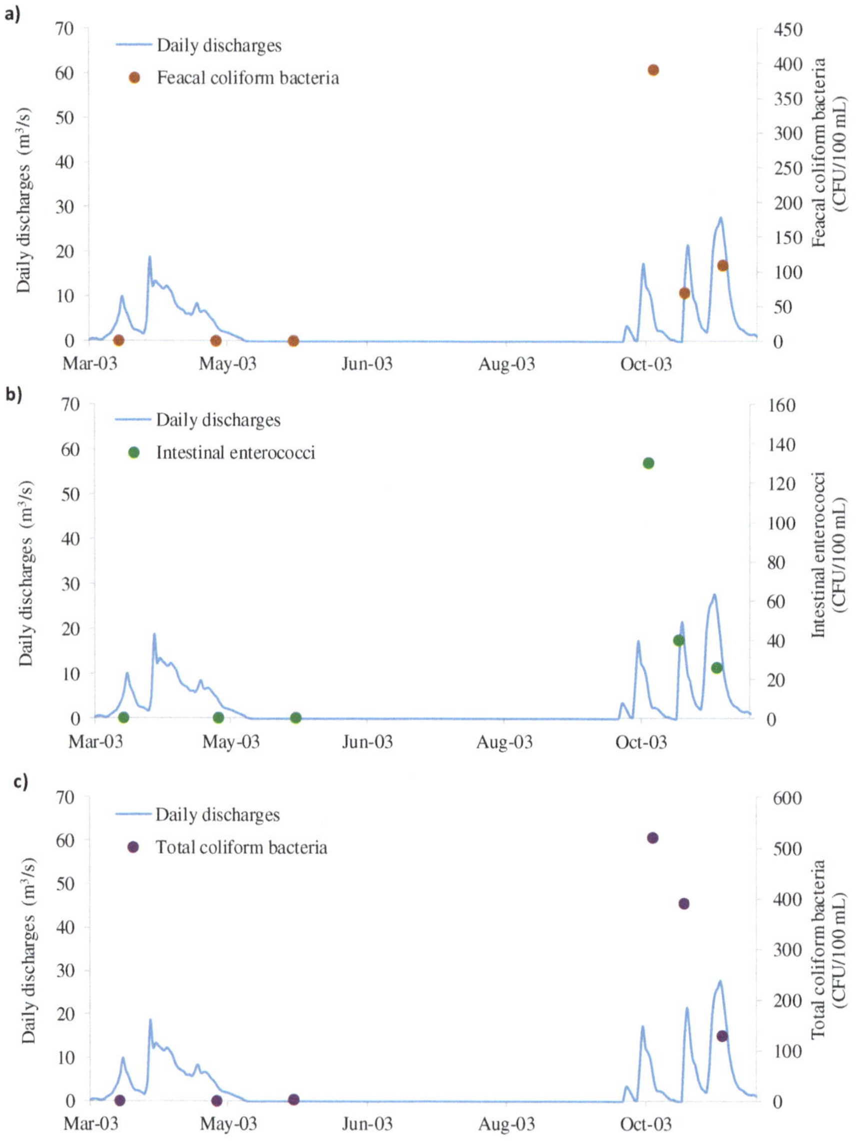

Fig. 9.5 Microbiological contamination at the Rječina spring after the occurrence of higher water levels following a long summer drought in 2013. **a** feacal coliform bacteria, **b** intestinal enterococci, and **c** total coliform bacteria

Water turbidity consists of suspended particles that may originate from the spring itself or be mobilised by the resuspension of sediments while increasing the flow velocity of water in the distribution network. In addition, turbidity may occur due to the presence of inorganic particles in the groundwater or the breakup of biofilm elements within the distribution system. Turbidity particles can provide a basis for microbial growth and development, and may reduce the efficiency of disinfection. A prerequisite for effective disinfection is low water turbidity. Water with turbidity higher than 4 NTU is non-compliant according to the legal requirements (OG 125/2017 and 39/2020, 2017), although it does not pose an immediate risk to human health.

The testing results of a total of 419 spring water samples from springs in the vicinity of Rijeka (203 water samples from the Rječina spring and 216 water samples from the Zvir I spring) during the period 2000–2021 were analysed. The turbidity of the Rječina spring water ranged from 0.2–10.3 NTU, and the number of samples with turbidity values > 4 NTU was 6 (3.0%). The turbidity of the spring water of Zvir I varied from 0.1 to 16.5 NTU, and in 4.2% of the samples the turbidity was > 4 NTU. Turbidity of water in both springs was positively correlated with all studied microbiological indicators (HPC/22, HPC/36, EN, EC, TCB, FCB, $r_s = 0.301$—0.462, $p < 0.05$) and with dissolved Fe ($r_s = 0.314$–0.421, $p < 0.05$). In the Zvir I spring, turbidity was positively associated with Total Suspended Solids—TSS ($r_s = 0.416$, $p < 0.05$) and in the Rječina spring, with phenol and PCBs concentrations ($r_s = 0.233$, $p < 0.05$ and $r_s = 0.332$, $p < 0.05$, respectively).

The trend of water pollution of the studied springs with faecal pollution indicators can be tracked in Fig. 9.6a and b. In both sources, there has been a positive trend in the number of *E. coli* and a negative trend in the number of enterococci over a 22-year period. The water of the Zvir I spring has been significantly more microbiologically contaminated when compared to the Rječina spring (Mann–Whitney U test, $Z = -6.476$, $p < 0.001$). In the Rječina source, the number of intestinal enterococci and *E. coli* ranged from 0 to 130 CFU/100 mL and 0–100 CFU/100 mL. In the water of the Zvir I spring, the number of EN and EC varied 0–520 CFU/100 mL and 0–640 CFU/100 mL.

Although increased water turbidity indicates the possible presence of microbiological and chemical pollution, in the water distribution network of the City of Rijeka during the study period it has been found that water turbidity exceeded the threshold of 4 NTU only in 1.6% of the analysed samples ($N = 105/6692$), with all HPC/22/37 analysis without any faecal indicators. It should be also emphasized that, in this case, the reason for the non-compliance of the samples was given to the increased HPC value at an incubation temperature of 37 °C. It is not considered as a risk to human health, it is more the parameter indicating the hygienic condition of the distribution system. However, water turbidity is not only a potential pollution problem, but also an aesthetic issue for consumers, who perceive water turbidity as a health risk. EC and EN, as indicators of faecal contamination, have not been detected in any sample during the observed period.

In the future, climate change is expected to manifest itself in two extremes within the wider area influencing the City of Rijeka. On the one hand, over the longer dry

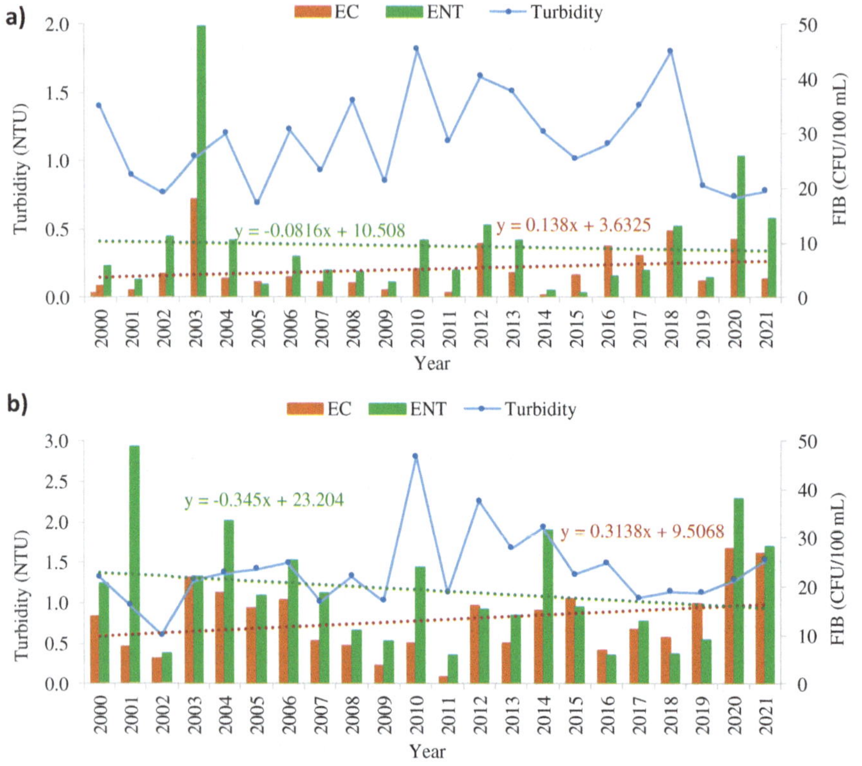

Fig. 9.6 Turbidity and faecal indicator bacteria (FIB) at **a** Rječina and **b** Zvir I springs during the period 2000–2021 (year average)

periods and, on the other hand, during the occurrence of more intense short-term precipitation. According to the analyses conducted within the RAINMAN project, and depending on their duration, these could increase in the range between 5 and 15% for the average scenario of possible changes, or even twice as much in case of applying the envelope of possible maximum changes (Faculty of Civil Engineering in Rijeka and Croatian Meteorological and Hydrological Service 2019). Such changes in precipitation intensity may lead to a more pronounced occurrence of floods with maximum discharges exceeding even the revealed percentage of increase in precipitation intensity, and thus also the expected effects of more intensive washout of pollution from the catchment area and its transfer to the water springs. Since there are no drinking water treatment plants in the area of Rijeka (only disinfection of drinking water is carried out), one of the possible solutions is the construction of filtration plants for water treatment in case of increased turbidity of the spring water.

Šibenik, Croatia—Coastal Water Quality

The coastal zone and its karstic hinterland of urban areas are linked by a series of natural and anthropogenic processes. The quality of the coastal sea depends largely on the dynamics of the interactions of these processes (Elliott et al. 2014). Very high sensitivity of these systems has been observed, as well as the changes in sea quality due to negative impacts from the mainland (Ilter Turkdogan Aydinol et al. 2012). Climate change increases the frequency of extreme rainfall events, which, together with the simultaneous overflow of rivers and sewage or poorly functioning sewage treatment systems, lead to microbiological contamination of bathing areas (Penna et al. 2021). The deterioration of bathing water quality under conditions of increased storms is largely due to land-based pollution, either from surface water and drainage systems or from groundwater. Deterioration of sea water quality after heavy rains usually lasts 3–7 days. The local waters of Šibenik are one of the areas sensitive to climate change, where the waters of the Krka River also have a strong impact on the quality of the coastal sea. It is especially expressed in the part of its flooded estuary up to its confluence with the Šibenik Channel.

The monitoring of the coastal water quality on the beaches of the Croatian part of the Adriatic Sea is carried out in accordance with the EU Bathing Water Directive (EU BWD 2006/7/EC, 2006). Bathing water quality is assessed on the basis of two microbiological indicators: intestinal enterococci (EN) and *Escherichia coli* (EC). In the studied area of Šibenik-Knin County, the quality of coastal waters is monitored at 100 sites. The results from 2009 to 2020 were analysed, and based on the data from previous observations, 10 stations (Fig. 9.7) were selected for more detailed analysis. These sites were assumed to be more directly influenced by inland waters, from the immediate catchment area. In 2017–2020, as many as 97% of the beaches in Šibenik-Knin County were rated as "excellent" and 3% of beaches were rated as "good". These were the beaches of Zaton, Crnica—Banj and Crnica—Banj End. All of them are located in the flooded estuary of the Krka River, where there is less water exchange, and in near proximity to the urban areas of Šibenik and Zaton. The following data sources were used: Croatian Meteorological and Hydrological Service, meteorological stations Šibenik, Vodice, and Skradin, and hydrological station Skradinski Buk Gornji.

The results of conducted analyses of coastal water quality show that the highest mean value for both indicators was measured on the beach in Skradin, followed by the beaches Crnica—Banj, Crnica—Banj near Šibenik and the beach in Zaton. Lower concentrations of the observed microbiological indicators occurred on all six beaches in the vicinity of the Municipality of Vodice. It was found that there were no pronounced trends of increasing maximum concentrations of the observed microbiological indicators that would indicate certain systematic changes in the quality of the coastal sea at the selected locations. An exception was the beach of Crnica Banj, where a clear trend of increase of EC was observed. It was most likely caused by the reconstruction of the drainage system of the town of Šibenik and the discharge of urban wastewater into the catchment area of this beach. It was

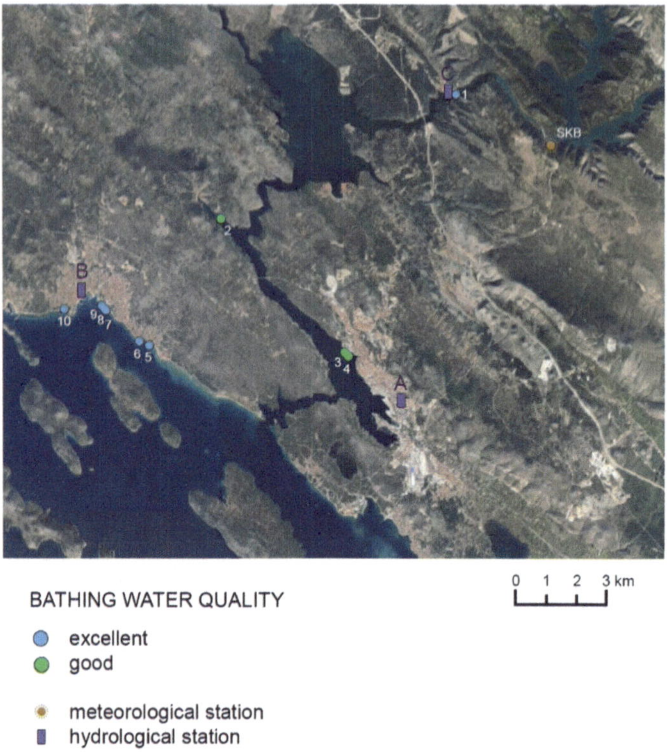

Fig. 9.7 Map showing the selected locations where coastal water quality monitoring is conducted (1) Skradin, (2) Zaton, (3) Plaža Crnica-Banj End, (4) Beach Crnica-Banj, (5) Srima, (6) Srima middle, (7) Hotel Olympia End, (8) Hotel Olympia, (9) Hotel Olympia beginning, (10) Plava plaža, meteorological station. **a** Šibenik, **b** Vodice and **c** Skradin), and hydrological station Skradinski Buk Gornji (SKB)

also noted that, with the increase of freshwater content in the sea samples at some beaches, the quality of the sea deteriorated (Fig. 9.8a–c).

Therefore, the Skradin beach was selected for modelling the impact of climate change on water quality in the coastal zone. This location is most directly influenced by the inflow of the Krka River, for which freshwater quality analyses were also conducted at the lowest station Skradinski Buk. The modelling was done with machine learning/artificial intelligence methods, using neural networks and regression decision trees. When modelling the occurrence of EN, it was found that the first three ranking influential attributes were the precipitation that felt on that day, the summary of precipitation that had fallen 1–3 days before, and salinity. On the other hand, when modelling the occurrence of EC, the order of ranking variables changed and the salinity of the sea dominated, then came the amount of precipitation that had fallen 1–3 days before, and finally the flow of the Krka on Skradinski Buk.

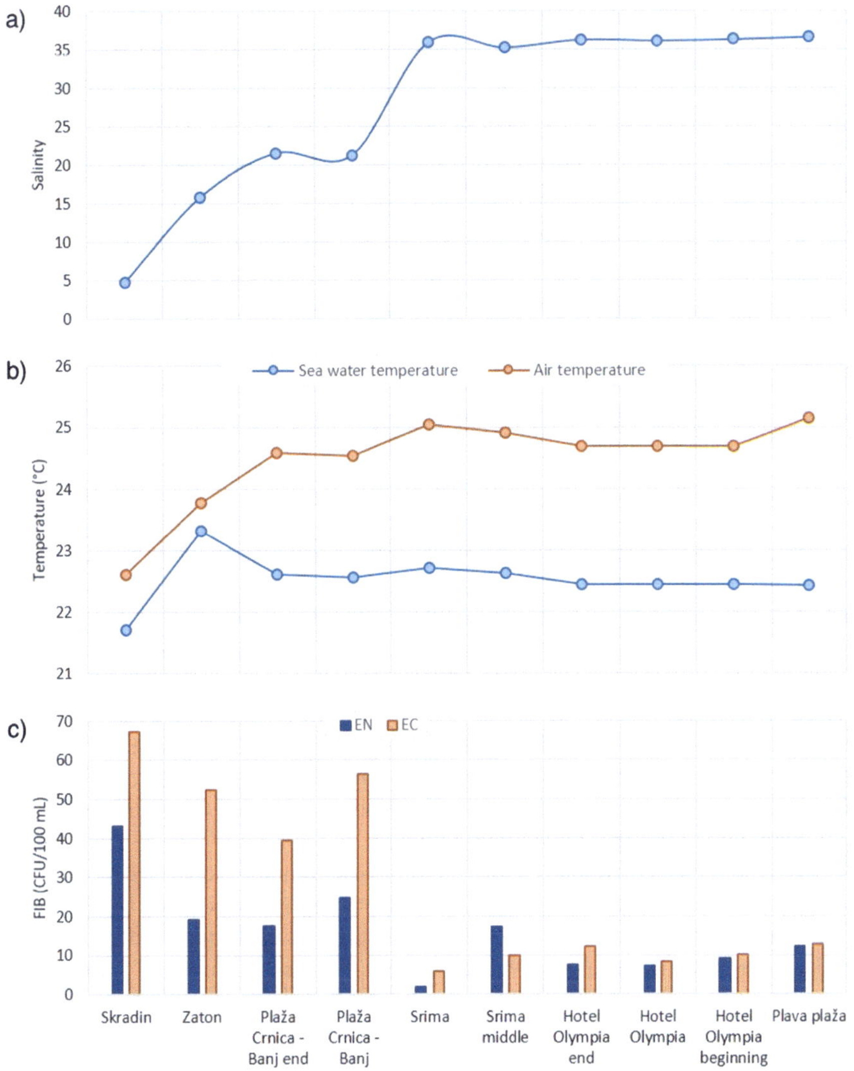

Fig. 9.8 Comparative presentation of mean. **a** salinity values, **b** mean air and sea temperature, and **c** mean FIB values (intestinal enterococci—EN and *Escherichia coli*—EC), at selected sites (Ružić et al. 2021)

Based on the trained neural network model, which gave acceptable results for predicting the occurrence of intestinal enterococci, changes were predicted under possible climate change conditions (Fig. 9.9). All input attributes were used as in the trained model, with air temperatures increased by 2, 2.5, and 3 °C, which is the range of the predicted changes in air and water temperature of the Krka River. Consequently, they were adopted for the changes in the sea temperature. Since precipitation

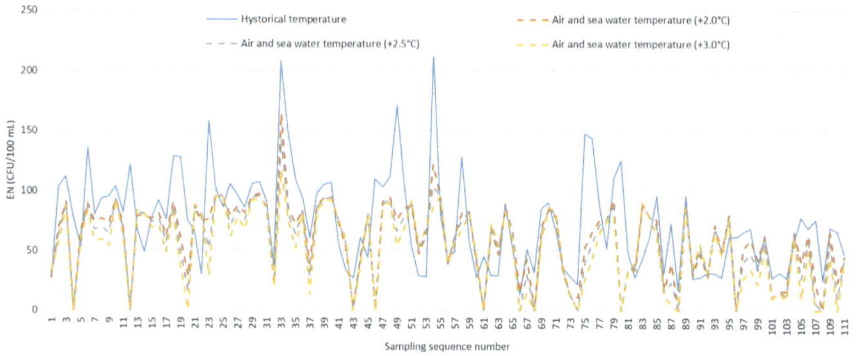

Fig. 9.9 Modelled values of intestinal enterococci (EN) under existing conditions and under conditions of potential climate change (changes in air and sea temperature of 2, 2.5, and 3°C) (Ružić et al. 2021)

possesses different possibilities of change in different climate scenarios (decrease and increase) compared to air temperatures, and for which there is a clear signal of continuation and strengthening of their increasing trend, their changes were not included in the modelling. It was found that, with an increase in the air and water temperature in the range between 2 and 3 °C, the microbiological load (EN) on the Skradin beach decreases by 15 to 28%, when the limit of 60 CFU/100 mL is exceeded. It decreases even more (45–64%), when the limit of 100 CFU/100 mL is exceeded. The reason for this is the aforementioned effect of shorter survival time of bacteria under more intense solar radiation when the air temperature is generally higher. A higher temperature of seawater also leads to a faster decrease in the number of bacteria due to increased metabolic activity.

Given the projected climate change, coastal water quality is expected to continue to deteriorate in the future, during and after heavy rainfall. In addition to the proper functioning of wastewater and stormwater drainage systems, the development of a predictive model for the microbiological quality of seawater will be of great importance (Vukić Lušić et al. 2017). These models will make it possible to predict the coastal pollution before it occurs, thus preventing the hazards to public health.

Municipality of Maglaj, Bosnia and Herzegovina—Water Supply During Flood

Floods are natural disasters, which in practice are considered one of the greatest natural disasters in terms of mortality and financial losses (Javadinejad 2022). In developing countries, flooding is mostly the result of climate change, excessive rainfall, construction on waterways, rising sea levels, soil moisture, uncontrolled rapid population growth, inadequate preparedness, and lack of political will (Umar and Gray 2022).

World Health Organization guidelines (WHO) state that climate change may affect the quality of water for human consumption. This is primarily due to increasing temperature extremes and more frequent and intense precipitation, leading to increased sedimentation and changes in chemical composition, total organic carbon content, and microbiological safety. Floods have direct health effects such as drowning, injuries, diarrheal diseases, vector-borne diseases, respiratory infections, skin and eye infections, and mental health problems. Heavy rainfall/flooding can lead to flooding of wastewater treatment plants, runoff of animal faeces and manure, and thus increased contamination of surface water and soil (WHO 2011). Waterborne diseases are among the greatest threats to public health worldwide, despite significant advances in the technical treatment of drinking water and wastewater (WHO 2017). After floods, intestinal infectious diseases are the greatest threat because the living space is contaminated with sewage and tap water may also be contaminated. For this very reason, the most common and significant health risk related to drinking water is microbiological contamination, the consequences of which determine that its control must always be of paramount importance (Bešić et al. 2019).

The floods that hit BiH in mid-May 2014 were the worst in the last 120 years. This natural disaster affected a quarter of the country's territory and about one million people. This represents about 27% of BiH's total population of 3.8 million. The total estimated economic damage amounted to more than EUR 2 billion, or 15% of the country's total gross domestic product (GDP) in 2014 (Jeleč 2015). The municipality of Maglaj in the Canton of Zenica-Doboj, with about 25,000 inhabitants, was one of the places most affected by the flood. The population and economy of the Municipality of Maglaj are supplied with drinking water through one public and several local water supply systems in the municipality, mainly in rural areas. Only 31.2% of the total population and public facilities in the Municipality of Maglaj are connected to the public water supply system (The Municipality of Maglaj 2012).

This means that 68.8% of the total population of the municipality from rural areas is supplied with drinking water from local water supply facilities that do not fall under the responsibility of the municipal public water supplier. Due to the lack of proper regulations for the management and monitoring of local rural water supply systems and water supply based on residents' own wells, there is no established continuous sanitary monitoring. Under these circumstances, it is impossible to take preventive action in the event of natural disasters or incidents that could affect the quality or abundance of water sources. In addition, the boundaries of the aquifers are not known, nor is information about possible negative impacts.

The current situation proved to be extremely unfavourable and dangerous during the natural disaster in May 2014. At that time, the floods and landslides caused flooding and contamination of springs, ruptures of water pipes, collapses of water facilities, and given the poor regulatory system and organization, qualitative intervention and management of the control and warning process was not possible. In order to determine the impact of flooding on drinking water in the municipality of Maglaj, a study was conducted (Institute for Public Health of the FBiH; Hydro-Engineering Institute Sarajevo 2017) in which 100 facilities belonging to the water supply network were evaluated, with 108 water samples collected. The selection of the exact facilities was influenced primarily by the number of inhabitants supplied with drinking water by a given facility. The results of previous water quality analyses and the location of the water source itself also played a role.

Of the total of 108 water samples tested for microbiological analysis, 93 (86.1%) did not comply with the reference values of microbiological parameters specified in the Ordinance on the Health Safety of Drinking Water (BiH 2010). Only 15 (13.9%) of the samples complied for human consumption. In 28 out of 93 samples (30.1%) that did not meet the requirements of the regulation, the causes of non-faecal pollution, saprophytic bacteria and coliform non-pathogenic bacteria were detected. In 50 (53.8%) of the 93 non-compliant samples, the faecal coliform bacterium *Escherichia coli* was isolated, while in 19 samples (20.4%) it was analysed in a number greater than 10^2. Only in 4 samples out of these 50, less than 10 bacteria were isolated. Faecal bacteria from the *Enterococcus faecalis* group were detected in 48 (51.6%) of 93 samples, while in 12 samples (25%) the values exceeded 10^2 and only in one of them less than 10 bacteria from this group were isolated.

A total of 12 samples (12.9%) were positive for the presence of bacteria belonging to the sulphite-reducing clostridia group. All 10 samples tested for the presence of *Legonella* spp. were negative. The levels of α-, β-, and γ-radiation in all water samples analysed were negative for radioactivity. It should be noted that most of the water samples analysed were raw (untreated) water and the present results reflect the actual water quality status at the source. However, from the microbiological point of view, the water quality can be much more favourable after the application of appropriate disinfection methods (Institute for Public Health of the FBiH; Hydro-Engineering Institute Sarajevo 2017).

Given the results presented, it was reasonable to expect an increase and more extensive occurrence of water and food epidemics. However, a look at the seven-year incidence trends of acute enterocolitis in BiH shows that there is a general trend of a slight increase in the rate, but without a large jump in 2014, when BiH was affected by major floods (Fig. 9.10). *Enterocolitis acuta* occurs most frequently in areas where the population is supplied with drinking water from individual water supply systems (wells, cisterns, untapped springs) that are not under the control of public health institutes and public water suppliers.

There is growing certainty that climate change will lead to changes in flood risk, so it is necessary to develop and improve public health preparedness and response plans to extreme events in countries at risk. The chances of effective response to health risks are greatly enhanced by reliable information on the health needs of the population and the rehabilitation needs of health, water and sanitation systems damaged by floods.

Fig. 9.10 Trends in the rate of acute enterocolitis per 100,000 inhabitants in Bosnia and Herzegovina, 2013–2019 (*Source* Reports on the Health Status of the Population)

Conclusions

As case studies elaborated above, it has become clearly obvious that climate change affects all areas of human activity and, as a consequence, the human health, too. The case studies presented highlight the impact of climate change on the quality of water resources. The intrusion of seawater into drinking water resources as a result of the lowering of the water level in the lake, as well as the increasing turbidity of drinking water after heavy rains, lead to a deterioration of drinking water quality. Heavy rainfall and non-functioning parts of the wastewater drainage system lead to the pollution of the coastal sea. Finally, extreme rainfall also leads to flooding, which drastically worsens sanitary living conditions and the microbiological and chemical safety of drinking water. Given the expected increase in extreme weather events in the future, a systematic preparation is needed to manage probable negative impacts. One of the important factors for monitoring and timely response is definitely taking into account the following elements:

- To increase the number of samples as well as monitored indicators;
- Based on monitoring results, to develop and use models to predict the potential impacts of climate change on water systems;
- To put in practice a database combining the results of local and national studies for the purpose of rationalization of the use of available financial resources and to maximize the effect of the available data.

This approach evidently requires a multidisciplinary approach with different types of experts involved, belonging both to technical and health sectors. In addition, since climate change knows no borders, interregional and regional cooperation is essential for any successful action.

Acknowledgements This work was financially supported by the: (1) University of Rijeka as part of the research project "Adaptation to the new EU Bathing Water Directive" (project code: uniri-biomed-18-292); (2) University of Rijeka as part of the research "Assessment of climate change impacts on hydrological stability of protected lake systems in Croatia's karst region" (project code: uniri-technic-18-298); (3) Management of Karst Coastal Aquifers Endangered by Climate Change (UKV), KK.05.1.1.02.0022; and (4) KLIMOD—Computer model of pollution, flooding and spreading of pollution in rivers and coastal sea areas, KK.05.1.1.02.0017.

References

Aydinol IT, F., Kanat, G., & Bayhan, H. (2012) Sea water quality assessment of Prince Islands' beaches in Istanbul. Environ Monit Assess 184(1):149–160. https://doi.org/10.1007/s10661-011-1954-5

Bešić A, Karakač S, Dautbegović A (2019) Evaluation of public health risk factors in water supply systems in Maglaj Municipality. Int J Manag Cases 21(1):67–82

Bosnia and Herzegovina (2010) Regulation on health safety of drinking water. Official Gazette of the Federation of Bosnia and Herzegovina 40/2010, 43/2010 and 30/2012.

Bubnová R, Hello G, Benard P, Geleyn J (1995) Integration of fully elastic equations cast in the hydrostatic pressure terrain-following coordinate in the framework of ARPEGE/ALADIN NWP system. Mon Wea Rev 123:515–535

Chowdhury S (2022) Effects of seawater intrusion on the formation of disinfection byproducts in drinking water. Sci Total Environ 827:154398. https://doi.org/10.1016/j.scitotenv.2022.154398

Croatian Bureau of Statistics (2022) Statistics. https://podaci.dzs.hr/en/

Croatian Institute of Public Health (2021) Comparison of indicators on leading public health problems in the Republic of Croatia and the European Union. https://www.hzjz.hr/periodicne-publikacije/objavljena-publikacijausporedba-pokazatelja-o-vodecim-javnozdravstvenim-problemima-u-republici-hrvatskoj-i-europskoj-uniji/

De Weist R (1965) Geohydrology. Wiley, New York

Elliott M, Cutts N, Trono A (2014) A typology of marine and estuarine hazards and risks as vectors of change: a review for vulnerable coasts and their management. Ocean Coast Manag 93:88–99

European Environment Agency (2017) *Climate change, impacts and vulnerability in Europe 2016: an indicator-based report*. Publications Office.

European Union (2006) *Directive 2006/7/EC of the European Parliament and of the Council of 15 February 2006 concerning the management of bathing water quality and repealing Directive 76/160/EEC*. European Union, pp 37–51

Faculty of Civil Engineering in Rijeka and Croatian Meteorological and Hydrological Service (2019) Precipitation analysis in pilot areas in Istria and Zagreb (RAINMAN project). Project Manager RubinićJ; Rijeka—Zagreb, Technical Documentation Fund

Hall M, Frank E, Holmes G, Pfahringer B, Reutemann P, Witten I (2009) The WEKA data mining software: an update. ACM SIGKDD Explor Newsl 11(1):10–18

Hoque MA, Scheelbeek PFD, Vineis P, Khan AE, Ahmed KM, Butler AP (2016) Drinking water vulnerability to climate change and alternatives for adaptation in coastal South and South East Asia. Clim Change 136:247–263. https://doi.org/10.1007/s10584-016-1617-1

Horvat B, Rubinic J (2006) Annual runoff estimation—an example of karstic aquifers in the transboundary region of Croatia and Slovenia. Hydrol Sci J 51(2):314–324. https://doi.org/10.1623/hysj.51.2.314

Institute for Public Health of the Federation of Bosnia and Herzegovina; Hydro-Engineering Institute Sarajevo (2017). Elaborat Cadastre of water supply facilities in the municipality of Maglaj, with a proposal for a system of organization, monitoring and response in crisis situations. Sarajevo

Javadinejad S (2022) Causes and consequences of floods: flash floods, urban floods, river floods and coastal floods. Resour Environ Inf Eng 4(1):156–166. https://doi.org/10.25082/REIE.2022.01.002

Jeleč N (2015) Crisis management—an example of the floods in BIH in 2014 with reference to the readiness of BIH to take in more refugees. https://www.parlament.ba/Publication/Read/4728?title=upravljanje-krizom---primjer-poplava-u-bosni-i-hercegovini--2014.-sa-osvrtom-na-spremnost-bosne-i-hercegovine-za-prijem-veceg-broja-izbjeglica-&pageId=0

Katalinić A., Rubinić J, Bušelić G (2017) Hydrology of two coastal karst cryptodepressions in Croatia: Vrana lake vs Vrana lake. In *Proceedings of taal 2007: The 12th world lake conference, 2007*, pp 732–743

Langbein W (1962) The water supply of arid valleys in intermountain regions in relation to climate. IAHS Bulletin 7(1):34–39

Ministry of Economy and Sustainable Development (2022) Coastal water quality in Šibenik-Knin County. https://vrtlac.izor.hr/static/plaze/

Official Gazette of the Republic of Croatia 125/2017 and 39/2020 (2017) *Ordinance on compliance parameters and methods of analysis, monitoring and water safety plans of water for human consumption and the manner of keeping a register of legal entities performing public water supply activities*. Ministry of Health, Zagreb, Croatia

Pal JS, Giorgi F, Bi X, Elguindi N, Solmon F, Gao X et al. (2007) Regional climate modeling for the developing world: the ICTP RegCM3 and RegCNET. Bull Am Meteorol Soc 1395–1409

Penna P, Baldrighi E, Betti M, Bolognini L, Campanelli A, Capellacci S et al (2021) Water quality integrated system: A strategic approach to improve bathing water management. J Environ Manage 295:113099. https://doi.org/10.1016/j.jenvman.2021.113099

Petrik M (1961) Available quantity of water in Vrana Lake on the Cres Island. 13(4):93–98

Rubinić J, Kuhta M, Brkić Ž (2011) Test area Lake Vrana—Cres Island (Croatia). EU Project CCWaterS. Croatian Geological Survey, Zagreb

Rubinić J, Mihovilović M, Radišić M, Rogić A, Vodanović F (2017). Water protection of Lake Vrana on the island of Cres—yesterday, today and tomorrow. In *Scientific conference Management of lakes and reservoirs in Croatia, Biograd na moru, 2017*

Rubinić J, Sarić M (2005) Water resources hydrology of the rječina catchment area. In Linić A (ed) *Professional and business conference with international participation—the past, the present and the future of water supply and drainage—experiences and challenges, Rijeka, 2005*, pp 199–208

Ružić I, Krvavica N, Rubinić J, Radišić M, Tadić A., Lončar G et al. (2021) The analysis of coastal infrastructure vulnerability due to climate change and definition and derivation of data for vulnerability analysis—ADRIADAPT project

Semenza JC, Tran A, Espinosa L, Sudre B, Domanovic D, Paz S (2016) Climate change projections of West Nile virus infections in Europe: implications for blood safety practices. Environ Health 15(Suppl 1):28. https://doi.org/10.1186/s12940-016-0105-4

The Municipality of Maglaj (2012) The development strategy of Maglaj municipality 2012–2020. https://www.maglaj.ba/slike2/dokumenti/Strategija%20razvoja%20op%C4%87ine%20Maglaj%20%202012-2020.pdf

Umar N, Gray A (2022) Flooding in Nigeria: a review of its occurrence and impacts and approaches to modelling flood data. Int J Environ Stud 1–22. https://doi.org/10.1080/00207233.2022.2081471

Verruijt A (1968) A note on the Ghyben-Herzberg formula. Int Assoc Sci Hydrol Bull 13(4):43–46

Vukić Lušić D, Ćuzela-Bilać D, Lenac D, Lušić D, Linšak Ž, Mićović V (2011) Drinking-water safety during the occurrence of turbidity in water sources of city of Rijeka. In *XV. scientific-professional conference "Water and public water supply", Vinkovci, Crotia, 2011*. Croatian Instutute of Public Health

Vukić Lušić D, Kranjčević L, Maćešić S, Lušić D, Jozić S, Linšak Ž et al (2017) Temporal variations analyses and predictive modeling of microbiological seawater quality. Water Res 119:160–170. https://doi.org/10.1016/j.watres.2017.04.046

WHO (2011) Guidance on water supply and sanitation in extreme weather events,. https://www.euro.who.int/__data/assets/pdf_file/0016/160018/WHOGuidanceFVLR.pdf

WHO (2017) Guidelines for drinking-water quality: fourth edition incorporating the first addendum. World Health Organization. Licence: CC BY-NC-SA 3.0 IGO.

Witten I, Frank E (2000) Data mining: practical machine learning tools and techniques with Java implementations. Morgan Kaufmann, San Francisco, Calif

Darija Vukić Lušić Master of Environmental Public Health. She studied at the Faculty of Medicine, University of Rijeka, Croatia, graduated as a Sanitary Engineer (2000) and received her Ph.D. in Oceanology from the Faculty of Natural Sciences, University of Zagreb, Croatia (2009). Since 2001 she has been working at the Teaching Institute of Public Health, Primorje-Gorski kotar County, Rijeka, Croatia and since 2008 at the Faculty of Medicine, University of Rijeka, Croatia, Department of Environmental Health (since 2019 as an Associate Professor). Since 2020, she has been the head of the Ph.D. programme Health and Ecological Engineering at the Faculty of Medicine, University of Rijeka, Croatia. She conducts research on the impact of various environmental factors on population health. The narrower area of her scientific interest is the quality of water resources, the safety of drinking water and the quality of bathing waters. She has made a special contribution in the field of improving the monitoring of seawater quality.

Daniel Maestro Master of Public and Environmental Health. He studied at the Faculty of Health Science, University of Sarajevo, Bosnia and Herzegovina, graduated as a sanitary engineer (2010) and received his Master's Degree in the field of Health and Ecology from the same University (2014). Since 2002, he has been working at the Institute for Public Health of Federation Bosnia and Herzegovina, Department of Environmental Health. The topic area of his scientific interest is water and food safety and quality, environmental engineering and quality assurance. He is currently completing his Ph.D. in the field of health sciences.

Arijana Cenov Master of Environmental Public Health (1998) at the Faculty of Medicine, University of Rijeka. She obtained her Master of Science (2010) and Ph.D. (2017) at the Faculty of Sciences, Department of Biology, University of Zagreb, Croatia. Since 2001, she has been working at the Teaching Institute of Public Health, Primorje-Gorski Kotar County, Rijeka, Croatia. She was elected Assistant Professor (2020) at the Department of Health Ecology, Faculty of Medicine, Rijeka. She is involved in various projects related to monitoring the impact of environmental factors on human health. A narrow area of her scientific interest is the quality of all water resources, the health of drinking water and the quality of recreational waters. She has made a special contribution in the field of improving microbiological methods for determining the quality of aquatic media.

Maja Radišić obtained her Master's Degree in civil engineering in 2017 at the Faculty of Civil Engineering of the University of Rijeka. Since 2014, she has been employed by the Department of Hydraulic Engineering of the aforementioned Faculty as the project assistant on several national and international research projects related to hydrological studies and the analysis of the climate change impact. Since 2019 she has been engaged in the project Management of the Karst Aquifers. She is a Ph.D. student at the Faculty of Civil Engineering of the University of Rijeka. Her particular contribution was in conducting the hydrological analysis and modelling.

Dražen Lušić Ph.D., Associate Professor. He graduated at the Faculty of Medicine, University of Rijeka with a Master's Degree of Environmental Public Health (1996). He obtained his Master of Science (2007) and Ph.D. (2011) at the Faculty Food Technology and Biotechnology, University of Zagreb in the field of Food Science. Since 1997, he has been working at the Faculty of Medicine of the University of Rijeka, Croatia. In 2021, he also received the position of Associate Professor at the Faculty of Health Studies of the University of Rijeka. During his professional career he has been involved in several scientific and technical projects that dealt with the issues of food and water quality and safety with special emphasis on their impact to human health. Presently, he is the Head of the "Clinical Nutrition" University course of Study of the University of Rijeka.

Marin Glad Master of Environmental Public Health. He studied at the Faculty of Medicine, University of Rijeka, Croatia, and graduated as a sanitary engineer (1999). In 2017, he received his Ph.D. in Biology from the Faculty of Sciences, University of Zagreb, Croatia. He conducted the experimental part of his Ph.D. at the Ruđer Bošković Institute, the Centre for Marine Research, Rovinj, Croatia. Since 2001 he has been working at the Teaching Institute of Public Health, Primorje-Gorski Kotar County, Rijeka, Croatia, where he has been the Head of the Department of Environment Health since 2021. Also, since 2021, he has been working as an Assistant Professor at the Faculty of Medicine, University of Rijeka, Croatia. He has contributed to the development of research in the field of environment and health, with special interest in water environment, ecology and microbiology. The narrower area of his scientific interest is water quality.

Željko Linšak Master of Environmental Public Health at the Faculty of Medicine of the University of Rijeka (MEDRI), graduated (2002) sanitary engineering at the University of Rijeka (UNIRI). Master's degree (2010) in natural sciences (Biology—Toxicology) at the Faculty of Science in Zagreb. Ph.D. (2012) in natural sciences (Biology—Ecology) at the Faculty of Science in Zagreb. Since 2004, he has been working at the Teaching Institute of Public Health, Primorje-Gorski Kotar County, Rijeka in the environmental area, and since 2020 at the Faculty of Medicine, University of Rijeka, Croatia, Department of Environmental Health (as an Assistant Professor). He is a researcher of environmental impact on public health. He has experience in Environmental Education, Public Health management, strategic planning and management, narrower area of interest is air pollution as well as heavy metals in environment. He currently works as deputy director of the Teaching Institute of Public Health, Primorje-Gorski Kotar County, Rijeka.

Nerma Maestro Master of Biology / Microbiology. She studied at the Faculty of Sciences, University of Sarajevo, Bosnia and Herzegovina, graduated as a sanitary engineer at the Faculty of Health, University of Sarajevo, Bosnia and Herzegovina. Since 2016, she has been working at the Institute for public health of Canton Sarajevo, Department of Hygiene, Bosnia and Herzegovina and since 2019 at the Faculty of Health Studies. The narrower area of her scientific interest is water resources quality, drinking water safety and bathing water quality.

Dijana Tomić Linšak Master of Environmental Public Health at the Faculty of Medicine of the University of Rijeka (MEDRI), graduated (2002) sanitary engineering at the University of Rijeka (UNIRI). Master's degree (2010) in natural sciences (Biology—Toxicology) at the Faculty of Science in Zagreb. Ph.D. (2012) in natural sciences (Biology—Ecology) at the Faculty of Science in Zagreb. Since 2004, she has been working at the Teaching Institute of Public Health, Primorje-Gorski Kotar County, Rijeka in the environmental area, and since 2018 at the Faculty of Medicine, University of Rijeka, Croatia, Department of Environmental Health (as an Assistant Professor). She is a researcher of environmental impact on public health. The area of her scientific interest is environmental health impact on animals and humans, heavy metals in the environment, pesticides and ecological sanitation. She currently coordinates the area of public health and ecological sanitation at the Teaching Institute of Public Health, Primorje-Gorski Kotar County, Rijeka.

Josip Rubinić Master of Civil Engineering. He studied at the University of Split and the University of Rijeka where he obtained his Ph.D. in 2017. After 20 years of activity in the field of water management and hydrological research, since 2002 he has been professionally engaged at the Faculty of Civil Engineering of the University of Rijeka. Presently, he holds the position of Associate Professor at the Department of Hydraulic Engineering. He has been mainly involved in the hydrological studies of the karst and lake waters, the ecohydrology as well as the studies of the impact of the climate change on water systems. His particular contribution was in the analysis of the hydrological aspects of the climatic change impacts.

Chapter 10
Unawareness About Vector-Borne Diseases Among Citizens as a Health Risk Consequence of Climate Change—A Case Study on Leishmaniosis in Northwest Portugal

Teresa Letra Mateus, Solange Moreira, and Rui Leandro Maia

Abstract An unprecedented change in climate and consequently an increase in the number and geographic dissemination of vectors have occurred in the last decades worldwide as well as in Portugal. Vector-borne diseases are infections transmitted by the bite of infected arthropod species. Among them, the incidence of leishmaniosis/leishmaniasis (transmitted through the bite of phlebotomine insects) is increasing in Portugal both in animals and humans. In this chapter, the authors want to present the results of a case study aiming to assess the leishmaniosis awareness of citizens in the Northwest of Portugal. Following the bibliographic review and the realization of two focus groups with citizens, a draft questionnaire was constructed, tested and personally administered to citizens ($n = 291$) for five months and three

T. L. Mateus (✉)
CISAS—Centre for Research and Development in Agrifood Systems and Sustainability, Escola Superior Agrária, Instituto Politécnico de Viana Do Castelo, Rua Escola Industrial e Comercial Nun'Álvares, 4900-347 Viana Do Castelo, Portugal
e-mail: tlmateus@esa.ipvc.pt

EpiUnit—Instituto de Saúde Pública da Universidade do Porto, Laboratory for Integrative and Translational Research in Population Health (ITR), Rua das Taipas, no. 135, 4050-091 Porto, Portugal

Veterinary and Animal Research Centre (CECAV), Associate Laboratory for Animal and Veterinary Sciences (AL4AnimalS), UTAD, Quinta de Prados, 5000-801 Vila Real, Portugal

S. Moreira
Escola Superior Agrária, Instituto Politécnico de Viana Do Castelo, Rua Escola Industrial e Comercial Nun'Álvares, 4900-347 Viana Do Castelo, Portugal
e-mail: moreira27@live.com.pt

R. L. Maia
UFP Energy, Environment and Health Research Unit (FP-ENAS), Universidade Fernando Pessoa, Praça 9 de Abril, 349, 4249-004 Porto, Portugal
e-mail: rlmaia@ufp.edu.pt

CITCEM—Centro de Investigação Transdisciplinar «Cultura, Espaço E Memória», Faculdade de Letras da Universidade Do Porto, Via Panorâmica S/N, 4150-564 Porto, Portugal

W. Leal Filho et al. (eds.), *Climate Change and Health Hazards*, Climate Change Management, https://doi.org/10.1007/978-3-031-26592-1_10

days per week in the district of Braga in the northwest of Portugal. Only a participants minority vaccinate their animals to prevent this disease, even if 62.5% had heard about the disease. More than half (56.0%) are unaware of the zoonotic potential of leishmaniosis. It seems that there is a need to inform but also to communicate as even those who seem to have knowledge still adopt risk behaviours. In the context of climate change, both disease and unawareness pose health risks that need to be mitigated. For this purpose, the authors suggest that the investment in health communication should be jointly done by the animal, human and environmental health professionals in a One Health perspective to address these hazards.

Introduction

It is said that we are now living the Anthropocene Epoch with accelerated anthropogenic changes being linked to several health risks such as the emergence of SARS-CoV-2 (Srivastava et al. 2021). These changes are seen as a huge threat to human health as they impact the structure and function of the global natural systems, known to be the base of planetary health. Resilience may be achieved by protecting natural systems including landscapes, ecosystems and biodiversity (Whitmee et al. 2015). But on the contrary, human activities are degrading the ecosystem services (Cui et al. 2021), leading to the paradox of improved health and natural systems deterioration (Whitmee et al. 2015), if one considers that we have now, after all, the best human health ever, with the highest life expectancy and the lowest child mortality rate (Sharrow et al. 2021; United Nations 2017). Nevertheless, this is not enough to reduce the risk of raising emerging zoonoses (Kuiken 2021) as anthropogenic activities have an enormous impact on hosts' diversity, composition and abundance as well as on pathogen prevalence (Johnson Pieter et al. 2015).

Disruptions—of seasons, species diversity, land use—and easy and fast cross-border and even cross-continental travels (tourism or trading, humans or goods) of the modern society have increased the potential exposition of animals and humans to (old or new) pathogens through the creation of new human to animal interfaces (Srivastava et al. 2021; Sánchez et al. 2021). Globalisation leverage the dissemination of pathogens and their vectors (Semenza and Suk 2018). Novel infectious diseases may appear in different locations as our world is now fully interconnected (Suk et al. 2014), as we could see recently with COVID-19.

Changes to land use such as deforestation, habitat fragmentation, agricultural and urban expansion have already been related to the increase in disease transmission (Gottdenker et al. 2014). Urbanization, and mostly the crowding in huge cities, are also linked to a higher risk of the emergence of zoonosis (Sánchez et al. 2021). Additionally, habitat encroachment and overlap between wildlife, humans and domestic animals promote spillover among species (Srivastava et al. 2021). Specifically, this change in land use has consequences in hosts but especially in vectors communities (Guo et al. 2019). That is why vector-borne diseases are the biggest winners of climate change. The temperature increase permitted the dispersion of vectors (in latitude and altitude) (McCain and Garfinkel 2021) and the host shift (Ruscio et al. 2015; Harvey and Malcicka 2015), as well as increases the development rate of pathogens, the number of vectors due to the reduction of their life cycle time, and the pathogen transmission season (Srivastava et al. 2021). This is particularly evident in Europe (Semenza and Suk 2018). So emerging vector-borne diseases may arise and interestingly Pinner et al. (2020) argue that these and the climate risks are both systemic (as their effects spread easily globally), nonstationary (their distribution is very volatile and unpredictable), nonlinear (have seemingly disproportionate socio-economic impacts), risk multipliers (highlight and reinforce less obvious risks) and regressive (they disproportionately affect the most vulnerable).

Millions of people worldwide are infected by diseases (zika, malaria, dengue, leishmaniases among others) transmitted by vectors (mainly insects like mosquitoes and flies). The awareness of the World Health Organization (WHO) that 80% of the world population is at risk of being infected at least by one vector-borne disease, prompted the release of the initiative the "2017–2020 Global Vector Control Response" (WHO 2017).

The rising temperatures may lead to a higher geographic range and abundance of vectors and, on the other side, the rainfalls can generate an unexpected increase in their number, promoting the emergence of vector-borne diseases (Sánchez et al. 2021). The interaction of other drivers and effects may also influence, and some authors argue that there is no consensus on the relation between climate change and mosquito-borne disease (Franklinos et al. 2019). The same happens concerning the measures to mitigate the known urban heat islands, and consequently control the insect communities: one size will not fit all the cities and strategies should be adapted to each specific local insect community (Ligsay et al. 2021; Medlock and Leach 2015).

Insects are ectothermic. They depend on environmental temperature to live (McCain and Garfinkel 2021). The urban heat island effect is more evident in big cities, where the majority of the world population lives. So, it is expected that mosquito-borne diseases will spread increasingly in the global population (Ligsay et al. 2021). Mosquitoes like *Aedes* and *Culex* are spreading in towns and cities where they found great habitats to proliferate next to humans (WHO 2017). These vectors are expected to invade more temperate regions, as they mostly exist in tropics and sub-tropics (Kraemer et al. 2019). That has already occurred with the human outbreak of dengue in Portugal (Madeira) caused by the mosquito *Aedes aegypti* known to have been established in 2005 there (Sousa et al. 2012). *Aedes albopictus* travelled from South East Asia to America and Europe. It may transmit several viral diseases like dengue, zika and chikungunya (Vega-Rúa et al. 2020). Medlock and Leach (2015) stated that the twenty-first century brought a new status for vector-borne disease in Europe. Among these diseases is leishmaniasis, caused by *Leishmania* spp., considered by WHO (2017) one of the major global vector-borne diseases of humans, and transmitted by Phlebotomine sandflies. When we refer to the disease in animals it is called leishmaniosis. Medlock et al. (2014) evidenced the risk of the establishment of these species on the European Atlantic coast and some central European countries like Austria, Switzerland and Germany (Medlock et al. 2014). Leishmaniasis is the most prevalent disease in Europe transmitted by sandflies (Semenza and Suk 2018). Portugal is considered an endemic country (WHO 2020). Leishmaniasis in Europe, caused mainly by *Leishmania infantum*, has two main clinical forms: visceral and cutaneous. The symptoms range from self-limiting dermatological problems to systemic and fatal, if not treated. Although other animals may be reservoirs, the domestic dog is the most important one and also may be severely sick (Burza et al. 2018). In Portugal, the seroprevalence of the disease in dogs ranges from 0.8 to 56.0%, which may be explained by several environmental and socioeconomic factors (Pires et al. 2019; Cortes et al. 2012). Between 2013 and 2016, the number of autochthonous visceral leishmaniasis cases officially reported in humans in Portugal was 35 (Gaspar et al. 2017), but the Portuguese national institute of health estimates that about 20 cases of visceral leishmaniasis are diagnosed per year in immunocompetent humans (Gargaté 2019). In Portugal and other countries of the European south, the co-infection with *Leishmania* and Human Immunodeficiency Virus is frequent (Monge-Maillo et al. 2014). Moreover, there are increased problems among drug addicts because they share contaminated syringes, thus not requiring the intervention of the vector for the disease transmission (Gargaté 2019). As mobilization and education of the community on a local basis is strongly suggested by WHO (2017) as a strategy to prevent this disease, the authors of this survey aimed to assess the awareness of citizens in the northwest of Portugal about leishmaniosis in order to be able to develop in the future targeted intervention measures on health education.

Material and Methods

The construction of the draft questionnaire began with a bibliographic review using PubMed to search for articles published between 2010 and 2020, with keywords like "One Health", "Leishmaniosis", "Leishmaniasis", "Vector-borne diseases", "Zoonosis", "*Leishmania*" and "Health Communication" between February and April 2020. Then, two focus groups were done with citizens of different ages, professions and socio-economic conditions. Based on the bibliographic review and two focus groups a draft questionnaire was constructed. That draft was then tested with ten persons and finally personally administered to citizens. The questionnaire included a total of 23 questions about socio-demographics (the place of residence, age, gender and number of years of studies), animal ownership (species owned, lifestyle, veterinary care, knowledge about veterinary prophylactic measures), and awareness about zoonotic parasites in general and *Leishmania* in particular, as well as willingness to receive information about this zoonotic disease. During five months and three days per week (two days on the street and one day at a veterinary clinic), citizens from the district of Braga in the northwest of Portugal were approached to inquire about their availability to participate in this study. This study was carried out as part of Solange Moreira's graduation thesis in Veterinary Nursing. At the time it was carried out, there was no ethics committee in the institution, but the final degree internship plan was approved by a specific committee that evaluated ethical issues at the time, before the study began. No data that could identify the participants was gathered and the participation was voluntary. All the participants were previously informed about the aims of the study and decided to participate freely. Oral consent was given. At the end of the questionnaire, a pamphlet with information about the disease and the prevention measures in animals and humans was given to all participants. Braga district was selected because this is a mixed district with a reasonably big city (named Braga, the third most populous city in the country) and several sub-urban and rural towns and villages. It is crossed by several rivers and waterways, has a temperate climate adequate for vector development and it is where the first cases of the disease in dogs were identified more recently when compared with the northeast of Portugal.

The statistical analysis was conducted using SPSS and focused on the construction of frequency tables, the chi-square test to evaluate the association between variables (considering the degree of error of less than 5%, $p < 0.05$) and the test Cramer's *V* to know the intensity of association between variables (ranging from 0 to 1, being noteworthy associations >0.300) (Pestana and Gageiro 1998).

Results

There were several (unfortunately) unaccounted refusals, but despite that, 291 answers have been collected, 171 on the street and 120 at the veterinary clinic.

The answers obtained about socio-demographics, animal ownership and awareness about zoonotic parasites and *Leishmania* and willingness to receive information are presented in Table 10.1.

Citizens were mostly between *21 and 40 years old* (52.2%, $n = 152$), *female* (65.6%, $n = 191$), attended *university* (46.0%, $n = 134$) and were *owners of one or more animals* (88.0%, $n = 256$), namely dogs (85.2%, $n = 248$). Reporting only to dog owners, 42.3% ($n = 105$) of these owners admitted that their pets live simultaneously *indoors* and *outdoors*. Also, 58 owners *walk their dogs on a leash* and 14 (24.1%) of them are in *risk areas* for the presence of insects near smooth waters. When asked about *the use of ectoparasiticides*, most affirm doing it for fleas and ticks (74.6%, $n = 185$) and only a minority for *Phlebotomus* (39.5%, $n = 98$). Concerning *vaccination*, most of the owners do it for rabies (74.6%, $n = 185$) which is mandatory in Portugal for dogs, and only 59 (23.8%) do it for leishmaniosis. Ticks are seen as *the most dangerous parasite for animals and humans* (73.9%, $n = 215$) with mosquitoes being the second choice but only for 32 (11.0%) persons. When asked "*have you heard about leishmaniosis?*" 62.5% ($n = 182$) said "yes" but 56.0% ($n = 102$) of these didn't know it as a *zoonosis*. This means that of overall respondents only 27.5% ($n = 80$) were aware of the zoonosis. Of the ones who knew the zoonosis, only 66.3% ($n = 53$) knew also its *vector-borne transmission* and 17.5% ($n = 14$) knew the possible *symptoms*. Citizens that have heard about leishmaniosis mostly (52.7%, $n = 96$) do not know how to prevent it. Finally, the majority (75.9%, $n = 221$) would

Table 10.1 Answers obtained about socio-demographics, animal ownership and awareness about zoonotic parasites and *Leishmania* and willingness to receive information ($n = 291$; otherwise, n value is presented)

Variables	Answer categories	Percentage (%)
Sex	Female	65.6
	Male	34.4
Age (years)	<20	13.7
	21–40	52.2
	41–60	26.2
	>60	7.9
Educational qualification	Did not go to school	0.3
	Basic Education	20.3
	High School	33.4
	University Studies	46.0
Pet owner	Yes	88.0
	No	12.0
Dog Life Style ($n = 248$)	Indoors and outdoors	42.3
	Outdoors	28.3
	Indoors but walks on leash	14.9
	Indoors	14.5
Dog walking time ($n = 58$)	Any hour	46.6
	Morning + at the end of the day	34.5
	At the end of the day	8.6
	During the day	6.9
	Morning	3.4
Dog walking place ($n = 58$)	Near the river	20.7
	Near the sea	3.5
	Near the lake	3.5
	None of the above	72.3
Dog deworming ($n = 248$)	Yes	65.7
	No	0.4
	I do not know	33.9
Dog mosquito prevention ($n = 248$)	Yes	39.5
	No	60.5
Dog vaccination ($n = 248$)	Yes	74.6
	No	25.5
Dog vaccination – leishmaniosis ($n = 248$)	Yes	23.8
	No	76.2
The most dangerous parasite for health	Tick	73.9
	Mosquito	11.0
	Do not know	10.0
	Fleas	3.4
	Lice	1.7
Have heard about leishmaniosis	Yes	62.5
	No	37.5

(continued)

Table 10.1 (continued)

Variables	Answer categories	Percentage (%)
Knowledge about zoonotic potential (n = 182)	Yes No	44.0 56.0
Knowledge about the routes of transmission (n = 182)	Mosquito Do not know Wrong answers	29.1 14.8 56.1
Knowledge about human symptoms (n = 80)	Do not know Dermatological Wrong answers	75.0 17.5 7.5
Knowledge about prevention (n = 182)	Do not know Vaccination Deworming Wrong answers	52.7 31.9 10.4 5.0
Willingness to receive information about the disease (n = 291)	Yes No	75.9 24.1
Preferred information route (n = 221)[a]	Veterinarian or veterinary nurse Family doctor or nurse Television Internet Flyers	61.5 34.4 29.9 28.5 20.4

[a]it was possible to choose more than one option

like to *receive information* about the disease mainly from the *veterinarian/veterinary nurse* (61.5%, n = 136) and the *family's doctor/nurse* (34.4%, n = 76).

The citizens ranging from *21 to 40 years old* seem to be better informed about the *transmission route* of this disease (p = 0.047, VC = 0.335) as well as the *preventive measures* (p = 0.016; VC = 0.292). Most of the citizens who attended *high school* (60.8%) and *university* (76.1%) *had heard about leishmaniosis* but only 35.6% of those who attended *basic education* had (p = 0.000; VC = 0.324). There is a statistically significant relation (p = 0.0001) of strong association (VC = 0.558) between the *hour at which owners walk the dog* (end of the day) and the interest in getting information by *flyers*. On the other side, those who walk the dog early in the morning prefer the *internet* (p = 0.046; VC = 0.409). Citizens that live in *urban parishes* are more informed about the disease than those living in sub-urban and rural parishes (p = 0.005; VC = 0.275). Pet owners tend to consider *ticks* the most dangerous parasite more frequently (76.6%) than those who are not pet owners (54.3%) (p = 0.0012; VC = 0.210). *Pet owners* are more enlightened about the disease *prevention* than those who are not pet owners (54.7 and 44.0% respectively know the disease) (p = 0.005; VC = 0.359). Of the respondents that consider the *mosquitoes* the most dangerous ectoparasite, most (81.3%) know leishmaniosis (p = 0.035) and is aware of its *zoonotic potential* (73.1%) (p = 0.002; VC = 0.287).

Discussion

It is known that pets are present in an increasing number of households, so not surprisingly most of our respondents were pet owners. Anyway, it is not necessary to have a pet to be infected, because the vector is an insect. Indeed, Di Muccio et al. (2015) state that leishmaniasis increased in western Europe thanks to the explosion of ecotourism as it allows greater exposure to the vector (Di Muccio et al. 2015). Many pet owners admitted to having risk behaviours (like pets living outdoors or dog walking in risky places) and low sensitivity to the importance of the prevention measures to avoid contact with insects or vaccination, even if they have heard about the disease. This may be because on the one hand, having knowledge does not mean changing behaviours, and on the other, people in general neglect parasitic and vector-borne diseases. It seems that there is a need not only to inform but also to communicate. It is important to reinforce here that as dogs are reservoirs of the disease, preventing their disease is fundamental to protecting human health in a One Health perspective. Although Hathaway and Maibach (2018) perceived a general awareness among medical professionals about climate change gravity, we are not sure the same awareness is present in the general community (Hathaway and Maibach 2018). Concerning leishmaniasis in European Union, Gradoni (2013) concluded that awareness of the disease should be promoted among both medical professionals and citizens (Gradoni 2013). A survey among Spanish and French veterinarians also concluded that these professionals (especially the French ones) need to create more awareness of the zoonotic potential of the disease (Le Rutte et al. 2018).

It was also evident that people from suburban and rural areas over 40 years old are less literate concerning diseases in general and leishmaniosis in particular. This is worrying as the prevalence of the disease in dogs is higher in rural areas of Portugal (Campino and Maia 2010). But, on the other side, the prevalence of the disease is also increasing in sub-urban areas in Portugal, due to the unplanned urban development that promotes the production and accumulation of domestic waste that attract stray dogs and insects (Cortes et al. 2007). This urban crowding together with the climate change is a perfect scenario for the insect and disease spreading and apparently safe places like small gardens in the cities where citizens walk their dogs together with children and elderly can become perfect spots of infection particularly or immunocompromised persons, so awareness is really needed. Climate change and the consequent increase of the urban heat islands will prompt the increase of the urban greenspaces to mitigate this problem, which may increase the risk of exposure to vectors (Medlock and Leach 2015). Diseases like vector-borne diseases that were in the past linked to rural areas will emerge in urban areas due to several conditions related to urbanization, urban heat islands and crowded animal and human population (Ligsay et al. 2021).

Semenza and Suk (2018) highlight that the performance of risks associated with climate change depends also on the socioeconomic contexts. Social determinants can attenuate or amplify climate change's impacts on health (Leal Filho et al. 2022). Social determinants should always be taken into account in studies about zoonoses, namely in health promotion and education interventions that should be directed according to these settings. In addition to the integration of social sciences, integrated surveillance of humans, animals and the environment (namely insects' species presence) is the key of the so-called One Health (Semenza and Zeller 2014). Le Rutte et al. (2018) recommended the development of a common online platform where veterinarians and physicians could report confirmed cases of the disease. Unfortunately, Sánchez et al. (2021) identified the lack of integrated investigation among different disciplines as the major gap to deal with zoonotic pathogens. Furthermore, it is extremely urgent to change the narrative as it was brilliantly proposed by Kuiken (2021) from "focus on human health" to "equal attention to the health of ecosystems, animals, and humans", from "emphasis on the financial cost to society" to "equal emphasis on environmental impact". This author also argues the need to develop long-term solutions acting directly on the cause of the emergence of the disease, promoting in the long run, a sustainable world. Probably in the present context of the pandemic and war in Europe, the degradation of the ecosystem will be unavoidable as this occurs when it is not an international priority to take care of environmental concerns (Riahi et al. 2017). So, engaging and mobilizing communities as proposed by WHO is an urgent need (WHO 2017).

A *business-as-usual approach* to climate change as referred by Ligsay et al. (2021) will not bring global health for sure. In the context of climate change, both disease and unawareness pose health risks that need to be mitigated. For this purpose, we would like to suggest: first, a radical change of the planetary health narrative as referred by Kuiken (2021); then, a real investment in health communication jointly done by the animal, human and environmental health professionals in a One Health perspective to address these hazards.

Conclusions

Climate change is leading to an increase in the diversity, quantity and geographical dispersion of vectors responsible for transmitting a wide range of diseases to animals and humans, namely leishmaniosis. Nevertheless, the authors hypothesized that awareness of this disease could be low. To assess it, a questionnaire was developed and administered to citizens in Braga, a district in the Northwest of Portugal. Of the 291 participants, 62.5% referred to knowing the disease but only a minority vaccinated their dogs to prevent the disease and more than half were unaware of its zoonotic potential. Vector-borne diseases spreading in the context of climate change, and unawareness about them need to be mitigated. Health education and communication in a One Health perspective should be dynamized to address these hazards. This is a case study and it used a convenience sample, indicating trends only. In the future, a mixed methodology could be developed to understand why even when one has knowledge one does not behave more assertively to protect their pets and their own health.

Acknowledgements The participation of Teresa Letra Mateus was supported by the projects UIDB/CVT/00772/2020 and LA/P/0059/2020 funded by the Portuguese Foundation for Science and Technology (FCT). The authors wish to thank Katarzyna Loskot for her constructive feedback on earlier drafts.

References

Burza S, Croft SL, Boelaert M (2018) Leishmaniasis. Lancet 392(10151):951–970

Campino L, Maia C (2010) Epidemiology of leishmaniases in Portugal. Acta Med Port 23(5):859–864

Cortes S, Afonso MO, Alves-Pires C, Campino L (2007) Stray dogs and leishmaniasis in urban areas Portugal. Emerg Infect Dis 13(9):1431–1432

Cortes S, Vaz Y, Neves R, Maia C, Cardoso L, Campino L (2012) Risk factors for canine leishmaniasis in an endemic Mediterranean region. Vet Parasitol 189(2–4):189–196

Cui F, Wang B, Zhang Q, Tang H, De Maeyer P, Hamdi R, Dai L (2021) Climate change versus land-use change—what affects the ecosystem services more in the forest-steppe ecotone? Sci Total Environ 759:143525

Di Muccio T, Scalone A, Bruno A, Marangi M, Grande R, Armignacco O, Gradoni L, Gramiccia M (2015) Epidemiology of imported Leishmaniasis in Italy: implications for a European endemic country. PLoS ONE 10(7):e0134885

Franklinos LHV, Jones KE, Redding DW, Abubakar I (2019) The effect of global change on mosquito-borne disease. Lancet Infect Dis 19(9):e302–e312

Gargaté MJ (2019) Leishmaníase. In: Doenças associadas a artrópodes vectores e roedores (Leishmaniasis. In: Diseases associated with vector arthropods and rodents]) (ed: Núncio MS and Alves MJ). Portuguese National Institute of Health, p 198

Gaspar CC, Augusto GF, Albuquerque MJ, Nascimento MR, Olivença P Nogueira VP (2017) Doenças de Declaração Obrigatória em Portugal – Direção Geral de Saúde (Notifiable diseases in Portugal—general directorate of health] 2013–2016. Available at: https://www.dgs.pt/portal-da-estatistica-da-saude/diretorio-de-informacao/diretorio-de-informacao/por-nivel-de-informacao-908701-pdf.aspx?v=%3d%3dDwAAAB%2bLCAAAAAAABAArySzItzVUy81MsTU1MDAFAHzFEfkPAAAA. Accessed 7 May 2022

Gottdenker NL, Streicker DG, Faust CL, Carroll CR (2014) Anthropogenic land use change and infectious diseases: a review of the evidence. EcoHealth 11(4):619–632

Gradoni L (2013) Epidemiological surveillance of leishmaniasis in the European Union: operational and research challenges. Euro Surveill 18(30):20539

Guo F, Bonebrake TC, Gibson L (2019) Land-use change alters host and vector communities and may elevate disease risk. EcoHealth 16(4):647–658

Harvey JA, Malcicka M (2015) Climate change, range shifts and multitrophic interactions. https://doi.org/10.5772/59269

Hathaway J, Maibach EW (2018) Health implications of climate change: a review of the literature about the perception of the public and health professionals. Curr Environ Health Rep 5(1):197–204

Johnson Pieter TJ, de Roode JC, Fenton A (2015) Why infectious disease research needs community ecology. Science 349(6252):1259504

Kraemer MUG, Reiner RC, Brady OJ, Messina JP, Gilbert M, Pigott DM, Yi D, Johnson K, Earl L, Marczak LB, Shirude S, Davis Weaver N, Bisanzio D, Perkins TA, Lai S, Lu X, Jones P, Coelho GE, Carvalho RG, Van Bortel W, Marsboom C, Hendrickx G, Schaffner F, Moore CG, Nax HH, Bengtsson L, Wetter E, Tatem AJ, Brownstein JS, Smith DL, Lambrechts L, Cauchemez S, Linard C, Faria NR, Pybus OG, Scott TW, Liu Q, Yu H, Wint GRW, Hay SI, Golding N (2019) Past and future spread of the arbovirus vectors Aedes aegypti and Aedes albopictus. Nat Microbiol 4(5):854–863

Kuiken T (2021) Implications of transformative changes for research on emerging zoonoses. EcoHealth 18(3):275–279

Le Rutte EA, van Straten R, Overgaauw PAM (2018) Awareness and control of canine leishmaniosis: a survey among Spanish and French veterinarians. Vet Parasitol 253:87–93

Leal Filho W, Balasubramanian M, Purcell W, Paz S (2022) Handling the health impacts of extreme climate events. Environ Sci Eur 34(1):45

Ligsay A, Telle O, Paul R (2021) Challenges to mitigating the urban health burden of mosquito-borne diseases in the face of climate change. Int J Environ Res Public Health 18(9):5035

McCain CM, Garfinkel CF (2021) Climate change and elevational range shifts in insects. Curr Opin Insect Sci 47:111–118

Medlock JM, Hansford KM, Van Bortel W, Zeller H, Alten B (2014) A summary of the evidence for the change in European distribution of phlebotomine sand flies (Diptera: Psychodidae) of public health importance. J Vector Ecol 39(1):72–77

Medlock JM, Leach SA (2015) Effect of climate change on vector-borne disease risk in the UK. Lancet Infect Dis 15(6):721–730

Monge-Maillo B, Norman FF, Cruz I, Alvar J, López-Vélez R (2014) Visceral leishmaniasis and HIV coinfection in the Mediterranean region. PLoS Negl Trop Dis 8(8):e3021

Pestana HM, Gageiro JN (1998) Análise de dados para ciências sociais - a complementaridade do SPSS [Data analysis for social sciences—the complementarity of SPSS]. Lisboa Sílabo, Lda

Pinner D, Rogers M, Samandari H (2020) Addressing climate change Post-coronavirus | McKinsey. Available at: https://www.mckinsey.com/business-functions/sustainability/our-insights/addressing-climate-change-in-a-post-pandemic-world. Accessed 15 May 2022

Pires H, Martins M, Matos AC, Cardoso L, Monteiro F, Roque N, Nunes T, Gottstein B, Cortes H (2019) Geospatial analysis applied to seroepidemiological survey of canine leishmaniosis in east-central Portugal. Vet Parasitol 274:108930

Riahi K, van Vuuren DP, Kriegler E, Edmonds J, O'Neill BC, Fujimori S, Bauer N, Calvin K, Dellink R, Fricko O, Lutz W, Popp A, Cuaresma JC, Kc S, Leimbach M, Jiang L, Kram T, Rao S, Emmerling J, Ebi K, Hasegawa T, Havlik P, Humpenöder F, Da Silva LA, Smith S, Stehfest E, Bosetti V, Eom J, Gernaat D, Masui T, Rogelj J, Strefler J, Drouet L, Krey V, Luderer G, Harmsen M, Takahashi K, Baumstark L, Doelman JC, Kainuma M, Klimont Z, Marangoni G, Lotze-Campen H, Obersteiner M, Tabeau A, Tavoni M (2017) The shared socioeconomic pathways and their energy, land use, and greenhouse gas emissions implications: an overview. Glob Environ Chang 42:153–168

Ruscio BA, Brubaker M, Glasser J, Hueston W, Hennessy TW (2015) One health—a strategy for resilience in a changing arctic. Int J Circumpolar Health 74:27913

Sánchez CA, Venkatachalam-Vaz J, Drake JM (2021) Spillover of zoonotic pathogens: a review of reviews. Zoonoses Public Health 68(6):563–577

Semenza JC, Suk JE (2018) Vector-borne diseases and climate change: a European perspective. FEMS Microbiology Letters 365(2):fnx244

Semenza JC, Zeller H (2014) Integrated surveillance for prevention and control of emerging vector-borne diseases in Europe. Euro Surveill 19(13):20757

Sharrow D, Hug L, Lee S, Liu Y, You D (2021) Levels and trends in child mortality. United Nations Inter-agency Group for Child Mortality Estimation. Available at: https://cdn.who.int/media/docs/default-source/mca-documents/rmncah/unicef-2021-child-mortality-report.pdf?sfvrsn=7bbacc7d_1&download=true. Accessed 15 May 2022

Sousa CA, Clairouin M, Seixas G, Viveiros B, Novo MT, Silva AC, Escoval MT, Economopoulou A (2012) Ongoing outbreak of dengue type 1 in the Autonomous Region of Madeira, Portugal: preliminary report. Euro Surveill 17(49):20333

Srivastava S, Khokhar F, Madhav A, Pembroke B, Shetty V, Mutreja A (2021) COVID-19 lessons for climate change and sustainable health. Energies 14(18):5938

Suk JE, Van Cangh T, Beauté J, Bartels C, Tsolova S, Pharris A, Ciotti M, Semenza JC (2014) The interconnected and cross-border nature of risks posed by infectious diseases. Glob Health Action 7:25287

United Nations (2017) Population facts—life expectancy at birth increasing in less developed regions. Available at: https://www.un.org/development/desa/pd/content/life-expectancy-birth-increasing-less-developed-regions. Accessed 16 May 2022
Vega-Rúa A, Marconcini M, Madec Y, Manni M, Carraretto D, Gomulski LM, Gasperi G, Failloux AB, Malacrida AR (2020) Vector competence of Aedes albopictus populations for chikungunya virus is shaped by their demographic history. Commun Biol 3(1):326
Whitmee S, Haines A, Beyrer C, Boltz F, Capon AG, de Souza Dias BF, Ezeh A, Frumkin H, Gong P, Head P, Horton R, Mace GM, Marten R, Myers SS, Nishtar S, Osofsky SA, Pattanayak SK, Pongsiri MJ, Romanelli C, Soucat A, Vega J, Yach D (2015) Safeguarding human health in the Anthropocene epoch: report of The Rockefeller Foundation-Lancet Commission on planetary health. Lancet 386(10007):1973–2028
World Health Organization (2017) Global Vector Control Response 2017–2030. World Health Organization, Geneva, Switzerland. Available at: https://www.who.int/publications/i/item/9789241512978. Accessed 16 May 2022
World Health Organization (2020) Status of endemicity of cutaneous leishmaniasis. Available at: https://apps.who.int/neglected_diseases/ntddata/leishmaniasis/leishmaniasis.html. Accessed 17 May 2022

Teresa Letra Mateus graduated in Veterinary Medicine, holds a Ph.D. in Veterinary Sciences, a master degree in Food Science and Technology and post graduate studies in Food Safety, Health Education and Promotion, and Health Communication. She is in the Portuguese board of the Wild Animals Vigilance Euro-Mediterranean Society and she is a member of the Portuguese Society of the Veterinary Sciences. Her research area is One Health and especially parasitology, food safety, zoonosis, environmental health and health communication.

Solange Moreira graduated in Veterinary Nursing. Currently she is working as a veterinary nurse at a veterinary clinic. Her interests are parasitology and zoonoses.

Rui Leandro Maia holds a Ph.D. in Sociology and Fundamental Methodologies (2002). He is Associate Professor at Fernando Pessoa University. He is researcher at CITCEM/FLUP—Transdisciplinary Culture, Space and Memory Research Centre and at FP-ENAS—UFP Energy, Environment and Health Research Unit. His main research interests are in the filed of Demography, Fertility, Migration, Urban and Environmental Sociology.

Chapter 11
Climate Change and the Increase in Disasters Related to Hydrometeorological and Climatic Events in Mexico: Main Impacts on the Population

Gabriela Narcizo de Lima, Roberto Ariel Abeldaño Zuñiga, and Deysi Ofelmina Jerez-Ramírez

Abstract Mexico is highly vulnerable to climate change and extreme hydrometeorological and climatic events, mainly due to the rapid growth of urban centers, deterioration of rural areas, and its high levels of poverty. Between 1970 and 2021, 210 disastrous hydrometeorological and climatic events were registered in the country, which generated a total of 6,728 deaths, 3,176 injured people and 7,636,419 affected, being Veracruz, Oaxaca, Chihuahua, and Sonora the most impacted states. Despite the efforts made to create strategies aligned with the Sendai Platform, Mexico continues to present alarming indicators of vulnerability and exposure to different threats, including those of climatic and hydrometeorological origin. One of the main future challenges for the country is finding a way to encourage the participation of communities in the transformation of structures, linking the different social actors in the process of generating public policies on disaster risk and climate change.

Introduction

In the last few decades, climate change has been placed on the public agenda of many countries and organizations due to the fact that it is one of the main challenges that

G. N. de Lima
Faculty of Letters, Geography Department, Porto University, Via Panorâmica s/n, 4150-564 Porto, Porto, Portugal

R. A. Abeldaño Zuñiga
University of Sierra Sur, 70805 Oaxaca, Mexico
e-mail: rariel@unsis.edu.mx

D. O. Jerez-Ramírez (✉)
Research Institute in Risk Management and Climate Change, University of Sciences and Arts of Chiapas, 29039 Tuxtla Gutiérrez, Mexico
e-mail: deysi.jerez@unicach.mx

W. Leal Filho et al. (eds.), *Climate Change and Health Hazards*, Climate Change Management, https://doi.org/10.1007/978-3-031-26592-1_11

humankind is facing these days. Such a phenomenon is usually regarded as a global problem bringing about regional consequences; it is a situation which has changed, impacted and modified natural surroundings, and therefore, it has created an imbalance in ecosystems and natural resources, which in turn, has produced alterations in human systems both socially and economically.

In the case of Mexico, the risks related to the adverse effects of climate change are significant, since the country presents climatic and hydrological conditions, as well as a geographic location, highly vulnerable to changes in climatic patterns and a wide variety of hydrometeorological phenomena. Among them, the following stand out: winter storms, frosts, cold fronts, tropical cyclones, droughts, heat waves, dust storms, pluvial and river floods, hailstorms, snowfalls, among others (Prieto-González 2020; Neri and Magaña 2016).

Since the beginning of the twentieth century, more than 230 disasters have been triggered by natural hazards[1] (CRED 2021; Alcántara-Ayala 2019). Out of this total, 80% of them show a meteorological origin; most of which are linked to storms and floods that have caused impacts involving costs of over 31,000 dollars; this means using 82% of the overall expense designated for damages caused by disasters of this type.

According to data provided by the National Center for the Prevention of Disasters (CENAPRED, in its Spanish acronym) (CENAPRED 2021), the largest amount of money allotted to damages-associated to atypical rains, extreme floods, hurricanes, tropical storms, severe draughts, and landslides—between 2000 and 2021—was spent on the states located on the coasts of the Pacific Ocean and the Gulf of Mexico. Consequently, it can be noticed that these regions are remarkably exposed to these types of phenomena in Mexico. As reported by CENAPRED (2021), the states of Veracruz, Guerrero, Tabasco, Chiapas, and Oaxaca use a high proportion of their expenditure to meet the needs of rebuilding damaged or destroyed infrastructures because of hydrometeorological events.

Disasters give us an idea of the measure of our vulnerability to extreme climate conditions which not only do they impact on our health and affect us; but also, cause human losses, together with material and structural devastation. All of which entail national and international economic demands. Considerable efforts are needed to help to adapt to these events to reduce damages in the short term, regardless of any climate change in the long term (Magaña 2013).

In this respect, the aim of this chapter is to characterize the patterns of occurrence of emergencies, contingencies and disasters related to hydrometeorological events in Mexico in the period from 1970 to 2021, considering the trends by states, and to carry out a balance of the effects (deceased and affected people) according to the type of event. Furthermore, the main forms of strategic governance frameworks existing in Mexico as well as their efficacy when implementing policies of Disaster Risk Reduction (DRR) and climate action have been considered.

---

[1] EM-DAT only considers "disasters" those events that comply with one of the following four criteria: (1) at least ten dead people, (2) at least one hundred affected people, (3) declaration of the state of emergency, and (4) a call for international assistance.

So as to achieve such an objective, precipitation rainfall and temperature sets of data have been analyzed (historic data and climate change scenarios) along with occurrence records of tropical cyclones happening near Mexican coasts (Pacific and Atlantic Oceans). Information for the latter was provided by the Climate Change Knowledge Portal (Royal Netherlands Meteorological Institute 2021) and the National Meteorological Service (SMN, in its Spanish acronym) (SMN 2021). Then, emergency, contingencies and disasters reports produced by the occurrence of hydrometeorological, and climate events have been reviewed per state in the period between 1970 and 2021. This review has been carried out based on the DesInventar disaster information management system (UNDRR 2021) and information supplied by CENAPRED (2021). As a consequence of such events, the balance of deceased and affected people has been developed on information provided by the International Disaster Database (EM-DAT) (CRED 2021), from the Centre for Research on Epidemiology of Disasters de la Université Catholique de Louvain, Belgium.

Finally, a description of legal and institutional frameworks that deal with risk management issues and climate change in Mexico is given, in addition to an analysis of the actual extent of the governance approach used in reaching these objectives.

Climate Trends of Mexico

As a whole, the climate of Mexico is characterized by great regional differences considering its topography and geographical position. Being located between two oceans, in areas with complex topography, the country is predominantly exposed to extreme hydrometeorological events such as: tropical cyclones, frosts, extreme temperatures, and so forth (Conde et al. 2020).

According to the SMN (2021), in 2020, the temperature in the country rose 1.4 °C regarding the period being referred to: 1981–2010; whereas on a global level the increase was 0.8 °C during that same period. In other words, the temperature in Mexico has been rising faster than in the rest of the planet. If we observe data from the Climate Change Knowledge Portal (Royal Netherlands Meteorological Institute 2021) about temperature anomalies in Mexico from 1901 to 2019 (Fig. 11.1), it can be noticed that as of the mid 1980's, positive anomalies turned frequent, and escalated gradually until 2019. At the end of the 1990's, positive anomalies surpassed 0.5 °C in relation to the average temperatures in the twentieth century until the mid-2000's. It was then that records of anomalies higher than 1 °C, likely to increase the incidence of extreme events like heat waves, began to be registered. In 2017, a thermal anomaly of 1.59 °C was recorded, reaching the highest value of the period under analysis here.

Temperature anomalies refer to the diversion of the annual median temperatures concerning the average ones in the twentieth century all through the country. However, now it is possible to identify meaningful increases of the impacts in the different regions in Mexico in a separate way. Among them are a 9.9% increase in the

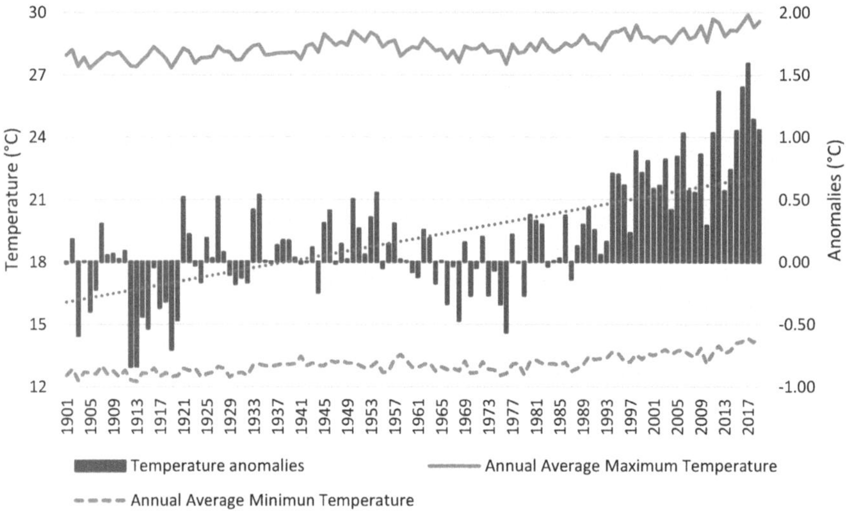

Fig. 11.1 Surface temperature anomalies, annual average maximum temperature, and annual average maximum temperature for México between 1901 and 2019 (*Source* Royal Netherlands Meteorological Institute 2021)

annual average of "hot days" and a 5.6% of "hot nights",[2] which affect the northern part of the country (Martínez-Austria and Bandala 2018).

In the period under analysis (1901- 2019), a positive constant ascending trend in maximum and minimum median temperatures can be observed around the country (Fig. 11.1), with totals of 2.5 °C and 2 °C, respectively. Positive upward trends have been more defined as of the mid 1970's mainly because of the rapid growth of urban centers in the diverse regions of the country during this period. The interest in studying the increase in temperatures, especially in relation to extreme maximums, derives from its effects on human health, particularly when heat waves are produced (Martínez-Austria and Bandala 2018).

Annual median precipitation in Mexico is approximately 750 mm (1960–2015). In the northern end of the country precipitations are lower than 50 mm per month throughout the year, while the southern and center high lands experience a humid season from June to October. On the other end of the country, the southern regions experience an average of 550 mm per month (SMN 2021). As stated by the Royal Netherlands Meteorological Institute (2021), annual rainfall between 1903 and 2019 had a statistically meaningless increase of around 7 mm per decade, being better defined by the data related to the rainy season and tropical cyclones in the country from May to November exclusively (Fig. 11.2). Moreover, the proportion of rain that took place during strong events increased in an average of 1.2% per decade. This shows that, even if total precipitations per year do not show a clear escalating trend,

[2] Hot days" or "hot nights" are defined by the temperature exceeded on 10% of days or nights in the current climate of that region and season.

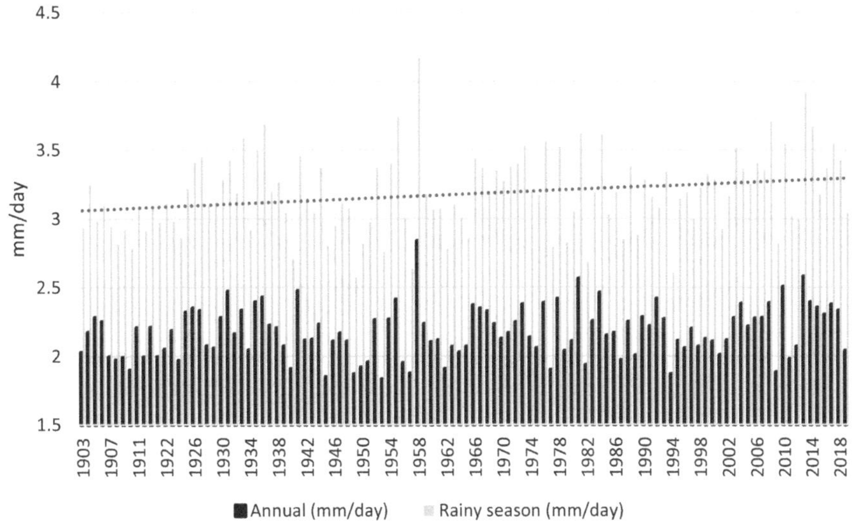

Fig. 11.2 Daily millimeters of rainfall for México between 1903 and 2019—Annual and Rainy season (*Source* Royal Netherlands Meteorological Institute 2021)

rains in the country are falling in a more concentrated way, which, in turn, also causes droughts to spread and increase.

Tropical cyclones are one of the primary causes of devastating disasters in Mexico. Conforming to data provided by the SMN (2021), between 1997 and 2020, a total of 834 cyclones happened (hurricanes, tropical storms, and tropical depressions) near Mexican coasts; out of which, 433 were registered as developing in the Pacific Ocean, and 401 in the Atlantic Ocean (Figs. 11.3 and 11.4). In both cases, we can see a clear rising trend in the total number of events in the last few decades. It is noteworthy that the number of hurricanes has been going up since 2010.

It should also be mentioned that in the last two decades, not only has the frequency of these phenomena increased, but their intensity too, due, principally, to ocean warming. By the same token, a high probability of the occurrence of diverse disasters is envisaged on the country coasts above all.

Future climate depends on how much the atmospheric composition and the properties of the earth change. Possible futures can be described through the various possible scenarios that may be constructed to study the potential consequences of climate change and explore the viable options to limit it. RCP's are the trajectories of the effects of greenhouse gas concentration adopted by the Intergovernmental Panel on Climate Change (IPCC). On its evaluation report (AR5) (IPCC 2014), it describes different climate scenarios, all of them considered possible, depending on the volume of effects produced by greenhouse gas emissions in the next few years.

As stated by scenario RCP4.5 (Fig. 11.5), a general warming is expected everywhere in the country by 2100, reaching values up to 2 °C, which are added to the increase that has already been registered in the last decades; especially in the central

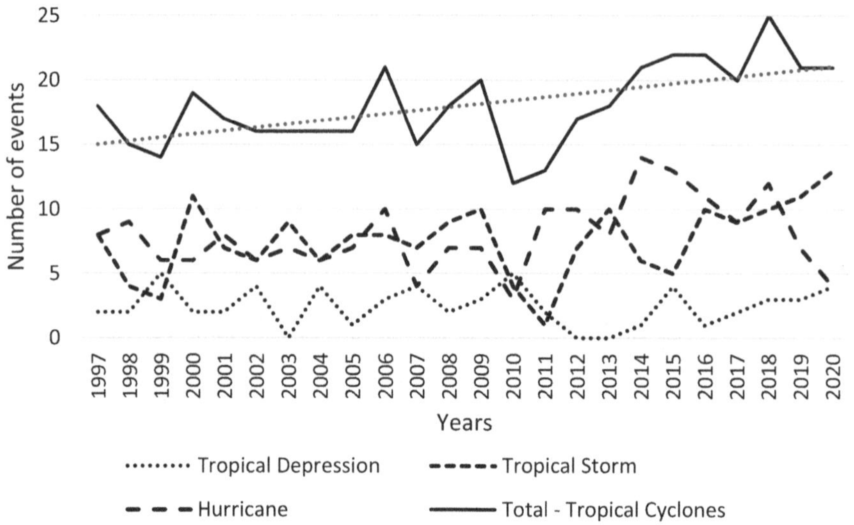

Fig. 11.3 Tropical cyclones in the Pacific Ocean between 1997 and 2020 (*Source* SMN 2021)

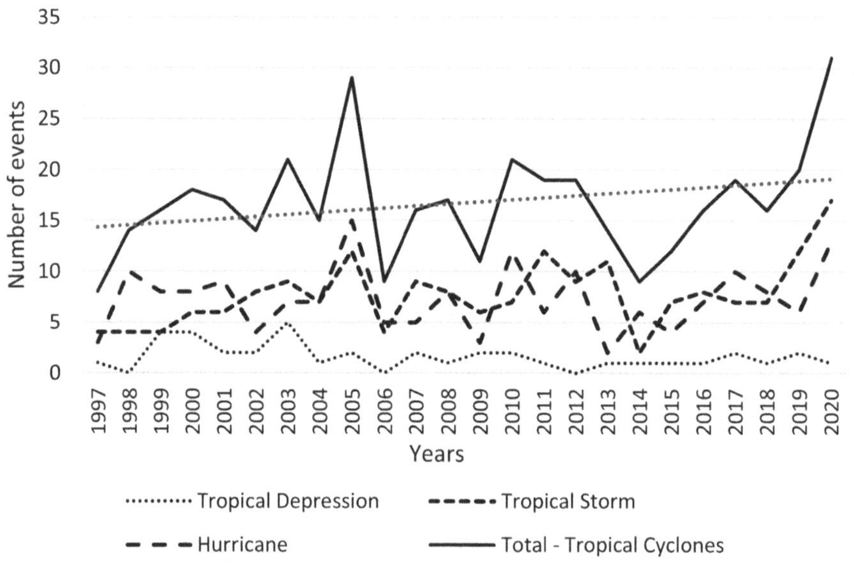

Fig. 11.4 Tropical cyclones in the Atlantic Ocean between 1997 and 2020 (*Source* SMN 2021)

and northern regions in Mexico and in a narrow strip on the border of Guatemala. In scenario RCP6.0 (Fig. 11.5), the same pattern repeats itself, but the zones with a higher expectancy of temperature rise will be very vast, reaching the Pacific and Atlantic coasts and regions spreading towards the south of the country.

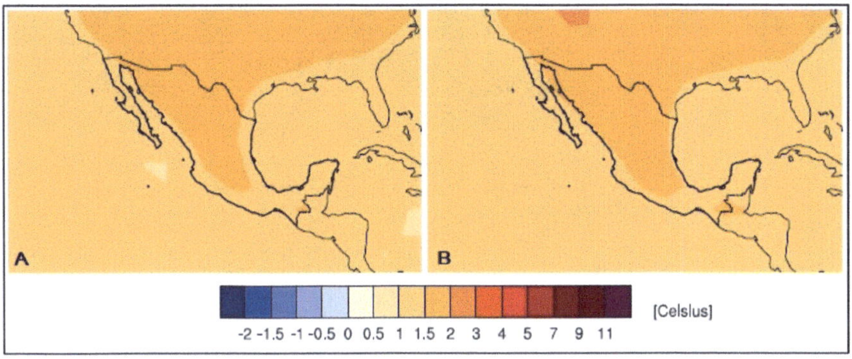

Fig. 11.5 Mean RCP4.5 (**A**) and RCP6.0 (**B**) temperature 2020–2100 minus 1986–2019 (January–December/AR5 CMIP5 subset)—México (*Source* Royal Netherlands Meteorological Institute 2021)

Rainfall predictions pose a considerable degree of uncertainty since they depend on factors related to the natural changeability of climate. It is predicted that rainfall variability will rise in the majority of regions in the country. In addition, it is very likely that moderate to severe draught events will be more frequent. On the other hand, there is an increasing trend of episodes of concentrated rainfall, which could produce serious floods, particularly during the rainy season, apart from recurrent tropical cyclones in the region (SMN 2021).

As reported by scenario RCP4.5 (Fig. 11.6), the whole country will be impacted by a decrease in precipitations by 2100 (in comparison to the period between 1986 and 2019). A drop of up to a 0.2 mm/day is expected, particularly, in the southeast and the Mexican Pacific coast. For the RCP6.0 scenario, there is more certainty in terms of the possible consequences caused by climate change in regard to precipitations in Mexico. Still, as a whole, zones that will have a precipitation decrease will be wider, while in the central plateau there will be a surge of around 0.1 mm/day by 2100.

Emergencies, Disasters and Contingencies Related to Hydrometeorological and Climate Events in Mexico

DesInventar (UNDRR 2021)—the management system of information about disasters- and CENAPRED (2021) are the most reliable sources of information in respect of disaster, emergencies and contingencies reports generated by natural threats on a local level in Mexico. DesInventar (UNDRR 2021) is a worldwide database, while CENAPRED (2021) owns data exclusively about Mexico. Both are systems that allow a homogeneous input and analysis of different events. They also offer a possibility for georeference information including a municipality level or something of the like. Both systems have institutional basis and information compiled

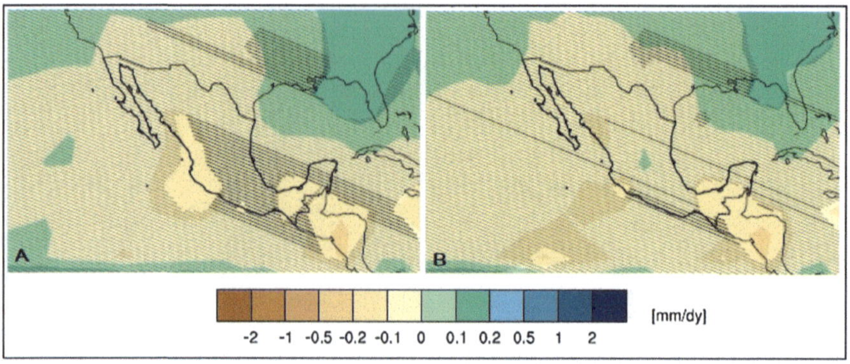

Fig. 11.6 Mean RCP4.5 (**A**) and RCP6.0 (**B**) precipitation 2020–2100 minus 1986–2019 (January–December/AR5 CMIP5 subset)—México (*Source* Royal Netherlands Meteorological Institute 2021)

through newspaper archives of free flow of information. Because of the methodology used in collecting information on both platforms, one same event seen as a threat might be reported in different municipalities for several days. For this reason, registers on these databases are more extensive than on the international disaster EM-DAT database (CRED 2021) used in the next section.

It is worth mentioning that data provided by DesInventar (UNDRR 2021) and CENAPRED (2021) were selected for analysis in this section because not only do they encompass disastrous events—where deceased and affected people are quantified—but they take emergencies and contingencies into account as well. The latter do not necessarily cause life losses or serious damages to infrastructures and society. Yet, they can become more dangerous if we consider that climate change suggests an increase in the frequency and intensity of extreme hydrometeorological and climate events that have been seen in other parts of the world other than Mexico, in the last few decades.

This section uses data from both sources for the period between 1970 and 2021. The compilation of information deals with events such as: draughts, low temperatures (including frosts, snowfall, and hailstorm) floods, storms (tropical cyclones, strong winds, rain, and tornadoes), and extreme temperatures (involving heat waves).

51% of the total number of reported events between 1970 and 2021, are related to severe storms (Fig. 11.7), normally characterized by tropical storms, hurricanes and intense rainfall in Mexico. All of these storms generate big impacts, chiefly on the Pacific coast and Gulf of Mexico, in states like Veracruz reporting 4,559 events, and Oaxaca reporting 4,667. The rising trends in the frequency of occurrence of tropical cyclones in the last twenty years, are an important reason for the rise of disasters and emergencies of this kind mostly in the poorest communities in the country, since they can barely meet the financial and operational costs of these phenomena.

Draughts and low temperatures are the second and third principal causes of disasters and emergencies in Mexico. They account for 18 and 17%, respectively, of

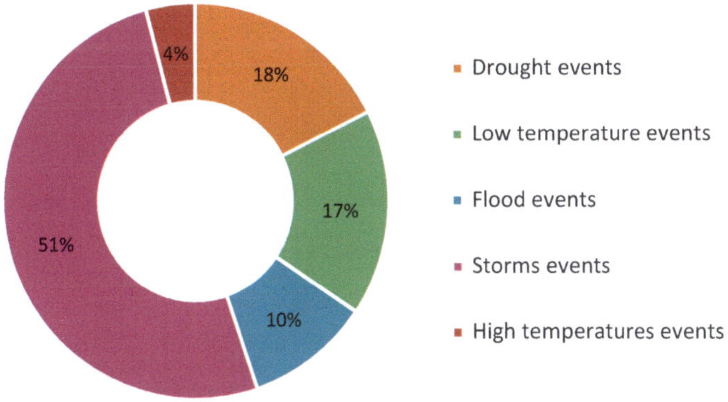

Fig. 11.7 Percentage of declarations by events in Mexico between 1970 and 2021 (*Source* UNDRR 2021; CENAPRED 2021)

all reported hydrometeorological and climate events in the period under analysis, primarily affecting the states of Chihuahua, Sonora, Zacatecas, San Luis Potosi and Oaxaca (Fig. 11.8). The increase in the occurrence of draught events in the last two decades is worrisome, in as much as approximately 71% of all collected records, occurred during this period, Oaxaca being the most impacted state, reporting 2,193 records.

The most important reported floods in Mexico between 1970 and 2021 were primarily due to the passage of tropical cyclones in the country. A recent example were the strong floods that started in October 2020 and mainly affected the southeast of Mexico, causing at least 28 deaths and thousands of affected people throughout the zone. The first major floods were reported in the states of Campeche, Yucatán and Quintana Roo in early October, due to the unusual occurrence of three almost consecutive cyclones, Tropical Storm Gamma and hurricanes Delta and Zeta. The most severe impacts were registered in November, affecting the states of Tabasco, Chiapas and Veracruz attributable to the interaction between Tropical Storm Eta and Cold Front number 11, which caused torrential rains, and overflowing of 10 rivers, including Usumacinta and Grijalva (SMN 2021).

Among the states most commonly affected by floods are Veracruz, Oaxaca, Chiapas and Tabasco on the Pacific and Atlantic coasts, and the State of Mexico and Mexico City in the center of the country (Fig. 11.8). The case of the last two states is relevant since floods are closely related to the occurrence of concentrated rain for short periods. Furthermore, these rains are becoming more frequent in this region of the country during the summer months, when urban infrastructure raises the vulnerability of the population to these events.

Temperature anomalies linked to extreme temperatures and heat waves are relatively frequent in Mexico. Likewise, they generally strike the northern states, where the highest temperature increases are expected in the next decades, according to the climate scenarios presented. Some of the states that reported events of this type, from 1970 to 2021, are Chihuahua, Sonora, Coahuila, Veracruz, Chiapas and Yucatan.

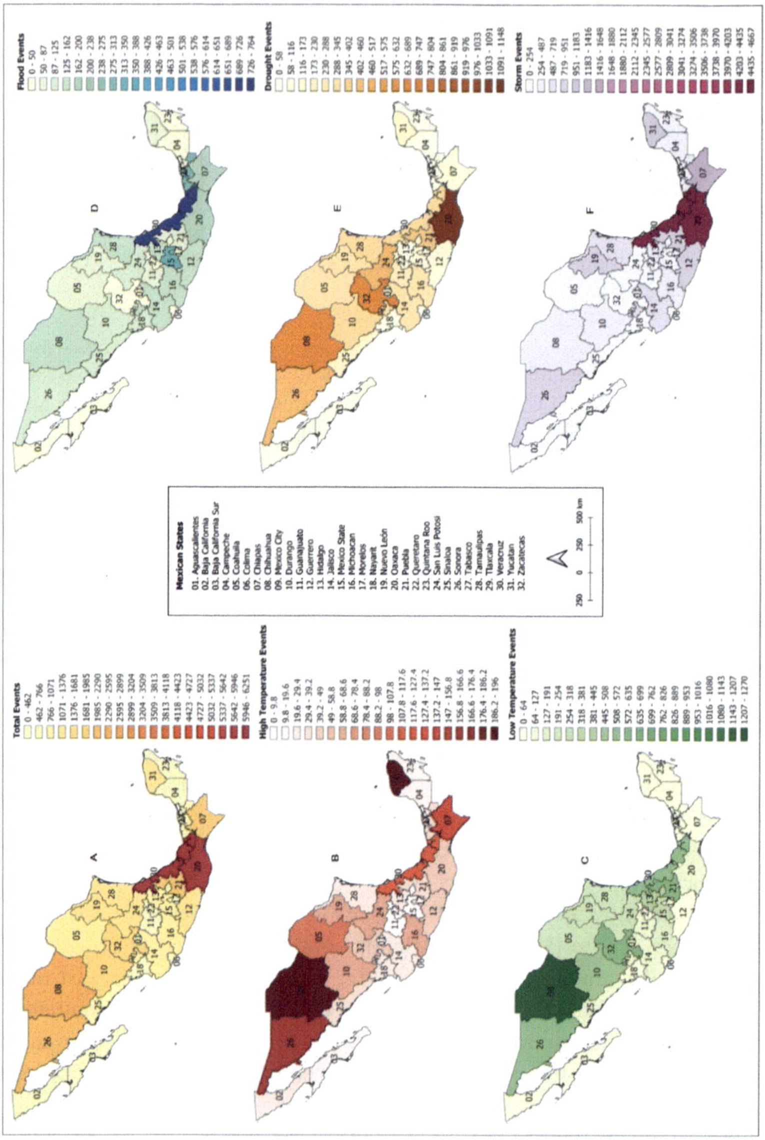

Fig. 11.8 Reports of disasters, emergencies and contingencies in Mexico between 1970 and 2021. **A** Total events, **B** High temperatures, **C** Low temperatures, **D** Floods, **E** Droughts and **F** Storms (*Source* UNDRR 2021; CENAPRED 2021)

In a special report on the impacts of 1.5 °C global warming, regarding pre-industrial levels, IPCC (2018) states that any increase in global warming will have an impact on human health, in a negative way for the most part. Studies show there will be fewer risks with 1.5 °C than with 2 °C concerning morbidity and lethality associated with heat; what is more, there is a high probability of a rise in vector-transmitted diseases such as malaria and dengue, which pose potential changes in terms of geographical scope.

Main Impacts of Disasters in Mexico

In this section, the most adverse disaster impacts derived from hydrological, climatological and meteorological threats on the population of Mexico are analyzed for the years between 1970 and 2021. Trends for longer periods of time in Mexico and threats from different causes have been examined in former studies (Abeldaño Zúñiga and González Villoria 2018) but have not been focused on the effects of the events from hydrological, meteorological and climatological origins.

In order to achieve the goals of this study, EM-DAT data sources have been consulted as they have been developed and kept current by the Center for Research on Epidemiology of Disasters (CRED) from the Université Catholique de Louvain, Belgium (CRED 2021). As reported by CRED guidelines, events can be classified in: (a) Hydrological events: identified by the occurrence, movement and either fresh and salt, or surface and ground water distribution, in addition to floods, landslides and surges of water; (b) Meteorological events: caused by risks of climate and atmospheric extreme conditions of short term, from micro to mesoscale, lasting from minutes to days; such phenomena include extreme temperatures, fog and storms; (c) Climatological events: caused by atmospheric processes of long duration, from mesoscale to macroscale; they range from intra-seasonal climatic variability to multiple decades, among which are mentioned draughts, glacier retreat and forest fires (CRED 2021).

EM-DAT data base has been consulted to identify the impacts of these types of events in Mexico during the period of interest of this chapter- and to recognize the most frequent events in Mexico as well as the most affected states. In Fig. 11.9, 210 events records have been identified within the period being analyzed in Mexico. 105 (50%) correspond to storms, followed by 69 (32.9%) floods, 15 (7.1%) extreme temperature events, and 10 (4.8%) landslides. It can be observed that storms and floods records have been rising decade after decade.

In Table 11.1, some effects of disasters on the Mexican population's health can be noted between 1970 and 2021. The total number of deaths by climatological events has been 50; the ones caused by hydrological events has been 2,435, while deaths caused by meteorological events have amounted to 4,243. Storms have caused the death of 3,145 people, whereas floods have killed 2,221 people. Regarding injured people, records should be taken cautiously because of data underrepresentation.

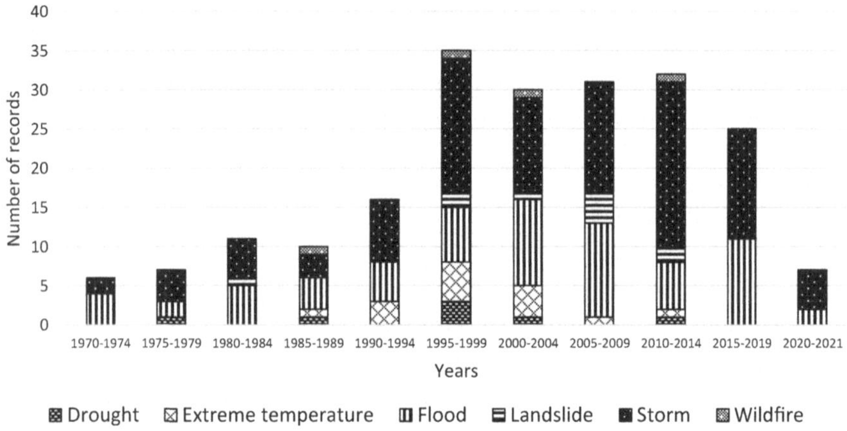

Fig. 11.9 Climatological, Meteorological, and Hydrological disasters recorded in Mexico. 1970–2021 (*Source* CRED 2021)

CRED accepts this variable. Nonetheless, 3,176 injured people have been reported in the whole period.

Significance of Governance in Disaster Risk and Climate Change Issues in Mexico

UN Special Representative of Secretary-General for Disaster Risk Reduction and head of the United Nations Office for Disaster Risk Reduction (UNDRR) has highlighted Latin America's commitment to be one of three regions where more than half of its states have informed the creation and start-up of strategies aligned with Sendai Framework (2015–2030) within the context of the commemoration for the International Day for Disaster Reduction in 2020.

Nevertheless, the regional evaluation report on disaster risks for Latin America and the Caribbean (UN 2021), presented in March 2021, pinpoints the limited effects of many propositions and actions for risk management that have been developed on the international cooperation agenda.

The region continues to pose alarming indicators of vulnerability and exposure to diverse types of natural threats like hydrometeorological and extreme climate events. They alter life, health, and the most susceptible and least prepared populations' means of livelihood in this zone.

DRR governance intends to deal with these systemic risks (UN 2019) that simultaneously have an effect on the three dimensions that maintain sustainable development: economic, social and environmental. Basically, this approach aim is to influence on the creation of organized webs for decision-making in regard to reduction and

Table 11.1 Total deaths, injured, affected people and homeless by disaster type in Mexico. 1970–2021

Years	Event group	Event type	Deaths	Injured	Affected	Homeless	Total affected
			Sum	Sum	Sum	Sum	Sum
1970–1974	Hydrological	Flood	202	97	250,000		250,097
	Meteorological	Storm	30				
1975–1979	Hydrological	Flood	6		67,000		67,000
	Meteorological	Storm	759	1400	525,000	100,000	626,400
1980–1984	Hydrological	Flood	129	200	125,000		125,200
		Landslide	50				
	Meteorological	Storm	604		310,000	32,500	342,500
1985–1989	Hydrological	Flood	76	12		200	212
	Meteorological	Extreme temperature	23				
		Storm	288		100,000	25,000	125,000
1990–1994	Hydrological	Flood	133	100	246,300	76,790	323,190
	Meteorological	Extreme temperature	460			16,000	16,000
		Storm	92		73,800	23,200	97,000
1995–1999	Climatological	Drought			65,000		65,000
		Wildfire	50				
	Hydrological	Flood	1086	160	571,200	86,000	657,360
		Landslide	24			120	120
	Meteorological	Extreme temperature	456				
		Storm	692	280	783,815	116,750	900,845
2000–2004	Hydrological	Flood	241	90	63,850	3000	66,940
		Landslide	10		200		200
	Meteorological	Extreme temperature	151				
		Storm	56	230	554,200	800	555,230
2005–2009	Hydrological	Flood	103	14	1,829,170		1,829,184
		Landslide	59				
	Meteorological	Extreme temperature	5				
		Storm	88		3,725,271	330,000	4,055,271
2010–2014	Climatological	Drought			2,500,000		2,500,000
	Hydrological	Flood	156		1,460,000	40,000	1,500,000
		Landslide	71				

(continued)

Table 11.1 (continued)

Years	Event group	Event type	Deaths	Injured	Affected	Homeless	Total affected
			Sum	Sum	Sum	Sum	Sum
	Meteorological	Extreme temperature	3		120,000		120,000
		Storm	334	135	1,508,525		1,508,660
2015–2019	Hydrological	Flood	59		54,442	6120	60,562
	Meteorological	Storm	160	458	51,702	2341	54,501
2020–2021	Hydrological	Flood	30		95,100		95,100
	Meteorological	Storm	42		4106		4106

Source CRED (2021)

prevention by means of reshaping the interaction between social and governmental agents.

As reported by Sendai instrument (UN 2015), climate change is classified as an intervening, underlying factor that drives risk disaster conditions all over the world in a significant way. Along these lines, the notion of governance has permeated the debate on "good government" for the management, mitigation and adaptation of anthropogenic climate variability. COP 2021 agreement for keeping global warming below 2 °C sets the pace for a new paradigm geared towards voluntary national actions[3] and climate governance processes (Córdova-Bojórquez and Romo-Aguilar 2015).

In the case of Mexico, public agenda on DRR and climate action are kept isolated (Jerez-Ramírez and Pinzón De Hijar 2017; Aragón-Durand 2010) so are attempts to include a governance plan for these purposes.

Decisions concerning risk reduction fall on the Ministry of Government (SEGOB, in its Spanish acronym) and are guided by the General Law of Civil Protection (Ley General de Protección Civil 2012), law-making framework that does not surpass the reactive approach and the technical rationality for disaster management. From 2019 onwards, CENAPRED has been assigned as a regulatory entity for Sendai framework in Mexico with the purpose of following up on the various goals for this international plan. However, conservative paradigm, which continues to sustain political action for DRR in the country, has hindered effective participation of different social agents and collective praxis for risk governance.

Management of policies and strategies in light of climate change, through a governance scheme, has proven to be a national challenge. Climate action -directed by the General Law of Climate Change (Ley General de Cambio Climático 2012) and the National Strategy of Climate Change (SEMARNAT 2013)—has not solved the necessity of practical procedures, in other words, those procedures that can enable

[3] This change in paradigm has been backed up since the failure of Kyoto Protocol and GHG emitters resistance to assume historic responsibilities.

an understanding among the various social and economic agents and governmental direction, for the purpose of adopting long term measures in terms of mitigation and adaptation (Rueda-Abad et al. 2017).

It is important to underline that the Mexican case is not alone in the regional context. Latin America and the Caribbean embody a complex territorial scope for the implementation of a governance model. Devised by external logics, this model ignores a series of tensions rooted in the relationships of power manifested in their territories. The intended "neutrality" observed on an international intervention for DRR, falls short in a region where risks and social vulnerability become part of the expected outcomes of a prevailing model of development. This way, evaluation of the actual accomplishments of risk governance for the continent and, specifically for Mexico, does not turn out positive in view of important social, economic, political, and institutional particularities which should frame the debate.

Conclusions

Climate change is a real phenomenon, and its impacts can be noticed in Mexico and the world. According to data shown in this chapter, positive anomalies over 1 °C were observed in the country as of the mid-2000's. Therefore, it is evident that the threat of irreversible impacts is a lot more immediate than what was expected several years ago.

- Between 1901 and 2019 a rise of 2.5 °C in the maximum median temperature and of 2 °C annual minimum median was recorded. By 2100, an increase between 1.5 °C and 2 °C is projected in the country's median temperature, which will add up to the values already registered.
- In the last few decades, it has been noticed that rains in Mexico are occurring in a more concentrated way during short periods. This event could spread and increase draughts. By 2100, a diminishing trend of precipitations with values up to 0.2 mm/day is predicted.
- From 1997 to 2020, a total of 834 tropical cyclones have been reported near Mexican coasts. Also, an upward trend in the frequency and intensity of these phenomena has been observed in the last 20 years, which increases the possibility of disaster occurrence throughout the country.
- It has been observed that climate patterns alterations in the country may generate an increase in the impacts by heat waves, on the one hand, mainly due to a higher frequency of extreme temperature events. On the other, an increment on the geographical scope of several vector-transmitted diseases, like malaria and dengue is foreseen.
- Most emergencies, disasters and contingencies triggered by hydrological and climate events have been recorded in the Mexican states of Veracruz, Oaxaca, Chihuahua and Sonora.

- Tropical cyclones, strong winds, rains, and tornadoes are the main cause for reports on emergencies, disasters and contingencies related to hydrometeorological and climate events in the country.
- Hydrometeorological and climate threats, along with an increase in vulnerability on the national territory, are key to understanding some recent disasters. Between 1970 and 2021, 210 disastrous events were recorded. These reports registered 6,728 deaths, 3,176 injured people and 7,636,418 impacted people.
- Despite efforts made for the creation of strategies in conjunction with Sendai Framework (2015–2030), the region of Latin America and the Caribbean, specifically, Mexico, continue to show alarming signs of vulnerability and exposure in view of diverse threats, such as those that stem from a climate and hydrometeorological origin.
- The creation of the General Law on Climate Change and the National Strategy of Climate Change has not influenced the development of practical instruments that generate the participation of different actors in the elaboration and implementation of mitigation and adaptation measures to climate change and its long-term impacts in the country. Thus, the construction of governance for DRR does not seem effective in national contexts, where there is a lack of participation instruments.

Future Prospects

The creation of effective strategies for the purpose of facing climate change and making a difference in DRR requires political will, a long-term vision, and social commitment. One of the main future challenges is to find a way to encourage the participation of communities in the transformation of structures and to link the different actors—civil society, the scientific community and decision makers—in the process of generating public policies.

References

Abeldaño Zúñiga RA, González Villoria AM (2018) Desastres en México de 1900 a 2016: patrones de ocurrencia, población afectada y daños económicos. Rev Panam Salud Pública 42. https://doi.org/10.26633/RPSP.2018.55

Alcántara-Ayala I (2019) Desastres en México: mapas y apuntes sobre una historia inconclusa. Investigaciones Geográficas 100:e60025. https://doi.org/10.14350/rig.60025

Álvarez-Miño L, Taboada-Montoya R (2021) Effects of climate change on public health 2015–2020: a systematic review. Rev Esp Salud Publica 95

Aragón-Durand F (2010) La adaptación al cambio climático en ciudades a través de la reducción del riesgo: Hacia un esquema articulador. In: Delgado GC, Gay C, Imaz M, Martínez MA (eds) México frente al cambio climático. Retos y oportunidades. Universidad Nacional Autónoma de México, pp 139–152

Centre for Research on the Epidemiology of Disasters—CRED (2021) EM-DAT, the international disaster database. https://public.emdat.be/data. Accessed 6 Apr 2021

Centre for Research on the Epidemiology of Disasters—CRED (2019) The international disaster database (EM-DAT). Sch Public Heal Univ Cathol Louvain. http://www.emdat.be. Accessed 18 Apr 2021

CRED C for R on the E of D (2021) Guidelines I EM-DAT. https://www.emdat.be/guidelines. Accessed 6 Apr 2021

Centro Nacional de Prevención de Desastres—CENAPRED (2000–2021) Sistema de Consulta de Declaratorias. http://www.atlasnacionalderiesgos.gob.mx/apps/Declaratorias/#. Accessed 10 de Apr 2021

Conde C, Luyando E, López-Díaz F et al (2020) Riesgos hidrometeorológicos en México. In Martínez-Austria PF, Corona-Vásquez B, Patiño-Gómez C (eds) Cambio climático y riesgos hidrometeorológicos, 1st edn. Fundación Universidad de las Américas, pp 47–62

Córdova-Bojórquez G, Romo-Aguilar ML (2015) Gobernanza climática: Actores sociales en la mitigación y adaptación en el estado de Coahuila, México. Nóesis. Revista de Ciencias Sociales y Humanidades 24(47–2):129–146

Intergovernmental Panel on Climate Change—IPCC (2014) AR5 Climate change 2014: impacts, adaptation, and vulnerability. Part B regional aspects. Cambridge University Press. https://www.ipcc.ch/report/ar5/wg2/. Accessed 10 de Apr 2021

Intergovernmental Panel on Climate Change—IPCC (2018) Global Warming of 1.5°C.An IPCC Special Report on the impacts of global warming of 1.5°C above pre-industrial levels and related global greenhouse gas emission pathways, in the context of strengthening the global response to the threat of climate change, sustainable development, and efforts to eradicate poverty. IPCC. https://www.ipcc.ch/site/assets/uploads/sites/2/2019/06/SR15_Full_Report_High_Res.pdf. Accessed 20 Apr 2021

Jerez-Ramírez D, Pinzón De Hijar J (2017) La reducción del riesgo de desastres y el cambio climático: Un abordaje epistémico. In: Rueda Abad JC, Gay García C, Ortiz Espejel B (eds) Gobernanza climática en México. Aportes para la consolidación estructural de la participación ciudadana en la política climática nacional, vol. II. Restos y opciones. Universidad Nacional Autónoma de México, PINCC, pp 165–186

Ley General de Protección Civil. 6 de junio de 2012. Diario Oficial de la Federación, 2–20. http://www.diputados.gob.mx/LeyesBiblio/ref/lgpc/LGPC_orig_06jun12.pdf. Accessed 5 Apr 2021

Ley General de Cambio Climático. 6 de junio de 2012. Diario Oficial de la Federación, 1–49. https://www.profepa.gob.mx/innovaportal/file/6583/1/ley_general_de_cambio_climatico.pdf. Accessed 5 Apr 2021

Lieber M, Chin-Hong P, Kelly K, et al (2020) A systematic review and meta-analysis assessing the impact of droughts, flooding, and climate variability on malnutrition. Glob Public Health 1–15. https://doi.org/10.1080/17441692.2020.1860247

Magaña V (2013) Guía metodológica para la evaluación de la vulnerabilidad ante cambio climático. INE/PNUMA, Ciudad de México, México

Martínez-Austria PF (2020) Riesgos hidrometeorológicos en México. In: Martínez-Austria PF, Corona-Vásquez B, Patiño-Gómez C (eds) Cambio climático y riesgos hidrometeorológicos, 1st edn. Fundación Universidad de las Américas, pp. 65–86

Martínez-Austria PF, Bandala ER (2018) Heath waves: health effects, observed trends and climate change. In: Sallis JP (ed) Extreme weather. IntechOpen, pp 107–123. https://doi.org/10.5772/intechopen.75559

Neri C, Magaña V (2016) Estimation of vulnerability and risk to meteorological drought in Mexico. Weather Clim Soc 8:95–110. https://doi.org/10.1175/WCAS-D-15-0005.1

Prieto-González R (2020) Riesgos hidrometeorológicos en México. In: Martínez-Austria PF, Corona-Vásquez B, Patiño-Gómez C (eds) Cambio climático y riesgos hidrometeorológicos, 1st edn. Fundación Universidad de las Américas, pp. 31–44

Royal Netherlands Meteorological Institute (2021) KNMI Climate Explorer. https://climexp.knmi.nl/start.cgi. Accessed 10 de Apr 2021

Rueda-Abad JC, Gay García C, Ortiz Espejel B (2017) La Gobernanza Climática en México: Aportes para la consolidación estructural de la participación ciudadana en la política climática nacional, vol. II. Retos y opciones. Universidad Nacional Autónoma de México, PINCC
Secretaría de Medio Ambiente y Recursos Naturales—SEMARNAT (2013) Estrategia Nacional de Cambio Climático visión 10-20-40. México. https://www.gob.mx/cms/uploads/attachment/file/41978/Estrategia-Nacional-Cambio-Climatico-2013.pdf. Accessed 20 Apr 2021
Servicio Meteorológico Nacional—SMN (2021) Información Histórica. https://smn.conagua.gob.mx/es/ciclones-tropicales/informacion-historica. Accessed 10 de Apr 2021
United Nations—UN (2015) Marco de Sendai para la Reducción del Riesgo de Desastres 2015–2030, Sendai. https://www.unisdr.org/files/43291_spanishsendaiframeworkfordisasterri.pdf. Accessed 18 Apr 2021
United Nations—UN (2019) Global assessment report on disaster risk reduction 2019, Geneva, Switzerland, United Nations Office for Disaster Risk Reduction. http://www.preventionweb.net/gar09. Accessed 18 Apr 2021
United Nations—UN (2021) Regional assessment report on disaster risk in Latin America and the Caribbean (RAR 2021), Geneva, Switzerland, United Nations Office for Disaster Risk Reduction. https://www.undrr.org/publication/undrr-roamc-regional-assessment-report-disaster-risk-latin-america-and-caribbean-rar. Accessed 18 Apr 2021
United Nations Office for Disaster Risk Reduction—UNDRR (2021) DesInventar Sendai . https://www.desinventar.net/DesInventar/results.jsp
United Nations Framework Convention on Climate Change—UNFCCC (2015) Paris Agreement of the United Nations Framework Convention on Climate Change. United Nations Framework Convention on Climate Change, Paris, France
Weilnhammer V, Schmid J, Mittermeier I et al (2021) Extreme weather events in Europe and their health consequences—a systematic review. Int J Hyg Environ Health 233:113688. https://doi.org/10.1016/j.ijheh.2021.113688

Gabriela Narcizo de Lima is Full-time Assistant Professor at the Department of Geography of the Faculty of Letters, University of Porto, Portugal. She holds Doctor of Science (Physical Geography) from the University of Sao Paulo (USP), Brazil, with a master's and bachelor's degree in Geography from the São Paulo State University "Julio de Mesquita Filho" (UNESP), Brazil. Between 2016 and 2018, Professor Gabriela Narcizo de Lima carried out two postdoctoral research stays in the Department of Territorial Planning and Geoprocessing of the Institute of Geosciences and Exact Sciences of Rio Claro/Sao Paulo (UNESP) and the Institute of Geography of the National Autonomous University of Mexico (UNAM). Her main lines of research are related to the impacts of anthropic activities on climate at various scales and issues associated with managing water resources and land use planning.

Roberto Ariel Abeldaño Zuñiga has a BSc in Nursing from the National University of Salta, Argentina. Also, he has a Magister in Public Health and a PhD in Demography from the National University of Córdoba, Argentina. He has been in academic stays at the Salvador Allende School of Public Health of the University of Chile and the University of Sao Paulo, Brazil. Currently, he works as a Full Professor at the University of Sierra Sur (Mexico) and Researcher at the National Research Council (CONACYT, Mexico). His main research interests are Public health and the environment. He has been a consultant for the United Nations Development Program, the United Nations International Organization of Migrations, and The World Bank.

Deysi Ofelmina Jerez-Ramírez is Full-time Research Professor at the Research Institute on Risk Management and Climate Change, University of Sciences and Arts of Chiapas, Mexico. Professor Jerez-Ramírez belongs to level I of the National System of Researchers of the National Council of Science and Technology in Mexico. She holds a PhD in Political and Social Sciences from the National Autonomous University of Mexico (UNAM), with a Master's degree in Social Work, and a Bachelor's degree in Social Work from the Industrial University of Santander (Colombia). Recognized with the Serge Moscovici 2019 research prize for the best international doctoral thesis on social representations. She has a research stay at the Pontifical Catholic University of São Paulo, Brazil. Her main lines of research are related to the socio-environmental problems of mining, climate change, social study of disasters and social implications of the exploitation of celestial resources.

Chapter 12
Climate Change and Mental Health in Bangladesh: A Cultural Variability Perspective

Nazwa Tahsin, Weronika Z. Gąsior, and Agnieszka Olszewska-Guizzo

Abstract Bangladesh has been increasingly facing emblematic consequences of climate change, such as salinity intrusion, rising sea levels and episodes of drought, floods and intense cyclones. The natural hazards in the region have a direct effect on the physical health of the population, which has been well documented in research. However, the mental health risks are understudied and provide a stronger challenge for researchers due to the cultural intricacies of the issues (gender-related, linguistic, ethnic and religious) and the intangible character of mental health itself. Additionally, traumatic events present different mental health challenges than those relating to long-term issues, further increasing the complexity. In this chapter, we discuss the impact of climate change on mental health in the Bangladesh context through the lens of cultural variability, showcasing the different social, cultural and political aspects at play which determine, not only the health and well-being of the community, but also the climate-change-related mental health research landscape in Bangladesh. We pay additional attention to the impact of the COVID-19 pandemic on the most vulnerable population and as a pivotal moment in the research perspectives and approaches. We focus on research available both in English and Bangla and consider further developments in this field.

N. Tahsin (✉) · W. Z. Gąsior · A. Olszewska-Guizzo
NeuroLandscape, 03-252 Warsaw, Poland
e-mail: n.tahsin@neurolandscape.org

W. Z. Gąsior
e-mail: w.gasior@neurolandscape.org

A. Olszewska-Guizzo
e-mail: a.o.guizzo@neurolandscape.org

W. Leal Filho et al. (eds.), *Climate Change and Health Hazards*, Climate Change Management, https://doi.org/10.1007/978-3-031-26592-1_12

Introduction

Climate change is an irrefutable fact, with clearly visible consequences on the planet and its inhabitants, despite remaining denial in politics and questionable pursuits of development activities in the global south (Schipper et al. 2021). Climate change presents risks around not only tangible factors, but also for less perceptible dimensions like mental health, well-being, happiness and hope. In continuation, we intend to understand the variables that often remain underestimated when analysing issues of mental health and well-being, specifically considering the cultural, historical and geopolitical context of Bangladesh. While Bangladesh is a "young" country achieving independence in 1971, the documented history and knowledge in different domains of the area dates more than 2500 years (BBS 2021a, b). In our discussion, we will focus on the significant elements of the society, history, architecture, philosophy, and geopolitics that affect religion, language, economy and, in result, disaster preparedness, climate adaptation and well-being.

According to most recent reports, nearly 30 million people have been displaced every year due to extreme climatic events at the end of 2020 (IEP 2021) and Bangladesh often leads the infamous list of climatic threats. The capital city of Dhaka alone has reportedly given refuge to over 8.3 million internally displaced climate migrants (Hossain 2022). Additionally, the country has been hosting over 700 thousand of Myanmar's Rohingya refugees in the largest refugee camp in the world. While the geographical position is a great contribution, this chapter will discuss other key factors that increase the climate change induced risks in Bangladesh that have a further effect on the overall well-being and mental health of the population, as shown in Fig. 12.1.

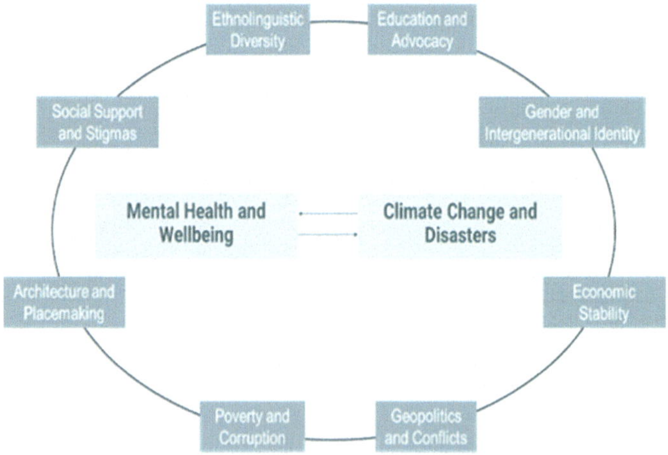

Fig. 12.1 Socio-economic drivers and links with mental health and climate change in Bangladesh

Research Approach

There are ample studies on climate change health hazards in the region of South Asia, which describe, analyse and discuss the issues within the disciplines of economics, sociology, epidemiology, urbanism, and many more. However, not many studies can be found which interpret the health risks from the cultural variability perspective. The aim of this chapter is to present an overview of the characteristics of Bangladesh in the light of climate change risks from a socio-cultural frame of reference. In doing so, we follow the methodology of literature review, available both in English and Bangla language in order to provide a comprehensive description and a critical evaluation of the relevant issues. We limit the discussion in terms of depth in order to provide a rather broad picture of the different points of influence and connect various disciplines (e.g. cultural studies, linguistics, etc.) that can inform the future directions of research and action in addressing climate-related health hazards in Bangladesh.

Physical Geography of Bangladesh

Bangladesh lies in the north-eastern part of South Asia, bounded by India, Myanmar, and the Bay of Bengal. The country has a very high population density of 1265 person/km^2 with a total beyond 160 mln. Most land of Bangladesh consists of flood-plain delta of the Ganges–Brahmaputra-Meghna River basin, with hilly areas mainly in the eastern terrain (BBS 2021a). There are 405 rivers and numerous other water bodies (Bangladesh Water Development Board 2020) (Fig. 12.2).

Natural Hazards and Climate Change

Just two decades ago, Bangladesh was described as having a moderate climate, with six distinct seasons. Now, the climate has shifted into a warm one, with only four seasons, which are also disappearing as summers elongate. The entire terrain of Bangladesh is prone to different types of floods. The frequency and magnitude of these are exacerbated by climate change and transboundary water conflicts (Islam et al. 2021a). Paired with man-made factors, floods also cause landslides in hilly areas (Ahmed 2021). Heatwaves, droughts and river erosion are now proven to be significantly increasing in Bangladesh due to anthropogenic climate change (Hossain 2022). It has also made rainfall erratic, causing drainage failures in urban areas, while introducing extreme salinity intrusion and super cyclones in the coastal region. All relevant studies indicate that the occurrence and intensity of the hazards is increasing rapidly, and they are becoming more erratic (Islam et al. 2021b) (Fig. 12.3).

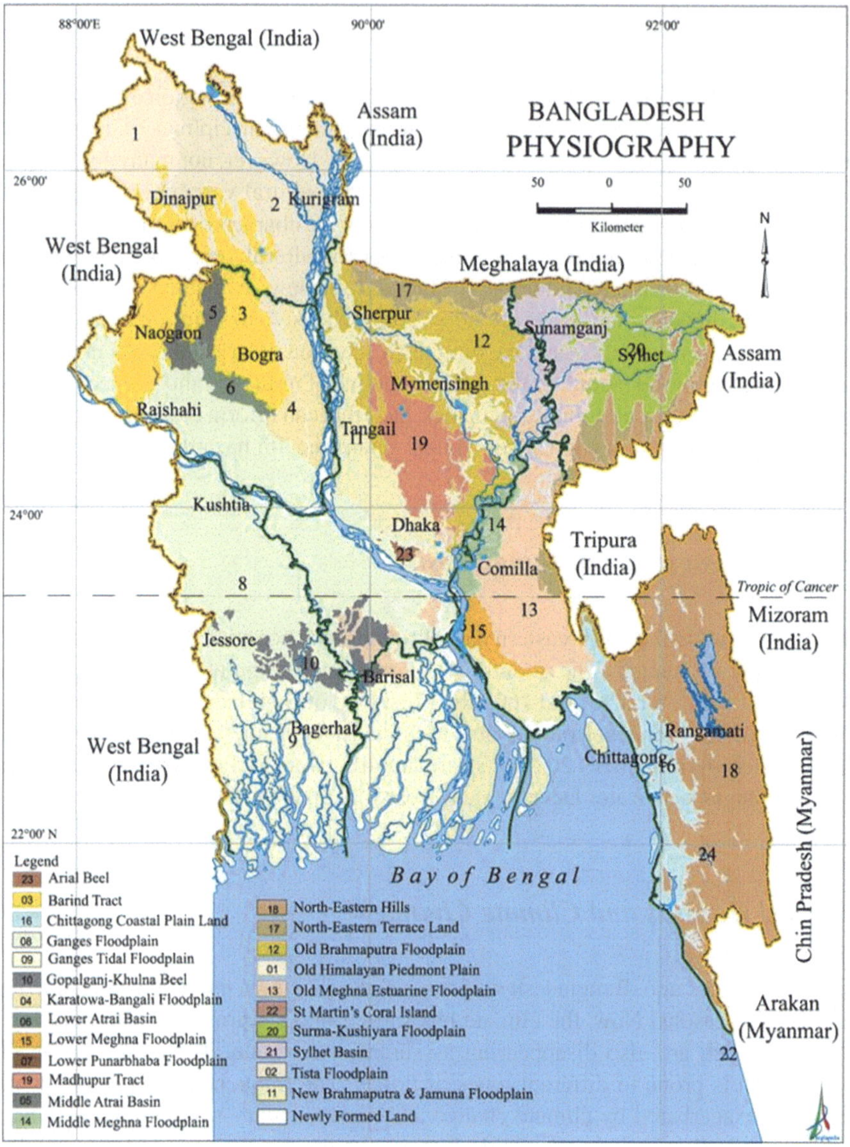

Fig. 12.2 Physiographic map of Bangladesh (Nasirkhan 2021)

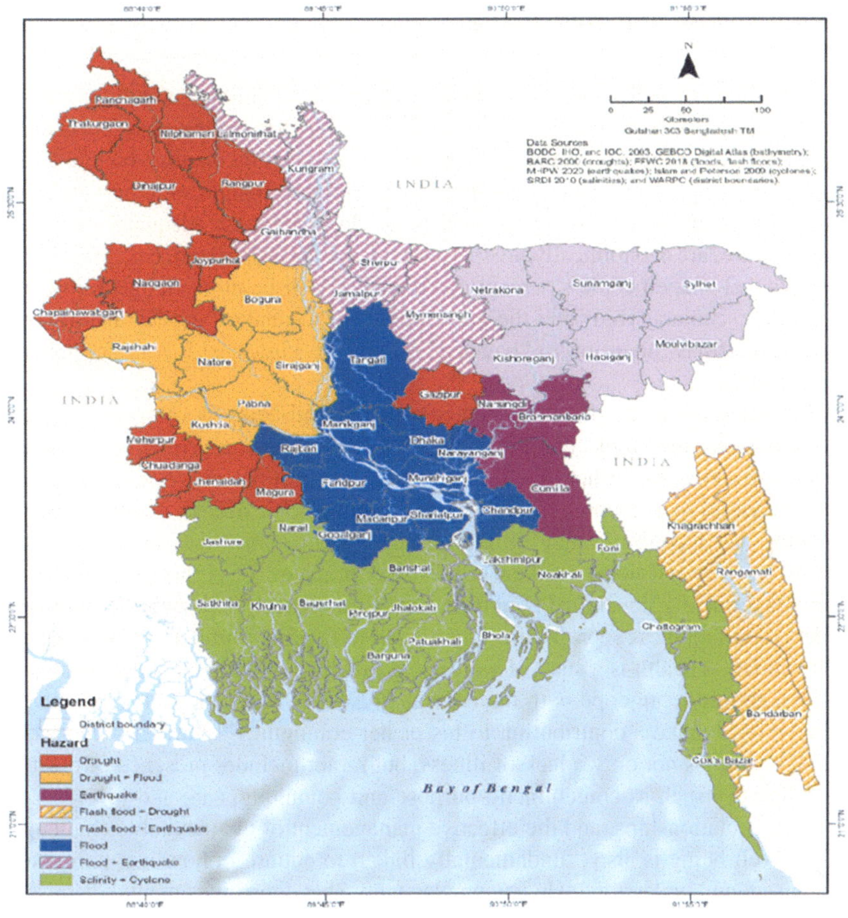

Fig. 12.3 Major hazards in different areas of Bangladesh (Asian Development Bank 2021)

Climate-change related natural hazards result in well-documented physical health risks, such as respiratory, dengue, water-borne diseases, malnutrition and growth stunting (Joarder 2021). Additionally, salinity intrusion results in limited access to drinking water and/or consuming saline water, leading to risk of hypertension, cardiovascular diseases and stroke (Shammi et al. 2019). The Bangladeshi government hence acknowledges climate change as a "health hazard threat" (Joarder 2021). Studies on mental health and well-being, however, are neither adequate nor comprehensive, which is shown in detail in the following sections.

Mental Health and Well-Being

The definitions of well-being and mental health are rather similar and intertwined. In Bangladesh, personal well-being is considered as an agglomeration of an individual's mind, body, environment, housing, happiness, economy and positive bonding with others. The scripts from ancient, mediaeval and modern Bangla suggest well-being is the highest in welfare societies, which are designed with adequate space and infrastructure, are economically prosperous and just towards all citizens. Most of the concepts that shape people's perception of mental well-being in Bangladesh come from vedic scholars, yogis and the philosophies of Chanakya and Khona. The Far Eastern traditions' attention to "balance" (similar to yin/yang) is also appreciated (Joarder 2021).

The contemporary academic concept of personal well-being, developed in the field of psychology, called subjective well-being (SWB), is "a person's cognitive and affective evaluations of his or her life" (Diener et al. 2002). A person's SWB comprises: life satisfaction (e.g. economic stability and meeting basic needs), positive feelings (enjoying life, having meaningful relationships) and negative feelings (having few chronic worries and being resilient). The main predictor of higher SWB is living in an economically developed country as struggle for food and shelter is minimal, while lower happiness score is strongly related to conflict. Similarly, following the WHO, "mental health is a state of well-being in which an individual realises his or her own abilities, can cope with the normal stresses of life, can work productively and is able to make a contribution to his or her community". Furthermore, "good mental health" is not only a lack of illness, but rather includes presence of positive characteristics such as contentment, purpose and belonging, care and love, having supportive relationships, and the effective management of the inevitable challenges of life. Well-being is also "fundamentally linked to culture in how it is expressed, recognised, and understood. How we understand what constitutes mental distress in any cultural group is important because this influences and determines how decisions are made in healthcare systems about interventions, and how these are measured to assess effectiveness" (BRAC 2021).

Both local and global concepts thus indicate climate change and disasters as a threat to the well-being of people by exacerbating poverty, inciting conflicts, displacements, and loss of spatiotemporal stability. However, the relation of well-being, happiness, placemaking and environment are still missing in modern West-centric research and the changes are slow in research on happiness. In the latest (2022) World Happiness Report, Bangladesh occupied place 94 (out of 146 countries). It is an important measurement for the country, since SBW is one of the indicators of UN SDG#3 regarding good health and Bangladesh aspires to prosper for global recognition. It has been slowly increasing its scores, oscillating around 5 (out of 10), with the goal being 7.6 threshold (Sachs et al. 2021). However, unlike the West, South Asian philosophy seeks for balance and harmony from a collectivist perspective and growth through criticism. Hence, research suggests that the cultural norms of modesty or self-criticism may lead Easterners/Asians to give *lower* ratings than

Westerners even when equally satisfied with life (Tov and Nai 2018). Recognition of place, environment and well-being in the concept of good health at global scale is a wisdom that the country offers to the rest of the world, elaborated in later sections.

Social Fabric of Bangladesh

Ethno-Linguistic Landscape

Bengalis are the ethnic majority of Bangladesh. The country also houses other ethnic groups, including *Adivasi* communities (indigenous ethnicities). There are two major strata of indigenous ethnicities—one residing in the Chittagong Hill Tracts (CHT) and are collectively the majority group of CHT; the other being ethnicities that reside in the plainlands and are the minority in that region. The latest government gadget has recognised 50 ethnic groups (*Cultural Establishment Act of Ethnic Minority Act 2010 (Amended)*). However, the census of 2021 was not updated, and the 2011 census acknowledged only 27 ethnicities, reporting 1.5 million of ethnic minorities, comprising 1.8% of the total population of the time. In fact, the International Work Group for Indigenous Affairs reports that the indigenous communities in Bangladesh claim to have a population around 5 million (Chakma and Chakma 2021). It is an important disparity, which reflects the political and cultural tensions within the country related to ethnic diversity (Fig. 12.4).

Right over land and language have been key conflicts among the Bengali majority and other ethnicities. The conflicts are rooted in property and land laws imposed

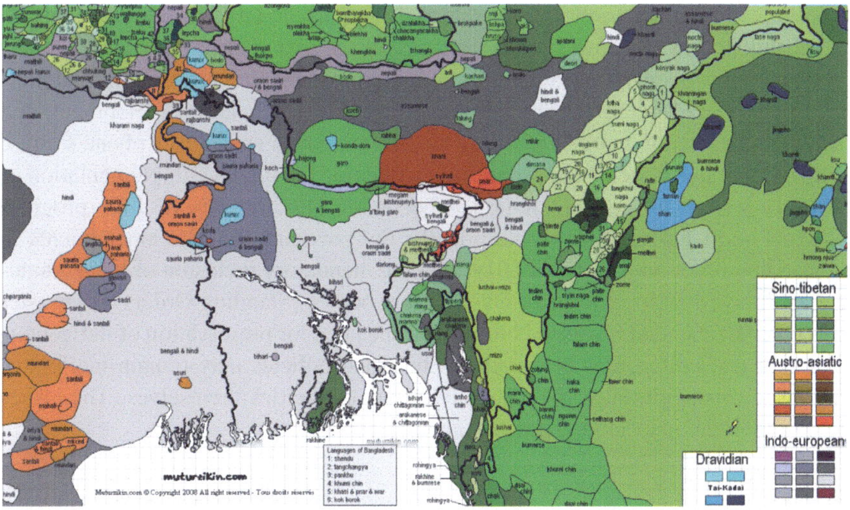

Fig. 12.4 Ethnolinguistic map of Bangladesh and neighbouring regions (Muturzikin 2008)

during the British colonial period, which did not align with the traditional mechanism followed by the indigenous communities, displacing many communities (Adnan and Dastidar 2011). As a result of this legacy, in the emerging context of prolonged poverty (since colonial destruction of economy in late 1700s) (Roy 2019) and land scarcity, exacerbated by overpopulation and climatic displacements, land grabbing by influential personnel and/or settlement of landless Bengalis became rampant, exploiting land rights of the indigenous people (Chakma and Chakma 2021).

Bangla has the status of the national language of Bangladesh, while little is done to preserve minority languages in the country. English has taken a similarly prestigious place in education and employment for global reach. Meanwhile, minority languages and local dialects of Bangla are reserved for home and family only. Consequently, younger generations are increasingly less proficient in their native languages/dialects, leading not only to language loss but also loss of local knowledge on environment, agriculture and climate. Minority languages are crucial carriers of local insights about climate adaptations and general ancestral knowledge, and great care must be taken to preserve them. Moreover, displacements caused by climate crises increase vulnerability of minority languages since languages die out when speakers relocate to different places where another language is spoken as a vernacular (Sultana 2021)—which is likely to occur in Bangladesh.

Indigenous communities in Bangladesh present different climate vulnerabilities in comparison to Bengali and non-indigenous groups, as the ecosystems of their settlements and the socio-political conditions are different. For instance, they are more likely to be landowners, but may endure greater illiteracy, as education in native language may be unavailable, lowering adaptability (Gunter et al. 2008). Beyond linguistics, ethnic minorities from plainlands experience prolonged discrimination in access to healthcare and job opportunities (e.g., higher bribes are expected from them after job quotas were eliminated in 2019) (IRI 2020). Ethnic representation in politics is minimal (only one member of the parliament) and the communities feel abandoned in the power structure to influence essential politics and policies. Discrimination against minority ethnic communities impede interventions and policies to provide adequate aid, infrastructure and services, health, development, and disaster management sectors. Hence, lack of trust in power and agency of one's future contributes to lower well-being scores. To overcome the distrust, implementation of the peace accord in the CHT region is becoming effective despite little progress (Chakma and Chakma 2021). Additionally, the government made amendments in the standing order on disaster in 2019, seeking mandatory inclusion of local ethnicities at all layers of local disaster management bodies (Standing Orders on Disaster 2019). In the amended Cultural Policy 2021, the need for preservation of indigenous languages and culture has been ratified, but such policies may remain ineffective without the recognition of the indignity of the ethnicities themselves. The issues therefore remain in a "complicated" status.

Religion

The main religion of the country is Islam, with minority presence of Hinduism, Buddhism and other local religions. In Bangladesh, the concept of *dharma* is used to signify more than religious affiliation of a person, meaning "to sustain, support, or uphold" and refers to the "proper cosmo-moral ordering of things" (Devine et al. 2019). A recent study into the connection between religion and happiness in Bangladesh found that people who live in communities where they are the majority religion (i.e. Muslim) are happier than the minority. This finding was further supported with higher happiness scores amongst those having a stronger political positioning for "having connections" to the in-group subordinate denomination (Devine et al. 2019). It is therefore imperative that research of religion, specifically in Bangladesh, distinguishes between intrinsic, spiritual aspects from extrinsically motivated religious beliefs which provide a stronger sense of structure, belonging and safety. Religion thus can be a double-edged sword which can play an important supportive or debilitating role in a person's and community's overall well-being, mental health, and happiness (Papaleontiou-Louca 2021). Finally, climate change may contribute to religious extremism. Despite the emergence of Bangladesh as a secular state in 1971, the partition of British ruled India based on the distribution of Hindus and Muslims succumbed the remaining Hindu populations in Bangladesh to becoming scapegoats of natural disaster, internal political tension and geopolitical appeasement (Mostofa and Brasted 2022). In face of calamities, Hindus feel insecure to migrate to Muslim dominated areas. Furthermore, poverty and unemployment, political corruption, external geopolitical support and poor local governance drove the rise of Islamic terrorism in the early 2000s (Hasan and Macdonald 2021), which was successfully eradicated later in the same decade with socioeconomic progress. Hence, extreme climatic events are likely to weaken the environment and economic stability, feeding into inequity and low well-being, fuelling the return of religious extremism, and making a positive feedback loop of regression, conflicts and displacements (Hossain 2022).

Gender

The region of Bangla from ancient to pre-colonial times consisted of multiple genders, reflected in various documentations in local languages. However, the contemporary health system is built for binary people, in terms of both service and infrastructure. Conservative Christian doctrines implemented during colonial times abolished the idea of gender diversity, which negatively affected the perception of well-being, livelihood, and human rights of locals (i.e., education, property rights, work permit), especially women and non-binary people (Khan 2021). Furthermore, the colonisers penalizing intersex, transgender (a part of *hijra*), and gay (*kothi*) identities through section 377, resulted in the discontinuation of their documentation, leading to further ostracism. Even after the recognition of the *hijra* communities and gender equity after independence, vague legal definitions, discriminations, and lack of understanding of gender fluidity remain, reflected in the design, distribution and structure of health infrastructure and services, beliefs and stigmas, and mental health service seeking behaviours (Khan 2021). This discrimination was also reflected in the negative experiences of these communities during the COVID-19 pandemic (Sifat 2020).

Regarding the status of women in Bangladesh, elaborate studies can be found. Bangladesh is now at the top in the South Asia region in terms of various gender indices, specifically for economic and educational progress (WEF 2021). However, Bangladeshi society has long shown preference for male offspring, since girls eventually leave the household when married (Asadullah et al. 2021) and, even in the recent past, financial struggles made families marry off young girls to reduce mouths to feed. Moreover, natural and economic collapse contributes amply to child marriage, especially for girls, who are more likely to suffer, as most ethnicities are patriarchal (except a few, i.e. Garo and Santals). Child brides are also highly susceptible to violence, illness and malnutrition (Mostofa Kamal and Aynul 2010; HRW 2015). Religious affiliations and settlement play a role in this regard too: Muslim and Hindu girls have the lowest median age for marriage in South Asia (16.7) (Saha 2016), while urban women tend to marry later than their rural counterparts (Mostofa Kamal and Aynul 2010; HRW 2015). However, daughters are considered good luck in Bangladesh and parents must offer shelter to daughters as required in pregnancies or facing violence. The country's economic growth enabled government and NGOs to use these positive sentiments towards daughters and marketed education as a panacea against poverty and violence in the long run. Educational reforms succeeded in building adjacent schools, providing stipends, free books and facilities exclusively in favour of girls. As a result, preference towards girls' education over boys' has been seen since 2011. Dropout of male students has been exceeding females in both primary and secondary education (BBS 2021b). Ironically, climate change consequences lead to higher risk of poverty and, therefore, gendered adversities on both male and female children. As salinity intrusion and land erosion multiplies, dropouts are expected to increase. Female children are less likely to attend classes just to collect water, while boys will be forced into labour (as seen during COVID-19) for retrieving household finances. A recent study has shown that children aged 7–12 years exhibit significant cognitive impairments due to increased consumption of saline water in the southwestern area (Akter 2019). As adults, women in Bangladesh exhibit a higher suicide rate than men (8.7 vs. 6.8 suicide deaths/100,000 population), with risks connected to pregnancy, malnutrition, illness in children, and domestic violence (WHO 2021). Recent statistics also show women are almost twice as likely to die from depression and other mental illnesses (1.13%) compared to men (0.6%) (BBS 2021a). Finally, a typical family in Bangladesh is multigenerational and provides the support for family members with mental health conditions through protection from any risks and participation in social activities and gatherings. On the other hand, the extended family may also be ostracised because of having a member with a mental illness (WHO 2021). In sum, despite progress regarding women empowerment and gender inclusion in Bangladesh, there is still a lot to be done, especially in the face of the climate and pandemic crises, which have been rather detrimental to the potential progress to be made in this area.

Cultural Strengths and Vulnerability in Bangladesh

According to Hofstede's classification of cultures, Bangladesh is traditionally a collectivist society (low score of 20 on Individualism scale), where it manifests in a close long-term commitment to the members of a group. Loyalty in a collectivist culture is paramount and overrides individual needs and other societal norms. The society fosters strong relationships where everyone takes responsibility for fellow members of their group (Hofstede 2022). This could explain why the COVID-19 pandemic lock-down and limitations on social contacts negatively impacted the Bangladeshi population, where research found that "about three in every four people in Bangladesh suffered from loneliness at any level of gravity during the lockdown period. Among them, 39% were suffering from moderate to severe loneliness" (Das et al. 2021). Furthermore, loneliness negatively affects overall well-being and can lead to depression (Fig. 12.5).

Power Distance dimension defines to what extent power inequalities are accepted in a culture. With a high score of 80 points, Bangladesh is a highly hierarchical society, where further justifications of social positioning are not sought out and accepted as they are. This score is also associated with highly centralised power structures, difficult to shift. A recent summary of the country's evolution regarding economic, political and governance transformations summarised its position as paying lip-service to democracy, but using legislation and beyond to actually maintain the status quo (BTI 2022). This also aligns with the dimension of Uncertainty Avoidance, which describes a nation's preference for action knowing that future is mostly unpredictable—and the anxiety this creates. With a high score of 60, Bangladeshi

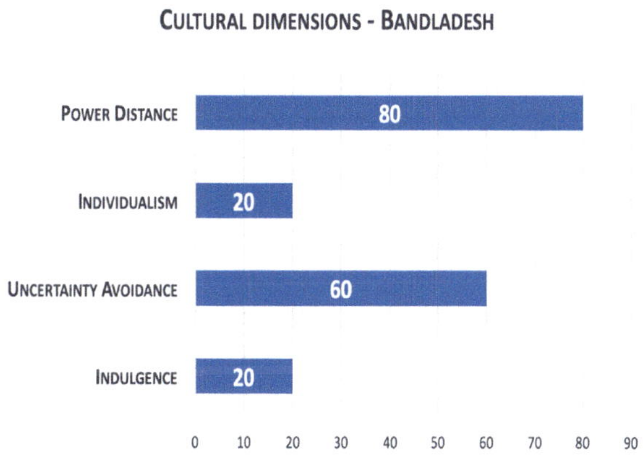

Fig. 12.5 Bangladesh cultural dimensions, adapted from Hofstede (2022)

culture leans towards avoiding uncertainty by maintaining loyalty to the existing social structures and for the safety of people (esp. of the same ethnicity), often putting the needs of others above personal. Finally, the Indulgence variable indicates the degree of attempt to control one's desires and impulses. With a low score of 20, Bangladesh is a restrained (not indulgent) country, where people feel controlled by social and religious norms and tend to put more emphasis on work over leisure. However, the context of Bangladesh is too understudied to interpret, as the social value of self-care and hobbies is paramount, but at the same time hard work and participation in collective festivities are expected.

The cultural dimensions of Bangladesh show potential health risks and solutions in the face of climate change. Firstly, having participatory welfare governance for most of history, a community-based approach where top-down rules are determined through bottom-up feedback seems essential. As such, the government must foresee the nature of challenges, which is expected to be communicated to the people for action strategy and determining the role of individuals and local implementers. At the same time, the guidelines must remain flexible enough to accommodate easy communication to intergenerational, cultural, religious and gender diverse populations. Secondly, regarding mental health, collective society provides intergenerational communications, offering knowledge on self-care. However, because the power over one's options in life is determined by the collective norms, economic status and organisation of the society, people (especially marginalised) may in fact feel powerless in taking care of their well-being and mental health individually. Adapting to societal expectations and roles leaves less space for putting one's individual needs first, which can hamper well-being. Lastly, context relevant knowledge of *vastushastra* (placemaking) to accommodate people's need of space for mental well-being amid climate crisis can be utilised. Stratified intergenerational housings are often economical and a great way to provide personal space while offering mental support and care to children and elders, especially for Bangladesh as the social services are still weak. The placemaking principles applied in traditional settlements often align with the climate, fulfil nutrition needs, and establish disaster and epidemic prevention, which can be studied to reform and utilise in rapidly urbanising parts of Bangladesh (Faruque 2018) (Fig. 12.6).

Given these cultural characteristics, the role of government, community organizations and NGOs is pivotal in providing services for communicating mental health distresses in changing climate by increasing social justice, eradicating discrimination and stigmas, reformation of placemaking for well-being, and access to appropriate and affordable mental health services.

Research on Mental Health and Climate Change

According to recent pre-pandemic data, the main mental health concerns in Bangladesh are depression, anxiety and high suicide rates (Naveed et al. 2020), paired with a very limited provision of mental health services, as well as social stigma related

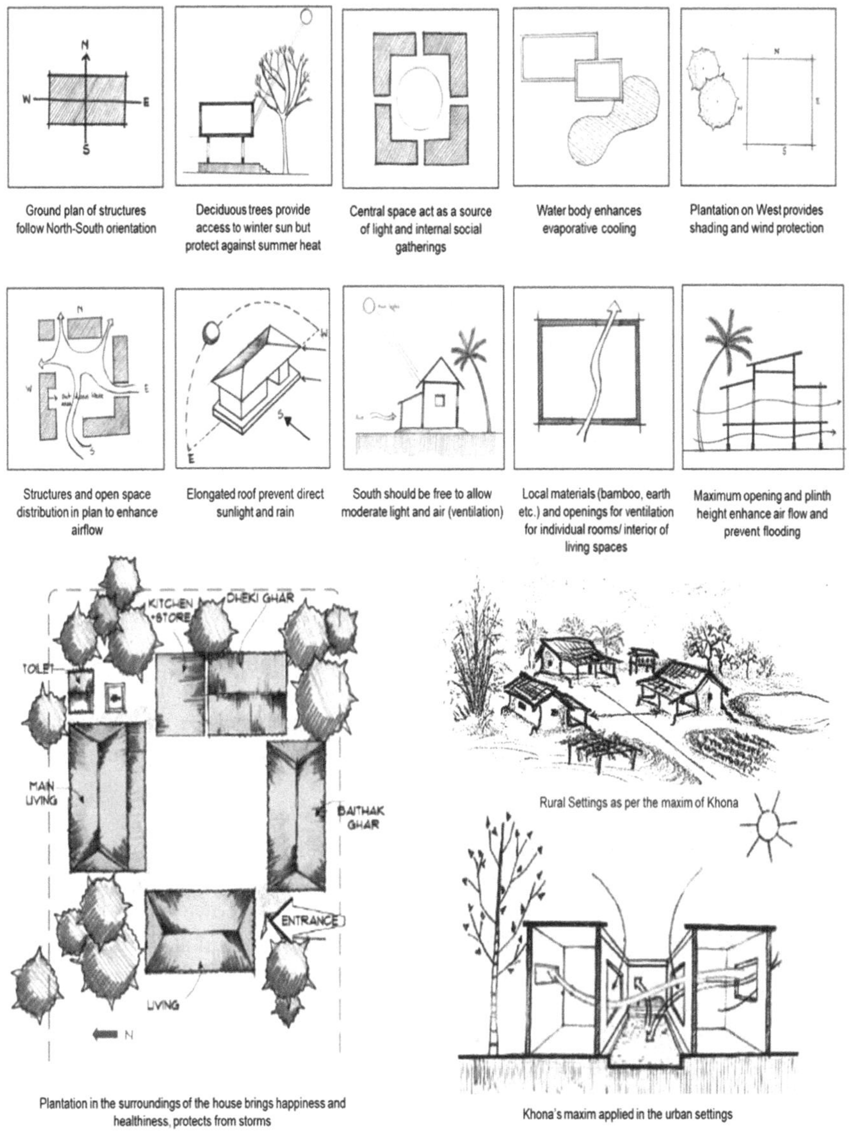

Fig. 12.6 Traditional principles of building a family residence for well-being, disaster and epidemic prevention through orientation, material selection, plantation, and bonding in rural and urban settings (Faruque 2018)

to mental health (WHO 2021). Substance abuse is also on the rise, with reported 8 million people using illicit drugs (Dhaka Tribune 2021). Mental health issues more specifically connected to climate change include PTSDs and sleep disorders related to natural disasters alongside stress, depression, anxiety and even "ecological grief" in chronic climate hardships (Hayward and Ayeb-Karlsson 2021). An astounding statistic in a national Mental Health Survey of 2018–2019 estimated that 92.3% of persons with diagnosable mental disorders were *not* receiving mental health treatment (WHO 2021). This is largely attributed, aside from the economic expense, to the stigma and subsequent lack of reporting. Because of the strong connection between mind and spirit, in many instances more spiritually oriented help is sought in traditional shamans and healers as opposed to medical help (ibid.). In addition, as the public health sector is not sufficiently subsidised for equitable healthcare distribution, most mental healthcare facilities are found in private, urban chambers and hospitals, which requires extensive out of pocket expenditure on health. The healthcare sector, therefore, is not structured to promote mental healthcare seeking behaviour (Joarder 2021).

The provision of mental health services in Bangladesh is run not only by national institutions, but also supported by international funds, NGOs, and private–public partnerships. However, the overall budget designated to mental health is minimal (0.05% of national health budget) and it is mostly (60%) dedicated to psychiatric hospitals. The country has very few mental health professionals, with an estimated 260 psychiatrists (0.16 per 100 k population), who work mostly in cities (WHO 2021). Nevertheless, a new Mental Health Policy, approved by the Ministry of Health in 2019, reflects a shift from a medical to a psychosocial treatment model with emphasis on decentralisation and community-based services and support for mental illness (WHO 2021). BRAC, the biggest NGO in Bangladesh, recognised in their Mental Health Strategy the need for mental health initiatives "grounded in an approach that espouses, endorses, and advocates for *culturally congruent*, respectful, and compassionate services" (BRAC 2021). It is difficult, however, to talk about culturally congruent strategies where one's identity and cultural self-determination is limited by law. Politicians from all major parties continue to deny the existence of indigenous ethnic groups in Bangladesh (Nafis 2021). It is clearly not a matter of semantics in a "tomayto—tomahto" way, because indigenous status relates so closely to a group's sense of belonging to ancestral land and the government's strategy is to prevent the possible interpretation of the ethnic minority status as a right to their land (ibid.). Consequently, the ethnopolitical issues intertwine with research and provision for mental health in Bangladesh.

COVID-19 in Bangladesh

The COVID-19 pandemic highlighted the interdependence of physical and psychosocial aspects of health in the human experience. A recent systematic review of prevalence of long-term mental health issues (depression, PTSD and sleep disturbance) in COVID-19 patients revealed that falling ill did not increase the long-term risks of the issues more than their predominance in the general public (Bourmistrova et al. 2022). Thus, any mental health deterioration could be more likely associated with the psychosocial effect of lockdowns and job-related factors rather than the virus itself (ibid.). In fact, the underlying determinant of high levels of anxiety, depression, and stress amidst the pandemic for the young population in Bangladesh was unemployment (Khan et al. 2021) and poor metabolic health (obesity), especially for females (Das et al. 2021). Consequently, the research suggests that the virus itself is not the actual culprit of mental health problems (Das et al. 2021; Khan et al. 2021; Bourmistrova et al. 2022), but it acts as a magnifying glass showing weak points in the social determinants of well-being. A similar metaphor has been used to describe COVID-19 not precisely causing, but exacerbating the existing disadvantages and retracting the improvements for instance in the realm of education in vulnerable societies around the world (Azevedo et al. 2021).

While overall poverty in Bangladesh has been steadily declining, the opposite is true for cities. This has been worsened by the COVID-19 pandemic, especially in the capital, as the city hosts 4 mln residents in slums, with inflow of climate migrants (Hasan and Macdonald 2021). Also, women are found to suffer more psychological distress than men when hospitalised due to COVID-19, which may be not only for medical reasons, but also due to the complex socio-psychological conditions, such as not being able to care for the family and missing out on income (also informal) while hospitalised (Kim et al. 2021). There have also been reported surges in rapes during the pandemic, with rural women being at higher risk of violence (Tasnim et al. 2021). The socioeconomic vulnerability of depression in women continues into older age, while higher levels of education have a protective effect (Rahman et al. 2020). The proportion of elderly population in Bangladesh has been increasing steadily and this trend is predicted to continue (BBS 2015). Luckily, the tradition of multigenerational families in Bangladesh, living under one roof, protects the elderly from the risks of poverty and loneliness, acting as a safety net also for their mental health, even in shortage of formal social support.

Discussion

Given the complexity of the phenomena presented, climate change highlights the power, ideology and identity struggles of communities, pertaining to their socio-cultural, ethnic, linguistic and geopolitical situation. The analysis of the different cultural dimensions influencing the mental health risks in the face of climate change in Bangladesh points towards a few clear key vulnerabilities. Firstly, the risks increase exponentially for females from ethnic minorities, living in poverty in the rural parts of Bangladesh. Given the strong influence of cultural dimensions that perpetuate these risks and inequalities, this cycle does not seem easy to break. However, Bangladesh has its own success stories with improving living conditions and working towards

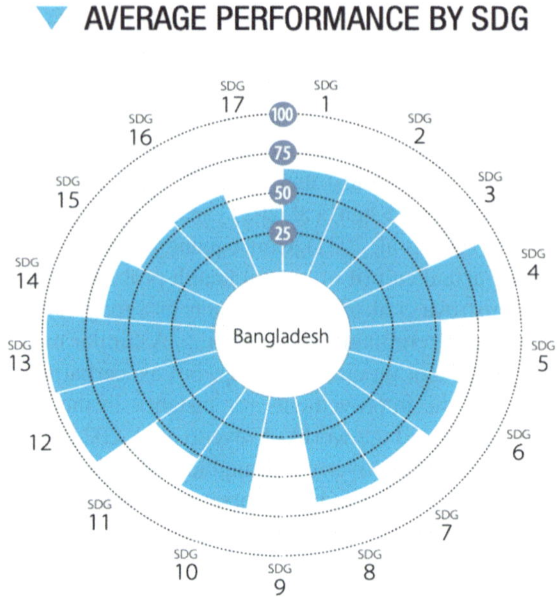

Fig. 12.7 SDG progress in Bangladesh (Sachs et al. 2021)

better quality of life, with the highest increase in SDG index scores since 2015 (Sachs et al. 2021) (Fig. 12.7).

Education and critical literacy have been called a key tool to eradicate helplessness and poverty (Falzon 2019). Literacy as a fundamental human right, provides access to education and employment, and lack of it leads to a lower quality of life (ibid.). Furthermore, lower literacy and skills are associated with poorer health and individuals seeing themselves as "objects rather than actors in political processes, and to have less trust in others" (OECD 2016) (trust is an important predictor of well-being and happiness). Bangladesh has achieved many milestones in education by significantly lowering drop-out rates, increasing quality and quantity of tertiary education, and addressing gender equality in access to education (BBS 2021b). While the progress may be commendable, the country still has not addressed the linguistic and cultural rights of indigenous ethnic communities (Sultana 2022).

Secondly, what impedes faster progress of Bangladesh in achieving SDGs, especially for climatic and health resilience is the combination of unwillingness to understand the context of the country by the international institutions, internal and external corruption, and illicit money practices (Sultana 2022). Bangladesh receives considerable international aid, but unfortunately a significant proportion of it is lost due to unscrupulous procedures, insufficient allocation for actual development work and overreliance on foreign over local professionals. The cultural theories of the origins of corruption link it to traditional authority structures and countries where the state plays the dominant role in the economy (Srirejeki 2020) as well as relationship-oriented cultures (Chandler and Graham 2010) and collectivism (Mazar and Aggarwal 2011). To address this, it may be imperative to understand the cultural motivations of tipping or gift-giving and segregating them from bribing to approach corruption in action plans from a less confrontational and more collaborative perspective.

Finally, another key hindrance in improving well-being and mental health in Bangladesh is lack of continuous and accurate data. Following the principle that we cannot improve what we do not measure, knowing the status quo of the various metrics that are key in a country's economic and social development, is paramount to design and implement the course of action for the improvement. Research must therefore take into account the sociocultural, linguistic and geographical peculiarities of the country in order to collect quality data to address climate-related risks.

Conclusion

As rightly pointed out in a recent systematic review of research into mental health and climate change in Bangladesh, the focus on mental (mostly ill) health from a western perspective is not enough and a well-being approach is rather more appropriate (Hayward and Ayeb-Karlson 2021), as it is a complex and rich concept, to a large extent cultural, existential, and context-specific. Moreover, climate change is challenging the status quo of how we perceive and measure mental health and well-being, as an overshadowing, lurky risk. Additionally, the COVID-19 pandemic has highlighted and worsened the issues already causing major mental health problems around the world and showing a need to re-examine the paradigm.

In the face of what appears to be an imminent climate crisis, for Bangladesh, addressing economic stability in the long run, gender equality and the provision of mental health services should be the key aspects to increase the well-being and mental health of the population. However, these actions must consider the unique social, cultural, religious, ethnic and linguistic dimensions of the Bangladeshi people and their land. Embracing the rich and diverse culture may be just what the country needs to really impel towards improved well-being and happiness, despite the risks of climate change it will inevitably face in the future.

References

Adnan S, Dastidar R (2011) Alienation of the lands of indigenous peoples in the Chittagong Hill tracts of Bangladesh. IWGIA, Copenhagen

Ahmed B (2021) The root causes of landslide vulnerability in Bangladesh. Landslides 18(5):1707–1720

Akter S (2019) Impact of drinking water salinity on children's education: Empirical evidence from coastal Bangladesh. Science of the total environment 690:1331–1341

Asadullah MN, Mansoor N Randazzo T Wahhaj Z (2021) Is son preference disappearing from Bangladesh? World Dev 140:e105353

Asian Development Bank (2021) Bangladesh climate and disaster risk Atlas: hazards, vol. I. https://doi.org/10.22617/TCS210518

Azevedo JP, De Rogers WF, Ahlgren H, Cloutier SE, Chakroun M-H, Chang B, Mizunoya G-Ch Reuge S, Brossard NJ, Bergmann M, Lynn J (2021) The state of the global education crisis: a path to recovery (English). World Bank Group, Washington DC

Bangladesh Water Development Board (2020) Analysis of water level along the Brahmaputra—Jamuna, Ganga-Padma & Surma-Meghna River System. http://www.hydrology.bwdb.gov.bd/img_upload/ongoing_project/757.pdf

BBS (Bangladesh Bureau of Statistics) (2015) Population monograph of Bangladesh—elderly population in Bangladesh: current features and future perspectives. Population Monograph, vol. 4. http://203.112.218.65:8008/WebTestApplication/userfiles/Image/PopMonographs/elderlyFinal.pdf

BBS (2021a) Report on Bangladesh sample vital statistics 2020, Ministry of Planning, Government of the People's Republic of Bangladesh http://bbs.portal.gov.bd/sites/default/files/files/bbs.portal.gov.bd/page/b343a8b4_956b_45ca_872f_4cf9b2f1a6e0/2021-06-28-07-11-c1784c6fe700cf52ff934e8dd7cf9147.pdf

BBS (2021b) Statistical yearbook Bangladesh. http://bbs.portal.gov.bd/sites/default/files/files/bbs.portal.gov.bd/page/b2db8758_8497_412c_a9ec_6bb299f8b3ab/2021-08-11-04-54-154c14988ce53f65700592b03e05a0f8.pdf

Bourmistrova NW, Solomon T, Braude P, Strawbridge R, Carter B (2022) Long-term effects of COVID-19 on mental health: a systematic review. J of Affective Disorders 299:118–125

BRAC (2021) BRAC Mental Health Strategy 2020–2030. http://www.brac.net/program/wp-content/uploads/2021/09/BRAC-mental_-health_strategy-2020-2030.pdf

BTI (2022) Bertelsmann Stiftung, BTI 2022 Country Report—Bangladesh, Bertelsmann Stiftung, Gütersloh

(Ministry of) Cultural Affairs of Bangladesh (2019) Cultural establishment act of ethnic minority Act 2010 (Amended)

Chandler JD, Graham JL (2010) Relationship-oriented cultures, corruption, and international marketing success. J Business Ethics 92(2):251–267

Chakma P, Chakma B (2021) The indigenous world 2021: Bangladesh, IWGIA, Dhaka

Das R, Hasan MR, Daria S, Islam MR (2021) Impact of COVID-19 pandemic on mental health among general Bangladeshi population: a cross-sectional study. BMJ Open 11(4):e045727

Deckers L (2013) Motivation: biological, psychological, and environmental, 4th edn. Routledge, New York

Devine J, Hinks T, Naveed A (2019) Happiness in Bangladesh: the role of religion and connectedness. J Happiness Stud 20:351–371

Diener E, Lucas RE, Oishi S (2002) Subjective well-being: the science of happiness and life satisfaction. In Snyder CR, Lopez SJ (eds) Handbook of positive psychology, Oxford University Press, New York, pp 63–73

Dhaka Tribune (n.d.) (2021) Home minister: all should work together against drug abuse. https://archive.dhakatribune.com/bangladesh/nation/2021/10/07/home-minister-all-should-work-together-against-drug-abuse

Falzon R (2019) Literacy and wellbeing. In Vella S, Falzon R, Azzopardi A (eds) Perspectives on wellbeing. Brill, Leiden, pp 79–95

Faruque F (2018) Geomantic study of homesteads in Bangladesh focusing on maxims of "Khona", MA Thesis, Bangladesh University of Engineering & Technology. http://lib.buet.ac.bd:8080/xmlui/handle/123456789/4972

Gunter B, Rahman A, Rahman AFM (2008) How vulnerable are Bangladesh's indigenous people to climate change? Bangladesh Development Research Center Working Paper

Haque M (2019) Indigenous knowledge and practices in disaster management: Experiences of the coastal people of Bangladesh. In: Zutshi B, Ahmad A, Srungarapati A (eds) Disaster risk reduction, Palgrave Macmillan, Singapore

Hasan M, Macdonald G (2021) How climate change deepens Bangladesh's fragility. United States Institute of Peace

Hayward G, Ayeb-Karlsson S (2021) 'Seeing with Empty Eyes': a systems approach to understand climate change and mental health in Bangladesh. Climatic Change 165:29
Helliwell JF, Layard R, Sachs JD, De Neve J-E, Aknin LB, Wang S (2022) World Happiness Report 2022, Sustainable development solutions network, New York
Hofstede G (2022) Hofstede insights—Bangladesh. https://www.hofstede-insights.com/country/bangladesh
Hossain A (2022) Visible climate impact: perspectives on Bangladesh (title translated from Bangla). Sharavan Prakashani, Dhaka
Hossain I, Mullick AR (2020) Cyclone and Bangladesh: a historical and environmental overview from 1582 to 2020. Intl Medical J 25(6):2595–2614
Human Rights Watch (2015) Child marriage: Bangladesh. https://www.hrw.org/reports/bangladesh_childmarriage_brochure_FINAL_web.pdf
Islam MN, Tamanna S, Noman M, Akhter E, Hasan MM, Rahman Rifat MA, Sakib EI, Yousuf BM, Khaled SM, Akangkha RA, Mohammad Riaz IS, Barua A, Mubashsara N, Abrar Jahin SM, Hossain F (2021a) Climate change impact and adaptation strategies in Bangladesh to strengthen regional cooperation. Springer Climate, pp 421–458
Islam MN, Tamanna S, vanAmstel A, Noman M, Ali MSS, Aparajita DM, Roy P, Tanha SR, Sarkar N, Ashiquzzaman M, Ghosh SK, Hasnat SR, Bhuiyan MF, Sayef M, Saha P, Hasan GR, Ghosh A (2021b) Climate change impact and comprehensive disaster management approach in Bangladesh: a review. Springer Climate, pp 1–39
IRI (International Republican Institute) (2020) The challenges facing plainland ethnic groups in Bangladesh: land, dignity and inclusion. https://www.iri.org/wp-content/uploads/2022/01/bangladesh-plainland-1_2.pdf
IEP (Institute for Economics and Peace) (n.d.) (2020) Over one billion people at threat of being displaced by 2050 due to environmental change, conflict and civil unrest (Report Press Release). Institute for Economics & Peace
IEP (n.d.) (2021) Ecological threat report 2021: Understanding ecological threats, resilience and peace, Sydney IEP. http://visionofhumanity.org/resources
Joarder T (2021) Autopsy of public health: COVID-19 and other issues (title translated from Bangla). The University Press Limited, Dhaka
Kim J, Kang H, Jhon M, Ryu S, Lee J, Kang S, Jung S, Shin I, Kim S, Stewart R, Kim J (2021) Associations between COVID-19 symptoms and psychological distress. Front Psychiatry 12:e721532
Khan KK (2021) Healthcare and the oppression of sexual and gender diverse communities in Bangladesh, PhD thesis, Massey University
Khan AS, Debnath S, Islam S, Zaman S, Ambia N-E, Barshan AD, Hossain MS, Tabassum T, Rahman M, Hasan MJ (2021) Mental health of young people amidst COVID-19 pandemic in Bangladesh. Helyon 7:e0717
Ministry of Disaster Management and Relief of Bangladesh (2019) Standing orders on disaster 2019, Ministry of Disaster Management and Relief. https://modmr.gov.bd/sites/default/files/files/modmr.portal.gov.bd/policies/7a9f5844_76c0_46f6_9d8a_5e176d2510b9/SOD%202019%20_English_FINAL.pdf
Mazar N, Aggarwal P (2011) Greasing the palm: can collectivism promote bribery? Psychol Sci 22(7):843–848
Mostafa Kamal SM (2010) Geographical variations and contextual effect on child marriage in Bangladesh. Pakistan J Women's Stud: Alam-e-Niswan 17(2):37–57
Mostafa Kamal SM, Aynul IMD (2010) Socio-economic correlates of malnutrition among married women in Bangladesh. Malaysian J Nutr 16(3): 349–359
Mostofa SM, Brasted H (2022) Islamic state's recruitment strategies: the case study of Bangladesh. In Ahmed I, Ahmed ZS, Brasted H, Akbarzadeh S (eds) Religion, extremism and violence in South Asia, Palgrave Macmillan, Singapore, pp 121–142
Munir MMH (2019) Bangladesh needs to fight against illicit finance flow and to achieve SDGs. https://www.socialwatch.org/sites/default/files/2019-BANGLADESH-Report-eng.pdf

Muturzikin (2008) Bangladesh & Bhutan—Linguistic map. https://www.muturzikin.com/cartesasiesudest/7.htm
Nafis H (2021) Bangladesh has no indigenous people. jamhoor.org, July 8. https://www.jamhoor.org/read/bangladesh-has-no-indigenous-people
Nasirkhan (2021) Physiography. https://en.banglapedia.org/index.php/Physiography
Naveed S, Waqas A, Chaudhary A, Kumar S, Abbas N, Amin R, Jamil N, Saleem S (2020) Prevalence of common mental disorders in South Asia: a systematic review and meta-regression analysis. Front Psychiatry 11:e573150
OECD (2016) Skills matter: further results from the survey of adult skills. OECD Publishing, Paris. https://doi.org/10.1787/9789264258051-1-en
Papaleontiou-Louca E (2021) Effects of religion and faith on mental health. New Ideas Psychol 60:e100833
Rahman MS, Rahman MA, Ali M, Rahman MS, Maniruzzaman M, Yeasmin MA, Ahmed NAMF, Abedin MM, Islam SMS (2020) Determinants of depressive symptoms among older people in Bangladesh. J Affect Disord 264:157–162
Roy T (2019) How British rule changed India's economy. Palgrave. https://doi.org/10.1007/978-3-030-17708-9
WHO (2021) Bangladesh: WHO special initiative for mental health situational assessment. https://www.who.int/publications/m/item/bangladesh---who-special-initiative-for-mental-health
Sachs JD, Kroll C, Lafortune G, Fuller G, Woelm F (2021) The decade of action for the sustainable development Goals. Sustainable Development Report 2021, Cambridge University Press, Cambridge
Saha, D (2016) 84% Of 12 million married children under 10 are Hindus, India spend (1 June). https://www.indiaspend.com/84-of-12-million-married-children-under-10-are-hindus-82446
Schipper ELF, Eriksen SE, Fernandez Carril LR, Glavovic BC, Shawoo Z (2021) Turbulent transformation: abrupt societal disruption and climate resilient development. Climate Devel 13(6):467–474
Sifat RI (2020) The effect of COVID-19 on hijra (third gender) people in Bangladesh. LANCET Psychiatry 7(12):1015–1016
Shammi M, Rahman MM, Bondad SE, Bodrud-Doza M (2019) Impacts of salinity intrusion in community health: a review of experiences on drinking water sodium from coastal areas of Bangladesh. Healthcare 7(1):50
Srirejeki K (2020) Corruption and culture: revising the claim of its relationship. In SHS Web of Conferences 86:e01037
Sultana F (2022) The unbearable heaviness of climate coloniality. Polit Geogr e102638. https://doi.org/10.1016/J.POLGEO.2022.102638
Sultana S (2021) Indigenous ethnic languages in Bangladesh: paradoxes of the multilingual ecology. Ethnicities. https://doi.org/10.1177/14687968211021520
Tasnim R, Sujan MSH, Hossain S, Amin M, Hasan ATMH, Islam MS (2021) An overview of the surge of rape incidents in Bangladesh during COVID-19 pandemic. CBMJ 10(1):43–49
Tov W, Nai ZLS (2018) Cultural differences in subjective well-being: how and why. In: Maddux JE (ed) Subjective well-being and life satisfaction. Routledge, New York, pp 50–73
WEF (World Economic Forum) (2021) Global Gender Gap Index Report 2021. https://www.weforum.org/reports/global-gender-gap-report-2021/
WHO (2013) Comprehensive mental health action plan 2013–2020. WHO, Geneva. http://apps.who.int/gb/ebwha/pdf_files/WHA66/A66_R8-en.pdf

Nazwa Tahsin is a Ph.D. candidate in hydrodynamics at Wageningen University, with MSc in Water Resources Management and Hydrology from the Bangladesh University of Engineering and Technology. She has worked with Neurolandscape in research on environmental health and equity with a focus on global south. Co-founder on a non-profit organization "Beyond Hills", working in digital empowerment of indigenous communities of Bangladesh; passionate about climate and environmental adaptation, improving health and justice

Dr. Weronika Gąsior is a Scientific Board member and Gender Equality Officer at NeuroLandscape. She received a Ph.D. in Languages and Cultural Studies from the University of Limerick, Ireland. As an expert in intercultural communication, she brings this perspective into qualitative research on topics relating to nature's role in mental health and wellbeing, and children and nature. She also applies her background in linguistics to analyse the discourse across the different disciplines relating to nature and health

Dr. Agnieszka Olszewska-Guizzo is a landscape architect and neuroscience researcher. She received her Ph.D. from the University of Porto, Portugal in Landscape Architecture and Urban Ecology. She has introduced and operationalized the concept of contemplative landscapes through the development of a tool called the Contemplative Landscape Model (CLM)—a quantitative expert-based scale for assessing and classifying landscape views based on their potential to generate mental health benefits in green space users through passive exposure. Most of her post-doctoral academic work has taken place in Singapore where she conducted research with the National University of Singapore and National Parks Board. In 2021 she was awarded a CUGE Fellowship from the Singapore National Parks Board for further development of the Contemplative Landscape Model. She is a Co-founder and Lead Researcher at NeuroLandscape NGO and a Fellow of the Center for Design and Mental Health.

Chapter 13
Impacts of Climate Change on Agriculture and Food Security in Africa and Latin America and the Caribbean

Gabriela Narcizo de Lima, Roberto Ariel Abeldaño Zuñiga, and Mina Margaret Ogbanga

Abstract Recent analyzes have determined that many countries have not made progress towards meeting the goal of eradicating hunger. The difficulties imposed by the COVID-19 pandemic added to the worsening of the climate crisis, which led to an increase in the occurrence of extreme events and the reduction of food production in many countries, have moved the world away from the path that leads to the end of hunger and malnutrition in all its forms by 2030. In 2020, almost one in three people in the world lacked access to adequate food, an increase of almost 320 million people between 2019 and 2020. If these data are analyzed by regions, in this same period approximately 46 million more people suffered from hunger in Africa and 14 million more in Latin America and the Caribbean (LAC). Therefore, this chapter aims to assess the main impacts generated by historical (1950–2021) and future (2050 and 2100) climate change trends on food production and its correlation with the prevalence of undernourishment and moderate and severe food insecurity in Africa and LAC between 2000 and 2022. The results of these analyzes will provide essential information that can be used to promote the development of sustainable food systems in the current context of the socio-sanitary and climate crisis, in accordance with SDGs 2, 13 and 17 of the 2030 Agenda.

Introduction

Recent analyzes have determined that many countries have not made progress toward meeting the goal of eradicating hunger worldwide (SDG 2) set by the Sustainable

G. N. de Lima
Geography Department—Faculty of Letters, Porto University, Via Panorâmica s/n, P.C. 4150-564 Porto, Portugal

R. A. A. Zuñiga (✉)
Faculty of Social Sciences, University of Helsinki, Helsinki, Finland
e-mail: ariabeldanho@gmail.com

PostGraduate Department, University of Sierra Sur, Oaxaca, Mexico

W. Leal Filho et al. (eds.), *Climate Change and Health Hazards*, Climate Change Management, https://doi.org/10.1007/978-3-031-26592-1_13

Development Goals (SDG) of the 2030 Agenda (Food and Agriculture Organization of the United Nations 2021a). The difficulties imposed by the COVID-19 pandemic added to the worsening climate crisis, led to an increase in the occurrence of extreme events and the loss and/or reduction of food production in many countries, have moved the world away from the path that leads to the end of hunger and malnutrition in all its forms by 2030.

Although the global prevalence of moderate or severe food insecurity has been growing slowly since 2014, the estimated increase in 2020 was equal to the sum of the previous five years. In 2020, almost one in three people (2.37 billion) worldwide lacked access to adequate food, increasing almost 320 million people between 2019 and 2020. If this data is analyzed by regions, in this same period, approximately 46 million more people suffered from hunger in Africa and 14 million more in Latin America and the Caribbean (LAC). In total, 33% of Africans and 8% of Latin Americans experienced some level of food insecurity in 2020 (Food and Agriculture Organization of the United Nations 2021a).

According to the definition of the Food and Agriculture Organization of the United Nations (FAO 2008) food security "exists when all people, at all times, have physical and economic access to sufficient safe and nutritious food that meets their dietary needs and food preferences for an active and healthy life." From this definition, four dimensions can be analyzed: availability, accessibility, biological utilization, and stability of the other three dimensions over time (Food and Agriculture Organization of the United Nations FAO 2008).

Availability refers to production, stock, and net trade mechanisms. Accessibility is an economic concept since food availability does not guarantee secure access, but is related to the income and expenditure of individuals and families, and food prices determined by markets. Biological utilization should be understood as to how the body uses the nutrients provided by food. Finally, stability can affect the other three dimensions. If, for example, there is a period of strong economic instability with market shocks and price increases (which would prevent access to food), extreme weather events that wipe out crops (e.g., an extreme hailstorm), or factors related to political instability (which would impact availability).

Climate variability and extremes are among the main drivers of the recent increase in global hunger and a major contributor to severe food crises. Climate variability and extremes negatively affect all dimensions of food security (Food and Agriculture Organization of the United Nations 2021b). However, it is necessary to consider that the effects of climate change are not equally distributed. It is likely that the greatest

---

M. M. Ogbanga
Indianapolis School of Social Work, Indiana University Purdue University-IUPUI, Indianapolis, IN, USA
e-mail: Mogbanga@iu.edu

CEO Institute for Partnership and Development Studies of CEDSI Nigeria, Building Exceptional Organizations for Effective and Sustainable Project Delivery, UN Sustainable Development Solutions Network, Port Harcourt, Nigeria

impacts will occur in populations already in a situation of vulnerability (Matondo et al. 2020).

Therefore, this chapter aims to assess the main impacts generated by historical and future trends of climate change on food production and the prevalence of moderate and severe malnutrition and food insecurity in Africa and LAC between 2000 and 2022. The analysis will be performed based on the following categories: climate change patterns; impact on crops (wheat, rice, and maize); and food insecurity. The data required for the analysis are available in the open-access datasets of the World Meteorological Organization (WMO) (Royal Netherlands Meteorological Institute 2022; World Bank Group 2022), FAO (Food and Agriculture Organization of the United Nations 2022) and the project "Effects of Climate Change on Global Food Production from SRES Emissions and Socioeconomic Scenarios" (Parrya et al. 2004; Iglesias and Rosensweig 2009).

The results of these analyzes will provide essential information that can be used to promote the development of sustainable food systems in the current context of the socio-sanitary and climate crisis, in accordance with SDGs 2, 13, and 17 of the 2030 Agenda.

Climate Change Patterns and Tendencies in Africa and Latin America, and the Caribbean

Reflecting on contemporary issues, it is clear that climate change is at the center of international political debates and agreements, involving a diversity of actors, interests, and disputes. On the one hand, meteorological models point to an increase in global average temperature over the twenty-first century and, consequently, an intensification in the occurrence of extreme weather events such as droughts, heavy precipitation, and heat waves. On the other hand, the threat of changes in climate patterns and the slowness of the necessary mitigation and adaptation actions put at risk urban structures, energy matrices, food production, and the very dynamics of cities, particularly coastal cities, which are threatened by rising sea levels (Intergovernmental Panel on Climate Change 2021).

The Intergovernmental Panel on Climate Change (IPCC) reports (Intergovernmental Panel on Climate Change IPCC 2018; Intergovernmental Panel on Climate Change 2021) show that although Africa and LAC account respectively for only 3.8% and 13% of global greenhouse gas (GHG) emissions, compared to 23% for China and 19% for the United States, these regions are among the most vulnerable to the impacts of climate change. These impacts include food insecurity, climate-related increases in internal migration, temperature extremes, and low labor productivity, mainly due to heat stress during the warmer seasons (United Nations Economic Commission for Africa UNECA 2008; The Centre for International Governance Innovation CIGI 2009; Economic Comission for Latin America and the Caribbean

2010; Intergovernmental Panel on Climate Change IPCC 2018; Abeldaño Zuñiga et al. 2021).

Latin America and the Caribbean

There has been a recent increase in extreme weather events in LAC and the number of people affected by extreme temperatures, forest fires, droughts, storms, and floods (López Cruz et al. 2021). It is estimated that more than 40 million people were affected by these events in the last decade, representing an increase of 12.5% since the 1990s. These events generated an estimated cost of US$ 40 billion in damages for the region (Bárcena et al. 2018; Centre for Research on the Epidemiology of Disasters—CRED 2018).

Take the Representative Concentration Pathway 4.5 (RCP4.5) scenario as an example, presented by the IPCC in its Fifth Assessment Report (Intergovernmental Panel on Climate Change 2014), which considers the stabilization of the level of radiative forcing at 4.5 W/m2 until the year 2100. By using strategies to reduce GHG emissions, an increase in average annual temperatures up to 2 °C above that in recent decades (1990–2020) is expected by the end of the century in South America, especially in the north-central part of the continent, while for Central America and the Caribbean this increase would be up to 1.5 °C (Fig. 13.1). It should also be noted that maximum temperatures would increase in the same proportions during the period analyzed, which would result in a greater frequency of days with extreme temperatures in a large part of the region.

Data taken between 1990 and 2020 (World Bank Group 2022) (Fig. 13.2), show that the countries that suffered the most significant increases in average annual temperatures during this period were Guatemala (1.9 °C), Guyana (1.9 °C), Venezuela (1.8°), and Belize (1.7 °C). This pattern is repeated concerning mean annual maximum and minimum temperatures, indicating conditions of lower thermal comfort for populations and possible impacts on food production in the different countries.

Countries such as Brazil, Mexico, and Argentina, which are among the largest emitters of GHGs in the region, presented an average increase of 1.07 °C in their average temperatures but suffered a significant increase in average annual minimum temperatures. This increase indicates that the number of cold and cool nights in these regions has been reduced over the last three decades.

Projections of changes in precipitation patterns present a high level of uncertainty and strong heterogeneity among countries and regions, depending on the emissions scenario considered. In this sense, the forecasts presented, based on multiple general circulation models and the main emissions scenarios, also show that changes in the rainfall regime will be differentiated in the region (Intergovernmental Panel on Climate Change 2014).

According to the IPCC RCP4.5 scenario (Intergovernmental Panel on Climate Change 2014), a reduction of up to 2 mm per day of precipitation will be observed

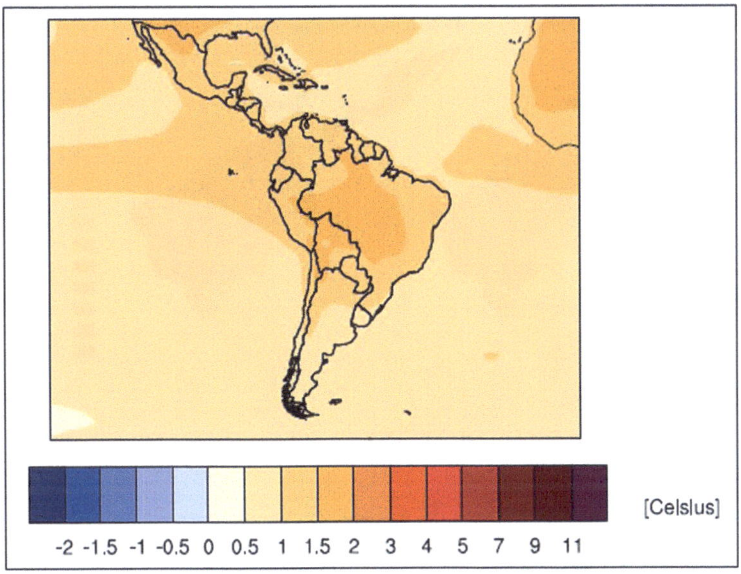

Fig. 13.1 Mean RCP4.5 temperatures 2021–2100 minus 1990–2020 (January–December AR5 CMIP5 subset)—Latin America and the Caribbean (*Source* Elaborated by the authors with data available from Royal Netherlands Meteorological Institute 2022)

until 2100 compared to the patterns observed between 1990–2020 in Mexico, part of Central America, and the Caribbean (Fig. 13.3).

In South America, within the same period, an increasing trend of up to 1 mm daily precipitation will be observed in the northwest of the continent, mainly affecting Colombia, Ecuador, and parts of Peru, and in the southeast, parts of Brazil, Uruguay, and Argentina. Conversely, a decreasing precipitation trend will be observed in almost the entire north-central part of the continent. This trend will mainly affect Venezuela, Guyana, Suriname, and the north-northeast of Brazil, areas that include a large part of the Amazon rainforest and play a key role in global and regional climate regulation.

Concerning historical data, from 1990 to 2020 (Fig. 13.2), some of the changes mentioned above can already be observed in LAC countries. In Central America and northern South America, there was a significant decrease in precipitation during the period analyzed, mainly in Suriname, Guyana, Venezuela, and Belize, which suffered reductions of 32%, 15%, 12%, and 11%, respectively, in their annual totals (Fig. 13.2). These significant changes in precipitation in the countries above impact water availability, aquifer replenishment, the maintenance of vegetation cover, and agricultural yields in the affected countries.

On the other hand, in several Caribbean islands, such as the Dominican Republic, Haiti, Dominica, and Saint Vincent and the Grenadines, there was a significant increase in annual rainfall totals during the same period, 12%, 7%, 6%, and 4%,

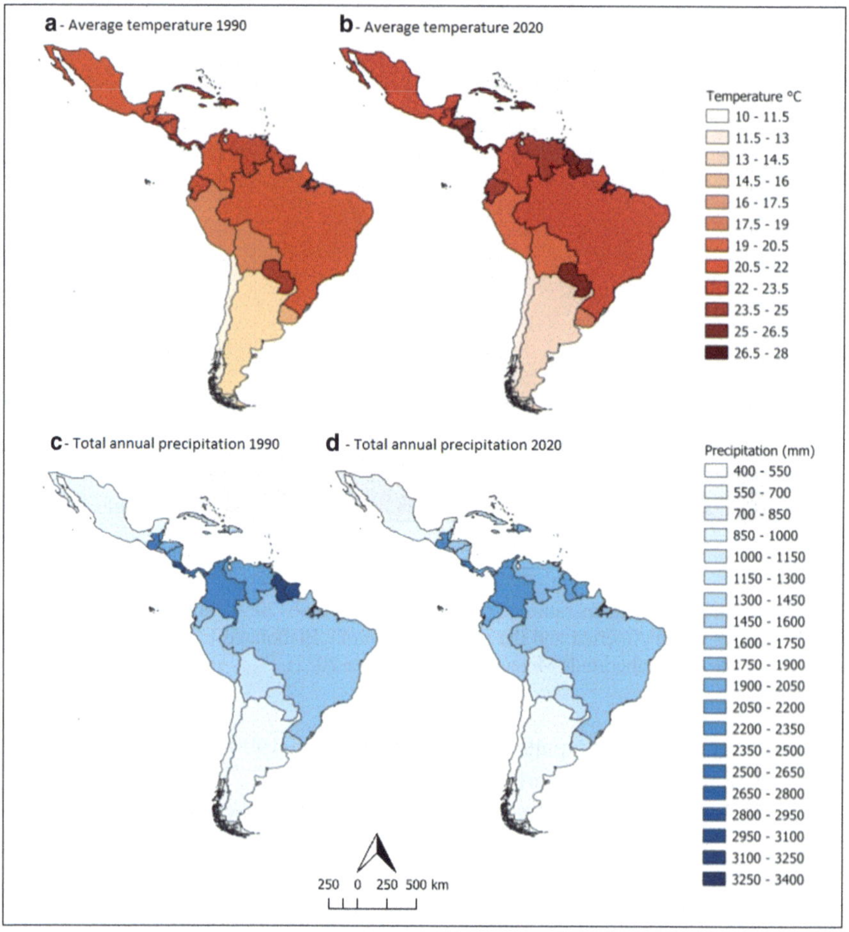

Fig. 13.2 Average annual temperatures and total annual precipitation in Latin America and the Caribbean—Values for 1990 and 2020 (*Source* Elaborated by the authors with data available from World Bank Group 2022)

respectively. This increase may be linked to a greater frequency of extreme hydrometeorological phenomena in the Atlantic Ocean, such as tropical storms, tropical depressions, and hurricanes (López Cruz et al. 2021).

Africa

Africa is highly vulnerable to the impacts of climate change. Temperatures in all countries of the continent are expected to rise faster than the global average rate.

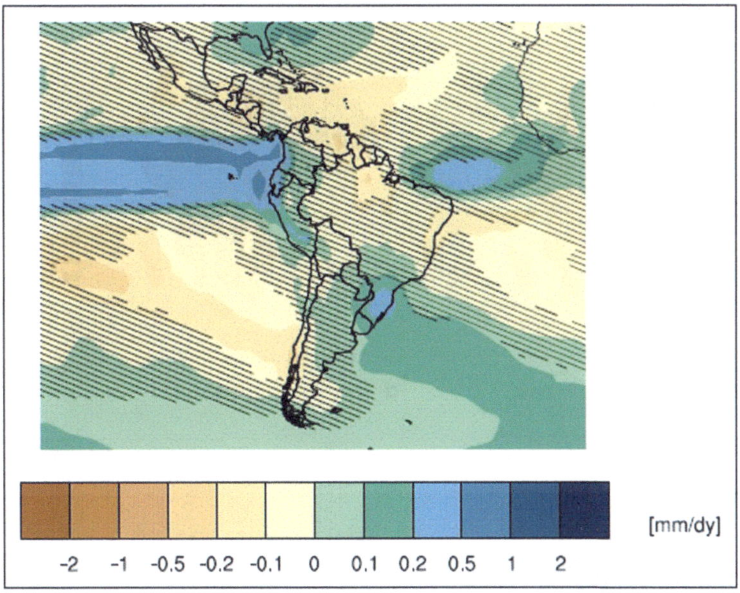

Fig. 13.3 Mean RCP4.5 precipitations 2021–2100 minus 1990–2020 (January–December AR5 CMIP5 subset). Latin America and the Caribbean (*Source* Elaborated by the authors with data available from Royal Netherlands Meteorological Institute [2022])

According to data from the Climate Change Knowledge Portal (World Bank Group 2022) (Fig. 13.4), an increasing warming trend has been observed since the beginning of the twentieth century, mainly in African regions where warm climates predominate. For the other areas, the increase in temperatures was not constant until the end of the 1970s (Abeldaño Zuñiga et al. 2022).

Looking at the data by country for the period 1990 to 2020, it is possible to observe a large variation in warming rates across the continent, with values above 1.5 °C in Tunisia, Namibia, Botswana, Lesotho, Zimbabwe, and South Africa and below 0.5 °C in Seychelles, Comoros, and Madagascar. But, in general, warming rates in most countries exceed 1 °C for this data period.

Concerning rainfall, a decreasing trend in annual totals is observed for 33 African countries from 1990 to 2020 (Fig. 13.4), generated mainly by the increase in intense and prolonged droughts in some regions of the continent. Libya, Zimbabwe, Mozambique, Zambia, and Namibia stand out as the countries with the largest reductions in their annual rainfall totals (63%, 53%, 36%, 35%, and 34%, respectively). For the other countries, the average reduction in rainfall is about 12%.

Yet, there has also been increased precipitation in recent decades due to the summer monsoons intensifying in West Africa. The increase in short rains in East Africa is due to the warming pattern of the Indian Ocean, and the increase in extreme precipitation caused by cyclones on land on the east coast of the continent (World

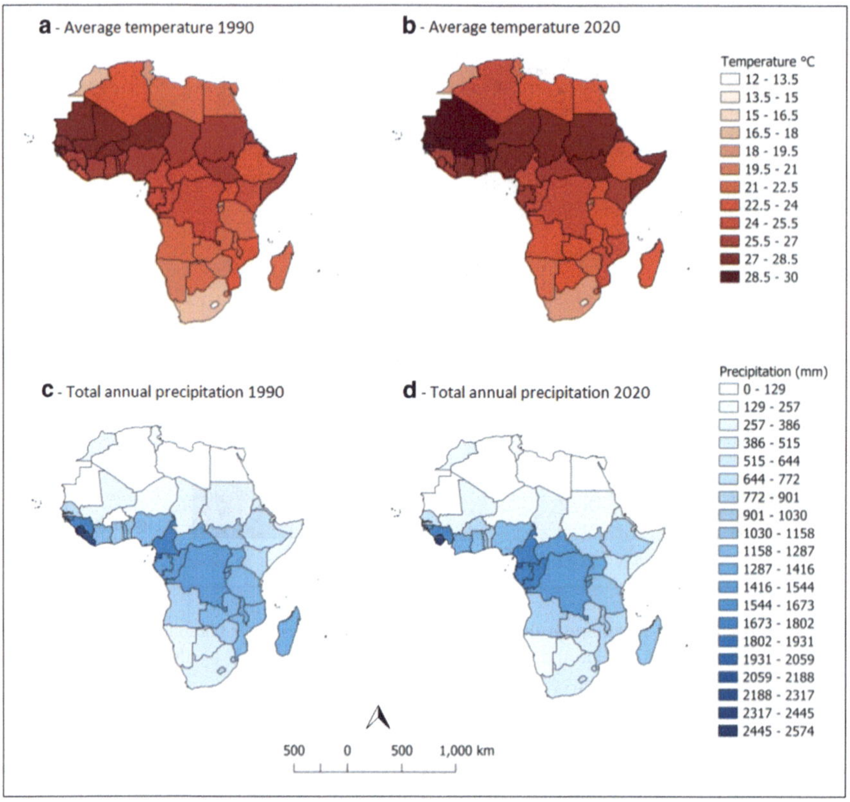

Fig. 13.4 Average annual temperatures and total annual precipitation in Africa—Values for 1990 and 2020 (*Source* Elaborated by the authors with data available from World Bank Group 2022)

Bank Group 2022). Among the countries that showed a significant increase in rainfall during the period analyzed were Cape Verde, on the Atlantic coast, and Djibouti, Somalia, and Ethiopia, in the Horn of Africa, with 58%, 34%, 22%, and 20%, respectively. The average increase in annual precipitation totals for the other countries analyzed is approximately 6% between 1990 and 2020.

Regarding future climate trends, according to the RCP4.5 scenario (Fig. 13.5), general warming of the African continent is expected by 2100, reaching values of up to 2 °C, especially in the arid and semi-arid zones of northern and southern Africa. The most affected countries will be Mauritania, Mali, Algeria, Niger, Chad, Libya, Egypt, Eritrea, Sudan, Zambia, Angola, Zimbabwe, Botswana, South Africa, and Namibia.

Future rainfall projections present a higher degree of uncertainty, as they also depend on factors related to natural climate variability. Rainfall variability is expected to increase in most African regions. There is a high probability of an increased frequency of high-intensity rainfall events, especially along the Atlantic coasts of

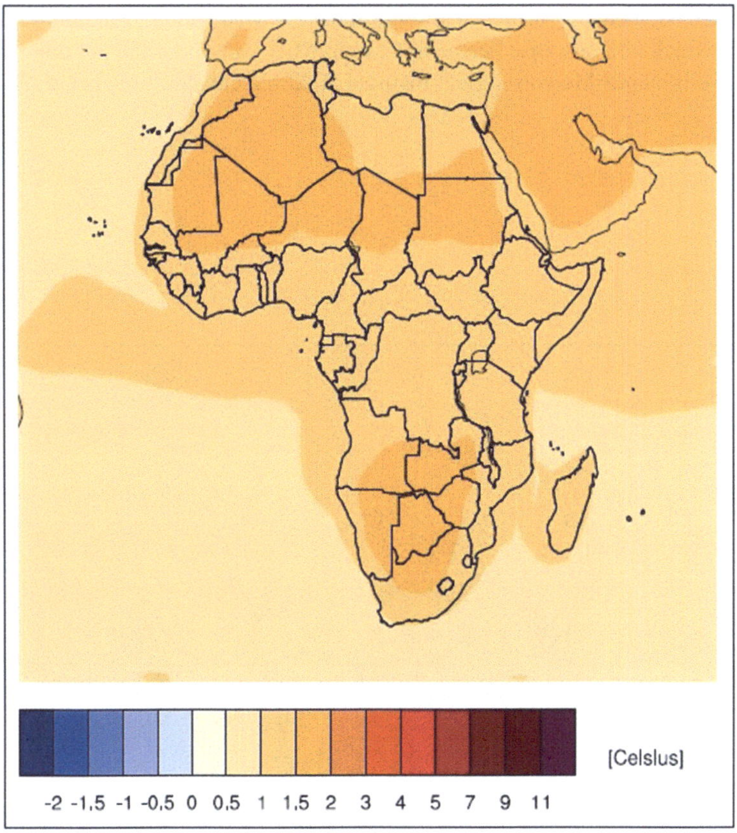

Fig. 13.5 Mean RCP4.5 temperatures 2021–2100 minus 1990–2020 (January–December AR5 CMIP5 subset)—Africa (*Source* Elaborated by the authors with data available from Royal Netherlands Meteorological Institute 2022)

central and western African countries and in parts of East Africa (Matondo et al. 2020; Abeldaño Zuñiga et al. 2022). Climate change could result in significant losses of endemic plant species, more than 50% of some bird and mammal species, and declines in the productivity of Africa's lakes by 20–30% by 2100 (Intergovernmental Science-Policy Platform on Biodiversity and Ecosystem Services (IPBES) 2018).

According to the RCP4.5 scenario (Fig. 13.6), almost all countries in tropical climate zones will be impacted by increased precipitation by 2100 (compared to 1990–2020). An increase of up to 0.5 mm per day is expected in Nigeria, Cameroon, Equatorial Guinea, and Gabon. On the other hand, for the dry climate zones, rainfall is expected to decrease further, with highlights for Madagascar, Mozambique, and Malawi, which will decrease between 0.2 mm to 0.5 mm of rainfall per day by 2100 (Royal Netherlands Meteorological Institute 2022; Abeldaño Zuñiga et al. 2022).

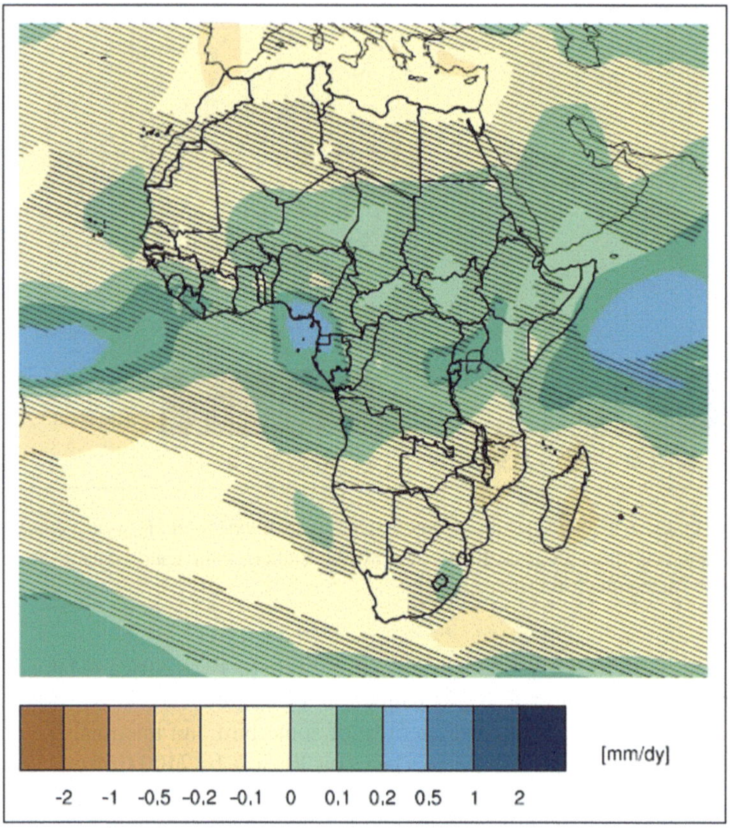

Fig. 13.6 Mean RCP4.5 precipitations 2021–2100 minus 1990–2020 (January–December AR5 CMIP5 subset). Africa (*Source* Elaborated by the authors with data available from Royal Netherlands Meteorological Institute 2022)

Climate Change Trends on Food Production in Africa and Latin America and the Caribbean

There is currently the talk of a circle of effects between agroindustrialization processes, climate change, and food security. Agroindustrialization processes have led to land degradation, water pollution, and GHG emissions (Secretaría de Medio Ambiente y Recursos Naturales 2019). These processes, in turn, have led to a decrease in crop yields in some areas, starting with what are called slow onset events associated with climate change (Abeldaño Zuñiga et al. 2021). Finally, this decrease in crop yields has impacted food security in several countries of the world.

On the other hand, extreme events related to climate change have also affected food security due to low crop yields, such as floods, hailstorms, frosts, and droughts (Food and Agriculture Organization of the United Nations 2018).

Previous studies suggest that, in the LAC region, slow-onset climate processes are closely related to poverty and food insecurity. This can be explained by the high economic dependence of this region on activities conditioned by environmental factors, such as agriculture, livestock, and fisheries (Economic Commission for Latin America and the Caribbean 2015; Gardi et al. 2015; Vargas et al. 2018; Castro-Llanos et al. 2019; Fernández et al. 2019; Jaramillo-Mejía and Chernichovsky 2019; Mahlknecht et al. 2020). In addition, population growth in most of the region's countries increases the demand for food production, water resources, and agricultural land (Abeldaño Zuñiga et al. 2021). In Africa, the situation is even more serious since changes in rainfall patterns, temperature increases, and desertification processes cause food insecurity in this region and provoke the spatial displacement of the population as a survival strategy (Africa Renewal 2021; Abeldaño Zuñiga et al. 2022).

To better understand the relationship between climate change and food production in both regions, this section assesses the potential impacts of climate change on staple crop production (wheat, rice, and maize) in LAC and Africa based on data generated by the project "Effects of Climate Change on Global Food Production from SRES Emissions and Socioeconomic Scenarios" (Parrya et al. 2004; Iglesias and Rosensweig 2009), while focusing on quantitative estimates of yield changes based on multiple climate scenarios.

The baseline for comparing the potential impacts of climate change on wheat, rice, and maize crops is the average 2000–2006 production (in tons) for different countries in Africa and LAC. The IPCC A1F1 and B1a scenarios (Intergovernmental Panel on Climate Change 2000) were taken as references because they represent divergent "possible futures," covering a significant part of the underlying uncertainties in the main driving forces, encompassing a wide range of key characteristics as demographic change, economic development, and technological change.

The A1 storyline and scenario family describe a future world of very rapid economic growth, a world population that peaks in the mid-twenty-first century and then declines, and the rapid introduction of new and more efficient technologies.

Special attention is given to subgroup A1FI, which describes future technological change based on an intensive use of fossil fuels.

On the other hand, the B1 scenario family describes a converging world with the same global population peaking in the mid-twenty-first century and then declining—as in the A1 scenario family—but with rapid changes in economic structures towards a service and information economy, with the introduction of clean and resource-efficient technologies. The emphasis is on global solutions for economic, social, and environmental sustainability.

Latin America and the Caribbean

Based on the A1FI scenario for the year 2050, which considers development on a global scale, with rapid economic and technological growth—based mainly on the use of fossil fuels—and a population that will reach its peak in that period, the impacts on wheat, corn and rice production in LAC would be considerable.

Twenty-two countries in the region would suffer losses of more than 1.5% in their corn production and 1.2% in their rice crops compared to the average from 2000 to 2006 (Fig. 13.7). Wheat production would be less affected under this scenario in most countries, with losses of approximately 0.4%. The cases of Ecuador, Panama, Suriname, and Mexico stand out, with losses of more than 7% in all crops considered under the A1FI scenario for 2050.

On the other hand, considering the important changes in weather patterns generated by the type of development expected in the A1FI scenario for South America—higher temperatures and increased precipitation totals—countries such as Argentina, Bolivia and Uruguay would see an increase in wheat and rice production of more than 10% by mid-century.

It is interesting to note that, in the case of LAC, even under scenario B1a, which also speaks of global development, but with a greater emphasis on solutions based on economic, social, and environmental sustainability, crop yields would not increase significantly by 2050. Most of the region's countries would continue to decrease their corn, wheat, and rice production. Argentina, Bolivia, and Uruguay would again record an increase in wheat and rice yields compared to the 2000–2006 average, but this increase would be smaller than expected for the A1FI scenario. Mexico is the only country that would register a significant improvement in the production of the three grains under a scenario of lower GHG emissions derived from a more sustainable economic and technological development.

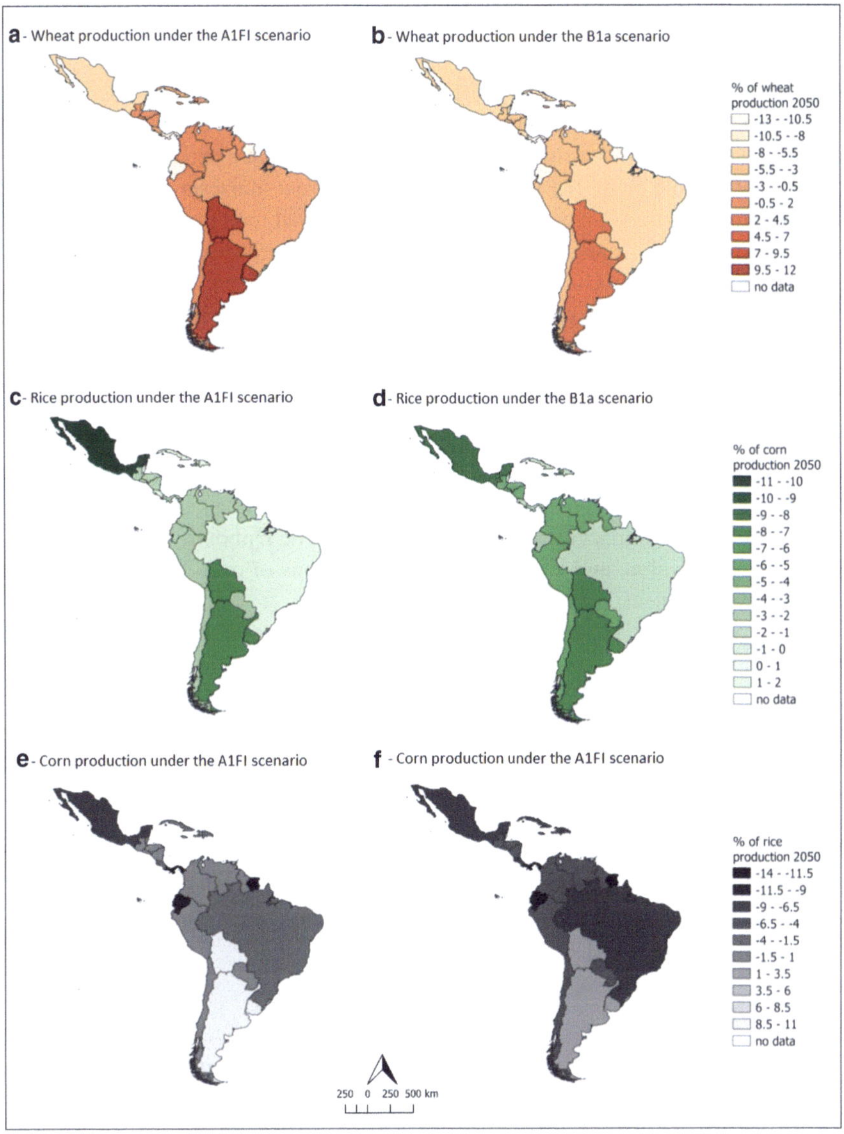

Fig. 13.7 The yield of corn, wheat, and rice crops in Latin America and the Caribbean under the A1FI and B1a scenarios for 2050 compared to the 2000–2006 average (*Source* Elaborated by the authors based on Parrya et al. 2004 and Iglesias and Rosensweig 2009)

Africa

According to the IPCC (Intergovernmental Panel on Climate Change 2014), Africa will experience a temperature increase that will be one and a half times higher than the average experienced worldwide. Evidence suggests that a persistent 1 °C rise on the continent would cause the loss of 65% of the current maize-growing capacity in producing countries and an estimated 20% drop in global food production (Research Program on Climate Change Agriculture and Food Security CCAFS (2011); Intergovernmental Panel on Climate Change 2014),). In addition, it is estimated that the price of wheat will increase by 120%, maize by 177%, and processed rice by 107% by 2030 (Research Program on Climate Change Agriculture and Food Security CCAFS 2011). This increase would put the most basic foods beyond the reach of many Africans and undermine their fundamental right to food.

Considering the expected future climatic conditions for the A1FI scenario, all regions of Africa would suffer significant grain crop losses by 2050, with averages of 2.5% for wheat, 3.5% for rice, and 10% for maize (Fig. 13.8). The countries with the most significant impact on food production would be Djibouti, Nigeria, Lesotho, Libya, Kenya, Zambia, and Zimbabwe, with reductions of over 4%, 5%, and 11%, respectively, for wheat, rice, and maize.

On the other hand, contrary to the pattern observed for LAC countries, which, even under a scenario of lower GHG emissions, would continue to register a significant reduction in crop yields, food production in Africa would benefit from more sustainable economic and environmental development.

Expected grain yields in 2050 under the B1a scenario would not reach the same levels as the average recorded between 2000 and 2006, still, the reduction would be less significant in most countries on the continent compared to the values predicted for the A1FI scenario. The average reduction in wheat, rice, and maize crops would be 1%, 2%, and 7%, respectively, across most of the continent, with countries such as Algeria, Congo, Egypt, and Gabon even registering a slight increase of 0.7% in wheat production by 2050.

Measuring the Impacts of Climate Change on Food Security in Africa and Latin America

The availability of data is fundamental for analyzing the impacts of climate change on food security. One of the most relevant data sources worldwide on food security issues is FAOSTAT (Food and Agriculture Organization of the United Nations 2022). The FAOSTAT site offers 30 indicators surveyed by FAO to analyze the food security situation at the country level. Of these 30 indicators, six correspond to the availability dimension, seven to the accessibility dimension, five to the stability dimension, ten to the biological utilization dimension, and two additional indicators of food energy requirements. From this list of indicators, this chapter analyzed the prevalence of

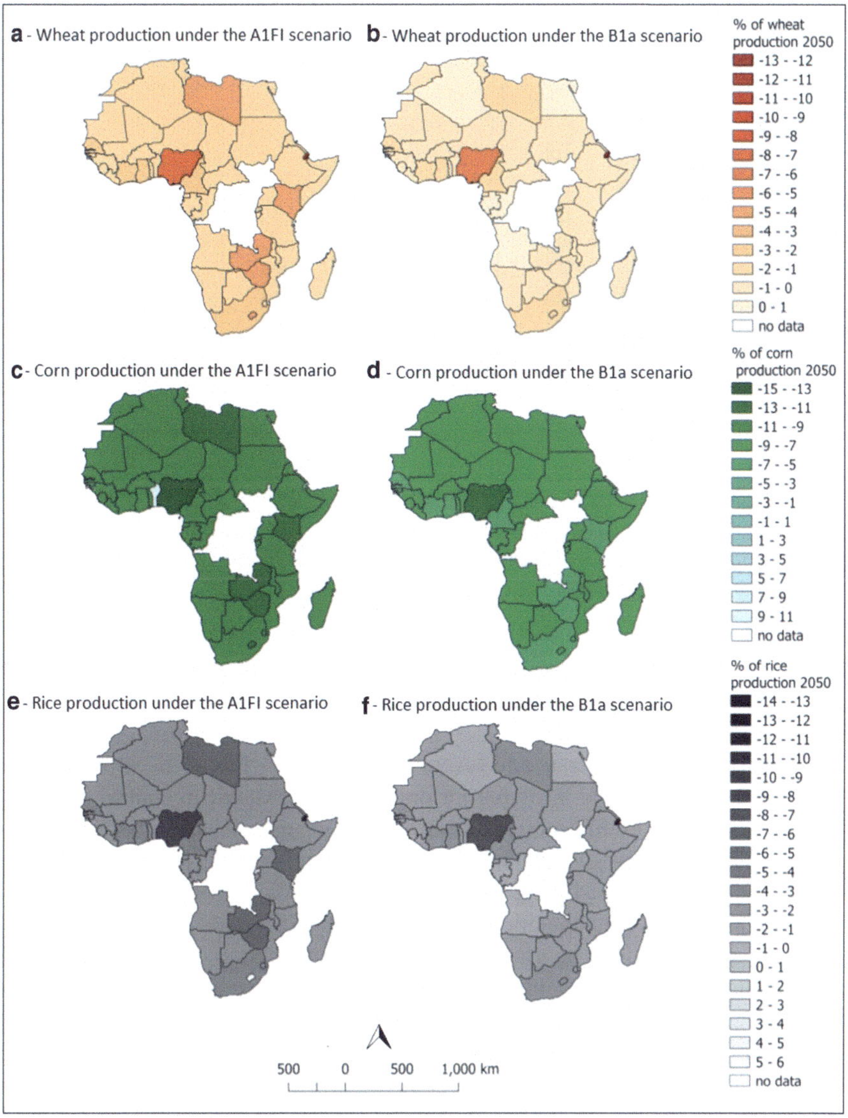

Fig. 13.8 The yield of corn, wheat, and rice crops in Africa under the A1FI and B1a scenarios for 2050 compared to the 2000–2006 average (*Source* Elaborated by the authors based on Parrya et al. 2004 and Iglesias and Rosensweig 2009)

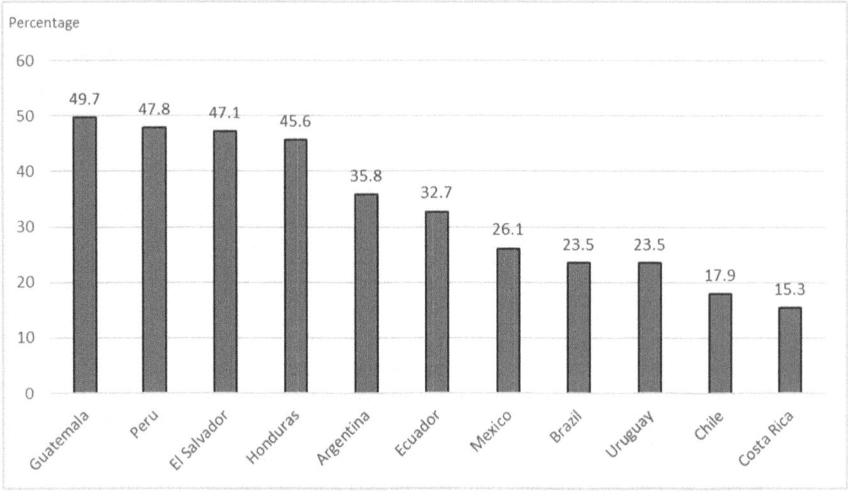

Fig. 13.9 Prevalence of moderate or severe food insecurity in the total population (percent) (3-year average) 2018–2020, in Latin American countries (*Source* Elaborated by the authors with data available from FAOSTAT 2022)

moderate to severe food insecurity, which is part of the food access indicators, and the prevalence of people with undernourishment, which is part of the biological utilization indicators.

The prevalence of moderate and severe food insecurity has been surveyed since 2014, and data is currently available for only 11 Latin American countries and 38 African countries. In Latin America, the most alarming prevalence of food insecurity is observed in Honduras, El Salvador, Peru, and Guatemala, where almost half of the people have difficulties accessing food (Fig. 13.9). In Africa, there are 27 countries where at least half of the population has difficulty accessing food (Fig. 13.10).

The prevalence of undernourished people has been reported since 2000, although there are currently no data available for 14 African countries and 9 Latin American countries. In seven Latin American and Caribbean countries, the prevalence of undernourished people exceeds 10% (Fig. 13.11). In 30 African countries, the prevalence of undernourished people exceeds 10% (Fig. 13.12).

In absolute numbers, 203.4 million people in Africa are undernourished (with 14 countries without data), and 43.6 million in LAC (with 9 countries without data). This means that it is very likely that this figure is underestimated due to the countries' populations that do not have data on this variable (Figs. 13.13 and 13.14).

These data can provide a basis for an approximate analysis of the reality of the countries; however, these data sources cannot yet have a comparative scope in which the situation of all countries can be analyzed. The FAO accepts that although the indicators are calculated consistently for all countries, there could be limited geographical comparability between countries due to differing methodologies and source data for the basic variables (FAOSTAT 2022).

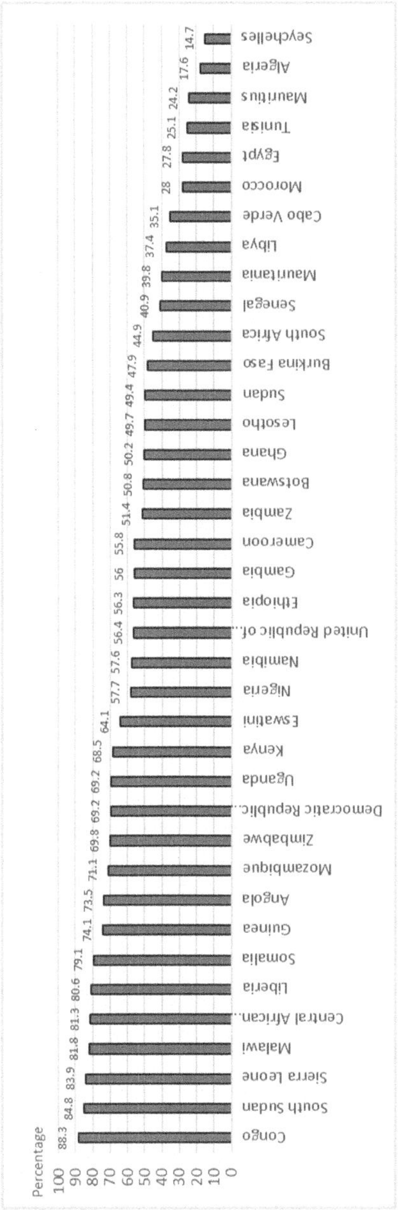

Fig. 13.10 Prevalence of moderate or severe food insecurity in the total population (percent) (3-year average) 2018–2020, in African countries (*Source* Elaborated by the authors with data available from FAOSTAT 2022)

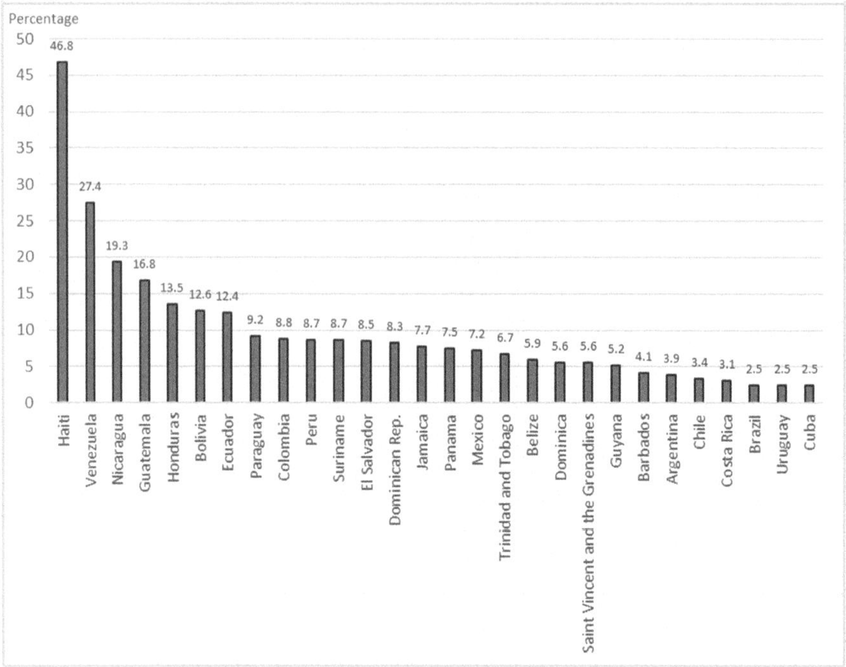

Fig. 13.11 Prevalence of undernourishment (percent) (3-year average) 2018–2020, in Latinamerican countries (*Source* Elaborated by the authors with data available from FAOSTAT 2022)

On the other hand, the Global Partnership for Sustainable Development Data initiative has suggested traditional data sources such as household surveys, agricultural censuses, local administrative data compiled by local ministries, and vital statistics. Then again, the same initiative has also pointed out some alternative data sources to analyze some dimensions of food security. These alternative data sources include satellite images for earth observation, data from road infrastructure and irrigation water networks, data generated through citizen science initiatives, sensors from official networks (climate), and data from the domestic and international trade balance (imports and exports) (Global Partnership for Sustainable Development Data 2021).

Based on the data analyzed, where 247 million people in different countries of the Global South are in a situation of undernourishment, a very complex scenario can be envisioned for the achievement of SDG 2 "Zero hunger" in these countries. In this regard, it is necessary to recall that, under the concept of climate justice (Oxfam International; The Mary Robinson Foundation—Climate Justice 2011; United Nations Sustainable Development 2019), climate change raises important questions about global justice. Climate justice refers to the fact that less developed countries generate less GHG emissions than more developed countries; however, the least developed countries suffer the most profound impacts of climate change.

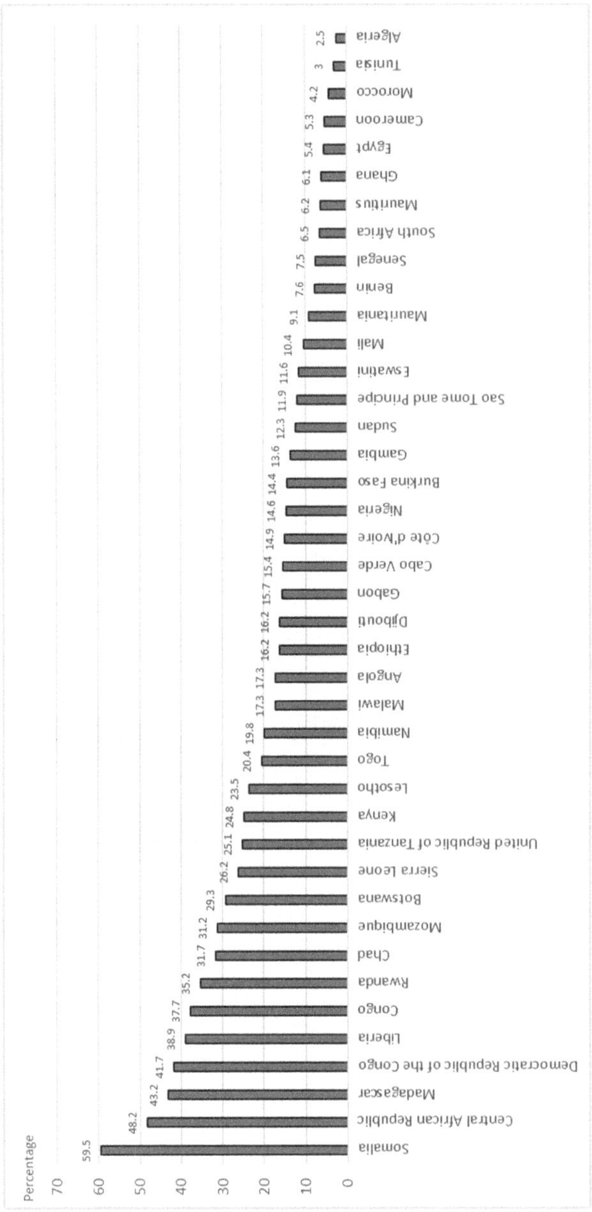

Fig. 13.12 Prevalence of undernourishment (percent) (3-year average) 2018–2020, in African countries (*Source* Elaborated by the authors with data available from FAOSTAT 2022)

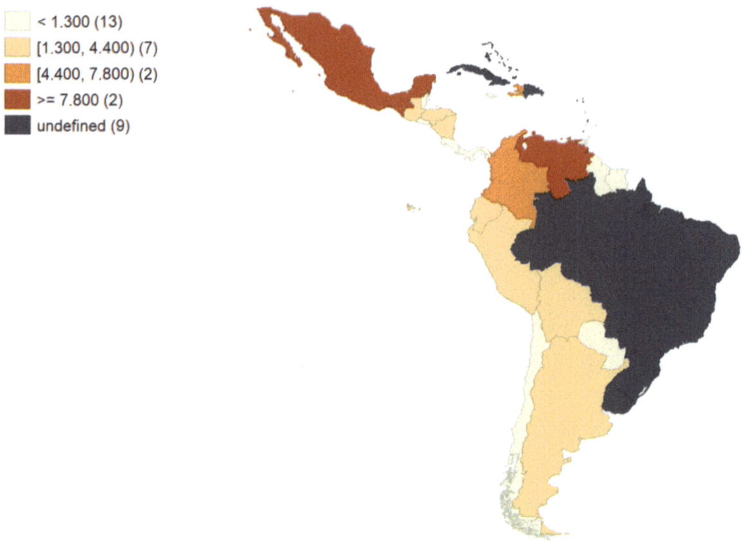

Fig. 13.13 Number of undernourished people (millions) (3-year average) 2018–2020, in Latin American countries (*Source* Elaborated by the authors with data available from FAOSTAT 2022)

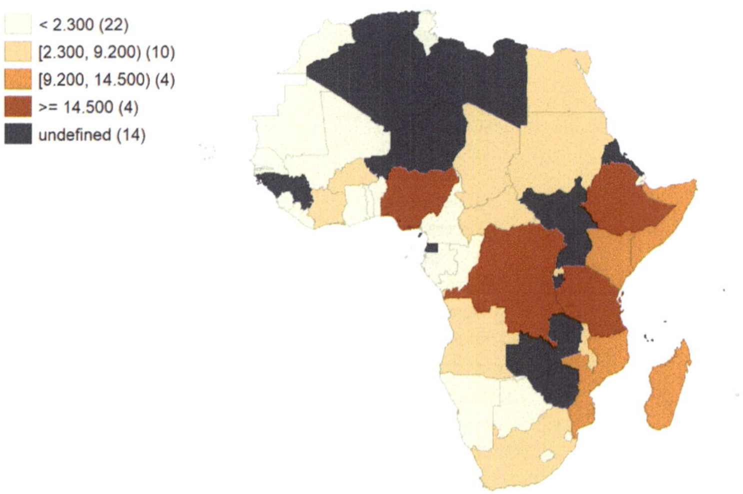

Fig. 13.14 Number of undernourished people (millions) (3-year average) 2018–2020, in African countries (*Source* Elaborated by the authors with data available from FAOSTAT 2022)

Conclusions

Since the early 1990s, the number of disasters related to extreme weather events has doubled. Harvests have decreased, and, as a result, food prices have increased. The climate crisis directly affects the food security of countries. Thus, with global food systems like the ones we have today, the impacts tend to be costlier for everyone, especially for the poorest regions, such as Africa and LAC.

Unfortunately, this trend will not disappear in the next few decades. Climate projections for LAC indicate a temperature increase of up to 2 °C by 2100, which will add to the 0.4 °C increase that has already been registered in several LAC countries in the last three decades. Concerning precipitation patterns, an uneven trend is observed in LAC. Central America and a large part of South America suffer a substantial reduction in annual rainfall and a more significant number of drought events, affecting water availability, yield agriculture, and plant covers in these areas. In contrast, parts of the Pacific and Atlantic coasts of South America will be affected by increased annual rainfall and a higher frequency of flood events.

Considering these trends, and based on the A1FI scenario for 2050, most LAC countries would lose more than 1.5% in their corn production and 1.2% in their rice crops compared to the average from the beginning of the twenty-first century. Countries such as Argentina, Bolivia and Uruguay would increase wheat and rice production by more than 10% by 2050. Interestingly, even under the climatic conditions expected for scenario B1a, most of the region's countries would continue to decrease their corn, wheat, and rice production. Mexico would be the only country that would register a significant improvement in the production of the three grains under a scenario of lower GHG emissions derived from a more sustainable economic and technological development.

Concerning the African continent, general warming is expected by 2100, which will reach values of up to 2 °C, especially in the arid and semi-arid zones of North and South Africa. Almost all countries in tropical climate zones will be impacted by increased rainfall by 2100 (compared to the 1990–2020 period). On the other hand, rainfall is expected to decrease even more in dry climate zones, especially in countries like Madagascar, Mozambique, and Malawi, which will decrease between 0.2 mm and 0.5 mm rainfall per day by 2100.

Under the conditions foreseen for the A1FI scenario, all regions of the African continent would suffer significant grain crop losses by 2050, with averages of 2.5% for wheat, 3.5% for rice, and 10% for maize. In counterpart, food production in Africa would benefit from a more sustainable economic and environmental development in the coming decades. The expected food production yields in 2050 under scenario B1a would not reach the same levels as the values recorded at the beginning of the twenty-first century. However, the reduction would be less significant in most countries of the continent.

The findings reported in this chapter are relevant for 87 countries of the Global South, with an approximate population of 2,191 million people living in LAC and

Africa. It has been identified that at least 247 million people are in a situation of undernourishment. Likely, this indicator is strongly underestimated because nine countries in Latin America and 14 countries in Africa do not have data on the prevalence of undernourishment. Beyond the potential underestimation of this data, it continues to show that climate change and the impact on crops represent a significant obstacle to achieving SDG 2 "Zero Hunger" in these 87 countries.

The data analyzed corresponds to the 2018–2020 period and does not consider the impact of the COVID-19 pandemic on food security. Hence, considering the impact of the COVID-19 pandemic, this figure is expected to have increased. Although the pandemic has aggravated the situation, hunger has risen since 2014. It is necessary to correct the profound vulnerabilities of agri-food systems in the face of climate change. Agricultural practices need to be sustainable and inclusive to deliver food where and when it is right where people need it.

Future Prospects

Based on the results observed in this study, two possible pathways are recommended to transform food systems to address the main determinants of food insecurity and malnutrition in the regions analyzed.

(a) Integration of humanitarian policies in areas with the highest rates of undernourishment as the basis for the fight against poverty and structural inequalities.
(b) Application of strategies to increase resilience to climate change in the different food systems and to strengthen the resilience of the most vulnerable population in the face of economic adversity.

References

Abeldaño Zuñiga RA, Lima GN, González Villoria AM (2021) Impact of slow-onset events related to climate change on food security in Latin America and the Caribbean. Curr Opin Environ Sustain 50:215–224. https://doi.org/10.1016/j.cosust.2021.04.011

Abeldaño Zuñiga RA, de Lima GN, Quiroga DE (2022) Climate change and population displacement in Africa. In: Odimegwu C, Adewoyin Y (eds) The Routledge handbook of African demography. Routledge, London, pp 510–532

Africa Renewal (2021) Climate change triggers food insecurity, poverty and displacement in Africa. https://www.un.org/africarenewal/magazine/november-2021/climate-change-triggers-food-insecurity-poverty-and-displacement-africa. Accessed 26 May 2022

Bárcena A, Samaniego J, Miguel Galindo L et al (2018) La economía del cambio climático en América Latina y el Caribe. Una visión gráfica

Castro-Llanos F, Hyman G, Rubiano J et al (2019) Climate change favors rice production at higher elevations in Colombia. Mitig Adapt Strateg Glob Chang 24:1401–1430. https://doi.org/10.1007/s11027-019-09852-x

Centre for Research on the Epidemiology of Disasters—CRED (2018) Natural Disasters 2017. Brussels, Belgium

Economic Comission for Latin America and the Caribbean (2010) La economía del cambio climático en América Latina y el Caribe. Economic Comission for Latin America and the Caribbean, Santiago, Chile
Economic Commission for Latin America and the Caribbean (2015) The economics of climate change in Latin America and the Caribbean: Paradoxes and challenges of sustainable development. Santiago, Chile
FAOSTAT (2022) Food Security Data. https://www.fao.org/faostat/es/#data/FS. Accessed 26 May 2022
Fernández FJ, Blanco M, Ponce RD et al (2019) Implications of climate change for semi-arid dualistic agriculture: a case study in Central Chile. Reg Environ Chang 19:89–100. https://doi.org/10.1007/s10113-018-1380-0
Food and Agriculture Organization of the United Nations (FAO) (2008) Food security information for action: Practical guidelines. www.foodsec.org/docs/concepts_guide.pdf. Accessed 5 May 2022
Food and Agriculture Organization of the United Nations (2018) Cambio climático y seguridad alimentaria y nutricional en América Latina y el Caribe
Food and Agriculture Organization of the United Nations (2021a) El estado de la seguridad alimentaria y la nutrición en el mundo 2021a. Transformación de los sistemas alimentarios en aras de la seguridad alimentaria, una nutrición mejorada y dietas asequibles y saludables para todos
Food and Agriculture Organization of the United Nations (2021b) El estado mundial de la agricultura y la alimentación (SOFA)
Food and Agriculture Organization of the United Nations (2022) FAOSTAT
Gardi C, Angelini M, Barceló S et al (2015) Soil Atlas of Latin America and the Caribbean. European Commission, Luxembourg
Global Partnership for Sustainable Development Data (2021) Data for food security. How can the international community drive transformative change? Global Partnership for Sustainable Development Data, New York
Iglesias A, Rosensweig C (2009) Effects of climate change on global food production from SRES emissions and socioeconomic scenarios
Intergovernmental Panel on Climate Change (2000) Summary for policymakers. Emissions scenarios—a special report of IPCC working group III
Intergovernmental Panel on Climate Change (2014) AR5 climate change 2014: impacts, adaptation, and vulnerability—IPCC. In: Intergov. Panel Clim. Chang.
Intergovernmental Panel on Climate Change (IPCC) (2018) Global warming of 1.5 °C. An IPCC Special Report on the impacts of global warming of 1.5 °C above pre-industrial levels and related global greenhouse gas emission pathways, in the context of strengthening the global response to the threat of climate change. Washington
Intergovernmental Panel on Climate Change (2021) Climate change 2021. The physical science basis
Intergovernmental Science-Policy Platform on Biodiversity and Ecosystem Services (IPBES) (2018) The IPBES regional assessment report on biodiversity and ecosystem services for Africa. Bonn, Germany
Jaramillo-Mejía MC, Chernichovsky D (2019) Impact of desertification and land degradation on Colombian children. Int J Public Health 64:67–73. https://doi.org/10.1007/s00038-018-1144-0
López Cruz DB, Abeldaño Zuñiga RA, de Lima GN (2021) Analysis of international climate change agreements in the context of Latin America and the Caribbean. In: Leal Filho W (ed) Handbook of climate change management, 1st edn. Springer International Publishing, Cham, pp 1–26
Mahlknecht J, González-Bravo R, Loge FJ (2020) Water-energy-food security: a Nexus perspective of the current situation in Latin America and the Caribbean. Energy 194. https://doi.org/10.1016/j.energy.2019.116824
Matondo J, Alemaw B, Sandwidi W (2020) Climate variability and change in Africa. Springer Nature Switzerland AG 2020, Cham

Oxfam International Food and Climate Justice. https://www.oxfam.org/en/food-and-climate-justice. Accessed 27 May 2022
Parrya ML, Rosenzweig C, Iglesias A et al (2004) Effects of climate change on global food production under SRES emissions and socio-economic scenarios. Glob Environ Chang 14:53–67
Research Program on Climate Change Agriculture and Food Security (CCAFS) (2011) Annual report 2011: groundwork for success. Copenhagen
Royal Netherlands Meteorological Institute (2022) KNMI Climate Explorer
Secretaría de Medio Ambiente y Recursos Naturales (2019) Cambio climático impacta la seguridad alimentaria por bajos rendimientos agrícolas y ganaderos. https://www.gob.mx/semarnat/articulos/cambio-climatico-impacta-la-seguridad-alimentaria-por-bajos-rendimientos-agricolas-y-ganaderos. Accessed 26 May 2022
The Centre for International Governance Innovation (CIGI) (2009) Climate change in Africa: adaptation, mitigation and governance challenges. Ontario, Canada
The Mary Robinson Foundation—Climate Justice (2011) Food security: a climate justice approach. https://ccafs.cgiar.org/news/food-security-climate-justice-approach. Accessed 27 May 2022
United Nations Economic Commission for Africa (UNECA) (2008) Climate change: African perspectives for a Post-2012 Agreement. Addis Ababa, Ethiopia
United Nations Sustainable Development (2019) Climate justice. https://www.un.org/sustainabledevelopment/blog/2019/05/climate-justice/. Accessed 27 May 2022
Vargas R, Cabrera M, Cicowiez M et al (2018) Climate risk and food availability in Guatemala. Environ Dev Econ 23:558–579. https://doi.org/10.1017/S1355770X18000335
World Bank Group (2022) Climate change knowledge portal

Gabriela Narcizo de Lima Assistant Professor at the Department of Geography of the Faculty of Letters—University of Porto—Portugal. Doctor of Science (Physical Geography) from the University of Sao Paulo (USP)—Brazil, with a master's and bachelor's degree in Geography from the São Paulo State University "Julio de Mesquita Filho" (UNESP)—Brazil. Between 2016 and 2018, Professor Gabriela Narcizo de Lima carried out two postdoctoral research stays in the Department of Territorial Planning and Geoprocessing of the Institute of Geosciences and Exact Sciences of Rio Claro/Sao Paulo (UNESP) and the Institute of Geography of the National Autonomous University of Mexico (UNAM). Her main lines of research are related to the impacts of anthropic activities on climate at various scales and issues associated with managing water resources and land use planning.

Roberto Ariel Abeldaño Zuñiga Ph.D. in Demography and Magister in Public Health from the National University of Córdoba, Argentina; Research Specialist from the University of São Paulo (USP), Brazil. His research topics are related to health, society, and environment; disasters, mental health and vulnerable populations; and public health in disaster situations. He has been a consultant for the United Nations Development Program, the World Bank, and UN Migration. He is a member of the National System of Researchers (SNI) in Argentina and Mexico. Currently, he is a full-time researcher at the University of Sierra Sur in Mexico.

Mina Margaret Ogbanga Dr Mina Ogbanga is an Associate Professor in the Faculty of Social Sciences River State University with demonstrated experience in Community Rights and Climate Justice. She is a Global Associate Scholar of the School of Social Work IUPUI a Ford Foundation and EADI Jump Fellow. Dr Ogbanga is a Graduate of several Local, National and International Universities including Harvard Kennedy School, United States, Cambridge University United Kingdom amongst others. Dr Ogbanga is a Multi award winning author and Development actor. Her strong skills in policy Influencing has led to the passage of several bills and policies that is impacting on several lives. Her fight for community rights has won ger several local, national and Global awards. She served as the national Coordinator of the United Nations Sustainable Development Solutions Network in Policy, Partnership and Projects amongst other leadership roles in the promotion of a more sustainable world.

Chapter 14
Gender Gaps in Food Security in Mexico, in the Context of a Changing Environment

Diana Guadalupe Antonio Sánchez, María Alejandra Sánchez Bandala, Nelly Victoria Gutiérrez Moguel, Gabriela Narcizo de Lima, and Roberto Ariel Abeldaño Zuñiga

Abstract In Mexico, the high social and economic inequality is reflected in the physical and economic difficulties that vulnerable groups face in accessing a healthy diet. Thus, in this chapter, an analysis of census data was carried out at the municipal level with the objective of identifying gender gaps in the food security of Mexican households. In general, it was observed that between 2015 and 2020 there was an improvement in the food security situation in the country. Meanwhile, it was noted that the levels of food insecurity throughout Mexico continue to be higher in households headed by women compared to households headed by men. Inequalities in access to and control of food not only affect the status of women, but also have significant implications for food security at the household and community levels.

Introduction

In recent decades, the region of Latin America and the Caribbean has made significant progress towards the effective exercise of the right to adequate food and the protection of the health of all its inhabitants (Food and Agricultural Organization of the United Nations [FAO] et al. 2019a). However, the current period of low economic growth, severe climatic phenomena, unsustainable modes of food production and consumption, the demographic, epidemiological, and nutritional transition, as well as the

D. G. Antonio Sánchez · M. A. Sánchez Bandala · R. A. Abeldaño Zuñiga (✉)
University of Sierra Sur, Guillermo Rojas Mijangos SN, Ciudad Universitaria, Miahuatlán de Porfirio Díaz, P.C. 70805 Oaxaca, Mexico
e-mail: ariabeldanho@gmail.com

N. V. Gutiérrez Moguel
Universidad de La Cañada, Carretera Teotitlán- San Antonio Nanahuatipán, K.M. 1.7 S/N, Teotitlán de Flores Magón, P.C. 68540 Oaxaca, Mexico

G. N. de Lima
Geography Department—Faculty of Letters, Porto University, Via Panorâmica s/n, P.C. 4150-564 Porto, Portugal
e-mail: glima@colef.mx

W. Leal Filho et al. (eds.), *Climate Change and Health Hazards*, Climate Change Management, https://doi.org/10.1007/978-3-031-26592-1_14

disparity between men and women in terms of rights, resources and opportunities, risk jeopardizing the advances made to date. On the other hand, current evidence has established a well-known link between climate change and food insecurity (Swinburn et al. 2019).

The latest available data on undernourishment in Latin America and the Caribbean show an increase in the last four years: 42.5 million people were hungry in 2018, representing an increase of 4.5 million people compared to the 38 million people reported in 2014 (FAO et al. 2019a). Food insecurity, understood as the limited or uncertain availability of nutritionally adequate and safe foods or the limited or uncertain ability to acquire acceptable foods in socially acceptable ways (FAO 1996), affects 187 million people in the Latin American region in its three levels: mild, moderate and severe food insecurity (FAO et al. 2019b).

The characteristics of food environments are essential to explain the different forms of malnutrition. For example, in Mexico the high social and economic inequality is reflected in the physical and economic difficulties that vulnerable population groups which include children, women, and certain ethnic groups all face in accessing a healthy diet. Additionally, according to the National Health and Nutrition Survey (ENSANUT) Mexico is in first place worldwide in childhood obesity (35.5%) and second in adult overweight/obesity (75.2%) (Secretaría de Salud de México 2018), which indicates that Mexican families are concerned about acquiring food in order only to satisfy their hunger, leaving aside the nutritional aspect of food (Aguirre Becerra et al. 2017).

Food environments have changed at an ever-increasing rate (Swinburn et al. 2019). Today, women experience growing participation in the labor market. However, in many cases, there is a lack of shared responsibility between men and women in domestic and unpaid work, which also influences food environments.

Recently, it has been emphasized that inside the family each member has inequitable control of food, other basic goods, and productive resources (Romero and Chávez 2013). Consequently, food security increasingly takes as a unit of analysis not the family as a whole, but each person, considering the different determinants of their vulnerability (gender, age, health condition). Households with persons under 18 years of age, with a female head in the household, lacking in schooling, where the head of the family and spouse speak an indigenous language or have a family composition of more than four members are the most vulnerable to being food insecure (Mundo-Rosas et al. 2013).

The conventional approach, which took the family as a compact unit, has been superseded by the finding that men and women have differences in terms of power, control of resources, perceptions, and priorities regarding family spending (Aguilar 2016). Women are generally discriminated against in the administration of resources and often also in access to food, a situation that is exacerbated in circumstances of food crisis (Romero and Chávez 2013).

Studies on household headship in developing countries has been dominated by efforts to estimate the greater incidence of poverty among female-compared to male-headed households at a point in time (Felker-Kantor and Wood 2012). Less common are studies that focus on the association between female headship and food insecurity.

Whereas it is hardly surprising that poor households are more likely to experience food insecurity compared to richer households, the aim of this chapter is to identify clusters of households with food insecurity in Mexico, and to explore the association between food insecurity and the gender of the household head.

Geographic and Gender Differences in Food Security

Geographically, all those territories where the majority of the population suffers from various forms of food insufficiency could be called "hunger zones". In this sense, entire populations around the world are being affected by food insecurity. However, according to the World Food Program, of all the starving people in the world, 60% are women and girls (Wu et al. 2021). Despite the fact that it is precisely women who produce most of the food grown in the world, because of gender issues, they rarely own the land they cultivate (Beckford and Campbell 2013). Therefore, it can be said that social inequalities are what create vulnerable populations.

Thus, in various parts of the developing world, women are less likely to own farmland and when they do, the land they have access to is often of inferior quality and in smaller plots. For example, in Kenya men's plots are on average three times larger. In Bangladesh, Ecuador and Pakistan, the average size of an agricultural plot of male-headed households is twice as large relative to female-headed households (Elum 2020). Various studies confirm that the gender division of labor in agriculture creates unequal power dynamics, responsibilities, and benefits within households. Similarly, their work tends to be more precarious (Adereti and Fasina 2017). In many cases they even face discrimination and sexual violence in the fields and factories where they work.

It is precisely the devaluation of what is feminine derived from the division of the sexes in regards to labor, which translates into the discrimination suffered by women due to their own gender condition: particularly in certain domestic life cycles (pregnancy, breastfeeding, old age), in the precariousness of employment, in wage inequalities between men and women, in the restriction of access to property and control of resources, and in the lack of autonomy over their own bodies in terms of sexuality and reproduction (Garcia and Wanner 2017). Current analysis reveals the vital role that women play in households as food providers, but often fails to recognize the extent to which they are sacrificing their own health and sometimes that of their daughters when access to food is undermined (Ruiz 2010).

In this way, women become real *buffers against the impacts of household food insecurity*. In fact, it has been documented that mothers in poverty spend more time and effort feeding their families (Mahadevan and Hoang 2016). This is contributed to by poor living conditions, intensive search for offerings in nearby environments, and pressure to make efficient decisions that avoid wasting resources. Evidence has also shown that the burden of this work on women leads them to sacrifice the quality of their own diets for the sake of privileging the consumption to children or household members who are gainfully employed.

Moreover, in female-headed households where no other adult is present, the female head of household is not only responsible for generating income but also for completing household chores (Felker-Kantor and Wood 2012). In other words, Western women may no longer be confined exclusively to the domestic field and participate in almost all activities of public life. But this does not mean that they have disengaged from activities in the domestic field, but rather that old and new practices are coexisting in a relationship of conflicting tension.

In this sense, care work, which includes everything from preparing food, cleaning the house, caring for the sick, elderly, and children, to work in the production of food and other resources for domestic use. These are activities usually attributed to women even when they are not only "homemakers" and have activities outside the home or paid jobs (Elum 2020). This additional load of activities often adds to the stress and duties that they are socially expected to absorb in order to manage their families' food security. Such as the constant pressure to provide sufficient and nutritious food for their families with less and less food and money available (Magaña-Lemus et al. 2016).

Research on household headship in developing countries has been largely dominated by efforts to estimate the greater incidence of poverty among female-headed compared to male-headed households at a point in time, and more controversially to document the "feminization of poverty," which refers to the increase over time in the proportion of the poor population that is female (Romero and Chávez 2013). Less common are studies that focus on the association between female headship and food insecurity. Most research on this topic has been carried out in the United States and in Africa (Felker-Kantor and Wood 2012). In Mexico, studies of female headship and food security have been either regional (Navarro 2010) or in the case of national approaches (Magaña-Lemus et al. 2016), have been based on cross-tabulations or economic models.

Measuring Food Insecurity in Mexico

This section proposes an analysis at the municipality level and the identification of gender gaps in the food security of Mexican households.

Several studies have analyzed food security in Mexico. Comparisons have been made between 2012 and 2018, also new integrated indicators have been proposed and some regional differences were identified (Mundo-Rosas et al. 2013, 2018; Shamah-Levy et al. 2013; Urquía-Fernández Robles 2013; Adereti and Fasina 2017; Ibarrola-Rivas and Galicia 2017; Lemos Figueroa et al. 2018; Torres and Rojas Martínez 2020). But so far, the possible gaps at the municipal level have not been identified for the whole country, nor have the possible gender gaps in this phenomenon been identified.

In Mexico there are two instances for measuring and monitoring household food security at the population level. One of these instances is the National Health and Nutrition Survey (ENSANUT), which is conducted every 6 years with the objective of monitoring risk factors for chronic noncommunicable diseases (NCDs), together with dimensions related to the diet of Mexicans (Instituto Nacional de Salud Pública 2021). This survey started in 2000, and then the 2006, 2012 and 2018 versions included a version of the Latin American and Caribbean Food Security Scale (ELCSA), in its Mexican version (Carmona et al. 2017).

The survey is nationally representative, and the food security scale used distinguishes households in which minors are present from those in which no minors are present.

In the 2006 and 2012 versions of the ENSANUT, its sample design allowed us to obtain analysis of the situation of all households by municipality. However, in the 2018 version this changed, and we can only obtain prevalences of household food security and insecurity at the aggregate level by state, not by municipality (Instituto Nacional de Salud Pública 2021).

For their part, the national population and housing censuses are a universal data source that are conducted every 10 years in all countries. In the case of Mexico, the 2010 National Population and Housing Census included 3 questions to investigate the food security situation of Mexican households. In order to monitor this situation in Mexican households, the intercensal survey of 2015 and the national population and housing census of 2020 already included a set of 6 questions of the Mexican Food Security Scale (EMSA) (Instituto Nacional de Estadística y Geografía [INEGI] 2016, 2021). These data sources also allow disaggregation at the level of municipalities; therefore, it is considered the best data source to represent the food security situation of all households in the almost 2,500 municipalities of the 32 states of Mexico (INEGI 2016, 2021).

The questions included in the Mexican Food Security Scale in these two data sources refer to each household, and were as follows:

- Due to lack of money or resources, they went without food.
- Due to lack of money or resources, what they ate was based on very little food variety.
- Due to lack of money or resources, they stopped eating breakfast, lunch, or dinner.
- Due to lack of money or resources, they ate less than they should have eaten.
- Due to lack of money or resources, they felt hungry, but did not eat.
- Due to lack of money or resources, they ate only once a day or stopped eating for a whole day.

The scale has a minimum score of zero and a maximum of 6. Zero is interpreted to mean a food secure household, while values of 1 and 2 points mean mild food insecurity, 3 and 4 mean moderate food insecurity, and 5 and 6 points mean severe food insecurity (INEGI 2016, 2021).

This section shows the analyses carried out with data from 2015 and 2020, stratified by gender of the head of household. In general, the 2015 data shows that the prevalence of food security in all Mexican households was 69% in male-headed households and 67.6% in female-headed households. In male-headed households, mild, moderate, and severe food insecurity was 17.2, 5.5 and 7.7% respectively. For female-headed households, mild, moderate, and severe insecurity was 16.4, 6.2, and 9.9% respectively.

By 2020, the situation has improved, registering a prevalence of food security of 70.7% in all households, being 71.5% in male-headed and 68.9% in female-headed households. Mild, moderate, and severe food insecurity affected 17.2, 5.4, and 6.0% of male-headed households respectively; while it affected 17.2, 6.2 and 7.8% of female-headed households respectively.

In 2015, the analysis of hierarchical clusters shows that 3 types of municipalities could be differentiated in Mexico. A cluster formed by 2,028 municipalities with a food security prevalence of 64%, another cluster formed by 403 municipalities with a food security prevalence of 35%, and a third cluster formed by 38 municipalities with a food security prevalence of 4%. This shows that 96 out of every 100 households in these municipalities had some level of food insecurity. These municipalities are mainly located in the Pacific-South region and in the Gulf of Mexico region (Fig. 14.1).

In the year 2020, the same patterns of geographic distribution of municipalities with lower prevalence of food security are observed; but in this case there is a cluster of 1484 municipalities with a food security prevalence of 71%, a second cluster of 915 municipalities with a food security prevalence of 47% and a third cluster of 70 municipalities with a food security prevalence of 20%. This means that the food security situation improved for all households, but the same territorial inequalities are still seen in the municipalities of the strip that goes from the Pacific-South to the Gulf of Mexico (Fig. 14.2).

When disaggregating by male-headed and female-headed households, important differences are observed that reaffirm what was stated in the previous paragraphs: female-headed households are in a situation of high vulnerability to food insecurity.

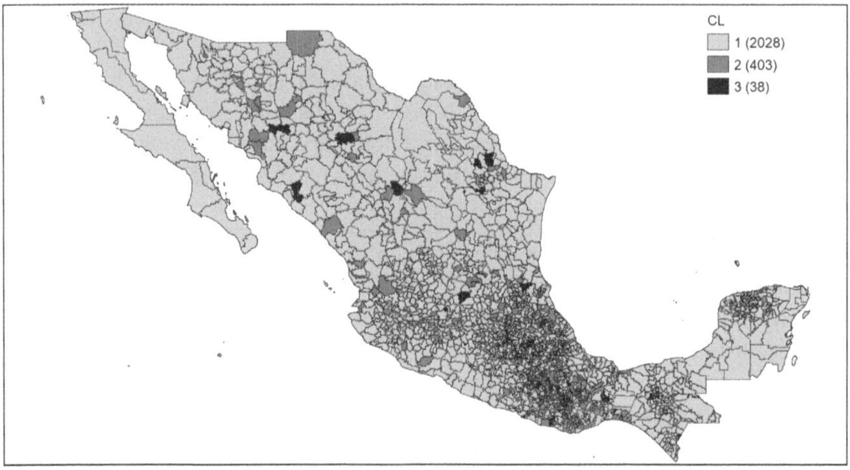

Fig. 14.1 Cluster analysis. Prevalence of food security in all Mexican households 2015 (*Source* Inter Census Survey 2015)

Figure 14.3 shows the cluster analysis of all female-headed households in 2015, by municipality. There is a cluster of 1,872 municipalities with a 64% prevalence of food security in female-headed households, a second cluster of 571 municipalities with a food security prevalence of 37%, and a third cluster of 26 municipalities with a 0% prevalence of food security. That is, municipalities where all female-headed households were food insecure (Fig. 14.3).

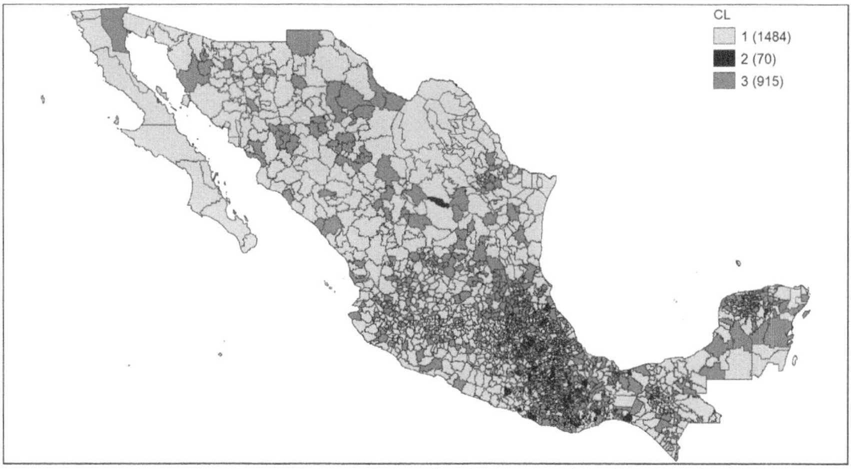

Fig. 14.2 Cluster analysis. Prevalence of food security in all Mexican households 2020 (*Source* Census of population and households 2020)

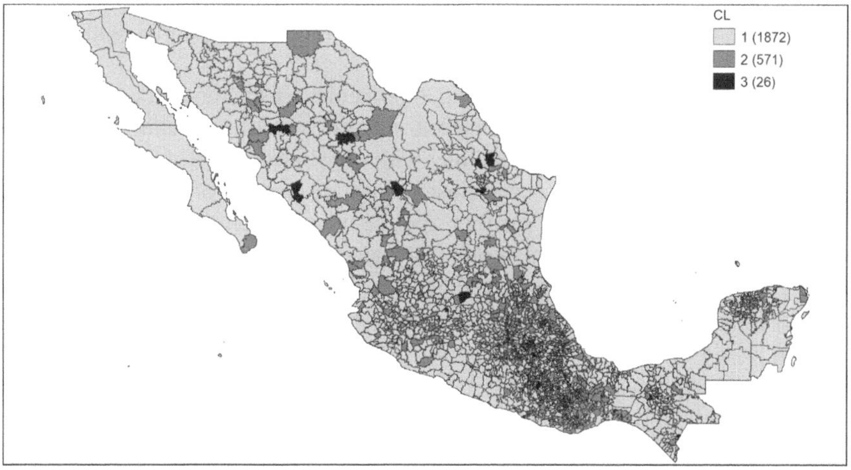

Fig. 14.3 Cluster analysis. Prevalence of food security in female-headed Mexican households 2015 (*Source* Inter Census Survey 2015)

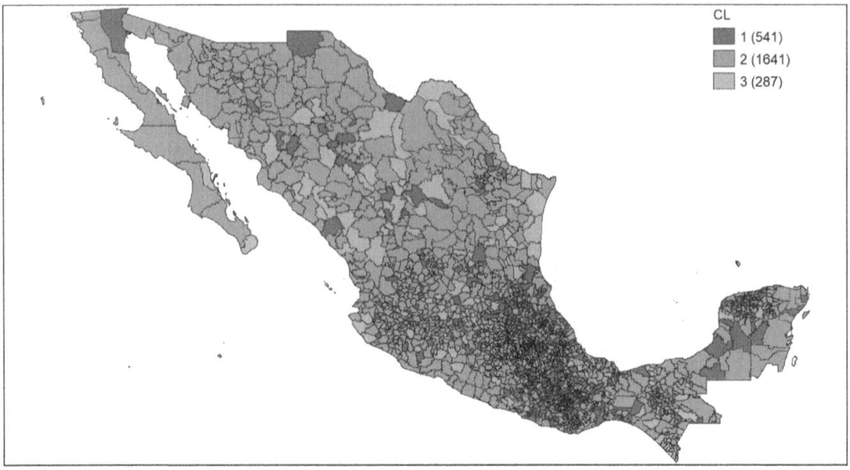

Fig. 14.4 Cluster analysis. Prevalence of food security in female-headed Mexican households 2020) (*Source* Census of population and households 2020)

The situation of female-headed households in 2020 improved substantially. The 2020 data allows us to identify a cluster of 541 municipalities with a food security prevalence of 83%, a second cluster of 1641 municipalities with a food security prevalence of 62%, and a third cluster composed of 287 municipalities with a food security prevalence of 35% (Fig. 14.4).

For its part, the analysis of hierarchical clusters in the male-headed households allowed identifying in 2015 a cluster made up of 1,931 municipalities with a food security prevalence of 66%, another made up of 496 municipalities with a food security prevalence of 38%, and a third cluster of 42 municipalities with a food security prevalence of 6% (Fig. 14.5).

In the 2020 data, on the other hand, an improvement in the food security situation in the male-headed households is identified, since an increase in the first cluster was identified, made up of 2,215 municipalities with a 65% prevalence of food security. A reduction was observed in the second cluster, this time made up of 248 municipalities with a prevalence of 31% food security, and the third cluster was also reduced to 6 municipalities with a prevalence of only 2% food security (Fig. 14.6).

These analyses were complemented with a risk estimation for both male-headed and female-headed households. An odds ratio (OR) estimate: 1.098 (95% CI: 1.096–1.100; $p < 0.001$) was obtained in the 2015 data. On the other hand, in the 2020 data, the estimate obtained was OR: 1.136 (95% CI: 1.134–1.137; $p < 0.001$). This implies that, although in 2020 the situation improved for all households, gender gaps still persist where female-headed households have a greater possibility of falling into a situation of food insecurity compared to male-headed households. The fact

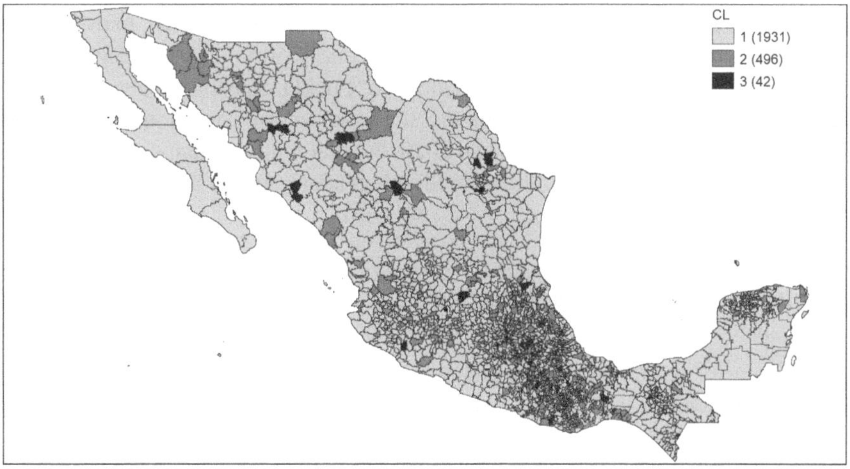

Fig. 14.5 Cluster analysis. Prevalence of food security in male-headed Mexican households 2015 (*Source* Inter Census Survey 2015)

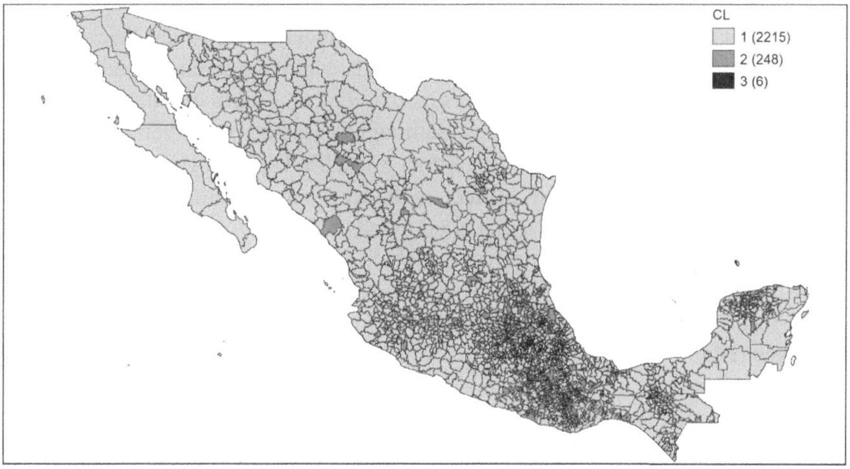

Fig. 14.6 Cluster analysis. Prevalence of food security in male-headed Mexican households 2020 (*Source* Census of Population and Households 2020)

that the risk for female-headed households has increased (albeit slightly) should be interpreted as meaning that the speed of change is slower in these households.

Table 14.1 shows this speed of change, calculated as the change in percentage points between the food security prevalence observed in 2015 and 2020, according to household headship. In Mexico City, Puebla, Sinaloa, Sonora, Tabasco, and Yucatan, food security prevalence for households registered a decrease in 2020. In the case of female-headed households, the decline in food security prevalence was registered in the states of Campeche, Mexico City, Guerrero, Puebla, Sinaloa, Sonora, Tabasco, Veracruz, and Yucatan. This reaffirms that the food security situation improved for Mexican households between 2015 and 2020, but that improvement did not have the same weight for female-headed households, so gender gaps persists.

Conclusions and Future Prospects

Food insecurity is a phenomenon exacerbated by inequitable global and national processes. According to the results in Mexico, the prevalence of food security from 2015 to 2020 had a considerable increase. Such variation could be due to a greater coverage of social and food aid programs across the country, as evidence points to the fact that an adequate design of cash and in-kind transfer programs generate positive effects on food security indicators of beneficiary households (Kairiza and Kembo 2019). However, in specific regions such as the South Pacific and the Gulf of Mexico region despite the improvement, the figures remain alarmingly low. A compelling finding is that regardless of the state of the republic in which one lives, levels of food insecurity are higher in female headed households compared to male headed households and even in 2020 when the numbers increased across the board, the existing gap in female headed households remained. Although it has been documented that when women are able to make decisions regarding household spending and consumption they tend to have a greater positive effect on the well-being of other family members (Felker-Kantor and Wood 2012), this good management is not enough to be able to ensure sustainable food for themselves and their families.

Inequalities in access to and control over food not only affects the status of women, but also results in significant implications for food security at household and community levels (Elum 2020). In relation to the latter, to recognize food insecurity as a gender justice issue is first and foremost to question the roles imposed and the social value placed on the work that women do. So as to finally be able to reflect on what types of interventions would really make a difference in terms of eradicating those practices that end up reproducing vulnerable populations.

The empirical evidence that female-headed households experience a higher risk of hunger confirms the importance of constant measurement and monitoring of food insecurity. As well as the study of gender inequality in households, which is presented as an opportunity to update the debate on the situation of female-headed households, which in turn will generate new knowledge on the determinants of quality of life. Estimates of gender differences in food insecurity also provide a point of reference with which to make diagnoses that will make it possible to prioritize certain regions and work, in the immediate future, on the design of a food security policy with a true gender perspective.

Table 14.1 Prevalence of food security in Mexican households by state and by head of households 2015 and 2020

State	Male-headed 2015 (%)	Male-headed 2020 (%)	Difference 2020–2015 (%)	Female-headed 2015 (%)	Female-headed 2020	Difference 2020–2015 (%)
Aguascalientes	69.47	72.21	2.7	67.02	68.55	1.5
Baja California	74.98	80.05	5.1	71.26	76.50	5.2
Baja California Sur	75.97	76.03	0.1	72.15	74.17	2.0
Campeche	62.27	62.61	0.3	61.59	59.12	−2.5
Chiapas	78.94	83.22	4.3	76.33	81.26	4.9
Chihuahua	64.45	69.03	4.6	59.35	63.35	4.0
Coahuila	63.66	63.77	0.1	62.36	59.89	−2.5
Colima	71.08	75.61	4.5	68.69	72.14	3.5
Mexico City	81.43	81.06	−0.4	78.82	77.97	−0.9
Durango	65.99	67.36	1.4	65.56	65.58	0.0
Guanajuato	65.90	73.77	7.9	64.99	71.75	6.8
Guerrero	47.22	47.81	0.6	46.94	46.61	−0.3
Hidalgo	60.32	62.15	1.8	58.67	60.11	1.4
Jalisco	70.41	76.58	6.2	67.52	72.72	5.2
México	67.56	68.45	0.9	65.23	65.22	0.0
Michoacán	54.67	65.84	11.2	52.99	62.80	9.8
Morelos	61.06	63.84	2.8	59.58	60.78	1.2
Nayarit	59.63	62.76	3.1	59.84	61.50	1.7
Nuevo León	82.68	84.11	1.4	80.43	82.07	1.6
Oaxaca	48.69	49.24	0.6	46.54	46.92	0.4
Puebla	56.06	56.00	−0.1	54.13	53.27	−0.9

(continued)

Table 14.1 (continued)

State	Male-headed 2015 (%)	Male-headed 2020 (%)	Difference 2020–2015 (%)	Female-headed 2015 (%)	Female-headed 2020	Difference 2020–2015 (%)
Querétaro	69.20	71.16	2.0	66.42	68.81	2.4
Quintana Roo	67.31	68.89	1.6	68.47	69.73	1.3
San Luis Potosí	58.98	65.84	6.9	59.49	65.03	5.5
Sinaloa	66.28	65.21	−1.1	65.68	62.79	−2.9
Sonora	76.35	73.83	−2.5	73.08	70.47	−2.6
Tabasco	59.62	51.39	−8.2	59.03	48.98	−10.0
Tamaulipas	74.42	77.41	3.0	73.59	74.66	1.1
Tlaxcala	63.44	65.45	2.0	59.55	61.53	2.0
Veracruz	57.05	57.08	0.0	55.27	54.62	−0.7
Yucatán	67.24	62.97	−4.3	64.68	59.68	−5.0
Zacatecas	64.52	71.67	7.2	64.49	69.73	5.2
Total (%)	69.60	71.50	1.9	67.60	68.90	1.3
Total (n)	15,555,958	16,986,783	1,430,825	6,168,833	7,657,913	1,489,080

Source Inter Census Survey 2015 and Census of Population and Households 2020

References

Adereti DT, Fasina OO (2017) Gender analysis of food security status of rural households in Ondo State, Nigeria. Russ Agric Sci 43:353–360. https://doi.org/10.3103/S1068367417040024

Aguilar L (2016) Mujeres jefas de hogar y algunas características de los hogares que dirigen. Una visión sociodemográfica [Women heads of household and some characteristics of the households they lead. A sociodemographic vision]. México

Aguirre Becerra H, García Trejo JF, Vázquez Hernández MC et al (2017) Panorama general y programas de protección de seguridad alimentaria en México [Overview and food safety protection programs in Mexico]. Rev Méd Electrón 39

Beckford CL, Campbell DR (2013) Women, agriculture, and food security in the caribbean. Domestic food production and food security in the caribbean. Palgrave Macmillan US, New York, pp 79–93

Carmona J, Paredes J, Pérez A (2017) La Escala Latinoamericana y del Caribe sobre Seguridad Alimentaria (ELCSA): Una herramienta confiable para medir la carencia por acceso a la alimentación [The Latin American and Caribbean food security scale (ELCSA): a reliable tool to measure the lack of access to food]. Rev Iberoam las Ciencias Soc y Humanísticas 6: https://doi.org/10.23913/ricsh.v6i11.118

Elum ZA (2020) Gender and poverty: its influence on household food security in Africa. In: Leal W, Azul AM, Brandli L et al (eds) Gender equality. Springer International Publishing, Cham, pp 1–13

FAO, PAHO, WFP, UNICEF (2019a) Regional overview of food security and nutrition in Latin America and the Caribbean. Santiago

FAO, UNICEF, FIDA et al (2019b) El estado de la seguridad alimentaria y la nutrición en el mundo 2019 [The state of food security and nutrition in the world 2019]

Felker E, Wood CH (2012) Female-headed households and food insecurity in Brazil. Food Secur 4:607–617. https://doi.org/10.1007/s12571-012-0215-y

Food and Agricultural Organization of the United Nations (1996) Report of the world food summit Rome, 13–17 November. http://www.fao.org/3/w3548e/w3548e00.htm. Accessed 4 Apr 2021

Garcia AS, Wanner T (2017) Gender inequality and food security: lessons from the gender-responsive work of the International Food Policy Research Institute and the Bill and Melinda Gates Foundation. Food Secur 9:1091–1103. https://doi.org/10.1007/s12571-017-0718-7

Ibarrola-Rivas MJ, Galicia L (2017) rethinking food security in Mexico: discussing the need for sustainable transversal policies linking food production and food consumption. Investig Geográficas 2017:106–121. https://doi.org/10.14350/rig.57538

Instituto Nacional de Estadística y Geografía (2016) Encuesta Intercensal 2015. Marco conceptual [Intercensus survey 2015. Conceptual framework]. Mexico City

Instituto Nacional de Estadística y Geografía (2021) Censo de Población y Vivienda 2020. Marco conceptual [Population and housing census 2020. Conceptual framework]. Mexico City

Instituto Nacional de Salud Pública (2021) Encuesta Nacional de Salud y Nutrición [National health and nutrition survey]. https://ensanut.insp.mx/. Accessed 14 Apr 2021

Kairiza T, Kembo GD (2019) Coping with food and nutrition insecurity in Zimbabwe: does household head gender matter? Agric Food Econ 7:24. https://doi.org/10.1186/s40100-019-0144-6

Lemos M, Baca J, Cuevas Reyes V (2018) Poverty and food insecurity in the mexican countryside: an unsolved public policy issue. Textual 71–105. https://doi.org/10.5154/r.textual.2017.71.004

Magaña D, Ishdorj A, Rosson CP, Lara- J (2016) Determinants of household food insecurity in Mexico. Agric Food Econ 4:10. https://doi.org/10.1186/s40100-016-0054-9

Mahadevan R, Hoang V (2016) Is there a link between poverty and food security? Soc Indic Res 128:179–199. https://doi.org/10.1007/s11205-015-1025-3

Mundo V, Shamah T, Rivera- JA (2013) Epidemiología de la inseguridad alimentaria en México [Epidemiology of food insecurity in Mexico]. Salud Publica Mex 55:S206–S213

Mundo-Rosas V, Vizuet-Vega NI, Martínez-Domínguez J et al (2018) Evolución de la inseguridad alimentaria en los hogares mexicanos: 2012–2016 [Evolution of food insecurity in Mexican households: 2012–2016]. Salud Publica Mex 60:309. https://doi.org/10.21149/8809

Navarro A (2010) ¿Mujeres proveedoras y jefas de familia? …Nuevas realidades rurales en localidades de la región zamorana [Women providers and heads of families? … New rural realities in localities of the Zamora region]. La Vent Rev Estud Género 4:139–171

Romero Y, Chávez Y (2013) Jefaturas femeninas: una aproximación a la feminización de la pobreza y de la responsabilidad en familias desplazadas por la violencia [Female-headed households: an approach to the feminization of poverty and responsibility for families displaced by violence]. Tabula Rasa 18:255–264

Ruiz H (2010) Nuestra raíz, nuestro futuro: el maíz. Una mirada a la legislación que regula el uso, conservación y diversificación de la semilla de maíz en México [Our root, our future: corn. A look at the legislation that regulates the use, conservation and diversification of corn seed in Mexico]. 1–13

Secretaría de Salud de México (2018) Encuesta Nacional de Salud y Nutrición 2018. Presentación de resultados [National Health and Nutrition Survey 2018. Main Results]. Mexico City

Shamah-Levy T, Mundo-Rosas V, A Rivera-Dommarco J (2013) La magnitud de la inseguridad alimentaria en México: su relación con el estado de nutrición y con factores socioeconómicos [The magnitude of food insecurity in Mexico: its relationship with nutritional status and socioeconomic factors]. Salud Publica Mex 56:79. https://doi.org/10.21149/spm.v56s1.5169

Swinburn BA, Kraak VI, Allender S et al (2019) The global syndemic of obesity, undernutrition, and climate change: the Lancet commission report. Lancet 393:791–846. https://doi.org/10.1016/S0140-6736(18)32822-8

Torres F, Rojas Martínez A (2020) Seguridad alimentaria y sus desequilibrios regionales en México [Food security and its regional imbalances in Mexico]. Probl del Desarro Rev Latinoam Econ 51: https://doi.org/10.22201/iiec.20078951e.2020.201.69521

Urquía-Fernández Robles N (2013) La seguridad alimentaria en México [Food securtiy in Mexico]. Salud Publica Mex 56:92. https://doi.org/10.21149/spm.v56s1.5171

Wu J, Zhang J, Ge Z et al (2021) Impact of climate change on maize yield in China from 1979 to 2016. J Integr Agric 20:289–299. https://doi.org/10.1016/S2095-3119(20)63244-0

Diana Guadalupe Antonio Sánchez Her academic background includes a degree in nutrition from the Universidad Veracruzana (Mexico) and a Master's in Public Health. She worked at the Food Security Observatory of Veracruz, where she became aware of one of the biggest problems we face as a society: hunger and inequality. Her main research topics are food security and climate change, studying both phenomena as global health problems in the Mexican context.

María Alejandra Sánchez Bandala She has a PhD and Master in Anthropology from the Center for Research and Higher Studies in Social Anthropology. She graduated in Anthropology from the Universidad Veracruzana (Mexico). Full-time Professor-Researcher at the Universidad de la Sierra Sur (Oaxaca, Mexico). Member of the Health, Society and Environment Study Group from the Institute for Research on Public Health. Her main research topics are related to the health needs of the vulnerable population and social response (emphasis on mental health, sexual and reproductive health, and chronic diseases), care trajectories and their social determination. She was awarded the Desirable Profile (PRODEP-SEP) and the National Researcher Candidate (SNI-CONACYT).

Nelly Victoria Gutiérrez Moguel She has a Degree in Nutrition, a Diploma in Pediatric Clinical Nutrition and a master's degree in Clinical Nutrition from the National Institute of Public Health. She has experience in the clinical area, providing nutritional consultation in the care units for workers of the Metro Transport System of Mexico City. She has made clinical stays in prestigious institutions in Mexico. She is a Full-Time Research Professor at the Universidad de la

Cañada (Mexico). She has published articles in nationally indexed journals related to the following topics: Nutritional medical therapy in patients with terminal chronic kidney disease; Consumption of energy drinks in university students; Overweight and obesity, anaemia and food insecurity in students of the Universidad de la Cañada; and medical nutrition therapy in hospitalized and critically ill patients with COVID-19.

Gabriela Narcizo de Lima Assistant Professor at the Department of Geography of the Faculty of Letters—University of Porto—Portugal. Doctor of Science (Physical Geography) from the University of Sao Paulo (USP)—Brazil, with a master's and bachelor's degree in Geography from the São Paulo State University "Julio de Mesquita Filho" (UNESP)—Brazil. Between 2016 and 2018, Professor Gabriela Narcizo de Lima carried out two postdoctoral research stays in the Department of Territorial Planning and Geoprocessing of the Institute of Geosciences and Exact Sciences of Rio Claro/Sao Paulo (UNESP) and the Institute of Geography of the National Autonomous University of Mexico (UNAM). Her main lines of research are related to the impacts of anthropic activities on climate at various scales and issues associated with managing water resources and land use planning.

Roberto Ariel Abeldaño Zuñiga He has a BSc in Nursing from the National University of Salta, Argentina. Also, he has a Magister in Public Health and a PhD in Demography from the National University of Córdoba, Argentina. He has been in academic stays at the Salvador Allende School of Public Health of the University of Chile and the University of Sao Paulo, Brazil. Currently, he works as a Full Professor at the University of Sierra Sur (Mexico) and Researcher at the National Research Council (CONACYT, Mexico). His main research interests are Public health and the environment. He has been a consultant for the United Nations Development Program, the United Nations International Organization of Migrations, and The World Bank.

Chapter 15
Assessing the Climate Change-Related Health Hazards in Africa

Newton R. Matandirotya, Tonderai Dangare, and Girma Berhe

Abstract The African continent remains extremely vulnerable to the negative impacts of climate change including health hazards. Climatic models are projecting that as the climate is changing so is the frequency and intensity of extreme weather events. The aim of the chapter was to assess the climate change-related health hazards that Africa faces and will face in the future through literature synthesis from various sources. Indications are that the burden of disease from climate change-related ailments is likely to increase for the majority of African countries moreso for those countries with already vulnerable health delivery systems, poor health policies, and poor health infrastructure. Some of the health-related challenges stemm from extreme weather events such as exposure to heat waves, an increase in vector-borne diseases as well increase in climate change-induced disasters.

Introduction

According to Chapman et al. (2022), Africa remains one of the most vulnerable continents to climate change yet it still lacks the adaptive capacity (Leal Filho et al. 2021; Matandirotya et al. 2021). Climate change is expected to contribute directly or indirectly to negative human health outcomes including an increase in the burden of

N. R. Matandirotya (✉)
Department of Geosciences, Faculty of Science, Nelson Mandela University, Port Elizabeth 6000, South Africa
e-mail: runyamore@gmail.com

Centre for Climate Change Adaptation and Resilience, Kgotso Development Trust, P.O. Box 5 Beitbridge, Zimbabwe

T. Dangare
Department of Electronics and Telecommunications Engineering, University of Zimbabwe, 630 Churchill Avenue, Mount Pleasant, Harare, Zimbabwe

G. Berhe
Haramaya University, Institute of Technology, School of Water Resource and Environmental Engineering, P. O. Box 138 Dire Dawa, Ethiopia

W. Leal Filho et al. (eds.), *Climate Change and Health Hazards*, Climate Change Management, https://doi.org/10.1007/978-3-031-26592-1_15

diseases. Broadly climate change is expected to affect humans through an increase in temperatures, an increase in extreme weather events, a rise in sea level and an increase in air pollution (Leal Filho et al. 2022). A rise in both indoor and ambient temperatures is also likely to contribute to an increase in heat-related human illnesses for example heat strokes, kidney diseases, diabetes and other chronic ailments. Climate change is therefore bringing a wide range of human health threats that are also further compounded by brain drain, weak health systems and infrastructure in Africa. The human health threats that the continent faces are multi-layered and can be viewed from a regional basis. This is so because as climate change impacts are not expected to be homogenous across the continent. Similar to other regions around the world the Southern Africa and the horn of Africa is expected to experience more extreme weather events characterised by drought, heat waves and floods while the central-eastern Africa will be more characterised by disease outbreaks, pandemics and vector population increase. On the other hand, droughts will result in a shortage of potable water leading to an increase in incidences of waterborne diseases including cholera, dysentery and typhoid. Drought will also see an increase in the widespread rise in cases of malnutrition and undernutrition. The aim of the chapter was to explore the human health threats that Africa faces because of climate change. Based on literature synthesis, the chapter is arranged in 5 sections; "Extreme High Temperature and Health in Africa" section gives an overview of extreme heat and human health, "Drought-Water Scarcity and Health Challenges in Africa" section gives an overview of drought and human health, "Floods-Cyclones and Health in Africa" highlights the flood and human health association, "Vector-Borne Disease and Health in Africa" section highlights the vector-borne disease proliferation and human health, "Air Pollution, Climate Change and Health in Africa" section highlights the air pollution and human health component, "Future Trends" section gives an overview of the expected future trends, "Policy Recommendations" section outlines the policy recommendations while "Conclusion" section gives the chapter conclusion.

Figure 15.1 summarises various pathways for climate change and human health challenges centred around an increase in global temperatures that will culminate in an increase in incidences of extreme weather events such as heat waves, floods, drought. These will also be accompanied by an increase in the burden of disease.

Extreme High Temperature and Health in Africa

One of the fundamental negative impacts of climate change is the increase in the incidences of heat waves characterised by warmer days/nights (Matandirotya et al. 2021). It is estimated that Africa is experiencing hotter and longer heat waves than in the last two decades (Russo et al. 2016). As the average global temperatures increase so are the prevalence of heat-related related ailments including diabetes, heat stress and high blood pressure (Chapman et al. 2022). Furthermore, exposure to non-optimal temperature has also significantly contributed to excess mortality and morbidity from different ailments (Zhao et al. 2021). Several studies have also found

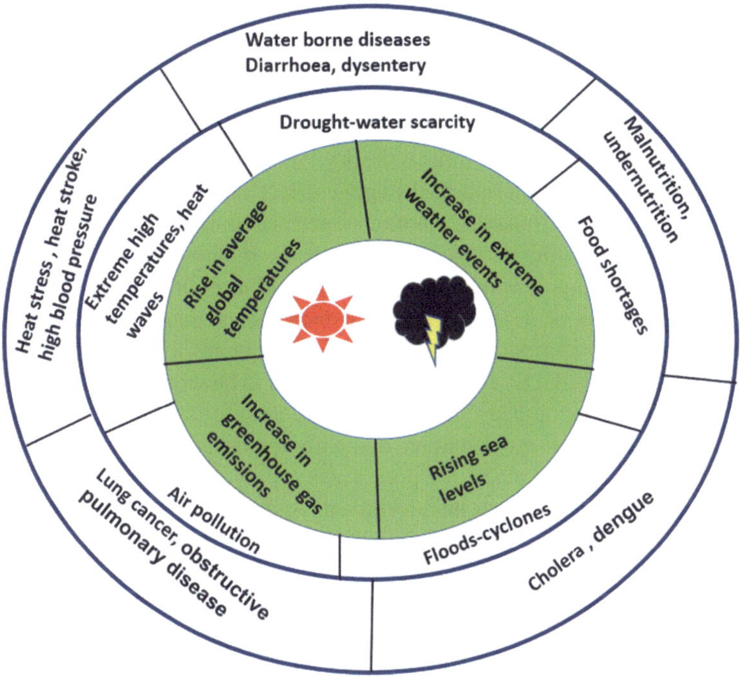

Fig. 15.1 Summarised climate change impacts on human health

strong evidence that links prolonged exposure to heat to short sleep duration (SSD) and in some instances disrupted sleep quality (Minor et al. 2022). Besides, exposure to indoor and ambient heat leads to thermal discomfort (Matandirotya et al. 2021) while on the other hand, SSD tends to increase the risk of cardiovascular disease, adverse mental health outcomes and in some instances type 2 diabetes decrease in immune functionality (Cappuccio et al. 2011). At a global scale, estimates are that heat-related mortality is around 500,000 (Zhao et al. 2021). Studies are indicating that nighttime (minimum daily) temperatures are increasing more rapidly than maximum daytime temperatures over large portions of the land surface. Due to their poor thermoregulatory abilities, old people and young children under the age of 5 are vulnerable (Chapman et al. 2022) as well as those residing in low-cost housing dwellings (Matandirotya et al. 2021). Vulnerability for children in low-income countries is often worsened by pre-existing conditions including undernutrition as well as poor healthcare facilities (Chapman et al. 2022). Strong empirical evidence exist on heat impacts on human health from around the world however very few studies have been conducted around Africa.

Drought-Water Scarcity and Health Challenges in Africa

Drought conditions are characterised by water scarcity that extends to a shortage of potable water for domestic use. The impacts of drought have a slow process to develop as they accumulate over time as the dry events continue (Rocque et al. 2021). As a result of these characteristics, drought's effect on society and related health challenges are easily overlooked. Climate change-related water scarcity is recognized when it affects the resilience of certain people by disrupting the socioeconomic environment which largely impacted the health condition of the society (Maskrey et al. 2011). Niang et al. (2014) stated that the adaptive capacity of low-income countries of Sub-Saharan Africa (SSA) to drought remains low and high mortality risk as a result of low GDP and weak governance (Maskrey et al. 2011). Nearly half of the SSA population do not have access to safe drinking water and uses unprotected sources to meet the water demands for drinking purpose (Matchawe et al. 2022). Further, inadequacy and poor water supply coverage in both urban and rural communities contribute to health risks in Africa (Oyebande 2001) including several waterborne ailments (Bryan et al. 2020). These water shortages become the major cause of death in Africa linked to waterborne diseases (Matchawe et al. 2022).

Asmall et al. (2021) reviewed the association between drought and adverse health effects in 9 SSA countries and they found scant evidence of a high prevalence of malnutrition, increased prevalence of anaemia, cholera, dengue, and increased incidence of child disabilities in the region. Simane et al. (2016) reviewed the climate change-related health problems in Ethiopia due to floods, heatwaves, vector-borne diseases and water-borne diseases among others increasing mortality and morbidity in the country. Lack of intersectoral collaboration and the absence of clear policy guidelines which interconnect climate change and health issues in a country are considered major challenges. In water-scarce arid and semi-arid regions of Ethiopia, the projected impacts of storing water as a result of water shortage may lead to creating favourable conditions for breeding sites of mosquitos and increase vector-borne diseases (Alcayna et al. 2021). Further, anxiety and stress are associated with water insecurity, particularly in drought-affected pastoral communities of Ethiopia (Alcayna et al. 2021). Ahmadalipour et al. (2019) quantified the future drought risk in Africa and found that it will increase the trend if no climate change adaptation policy is implemented. In West Africa, the future projection indicated that an increase in drought events will be recorded. Whereas, the projection in East Africa shows that the "dry gets drier and wet gets wetter (Ahmadalipour et al. 2019; Ayugi et al. 2022). Further, the Northern African countries (Morocco, Algeria and Tunisia) are projected to experience more severe drought at the end of the twenty-first century (Dai 2013; Prudhomme et al. 2014). Likewise, Southern African countries are also projected to experience a robust drying signal with varying magnitude (Maúre et al. 2018; Ayugi et al. 2022).

Food Insecurity

Food insecurity can be aggravated by the negative impact of climate change on agricultural production (Hasegawa et al. 2018). Food insecurity, poor health coverage and mass displacement of people due to flooding and drought in eastern Africa are posing a challenge to building long-term resilience and provision of livelihood support and protection for the displaced communities (Lindvall et al. 2020). According to the US Humanitarian Information Unit (HIU) report in 2022, persistent drought (four consecutive below-average rainfall) in eastern Africa since late 2020 resulted in food insecurity for more than 18 million people and it continues to rise in Ethiopia, Kenya and Somalia. In addition, more than 7 million children face acute malnutrition in the region (HIU 2022). Further, the effect of climate change impact on health is projected to increase malnutrition rates, stunted growth and impaired cognitive development In Sub-Saharan African countries (Niang et al. 2014; Serdeczny et al. 2017).

Talukder et al. (2021) studied on health impacts of climate change on global smallholder farmers in the world which covers 12% of the global agricultural lands (Ricciardi et al. 2018) and provides 70% of kilocalories to the people of Asia and SSA (Fanzo 2017). According to the 2014 Climate Change Vulnerability Index (CCVI) ranking, the SSA countries such as Guinea-Bissau, Sierra Leone, South Sudan, Nigeria, DR Congo, and Ethiopia in descending order experienced and will continue to experience frequent extreme weather events (Maplecroft 2014). Food insecurity due to drought in arid and semi-arid regions of Ethiopia affects pregnant and lactating women who are vulnerable to food unavailability and malnutrition. Besides, child marriage may increase due to drought to reduce the number of families to feed (Alcayna et al. 2021). However, this will expose children to different health risks. Recently, a severe decline in food production in West Africa is being reported due to the agricultural drought occurrence (Tsutsui et al. 2021).

Reed et al. (2022) also made an extensive study on the impact of flooding on food security across Africa. They estimated that ~12% of the modelled population's (5,671,657 people) food security status is impacted by floods during the 2009–2020 period. Further, the regional comparison finding suggested that a stronger relationship between flood and food security was observed in east Africa compared to west Africa and southeast Africa (Reed et al. 2022). Food insecurity due to drought in arid and semi-arid regions of Ethiopia affects pregnant and lactating women who are vulnerable to food unavailability and malnutrition. Besides, child marriage may increase due to drought to reduce the number of families to feed (Alcayna et al. 2021). However, this will expose children to different health risks. In South Africa, the mean annual temperature was increased by at least 1 °C during the past 50 years, which is approximately ~1.5 times the global average (Engelbrecht et al. 2015). Studies indicated that the temperature is expected to increase by 4–6 °C compared to the 1961–1990 baseline by the end of the twenty-first century if no adaptation or mitigation measures are made ahead ('business as usual scenario') (Engelbrecht et al. 2015; Maúre et al. 2018). This temperature rise exceeds the global average estimates. This temperature rise has a significant long-term impact associated with food insecurity due to deteriorating the quality and quantity of crop yields and other agricultural products which exacerbate the health conditions due to hunger and malnutrition (DEA 2013). In general, drought, water scarcity and its associated food insecurity increases the vulnerability of the African people to different health hazards.

Floods-Cyclones and Health in Africa

In contrast to drought, a flood is a sudden and high-impact event (Sheffield and Wood 2012) that has an immediate effect on societies. For example, the prevalence of water-borne diseases, malaria infestation, etc. are noted when flooding/cyclones are experienced in certain regions. Sub-Saharan countries are considered as least affected by flooding compared to South Asia, East Asia and the Pacific countries (Maskrey et al. 2011). However, their exposure to flooding/tropical cyclones induced a multi-hazard risk of mortality is growing most rapidly (Maskrey et al. 2011). Furthermore, Okaka and Odhiambo (2019) investigated the household perception of flood risk and health impact on informal settlers in the coastal city of Mombasa, Kenya and they found that severe flood risk with high negative physical and mental health impact. Lack of alternative places to settle, short-term mitigation measures, financial constraints, and lack of institutional support impacted the livelihood of the society.

The heavy seasonal rainfall in mid-January 2015 causes severe flooding and impacted a large number of people in southern African countries such as Malawi (638,000 people), Mozambique (373,026 people) and Zimbabwe (6,000 people). Likewise, in Madagascar, more than 10,000 people were affected due to flooding in February 2015. During this period, deaths and cholera were reported in all aforementioned countries (Bickton 2016). Further, tropical storms frequently hit Mozambique and are considered one of the most flood-exposed countries in Africa (Mushtaq et al. 2022). For instance, the recent Tropical Storm Ana (24 January 2022) and Tropical Cyclone Gombe (11 March 2022) cause severe flooding, damaged infrastructure, relocation of several people to temporary shelter, and the deaths of dozens of people. In addition, a total of 220,425 hectares of agricultural land is lost (Mushtaq et al. 2022). The flood affects 736,015 people during Tropical Cyclone Gombe with immediate (63 deaths) and direct (108 people injured) health impacts as a result of the natural disaster recorded (OCHA 2022).

Recurrent flooding in Nigeria mainly caused by anthropogenic influences, poor urban planning and flood disposal infrastructural limitations exacerbate the socioeconomic problems and marred the Sustainable Development Goals (SDGs) particularly in "Ensuring healthy lives" (SDG-3) set for the year 2030 (Echendu 2020). Besides, Nigeria suffers from flood-induced death and injuries which account for ~90% of the direct effects of flooding in developing countries (Zorn 2018; Echendu 2020). Reed et al. (2022) also investigated the relationship between the number of flood occurrences and durations with food security in Africa. These results revealed that the number of flood occurrences degrades food security while the duration is associated with improved food security in southern African regions (Malawi, Zimbabwe,

Mozambique and Zambia). However, a complex relationship between flooding and food security was noted in west Africa (Niger and Nigeria). For example, the 2012 flood incidence damaged crops, infrastructure and displaced people in one region, and good crop growth was experienced in some other regions in both countries (Reed et al. 2022). Due to this reason, it is difficult to associate flooding and food security in the west Africa.

Vector-Borne Disease and Health in Africa

Climate change is likely to increase the proliferation of vector-borne diseases across the continent of Africa due to increase in vector population including mosquitos. Already there has been some concern around resistance to chemical control for some vectors for example mosquitos. In Africa, a noticeable outbreak of the West Nile Virus has often coincided with summer drought conditions and high heat (Bryan et al. 2020). In water-scarce arid and semi-arid regions of Ethiopia, the projected impacts of storing water as a result of water shortage may lead to creating favourable conditions for breeding sites of mosquitos and increase vector-borne diseases (Alcayna et al. 2021). In addition, anxiety and stress are associated with water insecurity, particularly in drought-affected pastoral communities of Ethiopia (Alcayna et al. 2021). Boyce et al. (2016) investigated the interaction of severe flooding and malaria transmission in the flood-prone Western Ugandan Highlands. In the extreme flooding periods, the risk of an individual being affected by malaria was increased by ~30% compared to the post-flooding period and poses a significant health challenge to a regional malaria control program. Flood-prone areas of West Africa where aquatic habitats live, the transmission of Rift Valley fever Virus (RVFV) is common during wet years following the heavy drought incidence (Weaver and Reisen 2010).

Air Pollution, Climate Change and Health in Africa

Air pollution presents a big threat to health and development in Africa (el Morabet et al. 2021; Gwimbi 2017; Hystad et al. 2019, 2020; Khadija and Muhammad 2019; Stilianakis 2015; Shupler et al. 2020). Activities associated with sustained population growth, urban expansion, industrialisation and the use of fossil fuels in power generation, transport and domestic sectors are the major drivers of air pollution in most African countries (Atlas of Sustainable Development Goals 2018: From World Development Indicators 2018). Fossil fuel combustion contributes to the majority of airborne particulate pollution and almost all air pollution by oxides of nitrogen (NO_X), sulphur (SO_X) and carbon (CO_X) (Bulto 2020; Collett et al. 2010; Kasim et al. 2018). These pollutants have been linked to various respiratory ailments such as chronic obstructive pulmonary disease (COPD) which is associated with elevated levels of Ozone (O_3), acute lower respiratory illness (ALRI), cerebrovascular disease

(CEV), ischaemic heart disease (IHD) and lung cancer which is associated with high concentration levels of particulate matter (PM) especially those with diameters less than 2.5 microns (PM2.5) (HEI 2020).

Air pollution has also been linked with increased mental health problems such as dementia, schizophrenia, cognitive disabilities, psychotic experiences, anxiety and depressive disorders in adults while in children, $PM_{2.5}$ and NO_2 have been associated with IQ loss, increased risk of Attention Deficit/Hyperactivity Disorder (ADHD) (al Ahad et al. 2022). The cost of pollution and pollution-related diseases are not easily discerned in Africa because they are buried deep in health care costs and reduced productivity. The Health Effects Institute (2020) estimates that air pollution became the 4th leading cause of premature deaths on a global scale in 2019 overtaken by high blood pressure, poor diet and tobacco use.

More than 90% of African countries are in the low-income category according to World Bank classification by income levels (https://blogs.worldbank.org/opendata/new-world-bank-country-classifications-income-level-2022-2023) and heavily depend on unclean energy sources such as wood, cow dung, and fossil fuels such as coal, kerosine and natural gas for home heating and cooking This dependence results in potential high levels of inhouse air pollution exposure amongst women and children who spend the majority of their time in household proximity (Bulto 2020; Shupler et al. 2020). Air pollution was responsible for an estimated 1.1 million deaths across the African continent with household air pollution accounting for 697,000 deaths (63%) while ambient air pollution accounted for the remaining 37% (Fisher et al. 2021).

Although, indoor air quality has accounted for more deaths, the associated pollution levels have been decreasing while ambient air pollution has been on an increasing trend. This pattern is attributed to the nature of anthropogenic activities taking place in Africa which include expanding industrialization and the associated increase in power generation demand that rely on fossil fuel burning, increased rate of urbanization as well as increased vehicular emissions (Health n.d.). Fossil fuel combustion for either domestic or industrial purposes is a major source of greenhouse gases like carbon dioxide (CO_2) that drive climate change and negatively impacts the livelihood of nationalities in many different ways such as climate change-induced droughts, rising temperatures, extreme weather and associated diseases (Abera et al. 2021). The lack of robust health facilities in most parts of Africa further exacerbates the impact of air pollution on the poor and vulnerable groups of society that include the elderly and young children.

Future Trends

Based on the findings of a recent global agricultural economic model study, future climate change will lower major crop yield by 17% with increasing market price by 20% in 2050, after the adaptation of production across regions (Hasegawa et al. 2018). Even with adaptation mechanisms in place, the adverse effects of climate change will be felt most in low-income areas of SSA where there is already severe food insecurity, undernutrition and malnutrition. The future projection of drought in Africa shows that more severe drought events over the region in northern Africa, a robust drying with varying magnitude in southern Africa, and in east Africa drought/flooding will persist in drier/wetter areas respectively (Maúre et al. 2018; Ahmadalipour et al. 2019; Ayugi et al. 2022). On the other hand, Ahmadalipour et al. (2019) quantified the future drought risk in Africa and found that it will increase the trend if no climate change adaptation policy is implemented. In West Africa, the future projection indicated that an increase in drought events will be recorded. Whereas, the projection in East Africa shows that the "dry gets drier and wet gets wetter (Ahmadalipour et al. 2019; Ayugi et al. 2022). Further, the Northern African countries (Morocco, Algeria and Tunisia) are projected to experience more severe drought by the end of the twenty-first century (Dai 2013; Prudhomme et al. 2014). Likewise, the Southern African countries will experience a robust drying signal with varying magnitude (Maúre et al. 2018; Ayugi et al. 2022).

Policy Recommendations

Information on the sources, components and levels of pollution remains scanty in most African countries, a consequence of limited air quality monitoring and research that is imperative for gauging the nature of population exposure to air pollution. Conventionally air quality monitoring is done using Air Quality Monitoring Stations (AQMS) equipped with reference analysers that are recommended by various environmental regulatory authorities (Castell et al. 2017). However, standard AQMS are expensive, bulky, complicated to maintain and beyond the reach of financial means of most African countries (Laref et al. 2021). Several pieces of research have revealed the potential of low-cost sensors in air quality monitoring thus presenting a potential solution to this challenge (Al-Kaseem et al. 2020; Das et al. 2018; Nakamura et al. 2021).

Science-based control strategies such as rapid transition to green energy such as wind and solar will help Africa avoid becoming entrapped by fossil fuels and pursue smart development pathways while taking advantage of decreasing costs of clean energy sources (Fisher et al. 2021). Pollution prevention will retard climate change, improve human health, prevent disease, and keep nations on trajectories of social and economic development (Altieri and Keen 2019). The lack of intersectoral collaboration and the absence of clear policy guidelines which interconnect climate change and health issues in the majority of African countries are considered as major challenges. Africa can also take advantage of the high technological penetration to migrate and offer digital health. This can go a long way in assisting the poor or non-existent health infrastructure and the high brain drain that has bedelved the continent. Instead, skilled public health personnel can still be useful albeit digitally.

Conclusion

The chapter gave an overview of climate change human-health related challenges that face the African continent through synthesising various literature. Africa is likely to see a rise in climate change-related human health challenges that will be compounded by the various inherent infrastructure, brain drain, public health skills deficit and underbudgeting. As projected, drought events will increase that will culminating in a rise in incidences and prevalence of the water-borne disease. Meanwhile, rapid urbanisation and industrialisation will also contribute to air pollution and an increase in the burden of disease. In view of the aforesaid, Africa is presented with an opportunity to adapt to these various climate change health-related challenges.

References

Abera A, Friberg J, Isaxon C, Jerrett M, Malmqvist E, Sjöström C, Taj T, Vargas AM (2021) Air quality in Africa: public health implications. Annu Rev Public Health 42:193–210. https://doi.org/10.1146/annurev-publhealth

Ahmadalipour A et al (2019) Future drought risk in Africa: integrating vulnerability, climate change, and population growth. Sci Total Environ 662:672–686

al Ahad MA, Demšar U, Sullivan F, Kulu H (2022) Air pollution and individuals' mental wellbeing in the adult population in United Kingdom: a spatial-temporal longitudinal study and the moderating effect of ethnicity. PLoS One, 17(3 March). https://doi.org/10.1371/journal.pone.0264394

Alcayna T et al (2021) Climate change impacts on health: Ethiopia assessment

Al-Kaseem BR, Ahmed AF, Abdullah AM, Azouz TZ, Al-Majidi SD, Al-Raweshidy HS (2020) Self-powered 6LoWPAN sensor node for green IoT edge devices. IOP Conf Ser Mater Sci Eng 928(2). https://doi.org/10.1088/1757-899X/928/2/022060

Altieri KE, Keen SL (2019) Public health benefits of reducing exposure to ambient fine particulate matter in South Africa. Sci Total Environ 684:610–620. https://doi.org/10.1016/j.scitotenv.2019.05.355

Asmall T et al (2021) The adverse health effects associated with drought in Africa. Sci Total Environ 793:148500

Atlas of sustainable development goals 2018: from world development indicators. (2018). World Bank, Washington, DC. https://doi.org/10.1596/978-1-4648-1250-7

Ayugi B et al (2022) Review of meteorological drought in Africa: historical trends, impacts, mitigation measures, and prospects. Pure Appl Geophys 179(4):1365–1386

Bickton FM (2016) Climate change as the biggest threat to public health in southern Africa and measures to reduce its impacts. Malawi Med J 28(2):70–72

Boyce R et al (2016) Severe flooding and malaria transmission in the western Ugandan Highlands: implications for disease control in an era of global climate change. J Infect Dis 214(9):1403–1410

Bryan K et al (2020) The health and well-being effects of drought: assessing multi-stakeholder perspectives through narratives from the UK. pp 2073–2095

Bulto TW (2020) Impact of open burning refuse on air quality: in the case of "Hidar Sitaten" at Addis Ababa, Ethiopia. Environ Health Insights 14. https://doi.org/10.1177/1178630220943204

Cappuccio FP et al (2011) Sleep duration predicts cardiovascular outcomes: a systematic review and meta-analysis of prospective studies. Eur Heart J 32(12):1484–1492. https://doi.org/10.1093/eurheartj/ehr007

Castell N, Dauge FR, Schneider P, Vogt M, Lerner U, Fishbain B, Broday D, Bartonova A (2017) Can commercial low-cost sensor platforms contribute to air quality monitoring and exposure estimates? Environ Int 99:293–302. https://doi.org/10.1016/j.envint.2016.12.007

Chapman S et al (2022) Past and projected climate change impacts on heat-related child mortality in Africa. Environ Res Lett 17(7). https://doi.org/10.1088/1748-9326/ac7ac5

Collett KS, Piketh SJ, Ross KE (2010) An assessment of the atmospheric nitrogen budget on the South African Highveld. S Afr J Sci 106(5/6). https://doi.org/10.4102/sajs.v106i5/6.220

Dai A (2013) Increasing drought under global warming in observations and models. Nat Clim Chang 3(1):52–58

Das I, Pedit J, Handa S, Jagger P (2018) Household air pollution (HAP), microenvironment and child health: Strategies for mitigating HAP exposure in urban Rwanda. Environ Res Lett 13(4):045011

DEA (2013) Long-term adaptation scenarios flagship research programme (LTAS) for South Africa. Climate change implications for human health in South Africa. https://www.dffe.gov.za/sites/default/files/docs/human_health_bookV5.pdf

Echendu AJ (2020) The impact of flooding on Nigeria's sustainable development goals (SDGs). Ecosyst Health Sustain 6(1):1791735

el Morabet R, Barhazi L, Khan RA, Hejjaj A, Harrara R, Bouhafa S, Khan AR, Khan NA, Yeremenko S (2021) Study of seasonal variation and index based assessment of water quality and pollution in semi-arid region of Morocco. Ecol Ques. 23 September, 32(4):113–125 Accessed on 22 Feb 2023. https://doi.org/10.12775/EQ.2021.037

Engelbrecht F et al (2015) Projections of rapidly rising surface temperatures over Africa under low mitigation. Environ Res Lett 10(8):85004

Fanzo J (2017) From big to small: the significance of smallholder farms in the global food system. Lancet Planet Health 1(1):e15–e16

Fisher S, Bellinger DC, Cropper ML, Kumar P, Binagwaho A, Koudenoukpo JB, Park Y, Taghian G, Landrigan PJ (2021) Air pollution and development in Africa: impacts on health, the economy, and human capital. The Lancet Planetary Health 5(10):e681–e688. https://doi.org/10.1016/S2542-5196(21)00201-1

Gwimbi P (2017) Monitoring SO2 emission trends and residents' perceived health risks from PGM smelting at Selous Metallurgical Complex in Zimbabwe. Int J Equity Health 16(200):1–11. https://doi.org/10.1186/s12939-017-0696-6

Hasegawa T et al (2018) Risk of increased food insecurity under stringent global climate change mitigation policy. Nat Clim Chang 8(8):699–703

Health Effects Institute (2020) State of global air 2020. Special report. Boston, MA:Health Effects Institute

HEI (2020) A special report on global exposure to. www.stateofglobalair.org

HIU (2022) Horn of Africa: persistent drought increasing food insecurity, millions face threat of famine

Hystad P, Duong M, Brauer M, Larkin A, Arku R, Kurmi OP, Fan WQ (2019) Health effects of household solid fuel use: findings from 11 countries within the prospective urban and rural epidemiology study. Environ Health Perspect 127(5):1–10

Hystad P, Larkin A, Rangarajan S, Alhabib KF, Avezum Á, Burcu K, Calik T, Chifamba J (2020) Associations of outdoor fine particulate air pollution and cardiovascular disease in 157 436 individuals from (PURE): a prospective cohort study. Lancet Planet Health 4:235–245. https://doi.org/10.1016/S2542-5196(20)30103-0

Kasim OF, Abshare MW, Mukuna, TE, Wahab B (2018) Land use and ambient air quality in Bahir Dar and Hawassa, Ethiopia. Air soil Water Res 11:1178622117752138

Khadija BU, Muhammad I (2019) Assessment of the pollution extent of sulphur dioxide (SO2) nd nitrogen dioxide (NO2) in ambient air within Kano metropolis, Kano State, Nigeria. J Environ Sci Comput Sci Eng & Technol. https://doi.org/10.24214/jecet.A.8.4.39604

Laref R, Losson E, Sava A, Siadat M (2021). Empiric unsupervised drifts correction method of electrochemical sensors for in field nitrogen dioxide monitoring. Sensors 21(11). https://doi.org/10.3390/s21113581

Leal Filho W et al (2021) Impacts of climate change to African indigenous communities and examples of adaptation responses. Nat Commun 12(1):13–16. https://doi.org/10.1038/s41467-021-26540-0

Leal Filho W et al (2022) An analysis of climate change and health hazards: results from an international study. Int J Clim Chang Strateg Manage 14(4):375–398. https://doi.org/10.1108/IJCCSM-08-2021-0090

Lindvall K et al (2020) Health status and health care needs of drought-related migrants in the Horn of Africa—a qualitative investigation. Int J Environ Res Public Health 17(16):5917

Maplecroft (2014) World: climate change vulnerability index 2014.

Maskrey A et al (2011) Revealing risk, redefining development, global assessment report on disaster risk reduction. In: United Nations strategy for disaster reduction, pp 17–51

Matandirotya NR et al (2021) Risks of indoor overheating in low-cost dwellings on the South African Lowveld. In: African handbook of climate change adaptation, pp 1583–1600. https://doi.org/10.1007/978-3-030-45106-6_123

Matchawe C et al (2022) Water shortages: cause of water safety in Sub-Saharan Africa

Maúre G et al (2018) The southern African climate under 1.5 C and 2 C of global warming as simulated by CORDEX regional climate models. Environ Res Lett 13(6):65002

Minor K et al (2022) Rising temperatures erode human sleep globally. One Earth 5(5):534–549. https://doi.org/10.1016/j.oneear.2022.04.008

Mushtaq F et al (2022) A rapid geospatial analysis of the impact of tropical cyclone Gombe in Mozambique in 2022. FAO, Rome

Nakamura Y, Talusan JP, Mizumoto T, Suwa H, Arakawa Y, Yamaguchi H, Yasumoto K (2021). ProceThings: data processing platform with in-situ IoT devices for smart community services. ACM international conference proceeding series, pp 116–121. https://doi.org/10.1145/3427477.3429275

Niang I et al (2014) Africa. Climate change 2014: impacts, adaptation and vulnerability: part B: regional aspects: working group II contribution to the fifth assessment report of the intergovernmental panel on climate change, pp 1199–1266

OCHA (2022) Mozambique: tropical cyclone Gombe flash update no. 9 (As of 13 April 2022)

Okaka FO, Odhiambo BDO (2019) Households' perception of flood risk and health impact of exposure to flooding in flood-prone informal settlements in the coastal city of Mombasa. Int J Clim Chang Strateg Manage.

Oyebande L (2001) Water problems in Africa—how can the sciences help? Hydrol Sci J 46(6):947–962

Prudhomme C et al (2014) Hydrological droughts in the 21st century, hotspots and uncertainties from a global multimodel ensemble experiment. Proc Natl Acad Sci 111(9):3262–3267

Reed C et al (2022) The impact of flooding on food security across Africa. Proc Natl Acad Sci 119(43):e2119399119

Ricciardi V et al (2018) How much of the world's food do smallholders produce? Glob Food Sec 17:64–72

Rocque RJ et al (2021) Health effects of climate change: an overview of systematic reviews. BMJ Open 11(6):e046333

Russo S et al (2016) When will unusual heat waves become normal in a warming Africa? Environ Res Lett 11(5). https://doi.org/10.1088/1748-9326/11/5/054016

Serdeczny O et al (2017) Climate change impacts in Sub-Saharan Africa: from physical changes to their social repercussions. Reg Environ Change 17(6):1585–1600

Sheffield J, Wood EF (2012) Drought: past problems and future scenarios. Routledge

Shupler M, Miller-lionberg D, Jeronimo M, Arku RE (2020) Household and personal air pollution exposure measurements from 120 communities in eight countries: results from the PURE-AIR study. Lancert Planet Health 4(20):451–462. https://doi.org/10.1016/S2542-5196(20)30197-2

Simane B et al (2016) Review of climate change and health in Ethiopia: status and gap analysis. Ethiop J Health Develop 30(1):28–41

Stilianakis NI (2015) Susceptibility and vulnerability to health effects of air pollution: the case of nitrogen dioxide. Luxemburg: Publications office of the European Union

Talukder B et al (2021) Health impacts of climate change on smallholder farmers. One Health 13:100258

Tsutsui H et al (2021) Drought monitoring over West Africa based on an ecohydrological simulation (2003–2018). Hydrology 8(4):155

Weaver SC, Reisen WK (2010) Present and future arboviral threats. Antiviral Res 85(2):328–345

Zhao Q et al (2021) Global, regional, and national burden of mortality associated with non-optimal ambient temperatures from 2000 to 2019: a three-stage modelling study. Lancet Planet Health 5(7):e415–e425. https://doi.org/10.1016/S2542-5196(21)00081-4

Zorn M (2018) Natural disasters and less developed countries. In: Nature, tourism and ethnicity as drivers of (de) marginalization. Springer, pp 59–78

Newton R. Matandirotya is the Founder and Director of the Centre for Climate Change Adaptation and Resilience at Kgotso Development Trust, Zimbabwe and a Research Fellow at the Nelson Mandela University, South Africa. He holds a Ph.D. in Science with Geography and Environmental Management. His research interest include climate change adaptation, air quality monitoring and management, food and nutrition security, disaster management, indoor and outdoor thermal comfort. He is also very keen on the use of remote sensing techniques in various fields.

Tonderai Dangare is a lecturer in the department of Electronics and Telecommunications Engineering at the University of Zimbabwe. Currently he studying for a Ph.D. in Environmental Engineering and Instrumentation at the same institution. He received his Master of Science (M.Sc.) degree in Applied Physics, Bachelor of Science in Physics honours, B.Sc. degree in Math and Physics and an Advanced Diploma in Telecommunication Systems from the University of Zimbabwe. His research Interests are in Low-Cost Environmental Monitoring instruments development, Air quality monitoring, Internet of Things, Opto-electronics, R&D in Functional Materials and Devices.

Girma Berhe is an assistant professor, lecturer and researcher at Haramaya University, Haramaya Institute of Technology—School of Water Resources and Environmental Engineering. He received a B.Sc. degree in Agricultural Engineering and Mechanization from Hawassa (the then Debub University) in 2005 and M.Sc. in Irrigation Engineering from Haramaya University in 2011. His research work focused on hydrological modeling, catchment hydrology, water footprints, irrigation water management, water resource management, climate change studies on floods, droughts and adaptations, and GIS and remote sensing applications in land and water resources.

Part III
Climate Resilience and Adaptation to Climate Change

Chapter 16
Do the Resilience and Adaptive Capacity to Climate-Related Disasters Help Explain Human Health and Well-Being?

Miguel Alves Pereira, Alexandre Morais Nunes, and Diogo Cunha Ferreira

Abstract Climate change has been growing as one of the significant influencing factors of human health and disease. With the arrival of new circumstances and conditions, new health threats will emerge, and others will intensify. Their disruptive effect on physical, biological, and ecological systems already impacts public health, whose outcomes include increased cardiopulmonary diseases, injuries, and untimely casualties due to extreme weather events, modifications in the availability and distribution of food- and water-borne diseases, and mental health risks. Therefore, understanding the influence of such factors, especially those concerning the resilience and adaptive capacity to climate-related disasters, on human health and well-being is essential. By considering the countries most affected by extreme weather events, we use principal component analysis and multiple linear regression to ascertain if those climate change indicators affect human health. In the end, we conclude that the mortality associated with disasters and the quality of and actual adoption and implementation of disaster risk reduction strategies do not seem to be significant determinants of human health.

M. A. Pereira
INESC TEC, Rua Dr. Roberto Frias, 4200-465 Porto, Portugal
e-mail: miguel.a.pereira@inesctec.pt

CEGIST, Instituto Superior Técnico, University of Lisbon, Av. Rovisco Pais 1, 1049-001 Lisbon, Portugal

A. M. Nunes (✉)
Centre for Public Administration and Public Policies, Institute of Social and Political Sciences, Universidade de Lisboa, Rua Almerindo Lessa, 1300-663 Lisbon, Portugal
e-mail: anunes@iscsp.ulisboa.pt

D. C. Ferreira
CERIS, Instituto Superior Técnico, University of Lisbon. Av, Rovisco Pais 1, 1049-001 Lisbon, Portugal
e-mail: diogo.cunha.ferreira@tecnico.ulisboa.pt

W. Leal Filho et al. (eds.), *Climate Change and Health Hazards*, Climate Change Management, https://doi.org/10.1007/978-3-031-26592-1_16

Introduction

The health sector is relevant for citizens around the world because it presents itself as a sector that can promote or compromise other sectors, being understood as a factor that boosts well-being and quality of life since the health status of individuals affects not only their physical and psychological well-being but also their social and economic situation (Pförtner and Richter 2011). A healthy economy depends heavily on a healthy population (Betti 2017). However, this relationship does not depend only on itself, but also on other health factors and determinants that can have an impact on economic growth and are not necessarily related to health statuses, such as education, the environment, and transportation (Nunes and Ferreira 2022).

In the process of articulation between the health sector and the other sectors, it is essential that health entities understand the functioning and agenda of policies and the impact that health has on them and, in the opposite direction, assess the consequences on health in face of the decisions made in other sectors. To facilitate and promote positive development, the report of the international meeting on 'Health in All Policies' of the World Health Organization (WHO), many of them are at the base of climate change. Thus, the WHO recommends creating platforms that promote greater articulation between all sectors, evaluating the efficiency of joint work and the formulation of intersectoral policies, and the importance of involving other sectors in the fulfillment of a certain goal (e.g., objective examples of health policy that include environmental factors), which culminate in a joint action promoting better health conditions and well-being (World Health Organization 2010).

'Health in All Policies', in its transversal strategic essence to promote the health of the population, proposes an intervention at the level of health determinants (Sihto et al. 2006). There is a relationship between the reduction of inequalities and the improvement of the population's health status through the promotion of quality of life, work capacity/productivity, and the promotion of family and community economy that reduces social differences and levels of poverty. Based on this relationship, all over the world, governments have focused on economic development as the core of their respective political agendas (World Health Organization 2010). The SARS-CoV-2 pandemic put this idea to the test, given the negative impact of the great economic and social impact generated by the retraction of revenue and increase in State expenditure, even though, in environmental terms, they have reduced carbon emissions (Kershaw 2020; Sarkodie and Owusu 2021).

The social determinants of health embrace different origins, which can be biological, behavioral, educational, and economic, among others. They have a strong influence on the objectives of the process of implementing 'Health in All Policies' and are, therefore, addressed in this chapter, in line with Puska (2007) who states that intervening in these objectives can contribute to promoting the reduction of health inequalities: "An important consideration in the 'Health in All Policies' process is that addressing social determinants of health and disease can be a strong tool for reducing inequalities in health that is a major concern in most European countries".

Associated with socio-economic issues (such as employment and income, both in the case of citizens with more or fewer resources), a set of environmental determinants arise that are at the base of a set of chronic diseases, responsible for 80% of mortality in Europe (Nunes and Ferreira 2022). In addition to these, the issue of climate change has been growing as one of the factors that significantly influence human health and disease (Shankardass et al. 2018). Thus, key issues related to the effects of climate change on health and areas where a systematic approach to the problem is needed are presented. In addition, the challenges of climate change for societies and their impacts on human health are also presented, as well as some of the priority areas of action given the existence of interest groups.

Besides the theoretical issue at hand, we conduct a statistical analysis by means of principal component analysis (PCA) and multiple linear regression to understand the influence of the resilience and adaptive capacity to climate-related disasters on human health and well-being. In particular, we base our analysis on the United Nations (UN) Sustainable Development Goals (SDGs) framework in order to prove if the mortality associated with disasters and the quality of and actual adoption and implementation of disaster risk reduction (DRR) strategies are determinants of human health. Our study was performed in 23 countries and territories that suffered from climate-related disasters in 2020.

'Health in All Policies'

The WHO (1998) defines *health determinants* as a "range of personal, social, economic and environmental factors that determine the health status of individuals or populations". The idea of 'Health in All Policies' exists since the declaration of the WHO (1946) stating that "Governments have a responsibility for the health of their peoples which can be fulfilled only by the provision of adequate health and social measures". Since then, this view has been broadened over time, being present, for example, in the Declaration of Alma-Ata (1978), the Ottawa Charter (1986), the Treaty of Amsterdam (1997), several WHO conferences, the Lisbon Treaty (2007), the principles of the Commission on Social Determinants of Health (CSDH) (2008), and, since 2014, WHO's "Plan of Action on Health in All Policies" (Nunes and Ferreira 2022).

The current definition given by the WHO (2020) adds that the social determinants of health represent the circumstances in which people are born, grow, live, work, and age, which are shaped by the distribution of money, power, and resources at local, national, and global levels, are the main responsible for health inequities recognized as the basis of unfair and avoidable dissimilarities in health status observed between different countries. This involves a close link between health and other sectors such as education, economy, environment, agriculture, and transportation (Puska 2007). According to Nunes and Ferreira (2022), "This association is based on the principle that health status is determined by several factors, such as the workplace, eating habits, and consumption. These depend not only on individual options, but also on social, economic, environmental, and cultural factors, whose health policies can positively influence".

Intervention in terms of health determinants, namely those that, by their effects, are the cause or consequence of climate change, is a fundamental part of the health promotion strategy (Sihto et al. 2006). This finding was highlighted at several international conferences 35 years ago: (i) at the 1978 International Conference on Primary Health Care in Alma-Ata; (ii) in 1986, at the International Conference on Health Promotion in Ottawa; (iii) in 1997, at the WHO Global Conference on Intersectoral Action for Health; (iv) in 2005, at the Conference on Health Promotion in Bangkok. In all these conferences, the influence of a set of factors that could affect or promote the health status of the population was recognized.

State intervention to reduce health inequalities has been a priority all over the world, due, in part, to the pre-established relationship with economic development. In other words, only with a healthy population and with access to information, adequate living conditions, and health services is it possible to obtain greater productivity at work, well-being, and better quality of life (Greaves and Bialystok 2011).

Regardless of the definitions or the grouping of factors suggested by different authors, the model by Dahlgren and Whitehead (2007) depicted in Fig. 16.1 is internationally known as a reference since it considers that the social determinants of health influence the state of health since the social determinants of health influence the state of health and, consequently, the way we live due to a combination of several factors. Note the wide scope of environmental conditions, which are closely related to climate change and all the involvement they have on the health of citizens.

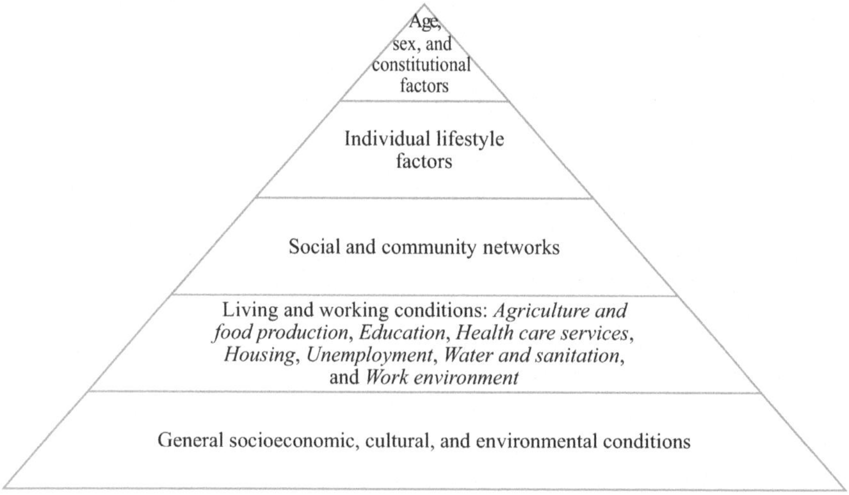

Fig. 16.1 The primary determinants of health. Adapted from Dahlgren and Whitehead (2007)

Climate Change, Society, and Health

The WHO (1998) has defined health determinants as a range of personal, social, economic, and environmental factors that determine individuals' or populations' health status. Meeting the definition adopted by the WHO for "determinant in health", Nunes and Ferreira (2022) consider the existence of multiple determinants influencing people's health status, highlighting environmental issues in the context of climate change.

Intervention in the field of fundamental health determinants, especially on factors that in some way may condition an entire society and limit well-being and quality of life, such as the environment and climate change that arise from the various aggressions caused by more developed societies, but which have greater effects in less developed countries, much due to their geographic location and greater health weaknesses (Sihto et al. 2006).

The global climate is changing and has several impacts on nature, biodiversity, and also on human well-being. Factors such as rising temperatures and extreme weather events have been devastating for the planet with serious consequences for the environment in a global way and, in particular, for human health (Leal Filho et al. 2016).

In addition to the causes of climate change, it is necessary to take into account that they arise as a result of economic activity and aggression toward the planet. It is known that growing energy needs have led to an increase in the consumption of fossil fuels and alarming greenhouse gas emissions (Saxe 2021). On this issue, questions about the future are increasingly being raised, as exemplified by Leal Filho et al. (2016): "how human species can survive and ensure the survival of future generations in the context of enormous changes in the functioning of natural systems? How can we live with dignity, health and ensure our wellbeing?" These are all valid interrogations.

In the literature, global warming appears as the factor that is at the base of the factors that promote climate change, associated with extensive use of fossil fuels—a problem that interferes at the soil level—, as well as the availability of essential resources, such as drinking water and the increase in seawater level associated with the melting of polar icecaps that can put at risk a quarter of the world's population and increase the acidity of the oceans, which compromises the marine environment (Leal Filho et al. 2016).

Unequivocally, we can claim that the health of populations is directly related to the environment due to the effect they have on the production of food, water, and the entire ecosystem (Smith et al. 2014). The consequences of climate change for human health are directly proportional to the exposure to them and the vulnerability of the exposed populations, where the most vulnerable communities and those who suffer from a greater degree of social inequalities gain greater significance, as discussed in the previous point. Hence, social factors, such as the lack of access to services, housing conditions, and culture and environmental and economic aspects, make some populations more vulnerable. This is clearly where the least developed countries stand out, which, even though they contribute the least to climate change, are the ones that suffer the most from its consequences (Smith et al. 2014).

In addition to the fact that climate change forces the displacement of populations, the destruction of fields for cultivation, the reduction of fish, and the limitation of the production of quality meat, the increase in the prevalence of diseases, including those of a mental nature caused by stress and anxiety can be added to the list (Leal Filho et al. 2016).

Climate change may underlie new health problems or exacerbate existing ones that are closely linked to cardiovascular disease, mortality, respiratory disease, the altered transmission of infectious diseases, and malnutrition due to crop failure (Patz et al. 2005) or even lead to death in extreme cases, such as those reported in India by Akhtar (2007) and Italy by Schifano et al. (2009). All this has an impact on other areas of activity, as mentioned when addressing the concept of 'Health in All Policies' and its relationship with the different health determinants. On this topic, Leal Filho et al. (2016), in an exhaustive survey, pointed out the main impacts of climate change on the health of the population in general (see Table 16.1).

In particular, according to the same authors, the effects of climate change are much more pronounced on more vulnerable groups, such as children, the elderly, pregnant women, and people who suffer from allergies (see Table 16.2).

Following the survey of diseases carried out by Leal Filho et al. (2016) and described in Tables 16.1 and 16.2, we approach the indicators with the causes of death associated with the diseases mentioned in the Organisation for Economic Co-operation and Development (OECD) countries. On average, in 2018, according to the OECD (2022), there were:

Table 16.1 Some of the impacts of climate change on health. Adapted from Leal Filho et al. (2016)

Impact	Consequence
Heat, heat waves, and drought	Discomfort, distress, cardiovascular and respiratory diseases, malnutrition, infectious diseases, malaria, and exacerbation of chronic diseases (e.g., diabetes, lung disorders, psychiatric disorders)
Extreme weather events	Discomfort, distress, and psychological problems due to loss of property
Reduced air quality and emission of gases	Breathing problems and higher incidences of allergies, skin cancer, ocular lesions, and impaired immune system
Floods, flooding, and proliferation of water-borne diseases	Malaria, dengue, transmissible diseases, stress, and psychiatric disorders
Cold	Respiratory infections, allergic reactions, cardiovascular and cerebrovascular diseases

Table 16.2 Overview of how climate change (focused on temperature increase) impacts some vulnerable groups. Adapted from Leal Filho et al. (2016)

Group	Influence of climate change due to increases in temperature	Impact
Children	Discomfort and distress	Apathy, reduced ability to concentrate, and lower learning performance
The elderly	Discomfort and variations in blood pressure	Less motivation to be outdoors and reduction in social contact
Pregnant women	Sudden changes in levels of blood pressure and sweat attacks	Lower ability to work and prolonged sick leave periods
Allergic people	Earlier release of pollen by some plants and longer pollination seasons	Worsening of allergy symptoms

- 284.9 deaths per 100,000 inhabitants from diseases of the circulatory systems (including cardiovascular and cerebrovascular diseases);
- 61.0 deaths per 100,000 inhabitants from diseases of the respiratory systems (including influenza virus, pneumonia, chronic obstructive disease, and asthma);
- 24.7 deaths per 100,000 inhabitants from diabetes mellitus;
- 2.95 deaths per 100,000 inhabitants from skin cancer;
- 27.55 deaths per 100,000 inhabitants from mental and behavioral disorders; and
- 13.1 deaths per 100,000 inhabitants from certain infectious and parasitic diseases (including tuberculosis and HIV).

These indicators, even though there are greater effects of climate change, have been attenuated because countries have been adopting WHO guidelines and establishing measures through priority health programs, which have contributed to the increase in average life expectancy at birth in OECD countries, whose average in 2020 was 80.44 years old (although it decreased from 80.98 years old).

The Targets of the United Nations' Sustainable Development Goal 13

The UN has planned the development of the SDGs aiming at a more sustainable future for all, based on the risks associated with environmental degradation, sustainability, climate change, and water security. To this end, it has set global goals for each of these areas, aspiring to leave no one behind and safeguarding an economic development that works for both the people and the planet.

As there is a consensus on the need to take urgent action to combat climate change and its impacts on health hazards, SDG 13 appears within the scope of this great plan in order to take urgent measures to combat climate change and its impacts. It is focused on the increased risk of global warming, knowing that, at present, to keep the increase in temperature below 1.5 °C and avoid the worst impacts of climate change, science urges that greenhouse gas emissions peak and then reduce by zero by 2050. Within this framework, the Committee for Development Policy (CDP) assessed the current context by looking at the global context, risks, opportunities, and benefits (see Fig. 16.2).

As immediate solutions to mitigate some of these effects, measures such as the following suggested by the Carbon Pricing Leadership Coalition (CPLC 2021) are strong possibilities:

- *Carbon tax.* Sets a direct price on emissions and requires economic agents to pay for each ton of emitted carbon pollution. In this way, it constitutes a financial incentive to reduce emissions, thus encouraging the adoption of more efficient processes or cleaner fuels, i.e., less pollution implies lower taxes);
- *Emission trading system (ETS).* Also known as a *cap-and-trade system*, it sets a cap on total direct emissions and establishes a market where emission rights (in

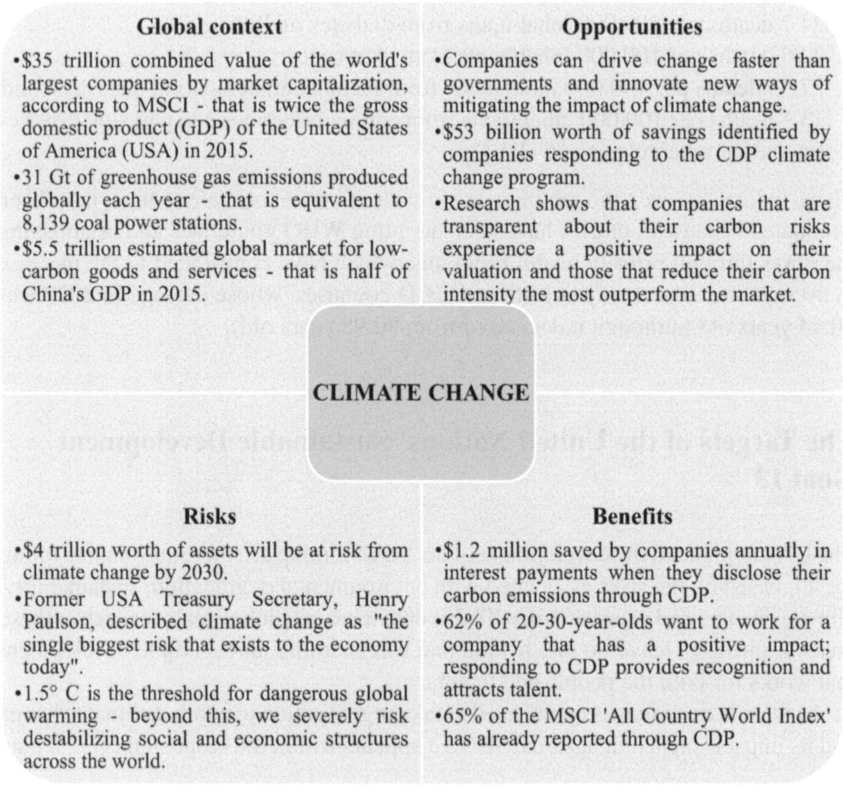

Fig. 16.2 Analysis of the CDP on climate change. Adapted from the CDP (2022a)

the form of licenses or carbon licenses) are traded. This approach allows polluters to meet emission reduction targets flexibly and at the lowest cost;

- *Crediting mechanism.* Emission reductions are the result of a project, promoted by a company, policy, or government, which can be negotiated. Entities seeking to reduce emissions can buy credits as a way of offsetting their actual emissions;
- *Results-based climate finance framework.* Entities receive funds when they achieve emission reductions, i.e., funding is linked to specific results, helps polluters to achieve climate goals, and encourages private sector investment; and
- *Internal carbon pricing.* Governments, companies, and other entities assign their own internal price to the use of carbon and take this into account in their investment plans/budget, which can occur in two ways: (i) assignment of a shadow price to carbon use to determine its hypothetical cost; (ii) create an internal carbon fee to voluntarily charge business units for their emissions.

According to 2020 data from the CDP (2022b), more than 2,000 companies (including almost half of the world's largest companies by market value) decided to implement an internal carbon price. It is also worth noting the European Commission's action in creating the European Climate Pact initiative to build a greener Europe offers a platform for sharing solutions for people, communities, and organizations to participate in climate and environmental actions (European Commission 2020). Within the scope of the pact, stakeholders undertake climate and environmental actions.

Methodology

To recenter our focus on the main goal of the study—analyzing the influence of climate change on human health—we need to resort to statistics. When we want to determine the intensity of the relationship between a variable that responds to changes in another variable and that variable, we are measuring the influence of the latter—the *independent variable*—on the former—the *dependent variable.* In statistics, the most common way to predict the value of the dependent variable based on the independent variable is by means of *linear regression.*

Linear regression models were the first type of regression models to be studied and to be used in extensive applications (Yan and Su 2009). They have a linear dependence between their parameters, which makes them easier to estimate than non-linear regression models that have non-linear dependences. This results in a more straightforward computation of the estimators' statistical properties.

In case we are dealing with one explanatory variable and one predicted variable, we resort to *simple linear regression.* In case there is more than one explanatory variable, but still one predicted variable, *multiple linear regression* should be used (Freedman 2005). In case multiple correlated dependent variables are predicted from a set of multiple predictors, *multivariate linear regression* is called for (Rencher and Christensen 2012).

Mathematically, given a set $\left\{y_i, x_{i1}, \ldots, x_{ip}\right\}_{i=1}^{n}$ containing the data of n units, the linear relationship assumed by a linear regression model can be modeled via Expression (16.1) as

$$y_i = \beta_0 + \beta_1 x_{i1} + \cdots + \beta_p x_{ip} + \varepsilon_i = \boldsymbol{\beta} \boldsymbol{x}_i^{\mathrm{T}} + \varepsilon_i, i = 1, \ldots, n \tag{16.1}$$

or in matrix notation as in Expression (16.2):

$$\boldsymbol{y} = \boldsymbol{\beta} \boldsymbol{X} + \boldsymbol{\varepsilon} \tag{16.2}$$

where $\boldsymbol{\beta} = \left\{\beta_0, \beta_1, \ldots, \beta_p\right\}^{\mathrm{T}}$ denotes the vector of regression coefficients (β_0 is the intercept term), $\boldsymbol{y} = \{y_1, y_2, \ldots, y_n\}^{\mathrm{T}}$ denotes the vector of observed values, $\boldsymbol{X} = \left\{x_1^{\mathrm{T}}, x_2^{\mathrm{T}}, \ldots, x_n^{\mathrm{T}}\right\}^{\mathrm{T}}$ denotes the matrix of regressor row vectors $\boldsymbol{x}_i$, and $\boldsymbol{\varepsilon} = \{\varepsilon_1, \varepsilon_2, \ldots, \varepsilon_n\}^{\mathrm{T}}$ denotes the vector of error terms (which attempts to capture the "noise" that influences the dependent variable beyond the effect of the independent variable).

Note that standard linear regression models tend to make some assumptions regarding their variables and their relationship. In general, when customary estimation techniques are used (e.g., ordinary least squares), these include weak exogeneity, linearity, homoscedasticity, error independence, and lack of perfect multicollinearity (Wooldridge 2013).

Nonetheless, when the dimensionality of the sample at hand is large, it can be reduced by means of a popular method known as *principal component analysis* (PCA). In essence, PCA summarizes the information of large amounts of data into a smaller set of key indices for the sake of visualization and analysis. In the case of the present study, PCA is used to reduce the dimensionality of the set of dependent variables by inferring from patterns in them (Syms 2008), thus transforming the initial multivariate linear regression into a multiple linear regression. For mathematical and additional details on PCA, the reader is directed to the work of Wold et al. (1987).

Bottom line, methodologically, since we will be handling multiple dependent and independent variables, we resort to a two-stage approach. First, we use PCA for dimensionality reduction purposes regarding the number of dependent variables. Second, we use multiple linear regression to understand the influence of several climate change predictors on human health.

Examples of simultaneous applications of PCA and multiple linear regression are abundant, ranging from the food industry (Liu et al. 2008) to the energy sector (Lam et al. 2008) and the environment (Tan et al. 2016) to pharmaceutics (Otsuka et al. 2011). However, to the best of the authors' knowledge, there are no such applications to climate change and human health and well-being. This fact further justifies the novelty of our study.

Case Study

This section characterizes the dependent and independent variables, as well as the countries considered as our units of analysis. It also details the software used to compute the results.

Variables

Among the 17 interlinked SDGs put forward by the UN (2015), 'Good health and well-being' and 'Climate action' (SDG 3 and SDG 13, respectively) are the ones that are more closely aligned with the aim of our study to understand the influence of climate change on human health. On the one hand, the former contains 26 indicators divided by 9 'Outcome' and 4 'Means of implementation targets; we have selected 17 indicators due to data unavailability in the UN's SDG Global Database (2022). On the other hand, the latter is comprised of 3 'Outcome' targets and 2 'Means of implementation' targets; from its total of 8 indicators, only 3 had no issues in terms of missing data according to the aforementioned database. Note that these 3 indicators are all related to the resilience and adaptive capacity to climate-related disasters.

Therefore, 3 SDG 13 indicators will be used as our independent variables to predict the 17 SDG 3 indicators that will be used as our dependent variables. Details on these variables will be addressed below.

Sample

23 UN Member States were chosen as our units of analysis among the organization's 191 Member States. Again, this was due to multiple instances of missing data, as well as the fact that we only considered countries and territories that were affected by extreme weather events, which also contributed to reducing the sample. Armenia, Australia, Bangladesh, Benin, Colombia, Costa Rica, Egypt, Fiji, Ghana, Guatemala, India, Kazakhstan, Liberia, Malawi, Malaysia, Mexico, Papua New Guinea, the Philippines, the Russian Federation, Togo, Trinidad and Tobago, Ukraine, and Vanuatu were the selected ones, comprising a set of 3 European, 4 Oceanian, 5 Asian, 6 African, and 5 American countries and territories.

Ultimately, we were left with a balanced panel data of 23 UN Member States and 20 SDG indicators regarding health and climate change in 2020—the year with the most recently available data. The key descriptive statistics of our sample are presented in Table 16.3.

Table 16.3 Key descriptive statistics of the dependent and independent variables

Type of variable	SDG indicator		Min	Max	Mean	Std. dev
Dependent	3.1.1 (y_1)	*Maternal mortality ratio*	6.00	661.00	141.30	168.90
	3.1.2 (y_2)	*Proportion of births attended by skilled health personnel*	53.00	100.00	87.00	15.57
	3.2.1 (y_3)	*Under-five mortality rate*	4.00	93.00	29.48	23.80
	3.2.2 (y_4)	*Neonatal mortality rate*	2.00	31.00	13.39	8.42
	3.3.2 (y_5)	*Tuberculosis incidence per 100,000 population*	6.60	554.00	117.03	143.40
	3.3.5 (y_6)	*Number of people (in millions) requiring interventions against neglected tropical diseases*	0.00	696.94	38.41	144.32
	3.4.1 (y_7)	*Mortality rate attributed to cardiovascular disease, cancer, diabetes, or chronic respiratory disease between age 30 and exact age 70*	9.10	30.60	21.13	5.68
	3.4.2 (y_8)	*Suicide mortality rate per 100,000 population*	2.70	31.00	9.48	7.29
	3.5.2 (y_9)	*Harmful use of alcohol, defined according to the national context as alcohol per capita consumption (aged 15 years and older) within a calendar year in liters of pure alcohol*	0.10	11.20	4.50	2.96
	3.6.1 (y_{10})	*Death rate due to road traffic injuries per 100,000 population*	5.60	35.90	18.29	7.50
	3.7.2 (y_{11})	*Adolescent birth rate (aged 15–19 years) per 1,000 women in that age group*	9.30	180.00	61.96	46.68
	3.9.1 (y_{12})	*Mortality rate attributed to household and ambient air pollution per 100,000 population*	8.40	249.60	106.97	69.20

(continued)

Table 16.3 (continued)

Type of variable	SDG indicator		Min	Max	Mean	Std. dev
	3.9.2 (y_{13})	*Mortality rate attributed to unsafe water, unsafe sanitation, and lack of hygiene (exposure to unsafe WASH*a *services) per 100,000 population*	0.10	59.70	11.60	16.53
	3.9.3 (y_{14})	*Mortality rate attributed to unintentional poisoning*	0.20	3.50	1.20	0.98
	3.b.1 (y_{15})	*Proportion of the population with diphtheria-tetanus-pertussis immunization coverage among 1-year-olds*	50.00	99.00	87.78	13.11
	3.c.1 (y_{16})	*Density of medical doctors per 10,000 population*	0.40	44.00	15.77	15.90
	3.d.1 (y_{17})	*Average of 13 IHR*b *core capacity scores*	17.00	99.00	57.00	25.84
Independent	13.1.1 (x_1)	*Number of deaths and missing persons attributed to disasters per 100,000 population*	0.07	110.96	13.21	25.28
	13.1.2 (x_2)	*Score of adoption and implementation of national DRR strategies in line with the Sendai Framework*	0.25	1.00	0.76	0.21
	13.1.3 (x_3)	*Proportion of local governments that adopt and implement local disaster risk reduction strategies in line with national disaster risk reduction strategies*	0.00	100.00	65.40	39.15

[a] Water, Sanitation, and Hygiene for all
[b] International Health Regulations

Application Details

The sample described in the previous subsections was used to populate the multivariate linear regression model previously described. IBM SPSS Statistics version 26 was the software used to implement the model and compute the results.

Results and Discussion

First, we look at the results of the PCA. The bivariate Pearson correlation testing among the 17 dependent variables returned satisfactory levels of correlation since 55 out of 136 correlation tests proved to be statistically significant, with approximately 71% of those being moderate to very strong correlations at the 0.01 level. Table 16.4 displays these results. Note that Bartlett's test of sphericity returned a p-value lower than 0.01, which means that the dependent variables are indeed significantly correlated.

Moreover, the total variance explained by PCA's initial solution is shown in Table 16.5. Since eigenvalues greater than 1 are traditionally seen as indicative of how many factors to retain, 5 components should be retained, which account for approximately 84% of the total variance of the sample, i.e., we can reduce the dimension of the dependent variable set from 17 to 5 with only an approximate 16% loss of information.

If we look at the communalities (Table 16.6), which indicate the amount of variance that is accounted for per variable, we can see that their moderate to very high values indicate that the extracted components represent the regressands well.

According to Krishnan (2010), the importance of components in the computation of a single score is not equal. Using the proportion of their partial variance explained to total variance explained, we can generate a single non-standardized dependent variable, which, after feature scaling, returns its value in the [0, 1] range. A similar procedure has been adopted previously in the literature (see, e.g., Antony and Rao 2007).

Second, we now address the results of the multiple linear regression considering 3 predictors and the PCA-based human health and well-being composite dependent variable. Taken as a set, the 3 predictors account for approximately 12% of the variance of the human health composite indicator ($R^2 = 0.12$). Additionally, the overall regression model was not significant: $F(3, 19) = 0.89$, $p = 0.47$, $R^2 = 0.12$ (see Table 16.7).

At last, Table 16.8 presents the multiple linear regression coefficient results. Since no predictor had a p-value lower than 0.05, no predictor's accounted for unique variance is statistically significant. This means that the *Number of deaths and missing persons attributed to disasters per 100,000 population*, the *Score of adoption and implementation of national DRR strategies in line with the Sendai Framework*, and the *Proportion of local governments that adopt and implement local disaster risk reduction strategies in line with national disaster risk reduction strategies* do not influence human health and well-being.

Table 16.4 Results of the bivariate Pearson correlation testing between the dependent variables

	3.1.1 (y_1)	3.1.2 (y_2)	3.2.1 (y_3)	3.2.2 (y_4)	3.3.2 (y_5)	3.3.5 (y_6)	3.4.1 (y_7)	3.4.2 (y_8)	3.5.2 (y_9)	3.6.1 (y_{10})	3.7.2 (y_{11})	3.9.1 (y_{12})	3.9.2 (y_{13})	3.9.3 (y_{14})	3.b.1 (y_{15})	3.c.1 (y_{16})	3.d.1 (y_{17})
3.1.1 (y_1)	1.00	–0.64**	0.89**	0.83**	0.35	0.02	–0.10	–0.24	–0.20	0.84**	0.65**	0.71**	0.90**	0.52*	–0.12	–0.61**	–0.37
3.1.2 (y_2)	–	1.00	–0.67**	–0.72**	–0.47*	–0.13	–0.08	0.32	0.48*	–0.40	–0.39	–0.67**	–0.61**	–0.30	0.22	0.68**	0.33
3.2.1 (y_3)	–	–	1.00	0.95**	0.32	0.08	0.10	–0.27	– 0.36	0.70**	0.57**	0.84**	0.97**	0.63**	–0.24	–0.72**	–0.50*
3.2.2 (y_4)	–	–	–	1.00	0.41*	0.27	0.16	–0.31	– 0.43*	0.64**	0.56**	0.88**	0.89**	0.57**	–0.21	–0.75**	–0.47*
3.3.2 (y_5)	–	–	–	–	1.00	0.18	0.34	–0.22	– 0.07	0.16	0.10	0.51*	0.21	0.08	–0.51*	–0.44*	–0.29
3.3.5 (y_6)	–	–	–	–	–	1.00	0.09	0.17	0.05	0.12	–0.23	0.28	0.10	0.24	0.01	–0.13	0.18
3.4.1 (y_7)	–	–	–	–	–	–	1.00	0.19	– 0.20	–0.17	–0.19	0.40	–0.03	0.20	–0.27	–0.16	–0.23
3.4.2 (y_8)	–	–	–	–	–	–	–	1.00	0.66**	–0.07	–0.28	–0.24	–0.17	0.41*	–0.04	0.60**	0.50*
3.5.2 (y_9)	–	–	–	–	–	–	–	–	1.00	–0.19	–0.17	–0.41	–0.27	0.01	–0.11	0.65**	0.57**
3.6.1 (y_{10})	–	–	–	–	–	–	–	–	–	1.00	0.46*	0.53**	0.76**	0.62**	0.03	–0.44*	–0.31
3.7.2 (y_{11})	–	–	–	–	–	–	–	–	–	–	1.00	0.26	0.53*	0.14	0.04	–0.29	–0.30
3.9.1 (y_{12})	–	–	–	–	–	–	–	–	–	–	–	1.00	0.76**	0.53**	–0.29	–0.77**	–0.44*

(continued)

Table 16.4 (continued)

	3.1.1 (y_1)	3.1.2 (y_2)	3.2.1 (y_3)	3.2.2 (y_4)	3.3.2 (y_5)	3.3.5 (y_6)	3.4.1 (y_7)	3.4.2 (y_8)	3.5.2 (y_9)	3.6.1 (y_{10})	3.7.2 (y_{11})	3.9.1 (y_{12})	3.9.2 (y_{13})	3.9.3 (y_{14})	3.b.1 (y_{15})	3.c.1 (y_{16})	3.d.1 (y_{17})
3.9.2 (y_{13})	–	–	–	–	–	–	–	–	–	–	–	–	1.00	0.69**	–0.20	–0.63**	–0.41
3.9.3 (y_{14})	–	–	–	–	–	–	–	–	–	–	–	–	–	1.00	–0.37	–0.25	–0.16
3.b.1 (y_{15})	–	–	–	–	–	–	–	–	–	–	–	–	–	–	1.00	0.22	0.22
3.c.1 (y_{16})	–	–	–	–	–	–	–	–	–	–	–	–	–	–	–	1.00	0.64**
3.d.1 (y_{17})	–	–	–	–	–	–	–	–	–	–	–	–	–	–	–	–	1.00

Table 16.5 Total variance explained of the extracted components

Component	Initial eigenvalue		
	Total	Variance (%)	Cumulative variance (%)
1	7.69	45.22	45.22
2	2.34	13.79	59.00
3	2.05	12.04	71.05
4	1.21	7.11	78.17
5	1.03	6.08	84.25
6	0.69	4.07	88.32
7	0.55	3.24	91.55
8	0.44	2.58	94.14
9	0.36	2.11	96.24
10	0.24	1.43	97.67
11	0.15	0.85	98.52
12	0.10	0.60	99.12
13	0.07	0.43	99.54
14	0.05	0.26	99.81
15	0.02	0.12	99.92
16	0.01	0.06	99.99
17	0.00	0.01	100.00

Conclusion

Our study analyzed a set of 23 UN countries and territories to infer if the resilience and adaptive capacity to climate-related disasters influence human health and well-being, based on data from the UN's SDG framework from 2020, using PCA and multiple linear regression. To the best of the authors' knowledge, our contribution is unprecedented in the literature.

Globally, our approach revealed that the three indicators regarding the resilience and adaptive capacity to climate-related disasters (*Number of deaths and missing persons attributed to disasters per 100,000 population*, the *Score of adoption and implementation of national DRR strategies in line with the Sendai Framework*, and the *Proportion of local governments that adopt and implement local disaster risk reduction strategies in line with national disaster risk reduction strategies*) do not seem to predict human health and well-being in our sample. However, these findings have some limitations since the sample was comprised of only 12% of the UN's Member States and there were major issues of data unavailability concerning the remaining four targets of SDG 13. We believe that, by incorporating the indicators on the integration of climate change measures into policies and planning, the construction of knowledge and capacity to meet climate change, the implementation

Table 16.6 Rotated component matrix and communalities

SDG indicator	Component					Communality
	1	2	3	4	5	
3.1.1 (y_1)	0.91	–0.17	0.18	–0.20	–0.05	0.94
3.1.2 (y_2)	–0.56	0.44	–0.36	0.07	–0.19	0.67
3.2.1 (y_3)	0.91	– 0.28	0.18	0.08	– 0.02	0.95
3.2.2 (y_4)	0.85	–0.37	0.22	0.07	0.17	0.94
3.3.2 (y_5)	0.15	–0.21	0.86	0.03	0.18	0.84
3.3.5 (y_6)	0.12	0.07	0.05	0.08	0.91	0.86
3.4.1 (y_7)	–0.06	–0.13	0.26	0.83	0.07	0.78
3.4.2 (y_8)	–0.01	0.87	–0.12	0.34	0.12	0.89
3.5.2 (y_9)	–0.17	0.88	0.17	–0.20	0.00	0.87
3.6.1 (y_{10})	0.87	–0.03	–0.11	–0.09	0.02	0.77
3.7.2 (y_{11})	0.59	–0.17	0.05	–0.40	–0.38	0.69
3.9.1 (y_{12})	0.70	–0.38	0.32	0.30	0.26	0.89
3.9.2 (y_{13})	0.95	–0.15	0.08	0.03	–0.01	0.94
3.9.3 (y_{14})	0.77	0.33	0.03	0.46	0.08	0.91
3.b.1 (y_{15})	–0.12	–0.14	–0.79	–0.31	0.18	0.78
3.c.1 (y_{16})	–0.52	0.71	–0.26	–0.11	– 0.12	0.87
3.d.1 (y_{17})	–0.32	0.63	–0.15	–0.29	0.36	0.73

Table 16.7 Analysis of variance

Model	Sum of squares	df	Mean square	*F*	Sig
Regression	1,156.31	3	385.44	0.89	0.47
Residual	8,276.55	19	435.61		
Total	9,432.87	22			

Table 16.8 Multiple regression coefficients

Model	Unstandardized coefficients		Standardized coefficients	*t*	Sig
	B	Std. error	Beta		
(Constant)	−23.22	20.54		−1.13	0.27
13.1.1 (x_1)	−0.09	0.18	−0.11	−0.49	0.63
13.1.2 (x_2)	30.12	21.80	0.31	1.38	0.18
13.1.3 (x_3)	0.11	0.12	0.20	0.92	0.37

of the UN framework convention on climate change, and the promotion of mechanisms to raise capacity for planning and management, greenhouse gas emissions and climate expenditure could have some degree of influence on human health and well-being.

References

Akhtar R (2007) Climate change and health and heat wave mortality in India. Global Environ Res 11:51–57

Antony GM, Rao KV (2007) A composite index to explain variations in poverty, health, nutritional status and standard of living: use of multivariate statistical methods. Public Health 121(8):578–587. https://doi.org/10.1016/j.puhe.2006.10.018

Betti G (2017) What impact has the economic crisis had on quality of life in Europe? A multidimensional and fuzzy approach. Qual Quant 51(1):351–364. https://doi.org/10.1007/s11135-015-0308-8

Carbon Pricing Leadership Coalition. (2021). What is carbon pricing? https://www.carbonpricingleadership.org/what

Committee for Development Policy. (2022a). Climate change. https://www.cdp.net/en/climate

Committee for Development Policy. (2022b). Carbon pricing. https://www.cdp.net/en/climate/carbon-pricing

Dahlgren G, Whitehead M (2007) Policies and strategies to promote social equity in health: background document to WHO—strategy paper for Europe (No. 2007:14). Stockholm

European Commission (2020) European climate pact. https://ec.europa.eu/clima/eu-action/european-green-deal/european-climate-pact_en

Freedman D (2005) Statistical models: theory and practice. Cambridge University Press, Cambridge. https://doi.org/10.1017/CBO9781139165495

Greaves LJ, Bialystok LR (2011) Health in all policies—all talk and little action? Can J Public Health 102(6):407–409. https://doi.org/10.1007/BF03404187

Kershaw P (2020) A "health in all policies" review of Canadian public finance. Can J Public Health 111(1):8–20. https://doi.org/10.17269/s41997-019-00291-4

Krishnan V (2010) Constructing an area-based socioeconomic index: a principal components analysis approach. Alberta

Lam JC, Wan KKW, Cheung KL, Yang L (2008) Principal component analysis of electricity use in office buildings. Energy Build 40(5):828–836. https://doi.org/10.1016/j.enbuild.2007.06.001

Leal Filho W, Azeiteiro UM, Alves F (2016). Climate change and health: an overview of the issues and needs. In: Leal Filho W, Azeiteiro U, Alves F (eds) Climate change and health. Springer International Publishing, Cham, pp 1–11. https://doi.org/10.1007/978-3-319-24660-4_1

Liu J, Li Q, Dong J, Chen J, Gu G (2008) Multivariate modeling of aging in bottled lager beer by principal component analysis and multiple regression methods. J Agri Food Chem 56(16):7106–7112. https://doi.org/10.1021/jf800879v

Nunes AM, Ferreira DC (2022) Social inequity and health: from the environment to the access to healthcare in composite indicators, the Portuguese case. In: Leal Filho W, Vidal DG, Dinis MAP, Dias RC (eds) Sustainable policies and practices in energy, environment and health research. Springer International Publishing, Cham, pp 371–389. https://doi.org/10.1007/978-3-030-86304-3_21

Organisation for Economic Co-operation and Development (2022) Health status: causes of mortality. https://stats.oecd.org/index.aspx?queryid=30115

Otsuka T, Iwao Y, Miyagishima A, Itai S (2011) Application of principal component analysis enables to effectively find important physical variables for optimization of fluid bed granulator conditions. Int J Pharmaceut 409(1–2):81–88. https://doi.org/10.1016/j.ijpharm.2011.02.044

Patz JA, Campbell-Lendrum D, Holloway T, Foley JA (2005) Impact of regional climate change on human health. Nature 438(7066):310–317. https://doi.org/10.1038/nature04188

Pförtner T-K, Richter M (2011) Getting social: public health's increasing awareness of the social determinants of health. J Public Health 19(1):1–2. https://doi.org/10.1007/s10389-010-0392-0

Puska P (2007) Health in all policies. Eur J Public Health 17(4):328–328. https://doi.org/10.1093/eurpub/ckm048

Rencher AC, Christensen WF (2012). Methods of multivariate analysis, 3rd edn. John Wiley, Hoboken, NJ. https://doi.org/10.1002/9781118391686

Sarkodie SA, Owusu PA (2021) Global assessment of environment, health and economic impact of the novel coronavirus (COVID-19). Environ Dev Sustain 23(4):5005–5015. https://doi.org/10.1007/s10668-020-00801-2

Saxe JP (2021) A light bulb moment for cities: opportunities to improve residential energy efficiency outreach. In: Lackner M, Sajjadi B, Chen WY (eds) Handbook of climate change mitigation and adaptation. Springer, New York, NY, pp 1–44. https://doi.org/10.1007/978-1-4614-6431-0_173-1

Schifano P, Cappai G, De Sario M, Michelozzi P, Marino C, Bargagli AM, Perucci CA (2009) Susceptibility to heat wave-related mortality: a follow-up study of a cohort of elderly in Rome. Environ Health 8(1):50. https://doi.org/10.1186/1476-069X-8-50

Shankardass K, Muntaner C, Kokkinen L, Shahidi FV, Freiler A, Oneka G et al (2018) The implementation of health in all policies initiatives: a systems framework for government action. Health Res Policy Syst 16(1):26. https://doi.org/10.1186/s12961-018-0295-z

Sihto M, Ollila E, Koivusalo M (2006). Health in all policies: the wider context. In: Stahl T, Wismar M, Ollila E, Lahtinen E, Leppo K (eds) Health in all policies: prospects and potentials. Ministry of Social Affairs and Health, Helsinki, pp 3–20

Smith K, Woodward A, Campbell-Lendrum D, Chadee D, Honda Y, Liu Q, et al (2014) Human health: impacts, adaptation, and co-benefits. In: Field C, Barros V, Dokken D (eds) Climate change 2014: impacts, adaptation, and vulnerability. Part A: global and sectoral aspects. Contribution of working group II to the fifth assessment report of the intergovernmental panel on climate change. Cambridge University Press, Cambridge, pp 709–754

Syms C (2008) Principal components analysis. In: Jørgensen SE, Fath BD (eds) Encyclopedia of ecology. Elsevier, Amsterdam, pp 2940–2949. https://doi.org/10.1016/B978-008045405-4.00538-3

Tan KC, San Lim H, Jafri MZM (2016) Prediction of column ozone concentrations using multiple regression analysis and principal component analysis techniques: a case study in peninsular Malaysia. Atmos Pollut Res 7(3):533–546. https://doi.org/10.1016/j.apr.2016.01.002

United Nations (2015) Transforming our world: the 2030 agenda for sustainable development. New York, NY

United Nations (2022) Global SDG indicators data platform. https://unstats.un.org/sdgs/dataportal

Wold S, Esbensen K, Geladi P (1987) Principal component analysis. Chemometr Intell Lab Syst 2(1–3):37–52. https://doi.org/10.1016/0169-7439(87)80084-9

Wooldridge JM (2013) Introductory econometrics: a modern approach, 5th edn. South-Western Cengage Learning, Boston, MA

World Health Organization. (1946). Constitution of the World Health Organization. New York, NY

World Health Organization (1998) Health promotion glossary. Geneva

World Health Organization (2010) Adelaide statement on health in all policies: moving towards a shared governance for health and well-being. Adelaide

World Health Organization (2020). About social determinants of health. https://www.who.int/social_determinants/sdh_definition/en/

Yan X, Su XG (2009) Linear regression analysis. World Scientific, Singapore. https://doi.org/10.1142/6986

Miguel Alves Pereira is a post-doctoral researcher at the Institute for Systems and Computer Engineering, Technology and Science (INESC TEC) in Porto, Portugal, and an associated member of the Centre for Management Studies of the Instituto Superior Técnico of the University of Lisbon (IST-ULisbon; CEGIST). He has been an invited teaching assistant at the IST-ULisbon over the past 2 years, where he teaches in the field of management. His main research interests concern operational research / management science, econometrics, and applied economics (with an emphasis on the development of efficiency measurement and performance assessment models), and their applications to sustainable development and the health and water and sanitation sectors. He has published more than a dozen scientific articles in top international peer-reviewed journals and acted as an ad hoc referee in 21 of them. He has been involved as a member of various national and international scientific societies, and he has participated in different capacities in 4 national and international scientific projects. He has also been a visiting researcher at the Royal Melbourne Institute of Technology.

Alexandre Morais Nunes completed his Ph.D. in Health Administration (2016) at the Institute of Social and Political Sciences of the University of Lisbon (ISCSP-ULisbon). He is an Assistant Professor at ISCSP-ULisbon, where he teaches Administration, Management, and Health Policies. He is a researcher in the Centre for Public Administration and Public Policies (CAPP). This activity resulted in 100 scientific articles in peer-reviewed journals, 4 books (including 2 manuals), 2 book chapters, and 5 scientific merit awards over the last four years. In addition to this professional academic experience, he was in government roles in the Office of the Minister of Health (2015–2018), holding the position of Advisor for the Administration of Health Services. His research interests include: public policies; health policy; management; public administration; efficiency; satisfaction; access; financing; health spending; economic impacts; and environmental and social determinants in health.

Diogo Cunha Ferreira is an Assistant Professor at the IST-ULisbon, an Invited Assistant Professor at the ISCSP-ULisbon, and an Adjunct Associate Professor at the Institute for Public Policy and Governance within the Deputy Vice-Chancellor Research's Office (University of Technology Sydney, Australia). He has published over half a hundred papers in high-ranking journals in the past five years, mostly focused on management science (from theoretical to empirical ones). His research interests include performance assessment, benchmarking, sustainability, risk analysis, project management, healthcare, climate change, governance, public policy, and public administration.

Chapter 17
EU Funding to Promote Climate Change Adaptation and Risk Prevention and Management in Portugal: Potential Effects on Mitigating Health Hazards

Eduardo José Rocha Medeiros

Abstract Following the publication of the most recent (2018) National Spatial Policy Programme (Programa Nacional da Política de Ordenamento do Território—PNPOT), several critical development risks were identified for the Portuguese territory over the next few decades. These include expected rises in average temperatures and overall reduction of precipitation levels. This novel climate panorama can negatively impact human and environmental health, for instance via increasing numbers and intensity of forest fires, biologic desertification processes, and heatwaves. These predicted climate changes can also lead to high peaks of intensive rain, leading to catastrophic river floods which, associated with rising sea levels, in a coastal country like Portugal, can lead to increasing health hazards of all sorts in Portuguese territory over the coming decades. In this context, this article analyses the EU funding on promoting climate change adaptation and risk prevention and management in Portugal (2014–2020) via the Operational Programme for Sustainability and Efficient Use of Resources (POSEUR), and its potential effects on mitigating health hazards in the Portuguese territory. In parallel, it analyses existing Portuguese environmental development strategies to unveil the extent to which policy actions to mitigate health hazards are being considered.

Introduction

Living a healthy life is considered by the United Nations (UN) as a critical capability for human development. Hence a 'health-related indicator' has been used as part of the UN Human Development Index (HDI) since its inception (UN 2020). This chapter discusses the potential negative impacts of climate change on human health, through its direct and indirect effects of increasing average temperatures. It also examines whether and how the Portugal 2020 (the partnership agreement undertaken between

E. J. R. Medeiros (✉)
Instituto Universitário de Lisboa (ISCTE-IUL), DINÂMIA'CET - IUL, Avenida das Forças Armadas, Edifício Sedas Nunes, Sala 2W4-d, 1649–026, Lisboa, Portugal
e-mail: Eduardo.Medeiros@iscte-iul.pt

W. Leal Filho et al. (eds.), *Climate Change and Health Hazards*, Climate Change Management, https://doi.org/10.1007/978-3-031-26592-1_17

Portugal and the European Commission [EC], which brings together the five European structural and investment funds—ERDF, Cohesion Fund, ESF, EAFRD and EMFF—setting the programming principles which enshrine the economic, social and territorial development policy to be advanced in Portugal in the programming period between 2014 and 2020) and its environmental sustainability related Operational Programme (PO SEUR), has contributed to mitigating health hazards related to climate change in Portugal.

Critically, climate change is considered as s small-moving global crisis, entailing complex operative policy solutions in response to human-induced actions at the base of a global warming trend observed in past decades (Sachs 2015). Being aware of such policy challenges, the UN 2030 Agenda identified a specific strategic development goal (SDG 3), dedicated to ensuring healthy lives for all. In the end, with the implementation of related measures, the UN expects mortality rates to be significantly reduced by 2030, especially in less developed countries, and that universal health coverage should be a reality in all countries by 2030. As important are the SDG 3 sub-goals 3.3 (by 2030, end the epidemics of AIDS, tuberculosis, malaria and neglected tropical diseases and combat hepatitis, water-borne diseases, and other communicable diseases) and 3.d (strengthen the capacity of all countries, in particular developing countries, for early warning, risk reduction and management of national and global health risks), which are directly related to climate change-related effects (UN 2016).

In this context, this chapter draws on current literature on health hazards and climate change, and in particular on the effects of climate change on health hazards in Portugal in past years and over the coming decades. Mostly based on desk research, from a methodological standpoint, it sheds particular light on analysis of the approved projects database of the PO SEUR which directly and indirectly relate to mitigating the effects of health hazards. As regards the literature review, there are a limited number of analysis on the impacts of the PO SEUR on health impacts. The same goes for similar analysis for the Portuguese regions. In essence, this chapter addresses the potential positive effects of the PO SEUR on mitigating these hazards in Portugal. The remainder of the chapter is structured as follows. It begins by setting out a literature review on health hazards and climate change, to identify the main analytical dimensions or components which can guide analysis of the PO SEUR project database. The following section debates the potential climate change effects in health hazards in Portugal. The empirical analysis of the case study (Portugal) is provided in the next section and policy implications are discussed in the conclusion.

Health Hazards and Climate Change: A Literature Review

The fundamental idea behind the scientific and political alerts to global climate change in past decades (Sachs 2015) is mainly linked with mounting evidence showing global warming trends (Shao 2017). These, in turn, are recognized by the scientific community as potentially having direct and indirect negative effects on human health from, for instance, heatwaves, droughts, and floods (Johansen 2017). Directly contributing to these global warming trends, the burning of fossil fuels like oil and coal creates the conditions for health problems, including lung and heart diseases (including cancer), allergies, chronic bronchitis, asthma, chronic obstructive pulmonary disease, stroke, and chronic respiratory diseases (Epstein and Ferber 2011). Furthermore, rising temperatures can bring about a series of heat-induced diseases, including "dehydration, heat rash, heat cramps, heat exhaustion, heat fatigue, heat syncope/fainting, and heatstroke" (Schulte and Chun 2009: 544).

Indirectly, rising temperatures create optimal conditions for disease-spreading vectors like insects, with particular attention to mosquitoes, strongly associated with the spread of crippling and often fatal diseases like malaria, dengue, and Zika (National Geographic 2016). Indeed, "according to the Intergovernmental Panel on Climate Change's projections for human health, a rise in global average temperatures of 3–5 °C by 2100 could lead to 50 million to 80 million additional annual cases of malaria worldwide, primarily in tropical, subtropical, and poorly protected temperate-zone populations" (Johansen 2017: 40). Besides the aggravation of health symptoms resulting from the spread of mosquito-related diseases (headache, rash, joint pain, fever, encephalitis, meningitis, fatigue, paralysis, severe neurological problems, etc.) (Rivera-Amarillo and Camargo 2020), global warming trends also affect the escalation in numbers of termites, cockroaches, rats, spiders, and scorpions (Johansen 2017; Loko et al. 2016).

The rise of allergies and asthma can also result from global warming trends since this favour the growth of plants that release allergy and asthma-causing pollens (Singer 2013) and tend to increase with the rise of forest fires (Fearnside 2012). More worrying still is the expected rise in the number of deaths attributable to rapid global warming trends (at least 150,000 per year), which is expected to double by 2035 (Johansen 2017). Reflecting a long anthropologic legacy of global pollution, the spread of suspended particulate matter from outdoor urban exposures can increase their harmful health effects in rising temperatures (Cifuentes et al. 2001). On the other hand, many of these gas emissions which result from burning fossil fuels are also pollutants (Johnsson et al. 2019).

From another viewpoint, climate change has the potential to affect the health of humans by influencing the indoor environment, depending on the characteristics of the buildings. Here, the levels of poverty are a crucial "indicator of individual vulnerability to climate change and are related to marginalization and lack of resources" (IOM 2011: 52). Most influential in this regard are also natural hazards which can become ever more dangerous as a result of rising global temperatures like tropical storms, which may lead to increasing outdoor and indoor health hazards (Theodore and Dupont 2012). Among the potential impacts on health from direct and indirect consequences of climate change influencing the indoor environment are (IOM 2011):

- Increased mortality and decreased productivity;
- Exposure to chemical emissions from damaged materials;
- Water and vector-based diseases;
- Respiratory distress and illness;

- Allergen-mediated distress and illness;
- Exposure to excessive heat and cold;
- Exposure to CO_2 from backup electrical generators;
- Increasing numbers of insects and rodents (pesticide exposure).

In all, according to the Intergovernmental Panel on Climate Change (IPCC), there is mounting evidence of "human health effects directly affected by climate change (e.g., heat stress, death, or injury in floods and storms) and indirectly through changes in the ranges of disease vectors (e.g., mosquitoes), waterborne pathogens, water quality, air quality, and food availability and quality" (Schulte and Chun 2009: 542). Moreover, the same entity concludes that "health impacts will be largely influenced by local environmental conditions and socioeconomic circumstances and by the range of social, institutional, technological, and behavioural adaptations taken to reduce the full range of threats to health" (Schulte and Chun 2009: 542). It is also worth noting that numerous factors can influence the degree of health hazards provoked by climate change: (i) age; (ii) obesity; (iii) pre-existing disease; (iv) body size; (v) socio-economic condition; (vi) pregnancy; (vii) immunologic status; (viii) type of work clothing; and (ix) genetic characteristics (Schulte and Chun 2009).

In much the same way, Patz and Hatch (2013) conclude that climate change presents a significant challenge to global health. Similarly, they identify a myriad of areas in which health impacts can be observed as a result of increasing average temperatures, natural hazards, and air pollution. In turn, these trends can lead to increased illnesses and deaths, related to allergic and infectious diseases and malnutrition. But most importantly, these authors highlight the fact that, over the next decade, a disproportional impact on health hazards is expected due to climate change across the world with the "populations that are often least responsible for climate change that experience the greatest adverse impacts" (Patz and Hatch 2013: 1). This raises ethical and moral issues which need to be addressed globally when dealing with concrete global policy actions aiming at mitigating climate change (Graham 2015; Gudynas 2016; Lukasiewicz 2017) in the many policy areas covered by climate change-related health hazards (Fig. 17.1).

Potential Climate Change Effects in Health Hazards in Portugal

The most recent version of the Portuguese National Spatial Planning Programme (Programa Nacional da Política do Ordenamento do Território—PNPOT) manifests several challenges faced by Portuguese territories from potential climate change effects. These are especially concerning regarding the risks associated with the rise of the average temperatures, the reduction of rainfall, the rise of the average sea level (Fig. 17.2), and associated natural risks: increasing droughts, coastal and river floods, and forest fires (DGT 2018a, b, c). When contemplating the expected impacts of climate change in Portugal, the same document sets the stage for increasing extreme

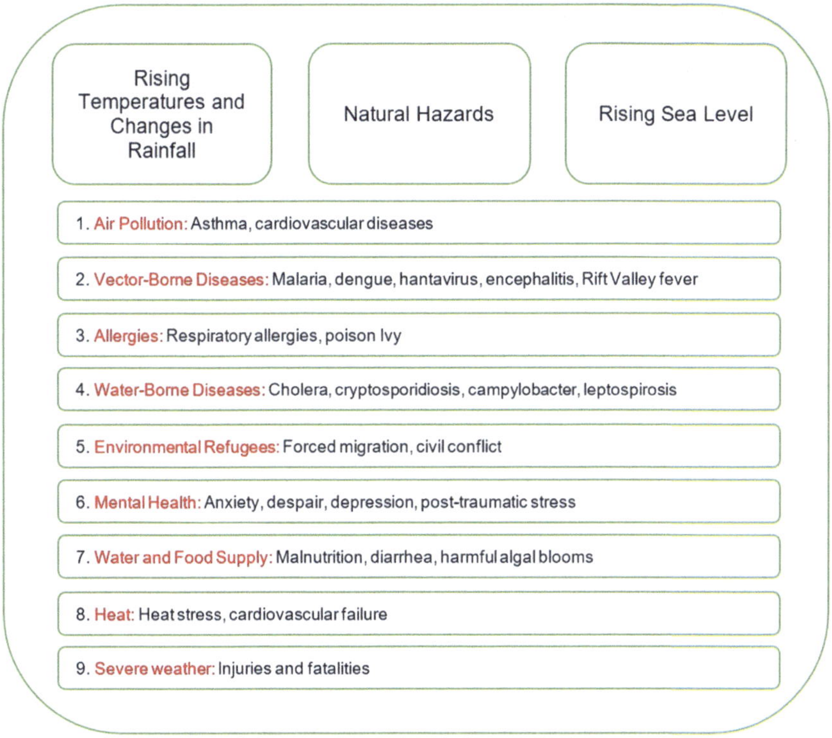

Fig. 17.1 Climate Change and Health hazards main analytical dimensions and components. Own elaboration based in—https://www.jkgeography.com/impacts-of-climate-change-on-people-and-places.html

meteorological events with different territorial levels, since the Portuguese weather conditions vary significantly from north to south and from east to west. Inspired by several international studies which predict potential climate trends in Europe, the PNPOT warns that, by the end of the twenty-first century, the average maximum temperature in the summer will increase between 0.5 °C along the coast and 2 °C in the inland territory on the mainland. In turn, respectively for the Madeira and Azores archipelagos, there is an expected rise of temperatures between 2 and 3 °C, and 2.5 and 3 °C.

In tune with several international scientific reports, the PNPOT predicts a reduction of the levels of rainfall for the end of this century in Portugal. In more detail, a significant reduction (between 20 and 40%) of the total annual rainfall is expected occurs in the mainland area of Portugal. Worst still, drought periods are expected be more recurrent and intense. Likewise, for the Madeira archipelago, a reduction of rainfall of around 30% is estimated by the end of the century. Conversely, for the Azores archipelago, a slight increase in rainfall levels of around 10% is expected during the winter, and a reduction during the summer (DGT 2018a).

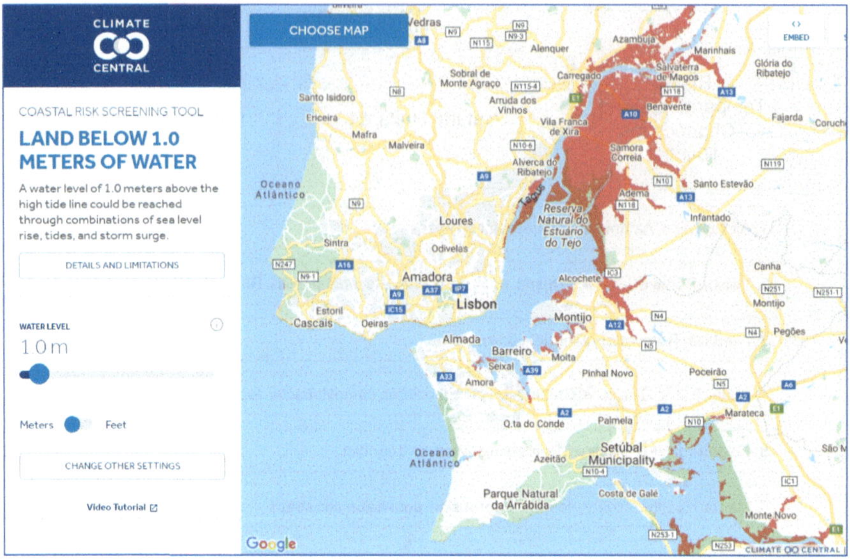

Fig. 17.2 Areas affected by mean sea level rise of 1 m in Lisbon Metropolitan Area (*Source* https://coastal.climatecentral.org/)

Finally, based on concrete measurements of the mean sea level made in the Portuguese city of Cascais for more than a century, the PNPOT draws attention to the fact that an increasing rise of the mean sea level can be observed in recent years in Portugal, when compared with past decades. In more detail, between 1992 and 2004 the sea rose on average 2.1 mm/year. Between 2005 and 2016 it rose 4.0 mm/year, which fits within the global trends, in which most estimates of global mean rise in the sea level this century will fall below 2 m (Kulp and Strauss 2019). For the Azores archipelago, this rise in the average sea level is expected to read 1 m by the end of this century. In all cases, this scenario poses great concerns in a country like Portugal, which has a vast coastline where more than 70% of the population lives, with particular concerns for several areas of the Lisbon Metropolitan Area (Fig. 17.2).

Crucially, based on the findings of the previous section, the predicted climate changes in Portugal can contribute to increasing the risks of various health hazards, including the spread of several diseases provoked by mosquitos and other animal pests, and other diseases associated with increasing temperatures and heatwaves. Furthermore, the reduction of available potable water and food production poses an alarming challenge to the country, with potential negative effects on the health of the Portuguese population. The same goes for the expected rise in extreme weather events, including floods and river and sea inundations, heatwaves and the occurrence of fires, with a serious impact on forested territories. Needless to say, Portuguese urban areas are particularly vulnerable to several of these climate change-related natural risks, and hence to related health hazards, thus requiring an adaptation of social and health infrastructure to respond to increasing heatwaves and potential mortality rates. In the end, the PNPOT identifies five main potential socio-economic impacts resulting from predicted climate trends in Portugal: (i) increasing financial effort due to emergencies; (ii) new needs to respond to social and health services; (iii) rising costs in infrastructure adaptation; (iv) rise of energy costs; and (v) a need to change economic models towards an efficient low-carbon economy (DGT 2018a).

Curiously, the Portuguese National Strategy for Climate Adaptation (Estratégia Nacional de Adaptação às Alterações Climáticas—ENAAC in Portuguese) is even more drastic in its climate change predictions for Portugal until the end of the century, as it predicts an average rise in summer temperatures between 3 °C in the littoral areas and 7 °C in the interior of the Portuguese continental territory. It is also interesting to note that the ENAAC expects the elimination of days with frost in most parts of the Portuguese territory. Hence, as previously seen, this scenario can provoke a rise in animal plagues (e.g. mosquitos), and a consequent rise in health hazards in Portugal (APA 2013). This report proposes a plethora of measures in several sectors to mitigate the predicted climate change-related impacts in Portugal. Only two were advanced for the health sector: (i) a Contingency Plan for Adverse Extreme Temperatures—Heat Module, and (ii) a National Programme for the Surveillance of Culicidae Vectors.

In a section dedicated to potential health hazards resulting from climate change in Portugal, the ENAAC highlights the potential introduction of novel risk factors to human health, especially related to the distribution of infectious and respiratory diseases transmitted via vectors like mosquitos and increasing pollution. Similarly, natural disasters like floods and heatwaves can have a direct and indirect physical and psychological impact on human health, for instance via injuries and the destruction of homes. By now, for instance, the Madeira archipelago is already dealing with a rise of dengue fever, due to rising average temperatures. On the other hand, the impacts on food production and the availability of potable water can contribute to increased risk of diseases associated with malnutrition and water contamination (APA 2013).

Added to that, there is a well-known association between intense precipitation phenomena and outbreaks of cryptosporidiosis. On the other hand, lower water flows in rivers and reservoirs, associated with higher temperatures, could increase human exposure to cyanotoxins produced by cyanobacteria, which could have serious consequences for human health. In the case of food-borne illnesses, the increase in air temperature is associated with a rise in cases of salmonellosis in different locations. Outbreaks caused by norovirus were also associated with cases of heavy rainfall and flooding that caused the overflow of wastewater. In sum, climate change and the expected effects on the distribution and prevalence of diseases in Portugal could lead to the emergence of new demands on health systems, requiring adaptation work that must be carried out as soon as possible to prevent and reduce the extension of the effects on the population (APA 2013).

Recent research has also provided a word of caution that should highlight the potential negative health effects of the rising sea level on the water and food supply. Indeed, on the Portuguese mainland, hundreds of thousands of inhabitants dwell near the Atlantic coast, which is also populated by key development infrastructure and economic activities. Hence, an extreme flood hazard scenario could provoke high levels of coastal erosion and rising coastal flooding risks (Antunes et al. 2019).

On the main expected impacts of heatwaves on the health of the Portuguese population, Dias et al. (2012) note the need to embrace measures with can effectively mitigate verified increased death rates from respiratory diseases in Portugal. Also noteworthy are the scientific studies which have demonstrated increased mortality rates in Portugal due to exposure to rising temperatures (Alcoforado et al. 2015; Almeida et al. 2010) causing severe respiratory diseases (Monteiro et al. 2016). It is expected that extreme heat events will have a negative profound effect, mostly in the Portuguese large metropolitan areas, and in Lisbon in particular (Cardoso et al. 2019). According to Forzieri et al. (2017) an increasing percentage of around 3000% of deaths is expected in Western Europe under a global warming scenario of a rise in average temperature of 2.8 °C by 2100. Here, the south of Europe and, in Portugal particular, is highly likely to see significant rises in the mortality rates from heat extremes (Naumann et al. 2020; Casimiro et al. 2006). Moreover, increasing ambient air levels of aeroallergens are expected to rise in the Portuguese territory in coming years (Schleussner et al. 2019), while an expected increase in forest fires in a warmer climate will tend to increase cardiovascular and respiratory-related illnesses (Vitolo et al. 2019). In tandem, vector-borne diseases such as leishmaniasis, Lyme disease, and Mediterranean spotted fever, are expected to rise in Portugal, as a consequence of the rising temperatures (Casimiro et al. 2006).

Contribution of the PO SEUR to Mitigate Health Hazards in Portugal: 2014–2020

As one of the 16 programmes created to implement the Portugal 2020 Strategy (Medeiros 2020), the PO SEUR (Programa Operacional Sustentabilidade e Eficiência no Uso de Recursos—Operational Programme for Sustainability and Efficient Use of Resources) is focused on supporting policy actions towards green and sustainable growth invoked in the Europe 2020 Strategy: promoting a sustainable development model based on a low-carbon economy and in adapting to climate change, and preventing risks, as well as supporting environmental protection and resource efficiency. The European Union (EU) Cohesion Policy is committed to financially supporting the Europe 2020 Strategy in the 2014–2020 programming period, which includes in its five main targets the core domain of the PO SEUR of combating climate change and supporting energy efficiency and sustainability. It was this EU context, favourable to the support of green policies towards sustainable growth, which culminated in the recent approval of the 'European Green Deal' (EC 2019) as a global plan to support the development of a green and sustainable economy in the European context, and in which the PO SEUR strategy was outlined.

On a national scale, the main objectives of the PO SEUR are aligned with various national development plans and strategies that support processes of growth and sustainable development (see Medeiros 2022), like the national Commitment to Green Growth. These, in turn, cover several areas of sustainable development, such as the support to: (i) the green economy (GDP, exports, employment, productivity, circular economy); (ii) more sustainable territories (urban rehabilitation, energy efficiency, public transport, water efficiency); (iii) a green energy transition (energy efficiency, renewable energies); (iv) combating natural risks, pollution and climate change (CO_2 emissions, air quality); and (v) the protection of nature (water bodies, biodiversity).

Specifically, the PO SEUR intends to contribute to the national sustainable growth and development priority, via the financial support to projects which address a low-carbon economy. It is fundamentally based on more efficient use of resources and wider promotion of greater resilience to climate-related natural risks and catastrophes. On the other hand, the PO SEUR acts as a policy-funding vehicle to materialise the national government goals of seeking a sustainable development trajectory, supported by a more competitive and resilient model of development. This entails lower consumption of fossil fuels and a simultaneous generation of new opportunities of employment and wealth generation in the green economy sector. This is particularly important in a country like Portugal which is still highly dependent on a tax of fossil fuels.

Therefore, the PO SEUR follows the EU guidelines to anticipate and adapt to expected global changes in the field of energy, climate change and more efficient use of resources, alongside a dynamic perspective that links competitiveness to sustainability. Indeed, Portugal is deeply committed to the structural transformation of its model of development, thus trying to create conditions for greater cohesion and convergence within the European context (Miguel et al. 2018). This national commitment towards a green economic structural transformation as a central model of development follows a remarkable positive change in fostering environmental sustainability policies in Portugal over the last 30 years (Cravo and Guerreiro 2019).

In concrete terms, the PO SEUR acts as the key financial package for supporting environmental sustainability policies in Portugal at all territorial levels, as complement to other main dimensions of territorial development and cohesion (see Medeiros 2016, 2019), such as economic competitiveness and social cohesion (Table 17.1). In detail, the PO SEUR is supported by three main intervention axes or strategic pillars: (i) support the transition to a low-carbon economy in all sectors; (ii) promoting climate change adaptation and risk prevention and management; and (iii) protecting the environment and promoting efficient use of resources. As seen, none is directly associated with the mitigation of health hazards. However, as discussed previously, it is possible to infer that many financed PO SEUR interventions are due to indirectly and directly contributing to mitigating the potential negative impacts of climate change in health hazards in Portugal.

The predicted financial provisions of the PORTUGAL 2020 earmarked for environmental sustainability-related policies were around 25% of the its total budget. By the end of 2021, however, the implemented projects in the field of environmental sustainability had only absorbed around 16% of the Portugal 2020 investments, in stark contrast with the 45% targeted at the domain of 'economic competitiveness'.

Table 17.1 Portugal 2020 main thematic domains

Main policy goal	Central thematic goal
Competitiveness and internationalisation	1. Strengthening research, technological development, and innovation; 2. Enhancing access to, and use quality of, information and communication technologies 3. Enhancing the competitiveness of small and medium-sized enterprises 7. Promoting sustainable transport and removing bottlenecks in network infrastructures 11. Enhancing institutional capacity of public administration efficiency
Social inclusion and employment	8. Promoting employment quality and sustainability and supporting labour mobility 9. Promoting social inclusion and combating poverty and discrimination
Human capital	10. Investing in education, training and professional training to acquire skills and in lifelong learning
	4. Supporting the shift towards a low-carbon economy in all sectors 5. Promoting climate change adaptation, risk prevention and management 6. Preserving and protecting the environment and promoting resource use efficiency

Source Own elaboration based on https://poseur.portugal2020.pt/en/po-seur/about-the-programme/

Within the PO SEUR, the bulk of the investment was aimed at finance water-related projects (mostly infrastructure) due to the need of many Portuguese localities to guarantee a clean water supply and to treat waste water (Fig. 17.3). As such, the PO SEUR policy interventions in the climate change domain only absorbed around 21% of its total budget (Fig. 17.4). Indeed, a cursory glance over the PO SEUR project database only reveals a few projects directly aiming at supporting public health issues or health infrastructure (Table 17.2).

Even so, the reading of the PO SEUR project database allows us to conclude that, directly or indirectly, many of its financed projects are providing a positive contribution, to mitigating climate change-related effects in Portugal, albeit with different levels of intensity. Indeed, there are several ongoing interventions aiming to: (i) correct coastal erosion due to the rise in sea level, for instance by depositing sediments and controlling floods; (ii) reduce air pollution, for instance by fomenting the use of renewable sources of energy, energy efficiency and the use of sustainable transports. These measures have also an indirect positive effect in mitigating rising temperatures and thus limit the spread of vector-based diseases and allergies, at least at the local level; (iii) treat residual waters, for instance by the construction of water treatment plants, with positive consequences for mitigating water-borne diseases; (iv) assure water and food supply, for instance by investing in water-related infrastructure;

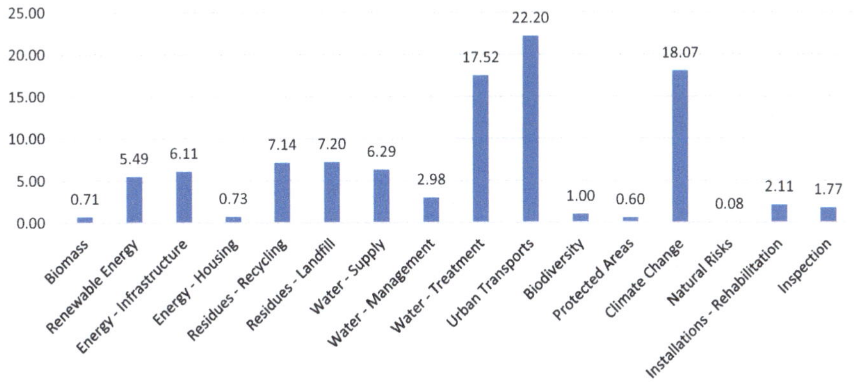

Fig. 17.3 PO SEUR investment (%) per main policy area of environmental sustainability (*Source* Own elaboration based on Portugal 2020 database)

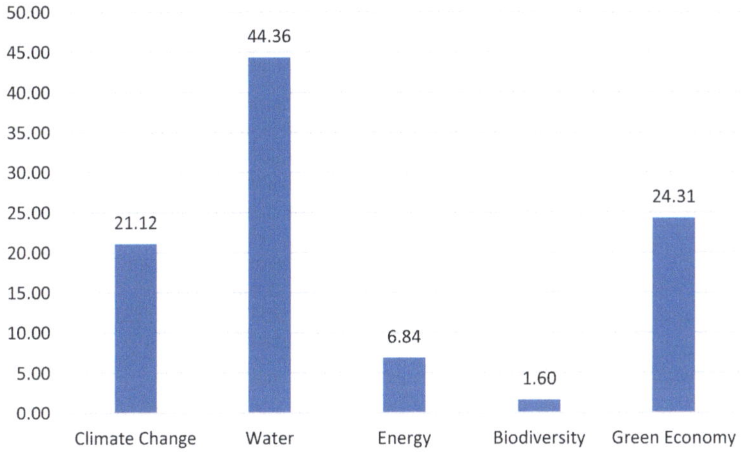

Fig. 17.4 PO SEUR investment (%) per main priorities (*Source* Own elaboration based on Portugal 2020 database)

and (v) protect the territory from severe natural disasters, for instance by artificially supplying sand to existing dunes along the coast.

Within this panorama, it is possible to conclude that the PO SEUR has had a positive impact in reducing the potential negative health hazard effects in several Portuguese regions and policy domains which are affected by climate change. How far these impacts can contribute to significantly reduce health hazards in Portugal resulting from climate change is hard to quantify and would require a more detailed ex-ante impact assessment exercise, demanding que collection and treatment of a vast pool of data which is out of the scope of this analysis. One thing is certain: the PO SEUR does not act in isolation. Indeed, other Portugal 2020 investments, for

Table 17.2 Examples of PO SEUR projects directly related Health Issues

Entity	Project	Health goal	€
Unidade Local de Saúde do Alto Minho, E.P.E	Energy efficiency in Viana do Castelo Hospital	Reconvert the energy system for improving efficiency	3,444,162
Instituto Nacional de Saúde Ricardo Jorge, I.P	Measures to increase energy efficiently via the use of renewable sources of energy	Implement measures to increase energy efficiency via the replacement of existing systems, changing interior and exterior illumination, promoting renewable sources of energy	2,332,202
Associação Humanitária dos Bombeiros Voluntários de Mondim de Basto	Amplification and requalification of the Mondim de Basto Fireman Edifice	Allow for a separation between the health parking space and the fireman parking spaces	479,993
Instituto Politécnico de Viana do Castelo	Energy efficient in the Health Superior School	Respond to climate change trends and environmental concerns	510,271
Instituto Politécnico de Viseu	Energy efficiency in the Superior Health School of Viseu	Optimize energy performance: guarantee indoor air quality, promote healthy spaces, increase thermal comfort, reduce consumption in Central Administration buildings, rationalize energy consumption and reduce greenhouse gases	235,330
Instituto Politécnico de Santarém	Improve energy efficiency in the Superior Health School	Using LED lights, replacement of ventilation systems, and installation of Installation of a photovoltaic solar system, application of thermal insulation in the pipe network to be reused in the hot water production system. Awareness Actions to promote the rational use of energy	213,856
Municipality of Espinho	New Urban Expansions of Basic Sanitation—Completion of the Water Supply Networks in the Municipality of Espinho	Minimizing the risks to public health associated with the consumption of water not coming from the public network	481,448

(continued)

Table 17.2 (continued)

Entity	Project	Health goal	€
Grândola Municipality	Water supply of Brejinho de Água	Provide the populations of Brejinho de Água with quality water supply, in quantity and with continuity, minimizing the risks to public health associated with the consumption of water not coming from the public network	216,318
Mirandela Municipality	Resolution of supply water quality problems with impact on human health	Carry out interventions at the level of Iron, Manganese and Arsenic removal filters	41,236
Funchal Municipality	Replacement of Water Distribution Networks—Western Sector of Funchal	The renovation of these networks aims to reduce the high losses that affect the network, increase the available flow to improve the quantity and quality of water distributed, thus rationalizing the consumption of drinking water, and replacing the material of the old ducts with another that complies with the regulations related to the risk to human health, also improving the hydraulic operation of the networks	1,487,595
Municipality of Viseu' Services	Closing of the Carvalhal Wastewater Sanitation Subsystem	Guarantee to the villages of Lamaçais, Santo Amaro and the Baixa de Figueiredo area of the Parish of S. Pedro de France a public service of quality wastewater sanitation, essential for the functioning of the social and economic fabric, as well as the protection and improvement public health and the environment	106,470

(continued)

Table 17.2 (continued)

Entity	Project	Health goal	€
Local Health Unit of the Alto Minho E.P.E	Energy efficiency in the Viana do Castelo Hospital	Reconvert energy systems to obtain significant savings in terms of energy consumption, while maintaining and improving service outputs and security of supply	3,444,162

Source Own elaboration based on Portugal 2020 database

instance in improving health systems and educational training programmes on the need to follow more sustainable development paths, can also contribute to responding to the predicted rise of health hazards due to climate change in Portugal in the coming decades. One good example is the EU Integrated Sustainable Urban Development Strategies, financed by EU Cohesion Policy since 2014–2020 (Medeiros and van der Zwet 2020a, b), which target European cities (108 in Portugal), as a concrete policy response to support green development in urban areas, which are known carbon footprint hubs (UN 2020).

Conclusion

Increasing temperatures and increasing humidity associated with climate change are expected to impact negatively on many policy domains, including health-related hazards. Among many other negative impacts on human health, climate change is expected to raise human mortality in many parts of the world, for instance via increasing exposure to water and air and vector-based diseases, the direct exposure to excessive heat and cold waves, increasing air pollution, and increasing natural hazards, with a potential rise in human injuries and fatalities.

As it is located in the south of Europe, all available studies demonstrate that Portugal is one of the European countries that is expected to have higher negative impacts of current global climate change trends, in particular a relatively high increase in average temperatures and reduced rainfall in the non-littoral areas. More detailed national studies, both on the mainland and in the two archipelagos (Madeira and Azores), are clear in pointing towards expected rise of temperatures between 3 and 7 °C (ENAAC), in a worst-case scenario. The consequences of such climatic dramatic trends on increasing negative health hazards effects are evident, with potential negative impacts in all the domains of the presented analytic dimensions of climate change-related health hazards.

At the very least, several national reports conclude that, by 2050, Portugal is expected to experience a significant rise in the average sea level, with negative consequences on the economic activities and the health security of thousands of inhabitants who reside near the sea, in a country like Portugal with a long coastal area. Running parallel to this, it is difficult to dispute that the expected rise in temperatures, especially high in non-littoral areas, will inevitably lead to increased vulnerability to vector and water-borne diseases like malaria and cholera, as well as respiratory/pulmonary diseases. The high possibility of increasing heatwaves could lead to increased cardiovascular failures and mental health-related symptoms (anxiety, depression, etc.), and also to rural migration towards the cooler littoral areas, in a country in which the bulk of the population already resides in these areas. The problem is that the large metropolitan area in Portugal (Lisbon) is expected to be largely and negatively affected by these increasing heatwaves, and may also suffer from reduced food production and clean water provision from other parts of the country due to climate change trends, which could ultimately contribute to health-related issues such as malnutrition.

In almost every way, the picture painted for the expected climate change trends in Portugal for the next decades is quite dark. Worse still, its potentially negative consequences for the health of the Portuguese population is not rosy at all. On the contrary, a wealth of negative health hazard-related effects is expected to occur in Portugal in the near future. So, it is important to know if and how the investments of the analysed PO SEUR are contributing to mitigating such negative trends. By detailed analysis of its 2022 approved projects until the end of 2021, it was possible to conclude that the PO SEUR has had a relatively positive impact on mitigating climate change trends in some policy areas, although it must be realised that climate change has a global character and, as such, each country can only do so much to invert current global climate change trends.

Despite the difficulty of convincingly relating the PO SEUR approved projects' funding to mitigating health hazards in Portugal, an in-depth project analysis allowed us to find that the projects directly associated with mitigating 'climate change' trends only received a little more than 20% of the total PO budget, so far. By itself, this number does not reveal the full scope of the potentially positive role of the PO SEUR in mitigating health hazards in Portugal. Indeed, in addition to financing projects which directly deal with the consequences of rising temperatures in Portugal, like the protection of the coast, the PO SEUR has also been contributing to adapting the Portuguese economy and its regions to a more sustainable development path by, for instance, financing projects in the area of production of renewable sources of energy and sustainable public transportation. Furthermore, the PO SEUR has placed great emphasis on improving water management and supply in all Portuguese regions, which can have a very positive impact on mitigating the potential negative impacts of climate change on food production and availability of clean water. In this stance, one can conclude that the PO SEUR has had indirect and direct positive effects on mitigating health hazards by financing projects which have contributed to: (i) reducing air and water pollution; (ii) lessening the vulnerability of the coastal areas to the rise of the average sea level; (iii) lowering the need to use of fossil fuels to produce energy and to urban mobility; and (iv) modernising some health-related infrastructure.

It goes without saying that combating climate change-related challenges requires a strong global response. In a global panorama, Portugal is just a relatively small piece of land in the west of a European peninsula. Hence, Portuguese-financed policies to combat climate change can only do so much to mitigate related health hazards. Even so, within the current global panorama, each country needs to deliver its pro-active policy measures to combat climate change. In the case of Portugal, it can be seen that there are several national strategies which predict, with detailed precision, the potential negative impacts of climate change to several health hazards for the coming decades. In this context, one would expect that the Portugal 2030 Climate Action and Sustainability Programme (2021–2027) can dedicate a specific line of investment to directly mitigate the health hazards resulting from predicted climate change trends in all Portuguese regions. Only time will tell, however, how far the EU funds have positively contributed to preparing the Portuguese territory to cope with climate change-related health hazards.

References

Alcoforado M, Marques D, Garcia R, Canário P, Nunes M, Nogueira H, Cravosa A (2015) Weather and climate versus mortality in Lisbon (Portugal) since the 19th century. Appl Geogr 57(2015):133–141

Almeida S, Casimiro E, Calheiros J (2010) Effects of apparent temperature on daily mortality in Lisbon and Oporto Portugal. Environ Health 9(12):1–7

Antunes C, Rocha C, Catita C (2019) Coastal flood assessment due to sea level rise and extreme storm events: a case study of the Atlantic Coast of Portugal's mainland. Geosciences 9(5):239

APA (2013) Relatório de Progresso Estratégia Nacional de Adaptação às Alterações Climáticas [Progress report national climate change adaptation strategy]. Agência Portuguesa do Ambiente, Amadora

Cardoso R, Soares P, Lima D, Miranda P (2019) Mean and extreme temperatures in a warming climate: EURO CORDEX and WRF regional climate high resolution projections for Portugal. Clim Dyn 52(2019):129–157

Casimiro E, Calheiros J, Santos F, Kovats S (2006) National assessment of human health effects of climate change in Portugal: approach and key findings. Environ Health Perspet 114(12):1950–1956

Cifuentes L et al (2001) Hidden health benefits of greenhouse gas mitigation. Science 293(5533):1257–1259

Cravo R, Guerreiro J (2019) Environmental sustainability: a 30-year progress assessment in Portugal. J Environ Prot 10(2019):1507–1535

DGT (2018a) PNPOT Alteração, Diagnóstico [National spatial planning programme—change, diagnosis]. DG Território, Lisboa

DGT (2018b) PNPOT Alteração, Estratégia e Modelo Territorial [National spatial planning programme—change, strategy and territorial model]. DG Território, Lisboa

DGT (2018c) PNPOT Alteração, Agenda para o Território [National spatial planning programme—change, agenda for the territory]. DG Território, Lisboa

Dias D, Tchepel O, Carvalho A, Miranda A, Borrego C (2012) Particulate matter and health risk under a changing climate: assessment for Portugal. Sci World J 2012:409546. https://doi.org/10.1100/2012/409546

EC (2019) The European green deal. European Commission, Brussels

Epstein P, Ferber D (2011) Changing planet, changing health: how the climate crisis threatens our health and what we can do about it. University of California Pres, Berkeley

Fearnside P (2012) Brazil's Amazon forest in mitigating global warming: unresolved controversies. Climate Policy 12(1):70–81
Forzieri G, Cescatti A, Silva F, Feyen L (2017) Increasing risk over time of weatherrelated hazards to the European population: a data-driven prognostic study. Lancet Planet Health 1(2017):200–208
Graham S (2015) Climate change, ethics and human security. Ethics Policy Environ 18(1):112–115
Gudynas E (2016) Climate change, the quadrilemma of globalization, and other politically incorrect reactions. Globalizations 13(6):938–942
IOM (Institute of Medicine) (2011) Climate change, the indoor environment, and health. The National Academies Press, Washington, DC
Johansen B (ed) (2017) Climate change: an encyclopedia of science, society, and solutions, vol III, ABC-CLIO, Santa Barbara
Johnsson F, Kjärstad J, Rootzén J (2019) The threat to climate change mitigation posed by the abundance of fossil fuels. Climate Policy 19(2):258–274
Kulp S, Strauss B (2019) New elevation data triple estimates of global vulnerability to sea-level rise and coastal flooding. Nat Commun 10(4844):1–12
Loko Y, Agre P, Orobiyi A, Dossou-Aminon I, Roisin Y, Tamo M, Dansi A (2016) Farmers' knowledge and perceptions of termites as pests of yam (Dioscorea spp.) in Central Benin. Int J Pest Manag 62(1):75–84
Lukasiewicz A (2017) Ethics, environmental justice and climate change. Austr J Environ Manag 24(3):332–333
Medeiros E (2016) Territorial cohesion: an EU concept. Eur J Spatial Develop 60. http://www.nordregio.org/publications/territorial-cohesion-an-eu-concept
Medeiros E (2019) Spatial planning, territorial development and territorial impact assessment. J Plan Lit 34(2):171–182
Medeiros E (2020) Portugal, 2020: an effective policy platform to promote sustainable territorial development? Sustainability 12(3):1126
Medeiros E (2022) The territorialisation of the UN Agenda 2030 and cross-cutting issues in energy, environment and health: the case of Portugal. In Leal Filho et al (eds) (2021) Sustainable policies and practices in energy, environment and health research. Springer, Cham, 393–411
Medeiros E, van der Zwet A (2020a) Sustainable and integrated urban planning and governance in metropolitan and medium-sized cities? Sustainability 12(15):5976
Medeiros E, van der Zwet A (2020b) Evaluating integrated sustainable urban development strategies: a methodological framework applied in Portugal. Eur Plan Stud 28(3):563–582
Miguel C, Mendes A, Madeira L (2018) An overview of the Portuguese energy sector and perspectives for power-to-gas implementation. Energies 11(12):3259
Monteiro A, Sá E, Fernandes A, Gama C, Sorte S, Borrego C, Lopes M (2016) Air quality over Portugal: now and in 2050. Sustain Dev Plan VIII(1):621–630
National Geographic (2016) Human health. National Geographic, Washington DC
Naumann G, Russo S, Formetta G, Ibarreta D, Forzieri G, Girardello M, Feyen L (2020) Global warming and human impacts of heat and cold extremes in the EU. Publications Office of the European Union, Luxembourg
Patz J, Hatch M (2013) Public health and global climate disruption. Public Health Rev 35(12):1–23
Rivera-Amarillo C, Camargo A (2020) Zika assemblages: women, populationism, and the geographies of epidemiological surveillance. Gend Place Cult 27(3):412–428
Sachs F (2015) The age of sustainable development. Columbia University Press, New York
Schleussner C, Menke I, Theokritoff E, van Maanen N, Lanson A (2019) Climate impacts in Portugal. Climate Analytics, Berlin
Schulte P, Chun H (2009) Climate change and occupational safety and health: establishing a preliminary framework. J Occup Environ Hyg 6(9):542–554
Shao W (2017) Weather, climate, politics, or God? determinants of American public opinions toward global warming. Environ Politics 26(1):71–96
Singer S (2013) Respiratory health and ecosyndemics in a time of global warming. Health Sociol Rev 22(1):98–111

Theodore L, Dupont R (2012) Environmental health and hazard risk assessment principles and calculations. CRC Press, New York

UN (2016) Transforming our world: the 2030 agenda for sustainable development. United Nations, New York

UN (2020) Human development report 2020. United Nations, New York

Vitolo C, Di Napoli, C, Di Giuseppe F, Cloke H, Pappenberger F (2019) Mapping combined wildfire and heat stress hazards to improve evidence-based decision making. Environ Int 127(June 2019):21–34

Eduardo José Rocha Medeiros is a Geography Professor (got the professionalisation in 2002) and an Integrated Research Fellow in DINÂMIA'CET-IUL, Instituto Universitário de Lisboa, Portugal. He has a Ph.D. in Geography—Regional and Urban Planning, and around 200 publications, including more than 50 published papers in international journals, 11 books and 16 book chapters. His research interests are focused on EU Cohesion Policy, Territorial Impact Assessment, Territorial Cohesion, Territorial Development, Territorial Cooperation, Environmental Sustainability and Spatial Planning. He is a DG REGIO (European Commission) an ESPON and URBACT III expert, and a Horizon 2020 evaluator. He is also a Regional Studies Association Fellow. He has coordinated several international policy evaluation projects and was a member of DG REGIO and ESPON projects. He was invited as a project adviser and to write reports and position papers by DG REGIO and the World Bank. He was already invited to be a keynote speaker by several International Universities and EU institutions (European Commission and Committee of the Regions). He is a member of the scientific and editorial committee of several journals and a peer reviewer of more than 35 international journals. He was a final Jury of the New European Bauhaus Prizes 2022.

Chapter 18
Increasing Healthcare Facilities Resilience to Hazards Resulting from Climate Change

Judith Harvey, Clemens Buter, and Dana van Alphen

Abstract In emergencies, countries need functioning health facilities to treat the sick and injured, despite the disaster taking place around them. In a project funded by the UK government Foreign Commonwealth and Development Office (FCDO), executed by the Pan American Health Organisation (PAHO) in collaboration with the local Ministries of Health, GBP 46.3 M is being spent on upgrading selected existing healthcare facilities to meet this challenge. The Smart Healthcare Facilities in the Caribbean project Phase 2 (SMART Project) was implemented in seven Caribbean countries. St. Lucia adopted a network approach, whereby almost half the stock of Wellness Centres was upgraded. Smaller health centres received basic upgrades to water and electrical systems so that they could re-open immediately after the emergency. Larger critical facilities, which must remain functioning during a hazard event, were upgraded to a higher standard of resilience. There is now improved access to healthcare island-wide in case of disasters and climate-related hazard events. Not only human health, but also environmental health benefited from the project, which raised the standard of greenness at each facility. This paper describes the SMART Project initiative by PAHO, and an approach based on using tools from PAHO's Smart Hospitals Toolkit for identifying how healthcare facilities can be strengthened to make them more resilient to climate change. It also describes some limitations of the methodology, the challenges of implementation, and level of success achieved.

J. Harvey (✉) · C. Buter · D. van Alphen
PAHO/WHO Office of Barbados & the Eastern Caribbean Countries, PO Box 508, BB11000 Bridgetown, Barbados
e-mail: harveyjud@paho.org

C. Buter
e-mail: butercle@paho.org

D. van Alphen
e-mail: vanalphd@paho.org

W. Leal Filho et al. (eds.), *Climate Change and Health Hazards*, Climate Change Management, https://doi.org/10.1007/978-3-031-26592-1_18

Climate Change and Human Health

The passage of extreme weather systems has a negative impact on the health and well-being of citizens living in the affected communities. Besides the psychological stress, there is limited access to clean water for sanitation and drinking, and reduced availability of the foods required to maintain good health. Another dimension is the switch from provision of routine primary healthcare services to focus on provision of emergency aid to casualties. People with chronic non-communicable diseases are often denied access to the medication and care needed to control their symptoms (Hassan et al. 2020).

Adaptation to Climate Change Enhancing Healthcare Provision

Adaptation to climate change is being addressed here as part of a disaster management strategy by the Pan American Health Organisation (PAHO) Health Emergencies Department. The focus is on healthcare provision that is resilient to natural hazards that cause disasters. Healthcare provision requires robust health infrastructure and medical staff that is prepared to operate in times of emergency.

During the actual emergency, common themes include the safety of the building and its surroundings, the functioning of critical building services, the functional readiness of the staff, and the linkages within the community that provide support to healthcare provision.

When planning all disaster managers incorporate a recovery plan. The first few days post disasters are dominated by the need for food, water and emergency care and treatment. Countries affected need to start the recovery process as soon as possible to reestablish all government services and functions.

Building back better after a disaster is not just repairing what was previously there but must include measures to increase resilience and reduce the impact of hazards driven by climate change. Additionally, all new buildings should be made safe and green, and therefore "SMART", as a protection against future hazard events. This is the goal of the SMART movement, which has developed from the Smart Healthcare Facilities in the Caribbean Project.

The SMART Project

The Smart Healthcare facilities in the Caribbean Project Phase 2 (SMART Project) achieved adaption to, and mitigation of, climate change by retrofitting over 50 facilities in seven Caribbean countries. The project is represented in Belize, Guyana,

Jamaica, Grenada, St. Vincent and the Grenadines, Dominica, and St. Lucia. Each country was allocated an approximate budget.

Facilities were selected based upon agreed criteria related to socio-economic measurements as well as cost effectiveness analysis developed by Florida International University (FIU). FIU created a Retrofitting Economic Support Tool (REST) (Arrieta et al. 2018). The choice of wellness centres was refined based on population statistics, accessibility, and centrality, as well as the services which could be provided to the community. The main deciding factors, however, were technical assessments embodied in the Hospital Safety Index (HSI) and Green Checklist (Pan American Health Organisation 2017).

Approximately GBP 46.3 M was provided by the UK's Foreign and Commonwealth Development Office (FCDO) to the Pan American Health Organisation (PAHO), which developed the concept and related standards and implemented the project. There was close collaboration throughout with the local Ministries of Health.

Execution of the project followed similar lines in all project countries, being administered by the same coordinating head office. To simplify this paper, St. Lucia will be used as an example throughout. This island demonstrated a network approach to the choice of retrofits. By performing small scale upgrades at twelve locations island-wide, there is a great probability that close to any given community there will be an operational wellness centre to attend to affected persons after the disaster. Such a centre can also continue to service the needs of those with chronic non-communicable diseases and is well placed to tackle potential outbreaks such as gastrointestinal disease.

In combination with small scale upgrades, three larger facilities were retrofit to a higher standard, achieving hazard resistance rather than mere resilience. These facilities required to be operational both during and after hazard events. Two of them provide residential care to vulnerable populations (Comfort Bay Senior Citizens' Home and New Beginnings Transit Home for Children) and one serves a critical node in the network at the southern tip of the island (Vieux Fort Wellness Centre).

The Aims of This Chapter

The SMART Project is based on concepts and standards that promote adaptation to climate change and mitigation of climate change in the context of healthcare provision in the public sector. A toolkit was developed incorporating a suite of documents that support the assessment of vulnerabilities, the establishment of base-line situations, and the provision of guidance on how to strengthen facilities and operations.

The first part of the chapter will focus on the Hospital Safety Index (HSI). This is primarily a tool to measure the safety of a facility related to various hazards. It will introduce the HSI components and explain how they were used to identify the tasks carried out under the project; the goal of these works being to strengthen the healthcare facilities against geophysical and meteorological hazards. The HSI effectively identifies adaptation measures both from an engineering and architectural standpoint, and an operational standpoint.

Mitigation of climate change is achieved by means of a Green Checklist scoring system, similar to the HSI. This checklist identifies works that can be done to reduce consumption of energy and water, to improve indoor air quality, to make sustainable procurement decisions and to optimize waste disposal. These tasks contribute to the standard of environmental health experienced by building occupants. A later section of the chapter will explain further.

The Smart Hospitals Project Utilises Other Tools

The Smart Hospitals Toolkit (Pan American Health Organisation 2017) and the PAHO website has freely available resource materials about climate change and hazards, and how they affect healthcare facilities. Two of the principal tools are discussed in this chapter. Other tools were developed under the SMART Project. For example, an illustrated guide for builders produced by PAHO, gives prescriptive guidance on the design of timber roofs to withstand category 5 windstorms.

The use of appropriate standards was emphasized throughout the project. For example, the Caribbean Basin Wind Hazard Maps were updated in 2008 and in 2019 to incorporate the most recent meteorological data. This hazard information is intended for use with relevant chapters of ASCE 7–16 to calculate wind loads on buildings (Mudd et al. 2019). However, the effects of climate change were not specifically built into the model used to generate these maps. Therefore, multipliers are used to factor up the wind speeds, depending on the category of the building and its location (Gibbs 2020).

The REST tool mentioned above is being calibrated post-retrofit by Florida International University. It calculates the cost effectiveness of investments in safety and greenness (Arrieta et al. 2018). The model predicts the expected damage to the facility and surroundings after a hazard event (primarily hurricane and earthquake). Retrofits are proposed for the facility along with estimated costs. The output predicts the reduction in damage and gain in life-years (QALYs) due to the retrofits proposed. This cost effectiveness analysis tool will be made available to Planners within the Ministries of Health in the target countries. This paper will not go into further detail about the REST model.

The Hospital Safety Index Tool

The Hospital Safety Index for Small-Medium facilities (Pan American Health Organisation 2015) was developed by PAHO to assess key parameters relating to the performance of healthcare facilities after hazard events (Pan American Health Organisation/World Health Organisation 2010). The index gives scores in three

Classification of the Hospital: B

Safety Index	Category	What should be done?
0 – 0.35	C	Urgent intervention measures are needed. The hospital's current safety levels are inadequate to protect the lives of patients and hospital staff during and after a disaster.
0.36 – 0.65	B	Intervention measures are needed in the short-term. The hospital's current safety levels are such that patients, hospital staff, and its ability to function during and after a disaster are potentially at risk.
0.66 – 1	A	It is likely that the hospital will function in case of a disaster. It is recommended, however, to continue with measures to improve response capacity and to carry out preventative measures in the medium- and long-term to improve the safety level in case of disaster.

Fig. 18.1 Definition of HSI safety categories and the scores associated with them (HSI Calculator spreadsheet PAHO Smart Health care facilities in the Caribbean Project 2016)

categories. 'A' scores are facilities that would be able to continue functioning during a hazard event and to operate at nearly full capacity immediately after the disaster. 'B' scores would be those that sustain slight damage, in such a controlled manner that they can be quickly restored to use after a major hazard event. 'C' scoring facilities would be expected to suffer severe damage due to the hazard event and could not be used for some time afterwards. The graphic below taken from the HSI calculator gives the official definitions Fig. 18.1.

The Smart Healthcare facilities in the Caribbean Project commenced with facility assessments at approximately 400 locations, looking at their HSI scores. This provided a ranking of the facilities according to their need for improved resilience.

The Hospital Safety Index assesses three areas of resilience: Structural, Non-Structural, and Functional. These three areas capture the building skeleton, the building services and building envelope, and the staffing and resources available to operate the facility during emergencies.

The HSI is a calculator tool. Weightings are given to different aspects, for example safety of the roof structure receives a greater weight than safety of the building handrails, although both are important. The weighting system places more emphasis on building integrity during a disaster, rather than operational safety on a normal day.

The main hazards incorporated in the Hospital Safety Index are hurricane, earthquake, fire and torrential rain. Of these, hurricane, fire and torrential rain are examples of climate-impacted hazards, expected to become more serious in the coming years (United Nations 2008).

Components of the HSI

This section looks in more detail at the HSI, breaking it down into its component parts. A vulnerability index is the opposite of the safety index and is less often quoted.

$$\text{Safety Index} + \text{Vulnerability Index} = 1$$

The format of the tool changed in 2022. Instead of a hardcopy table as depicted in the toolkit, there is a calculator spreadsheet where data can be entered directly in the field as shown in Fig. 18.2.

The Structural Section of the HSI

This section looks at the history of previous building modifications and the present building condition, including its configuration/basic shape to withstand hazard forces. In the context of hurricane, it specifically refers to the integrity of the roof structure, the strength of connections that must hold the elements together when wind forces would tear them apart, and the capacity of foundations to anchor lightweight structures against uplift.

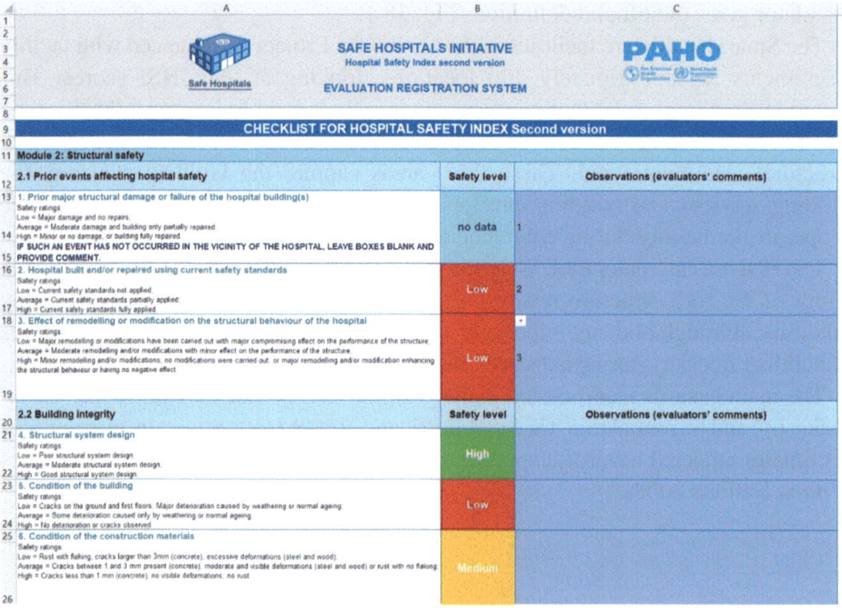

Fig. 18.2 Snapshot of the Hospital Safety Index spreadsheet for entry of data and automatic calculation of the HSI for any assessed facility (HSI Calculator spreadsheet PAHO Smart Health care facilities in the Caribbean Project 2022 [unpublished]).

The assessment is subjective with trained assessors ascribing a safety rating of 'low' 'medium' or 'high'. They make this judgement based on the printed guidance combined with technical know-how and based on precedent, as explained during their initial training (World Health Organisation and Pan American Health Organisation 2015). For example, a timber roof structure significantly affected by termites would have a low level of safety.

Resistance to the torrential rain hazard or other conditions local to the area, such as wave impact or sea level rise, are covered by the last item in this section. The HSI does not specifically differentiate between slow onset and rapid onset hazards, but there is a timeframe for the application of the assessment results, typically 5 years. Therefore, rapid onset extreme events are more relevant.

The Non-Structural Section of the HSI

This section looks at building services. Reserves are particularly important to continued functioning after a disaster occurs. This refers to the supply of water and electricity, and alternate communication systems. Medical gas and fuel reserves are also included in the assessment.

Systems of emergency maintenance and restoration of the building services are necessary for functioning after the passage of a hazard event. This applies especially to ventilation systems, without which the building would not be inhabitable.

The section captures solid waste collection, sewerage and stormwater drainage with emphasis on the security of solid waste containment and the efficiency of storm drainage. Solid waste, if not properly handled would contribute to flying debris during high winds. Climate change requires increasing storm drainage capacity as precipitation is expected to become more intense.

Protection of furniture, equipment and supplies is critical. The storage areas must be secure from high winds, flooding and fire, especially if there are separate warehouses and outbuildings.

Architectural components include the building envelope, which must maintain its integrity and remain watertight during a storm, for example. Elements of key importance include roofing, windows, and doors, along with external appendages that could loosen in high winds.

Access routes must remain open and pathways not be rendered unsafe by flood or landslide.

As temperatures rise, the fire safety of health facilities takes on a new significance. There is a greater likelihood of bush fires creating a threat to healthcare buildings. The fire safety component of the index looks at detection, fire suppression and the presence of safe, labelled, evacuation routes.

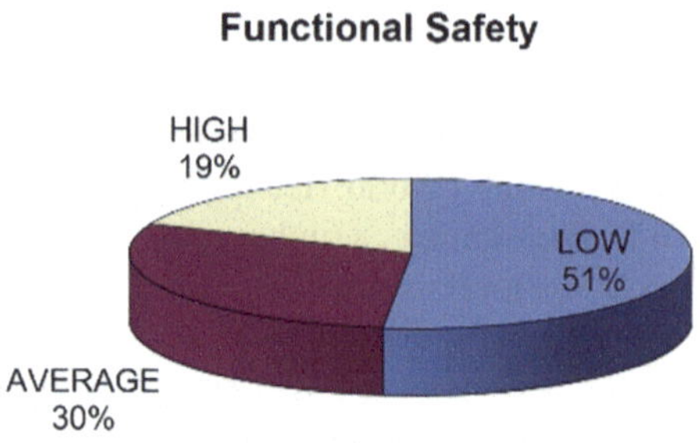

Fig. 18.3 Functional Safety graphic for one facility showing the percentage of high, low and average performance areas (HSI Calculator spreadsheet PAHO Smart Health care facilities in the Caribbean Project Output 2016)

The Functional Section of the HSI

Healthcare staff must be prepared for emergencies, including those linked to climate change. Provision of expanded space for casualties, stabilization of injuries and referral procedures are important after any disaster. Supplies of food and pharmaceuticals, functioning medical equipment and arrangements for transport of patients are necessary. These require co-ordination with the local disaster preparedness network, the health services network and volunteer community groups. It is too late when the hazard event occurs to consider and to put these measures in place. The HSI functional assessment examines the status of these and other aspects of the healthcare facility management and operational system, identifying weak and strong areas. For each HSI section a graphic is produced showing the percentage of high, low and average performance areas, as in the example Fig. 18.3.

Other Tools Measuring Parameters Similar to the HSI

In all the Small-Medium HSI has nearly 100 questions capturing the resilience of healthcare facilities to meteorological hazards that are becoming ever more serious. A large facility HSI formed the basis for the Small-Medium tool described here (World Health Organisation and Pan Amerian Health Organisation 2015). Also, there have been similar tools developed simultaneously by other organisations which deal with conditions specific to their countries and the hazards encountered there. Examples are cited in the Health Care Facility Climate Change Resiliency Workshop Primer (Tzaneva 2015), such as the Health Care Facility Climate Change Resiliency Toolkit developed in Canada (Paterson et al. 2014). Some tools also incorporate greening advice for private healthcare practices, and for large hospitals. Such tools are not directly applicable to the Caribbean where for example there is no winter season, and they will not be addressed further in this paper.

Success of Retrofitting Works Based on the Findings of the HSI

This section looks in more detail at the works carried out in accordance with the findings of the HSI assessments. As illustrated in Fig. 18.4, retrofitting substantially increased the HSI score at the facilities participating in the SMART project. Smaller health centres (HC) with lower budget allocations of the order of US$100 k, achieved 'B' grade safety, close to 0.5. 'A' grade safety, exceeding 0.65, was achieved by the larger facilities with greater budget allocations, of the order of US$1 M per facility.

Retrofitting Works Impacted Structural Safety

The Structural part of the HSI identified works needed to strengthen the building frame and foundations. Some examples of typical interventions are given below.

At selected facilities in St. Lucia, hurricane shutters were installed instead of replacing the windows to increase wind resistance. Window replacements were made in other countries. The decision was made depending on the condition of existing windows and the whether the building needed improved ventilation. For example, air-conditioned rooms might benefit from new tinted window glazing in addition to hurricane resistant laminated glass.

Structural safety was also enhanced by means of additional connectors between the roof and the walls, as well as metal brackets connecting individual timber elements to each other. These measures aimed at resisting the uplift forces on the timber roof structure during a hurricane. In other islands, where other considerations apply, new

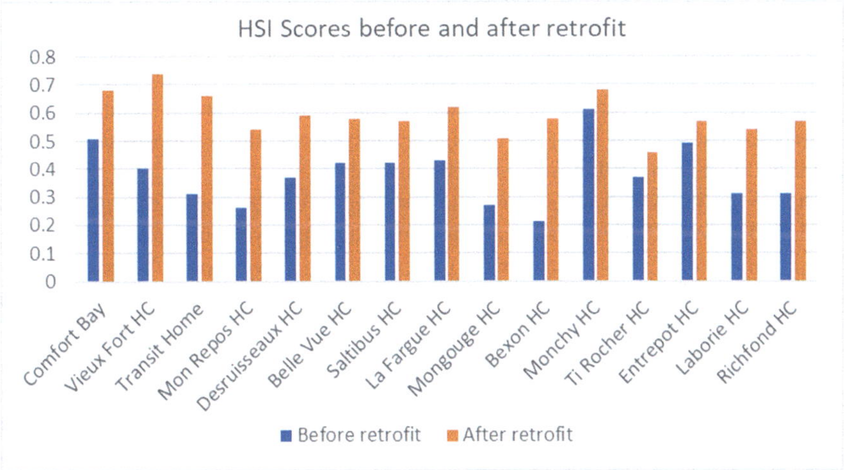

Fig. 18.4 Hospital Safety Index scores for St. Lucia Healthcare Facilities before and after retrofit

wind-resistant roof structures were designed and built of reinforced concrete. For example, in St Vincent, the new concrete roof withstood the weight of volcanic ash dumped on Chateaubelair Hospital during 2021.

Retrofitting Works Impacted Non-Structural Safety

The building services upgrade was the main part of the work done at St. Lucia's small facilities.

Improvement to electrical and water supply systems was both a greening and a safety measure. However, providing alternate electricity and water supplies is a key adaptation strategy. St. Lucia and other islands were impacted by Hurricane Elsa, 1–2 July 2021. In the aftermath of this event, all facilities were intact and fully operational, despite loss of mains power for hours and loss of public potable water supplies for some days. The project aims at a level of hazard resistance and resilience greater than that required by Hurricane Elsa, but this served as an initial test.

Rainwater harvesting was implemented with dual benefit: collection and control of runoff, and water supply in times of shortage. Rainwater must be treated for clinical purposes, but it is also used directly in laundry and sanitary fixtures, and for landscaping. A combination of gravity feed and pumped supply was installed. The stored water is managed in that a minimum of three days' supply is provided. Low flow fixtures maximise the use of the water available.

Diesel operated standby generators were installed at all facilities needing back-up power supplies, with the understanding that biodiesel would be procured as soon as it becomes feasible. Although photovoltaic electrical systems were installed at some facilities, these were grid-tied systems which would not serve as alternate sources of electricity during a power cut. For facilities with moderate power usage, and daytime operation, grid-tied systems were more efficient.

Fire safety upgrades were put in place, with a thorough overhaul of the existing systems of smoke detectors and fire extinguishers at small facilities. Larger establishments received fire suppression systems including hose reels and dedicated water supplies for use against Fire. Strobe lights and pull stations were installed where bigger buildings required more comprehensive alarm systems.

The doors and windows sealing the building envelope were strengthened, as mentioned above. Closure mechanisms and fastenings were upgraded, providing a minimum of three heavy duty hinges per exterior door, with tower bolts and deadbolt locks into a secure door frame. Hurricane rated doors were installed at selected facilities.

An important building component in the context of climate change is the rainwater guttering and roof covering. Special attention was given to fixings for the roof sheeting, repair of leaks and design flaws tending to leaks, as well as the slope and capacity of the rainwater collection gutters.

Specialised Training Impacted Functional Safety

Staff were trained to develop and keep updating Contingency plans for their individual facilities.

These plans documented responses to emergencies such as pandemics, fire, storms and flooding, power cuts and ingress of seawater. These hazards are relevant to climate change. Responsible staff and chains of command were identified, as well as assembly points during evacuation, provisions for vulnerable patients and sources of supplies and communication lines during and immediately following the event. In each plan the onset of the emergency was traced through the event to its conclusion, and restoration of services in the aftermath. As a next step staff are expected to perform regular simulation exercises which put segments of the plan into practice. These are envisaged to be held annually. For example, training was given by officers of the St. Lucia Fire Service in the correct procedure for operating the firefighting equipment at the wellness centres. This involved the staff combating a controlled blaze in the yard of the building.

How Greening Addressed Hazards to Environmental Health Due to a Changing Climate

The project implemented a Green Checklist to assess the level of greenness at each healthcare facility. This tool was similar to the HSI, with percentage scores assigned to each location based on consumption of water and electricity: the equipment used and their operational procedures promoting monitoring and conservation. The Green checklist also evaluated whether there were responsible sourcing practices, the use of chemicals, the disposal of waste, and provisions for ventilation. After retrofitting, nursing staff and clients using the buildings benefitted from improved environmental health indicators such as better ventilation, improved indoor air quality, and safer drinking water.

The previous section looked at benefits to human health at a community and national level by provision of emergency and routine care after a disaster. This section demonstrates that on a localized level, the environmental health standard at all healthcare facilities was also impacted by the project. This was particularly relevant at facilities with residential patients.

Challenges Associated with Improvements to Environmental Health

Ventilation improvements combatted expected rises in ambient air temperatures. These measures increased the natural air flow and provided more ozone friendly, energy efficient, air conditioning. At one facility, a system of solar extractor fans was installed to cool the ceiling air space and to reduce the load on the air conditioning. Enhanced air conditioning reduced internal humidity levels and reduced the amount of particulate air pollution entering the buildings, although it increased the electrical load. Increasing airborne dust is expected to be an outcome of climate change, especially the arrival of dust plumes coming from the Sahara. However, dust-proofing the facility and reduction in humidity was achieved at the cost of increases in maintenance and consumption expenditure.

Indoor air quality was the greatest challenge in the context of environmental health at many facilities. There were instances of mould, mildew and efflorescence developing on walls and in cupboards due to high humidity, condensation, and penetration of moisture through the walls. The project arranged for these to be investigated and rectified on an individual basis.

Mosquito breeding and spread of vector borne diseases tends to become more prevalent with higher air temperatures and more frequent precipitation. Mosquito screens in windows were included in the project scope. Measures against mosquito breeding also included the use of mesh at entry and exit points of rainwater harvesting systems. Treatment of rainwater using UV radiation was installed to ensure that it was safe against mosquito borne vectors, where the application required it.

Limitations of the Methodology

Despite challenges, using the Green Checklist and the HSI tools for guiding the scope of retrofits proved successful overall.

As described in Section "Other Tools Measuring Parameters Similar to the HSI" there are other tools which were found to be unsuitable for the Caribbean region and its unique combination of geophysical and meteorological hazards, when applied in the context of small-medium sized healthcare facilities. The tools described here were specially developed to meet this need. Nevertheless, the methodology employed had some limitations, and there were lessons learned for future projects implemented under the SMART movement.

During the interim between carrying out the assessments and the start of construction, the buildings underwent further deterioration. This had to be addressed as an unforeseen additional expense to the project. The Covid-19 pandemic also had implications for service provision that were unforeseen at the time of design. One example was the need to install sneeze guards on Pharmacy dispensing windows. This was a safety measure imposed by a hazard that was not directly climate induced. A retrofitting process such as this one cannot ignore critical safety measures, although not directly related to climate induced weather extremes.

Conclusion

The SMART Project has used adaptation to and mitigation of climate change to improve the physical and functional resilience of healthcare facilities in the Caribbean. This lessens the adverse effects of climate change on human health. Selected facilities were evaluated according to the Hospital Safety Index for Small-Medium facilities and based on this were retrofitted. Their structural, non-structural and functional components were upgraded. Thus, they were empowered to provide emergency, and routine but necessary health care during, and immediately after disasters are declared. Additionally, the standard of environmental health at the facilities benefitted from climate change mitigation measures based on the application of the Green Checklist.

The project retrofitting works come to an end in the year 2022. Good maintenance is therefore important, so that facilities remain SMART as changes in climate manifest themselves in the future.

Acknowledgements The authors thank Tony Gibbs for assisting with review of this document.

References

Arrieta A, Amick B, Adler J, Chen W (2018) The retrofitting economic support tool (REST) an economic evaluation of SMART retrofitting of healthcare facilities. Florida International University, Florida

Gibbs T (2020, April 1) Design wind speed maps for the Caribbean. Retrieved May 30, 2022, from https://www3.paho.org/disasters/index.php?option=com_docman&view=download&alias=2750-design-wind-speed-maps-for-the-caribbean&category_slug=smart-hospitals-toolkit&Itemid=1179&lang=en

Hassan S et al (2020) Management of noncommunicable disaeases after natural disasters in the Caribbean: a scoping review. Health Aff 39(12):2136–2143

Mudd L, Liu A, Vickery PJ (2019) Development of design wind speed maps for the Caribbean for application with the wind load provisions of ASCE7-16 and later. Pan American Health Organisation, Washington DC

Pan American Health Organisation (2015) Medium and small hospitals safety index. PAHO, Washington DC

Pan American Health Organisation (2017) Smart hospitals toolkit, 1st edn. PAHO, Washington DC

Paterson J, Berry P, Ebi K, Varangu L (2014) Health care facilities resilient to climate change impacts. Int J Environ Res Public Health 11(12):13097–113116

Pan American Health Organisation/World Health Organisation (2010) Hospital Safety Index adapted for small health facilities. PAHO/WHO Emergencies News, 113

Tzaneva V (2015) Healthcare facility climate change resiliency workshop primer, 1st edn. Pan American Health Organisation, Montreal
United Nations (2008) UNFCC Framework Convention on Climate Change. Physical and socioeconomic trends in climate related risks and extreme events and their implications for sustainable development. United Nations.
World Health Organisation and Pan American Health Organisation (2015) Hospital safety index: guide for evaluators, 2nd edn. WHO, Geneva

Cross References

Caribbean Basin Wind Hazard Maps Aid in Siting of Hospitals - PAHO/WHO I Pan American Health Organization (https://www.paho.org/en/health-emergencies/smart-hospitals/caribbean-basin-wind-hazard-maps-aid-siting-hospitals).
Hurricane resistant buildings: building CAT-5 resistant timber roofs - PAHO/WHO I Pan American Health Organization (https://www.paho.org/en/documents/hurricane-resistant-buildings-building-cat-5-resistant-timber-roofs).
Safe Hospitals - PAHO/WHO I Pan American Health Organization (https://www.paho.org/en/health-emergencies/safe-hospitals).
Smart Hospitals Toolkit - PAHO/WHO I Pan American Health Organization (https://www.paho.org/en/health-emergencies/smart-hospitals/smart-hospitals-toolkit).

Judith Harvey holds a Ph.D. in Engineering from Cambridge University and is a Chartered Engineer with the Institution of Civil Engineers, UK. She is registered as a Civil and Environmental Engineer in St. Lucia and is presently a Technical Consultant on PAHO's Smart Healthcare Facilities in the Caribbean Project.

Clemens Buter is the Planning and Operations Advisor on PAHO's Smart Healthcare Facilities in the Caribbean Project. One of his special areas is the coordination of Output 3, which is the dissemination of Project information to create a SMART movement, where others can replicate what has been done here. Most recently he represented the project by making presentations at COP 26.

Dana van Alphen is the SMART Project's Regional Coordinator and an Honorary Fellow of the Institution of Structural Engineers, UK. The title was conferred upon her for her contribution to the structural resilience of healthcare facilities through her work with the PAHO Health Emergencies Department. The vision for the SMART project was Dr. van Alphen's and she is primarily responsible for making it happen.

Chapter 19
Climate Change and Health Hazards: Mitigation Roles of Public Sectors (Ministry, Department and Agencies)

Umar Ibrahim

Abstract Addressing hazards to human and environmental health due to changing climate becomes a subject of debates and discourse among decision makers in the Ministries, Department and Agencies (MDAs), across the globe. The paper highlights the actual and potential roles of the MDAs in mitigating climate change related health hazards. In this regard, an extensive literature review was conducted, to search for publications on climate change issues and mitigation roles of the MDAs, focusing on air, land, water and food security among other climate change issues, to highlights the interdependence between environmental changes, climate change and health hazards, and how to mitigate them. Therefore, the narrative has emanate from the roles assumed by various MDAs in mainstreaming climate change policies, with which they take care of the climate change hazards confronting the global communities. Indeed, there is hardly an MDA that is completely insulated from climate change issues, as such the content of the paper can be adopted elsewhere, for similar results.

Introduction

Climate change is among the top most leading contemporary environmental challenges. Climate change occurred due to indiscriminate human activities, such as greenhouse gas emission, which exert pressure on the natural systems and the resultant consequences and catastrophic effects on human health. Indeed, climate change is one of the serious contemporary threat, confronting human health in the twenty-first century—it effects on health and wellbeing have manifested globally, in several forms, such as malnutrition, malaria, diarrhea, and heat stress among others, with Africa and Southeast Asia, as regions affected the most (Fawzy et al. 2020; Mordecai et al. 2020; Qazlbash et al. 2021). The World Health Organization (WHO) postulates

U. Ibrahim (✉)
Public and Environmental Health Department, Faculty of Basic Medical Sciences, College of Medicine and Allied Medical Sciences, Federal University Dutse (FUD), Ibrahim Aliyu By-Pass, PMB 7156, Dutse, Jigawa State, Nigeria
e-mail: umar.I@fud.edu.ng

W. Leal Filho et al. (eds.), *Climate Change and Health Hazards*, Climate Change Management, https://doi.org/10.1007/978-3-031-26592-1_19

that over 140,000 die annually due to climate change effects (Harris et al. 2017; Mordecai et al. 2020).

Climate change effects on health can be directly or indirectly. Climate change affects human health directly due to extreme weather impact, such as heat waves, floods and bushfires among others; and indirectly, due to, for example, the worsening of air quality caused by air pollutions that in turn lead to communicable and non-communicable diseases (Sutton et al. 2020). The effect of climate change on human health appears in several ways, depending on persons' demographic attributes like socio-economic status, age, and location (CDC 2022a). Other effects of Climate change include unemployment, food insecurity, and increase in social inequalities (Sutton et al. 2020). Climate change constitutes a threat to global environmental prosperity, in need of response from the public sectors across the globe (World Bank n.d.).

Government, otherwise known as public sector, play significant roles in mitigating the effect of climate change. Public sector, thus; Ministry, Department and Agencies (MDAs) protect the citizens by safeguarding their health and well-being from the effects of climate change at federal, states and local government levels. MDAS across the globe have taken steps toward addressing this public health menace (Hallegatte et al. 2011). The public sector roles has potentials of achieving goals number 3, 11 and 13 of the 17 Sustainable Development Goals (SDGs), through stronger and broader environmental sustainability practices (Stern and Fankhauser 2016).

This chapter presents the roles of public sector in mitigating CC effects on human health. The chapter was developed using extensive literature review, which was limited to the roles of public sectors in mitigating the effects of climate change related health hazards. The chapter provides a conceptual background of climate change and health hazards, focusing on the actual and potentials climate change effects' mitigation roles of the MDAs, across the globe. MDAs' roles evolved around policy making, coordination and implementation. The chapter, was as such, organized under these headings, thus; synopsis of the MDAs, overview of climate change and health hazards, roles of the MDAs in mitigating climate change and health hazards, challenges and the way forward. Therefore, the chapter aims to provide guide for policy brief, analysis and development, and mainstreaming it to the climate change mitigation practices and national health development strategies.

Synopsis of Ministries, Department and Agencies (MDAs)

Ministers, Departments and Agencies (MDAs) are responsible for the runs of public sector affairs. Public sectors across the globe are the state own organization that control the administrative landscape of every country. Public sectors are distributed in the forms of MDAs at national, state and local government levels of a given country. These MDAs conduct activities that significantly enhanced the landscape of the natural and physical environment and its sustainability mechanisms, which include, among others, health systems strengthening, capacity building and disaster preparedness (Harris et al. 2017; Mordecai et al. 2020).

Public sectors at the national level promote collaborative actions. Public sectors at federal levels placed state and local government at the center of service delivery, by working together in collaboration. For example, bureaucrats at state and local government are the one implementing the policies formulated at the national levels (Jackson and Lynch 2011).

Different MDAs performed different functions. Different public sectors delivered different services that ensued climate change mitigation practices. For example, Ministry of Land and Housing implement land use control duties; Ministry of Transport takes care of transport systems; Ministry of Water Resources look at water supply activities; Ministry of health deals with health and well-being; and Ministry of environment facilitates environmental management functions among others. These MDAs play major roles that shaped climate change and health hazards mitigation activities.

Overview of Climate Change and Health Hazards

Climate change is the Alpha and Omega of all environmental change issues. This is so, because greenhouse gas emission is emanating without boundary, from industries, farms, households in a great proportion, subjecting all and sundry to the associated hazards—which in turn affect health. Greenhouse emissions in the year 2000 alone were estimated to be 24 billion metric tons (Tol 2009). Consequential effects of climate change are numerous, diverse and pervasive, with various degrees of morbidity and mortality consequences on the population (UNHCHR 2016).

Climate change caused wide range of health hazards. Climate change disrupt ecosystems, altered food chain systems, destroyed water supply, settlements and infrastructures, resulting to loss of jobs, increased food insecurity, mental health, diseases and stress among others, depending on location, socio-economic, gender, age, disability status, and cultural background, among others. In this regard, the Intergovernmental Panel on Climate Change (IPCC), posit in its fifth assessment report that those who are marginalized socially, economically, culturally, politically, institutionally and otherwise, are particularly vulnerable to climate change related health hazards (UNHCHR 2016).

Climate change also escalates the effects of environmental determinants of health by acting as a threat multiplier. Furthermore, unmitigated warming is expected to reduce average incomes roughly by 23% by the year 2100, resulting into wide global income inequalities, in comparison to circumstances without climate change (Cimato and Mullan 2010). The scale at which climate change effects on health increase is alarming, as reflected in the annual rise in extreme weather events, like flood, draught and heat-waves among others (Abdeltawab et al. 2021).

Flooding occurs with much consequences. Flooding is among the climate change drivers that breed vector borne diseases. For example, mosquitoes find flooded areas conducive for breed and reproduction. Mosquitoes caused malaria, dengue, Zika, yellow fever and chikungunya disease among others (CDC 2022a). Therefore, the direct effect of flood is serving as a medium for the transmission of vector borne diseases.

Water quality is also affected by climate change. Climate change caused a broad range of water related public health issues (Sutton et al. 2020). For example, change in water temperatures may precipitate transmission of waterborne illnesses; algal toxins found in the water at different period of the year, can also be harmful in places where they were not previously a burden. In this regard, waterborne disease causing organisms, like fungi, parasites, bacteria and virus were found to escalate the quality of water (CDC 2022a).

Extreme weather altered normal life proceedings. Extreme weather events increased surge in storm, and vulnerability to communicable and non-communicable diseases, as a result of exposure to chemical substances and other toxins (CDC 2022b). Extreme weather events strain cities' shields, roads, drainage systems, power plants, ports, and other social amenities (Willems and Baumert 2003). Additionally, episodes of diarrheal diseases increase during a high temperature climate, often characterized by different patterns and intensity of the weather, depending on the location and the type of pathogen involved. Moreover, diarrheal diseases have also been found to occur more frequently during sporadic ice and snow melt, and increases in stream flow rates, due to warm climate (CDC 2022a, b).

Infectious diseases increase or decrease according to the weather variables. Infectious diseases spread based on climate suitability, as the temperature rise, so is the burden of the infectious diseases (Mordecai et al. 2020). Also, warm weather disrupts the physical, biological, and ecological systems, resulting in cardiovascular and respiratory ailments, physical injuries and premature deaths, related to climate change (CDC 2022b).

Climate change seriously affects mental health. Sound mental wellbeing depends upon conducive environment that support the society. As such, destruction of social support systems due to environmental changes like floods, droughts and temperature rise, confound the burden of mental health (Abdeltawab et al. 2021). Indeed, environmental disaster is mostly accompanied by mental health challenges among the people with no previous history of mental illness (CDC 2022b).

Air pollution is a major climate change and health hazards issue across the globe. Air pollution might be the most highly publicized climate change and health issue, related to fossil fuel use. Fossil fuel is known to have produced a range of pollutants that negatively affect indoor and outdoor air quality, with accompanied health cost. Air pollution have caused over seven million avoidable deaths in the year 2012, meaning one in every eight deaths is linked to air pollution. Air pollution is common in an environment whose major source of energy is fossil fuel like coal and fire woods among others. Air pollution caused about 29,000 premature deaths annually in the UK, with a total cost to EU health systems estimated to be over 42 billion Euros annually (Abdeltawab et al. 2021).

Indeed, climate change associated health hazards are linked to increase in ground level ozone. Ground-level ozone and particulate matters are key constituents of smog; smog is responsible for several climate change associated health hazards (CDC 2022a, b), such as cardiovascular and respiratory diseases, lung cancers, asthma, and increases in premature deaths among others (Abdeltawab et al. 2021; Sutton et al. 2020; CDC 2022a).

Roles of Ministries Departments and Agencies (MDAs) in Climate Change Mitigation

Ministries Departments and Agencies (MDAs) are found at the federal, states and local Government levels of every country. MDAs play vital climate change and health hazards mitigation roles as the public sector custodians. These public sectors conduct activities that mitigate climate change and health hazards, through the creation of appropriate and conducive environmental friendly mitigation policies and practices (Cimato and Mullan 2010; Gogoi et al. 2017). Public sectors' role in climate change mitigation comes in two ways, thus; adaptation and mitigation. Adaptation means preparation for the effects of climate change on health, while mitigation denotes addressing and/or decreasing the consequences of climate change on human health. This chapter focus on the mitigation roles of the MDAs.

In this regard, climate change mitigation roles of the MDAs are based on policy formation and implementation. MDAs in both global north and south formulate and implement environmentally friendly climate change mitigation policies and practices (Harris et al. 2017; Cimato and Mullan 2010, Gogoi et al. 2017). Public sectors across the world are responsible for environmental protection duties and activities that consolidate stakeholders' support, acceptance and compliance with climate change mitigation practices (Kulin and Sevä 2019). The policies were designed to provide qualitative, descriptive and suggestive mitigation guidelines (CDC 2016). For instance, formulation of climate change mitigation policy should start by identifying the required inputs, specifications, techniques and approaches to be deployed using available data. Policies developed using these approaches would help in removing the mitigation techniques implementation barriers (CDC 2022b). Also, policy implementation should prioritized vulnerable locations, monitoring and evaluation of the base line and end line mitigation projections.

Indeed, MDAs, have several actual and potential roles. This section of the chapter highlights the potential and actual roles of MDAs in climate change and health hazards mitigation, thus presented below under different headings and subheadings:

Corrupt Practices Discouragement

Corrupt practices encouraged environmental degradation activities. Sinha et al. (2019) in their study found that corrupt practices by the bureaucrats encouraged environmental degradation activities; it for instance, motivates negative renewable energy practices in countries, where development issues are not stringent. Therefore, mitigation policies should be robust and resilience in terms of corrupt practices. It should also prescribe how to eliminate or considerably reduce corrupt practices with focus on corrupt free mitigation practices (Sinha et al. 2019).

To achieve this feat, the MDAs should ensure that corrupt free mitigation practices are implemented. This implies that the roles of public sector in protecting the environment goes far beyond planning sound environmental friendly policies alone, to include actions that checked corrupt practices, hence ineffective and corrupt bureaucracy hinders public acceptance of critical environmental protection policies (Kulin and Sevä 2019). It is therefore, imperative for MDAs to developed and implement transparent and accountable mitigation policies that are capable of winning public support and confidence.

Climate Change Risk Assessment Prediction

Climate risk assessment guides better understanding of climate change patterns. Climate change assessment requires an in-depth analysis of weather and climate landscape of a giving location. Climate change assessment can guide the prediction of health effects associated with pollutants such as ground level ozone, as an example. In this regard, metrological agency as a public sector should conduct climate change assessment at regular intervals, so as to predict ground-level ozone percentage, for proper planning and implementation of mitigating strategies. So, reducing ozone related pollutants thrive on the control of other sources of greenhouse emissions, such as wildfires among others (CDC 2022a). Indeed, climate-driven increase in health hazards leads to increased hospital visits, loss of school and work days, and premature deaths among others. As such, climate risk assessment if done properly and timely can mitigates the climate change associated health hazards and other peculiarities.

Climate Change Mitigation Monitoring

Monitoring the effects of climate change mitigation outcomes thrives on data collection. Mitigation activities should be observed by the public sectors, in a context that rely on reliable data for replication in other settings. Evidence based data should be collected during policy implementation processes, for future references. Public sectors as a coordinating body of all the stakeholders in climate change mitigation practice have the advantage of acquiring in-depth, and qualitative information (Hepburn 2010). Furthermore, monitoring climate change mitigation practice has its prospects and constraints, it therefore, require a clear objective that explicitly harness appropriate indicators. Additionally, climate change monitoring needs complementary measures that are credible, stable and reliable, devoid of anticipatory inputs (Hepburn 2010; CDC 2022b). Finally, given the significance of climate change mitigation monitoring, one important function of the public sector is the production of a distinct monitoring methodology.

Food Security

Climate change affects food production quality and quantity. Quality and quantity of food, supply chain, cost and distribution were all affected by climate change. Food production capacity declined due to the combined impact of climate change (CDC 2022a, b). The decline has profound consequences on the nutritional status, therefore, food security measures should include mitigation technique that can reduce the effects of food production challenges (CDC 2022b).

Additionally, after climate change anticipatory reports the government should have food security policy in place, to prepare for the adverse effects of climate change. The prediction reports should galvanized the MDAs towards formulating approaches that can address the anticipated food shortage, to avert pandemic situation of malnourished citizens. Indeed, food security is the approach that ensued access to safe, nutritious and sufficient food that meets the daily sustenance requirement, for productive and healthy life, at all the times (World Food 1996).

Carbon Taxes

Carbon taxes could encourage the acceptance of alternative sources of energy. Alternate source of energy utilization produced lower carbon emission. Achieving this aim can be reached through creation of carbon tax, in the form of per-unit tax, according to the emitted pollutant by individuals or organizations. Carbon tax could increase the cost of using carbon-based energy, thereby given consumers options of using other source than fossils, which can in turn reduce the pressure of carbon emissions on the environment (Harris et al. 2017).

Several countries around the globe have established carbon taxes. For example, nationwide carbon taxes was instituted on coal use in India in the year 2010; new vehicles taxes according to their carbon emissions ability was established in South Africa in the year 2010; fuel carbon tax in Costa Rica was initiated in 1997), and carbon taxes was introduced in the Canada's Quebec, British Columbia and Alberta provinces (Harris et al. 2017). Carbon taxes policies enacted in these countries was a catalyst to the mainstreaming of climate change mitigation practice that could leads to the realizing of the expected outcomes with sustain results (UNDP-UNEP 2011, Gogoi et al. 2017). Carbon taxes can help ensure that climate change mitigation is factored in, in a way that reduces carbon emissions and help achieve Sustainable Development Goals number 13 and other 2030 agendas.

Insurance Coverage

Payment of insurance premium help in sustaining mitigation practices. Insurance is a key catalyst of mitigation practices; it encourage positive actions, because knowing that if one becomes negligent, losses may be incurred, which means losing of insurance benefits. Hence, premiums based on hazards provide the insured victims with the incentives to adopt mitigation practices, knowing once a climate change related disaster occurred, not due to carelessness; the involved risk will be covered by the said insurance. In this regard, the insurance coverage will take care of the potential losses. Example of climate change related insurance coverage is the type offered in the UK, in which insurance brokers covered floods related disasters (Cimato and Mullan 2010). Climate change related health hazards insurance coverage should be encourage by the public sector as a means of improved mitigation practice.

Provision of Mitigation Infrastructure

Implementation of mitigation practices thrives on the use climate change resistance infrastructure. Infrastructure is essential aspects of climate change mitigation practices, without which mitigation activities will be stunted. Infrastructures provide the MDAs a means of service regulation, coordination and implementation. As such, climate change resistance infrastructure is an essential tool that ensure that climate change related hazards mitigation received the right incentives (Cimato and Mullan 2010).

Allocation of Building Code and Standard

Allocation of building code should be made a priority. Building occupied high place in physical environment conservation. Government in the United State have adopted green building codes programs through which heat reduction strategies and building standards is initiated and maintain through the use of standard codes that indicate compliance. Standard building code is among the strategies that ensure that construction, adjustment, and maintenance of buildings and other structures follow standard that reduced heats in the buildings. Consequently, green building codes enables inclusion of other building heats' reduction requisites, such as cool roofing code, for energy saving during energy peak (CDC 2022b).

Health Promotion

Health promotion provides solid foundation for climate change and health hazards mitigation practice. Ideal health promotion practices should be utilized, through the application of behavior change principles. Health promotion principles denote a solid foundation upon which mitigation practices and actions that address climate change are implemented. Health promotion practices engaged stakeholders from within and outside the health sector, through transdisciplinary approaches using both generic and specific disciplines for actions that mitigate climate change and health hazards at community levels (Patrick et al. 2012). Health promotion practitioners make use of disciplines with stake in climate change mitigation, depending on the situation and need. Indeed, health promotion field presents a clear vision for health promotion practices suitable to climate change and health hazards mitigation (Patrick et al. 2012).

Surveillance

Climate change surveillance help in mitigation preparation. Surveillance provide data on climate change and associated ailments, such as diarrhea, vector borne diseases (malaria) and respiratory tract diseases among others, over a long period of time within a given location. The surveillance data enable comparison and understanding of climate patterns, and eventual mitigation plans, design and implementation. Use of certain indicators during the surveillance could identify evidence based climate-change effects on human health. Additionally, surveillance of climate change and health hazards related impact should encompass variables that can establish relationships between climate changes and health hazards.

Surveillance of climate change provide information on the patterns, frequency and/or the nature of the related health hazards occurrence. Climate change surveillance are supported by several factors, like early warning systems which can help in the preparation of mitigation measures (Georgeson et al. 2017). Consequently, Climate change surveillance require data collection expertise, and analytical skills.

Coping with Extreme Heat

Increased in temperature was witnessed over the last few decades. An average increased in heat waves have become intense, hotter and more frequent globally. The trend is expected to be sustained and to stay safe from increased heat threats, the following measure should be adopted, thus:

- Drinking of 2–3 L of water daily
- Avoid sitting in a non-moving car
- Resting in cool environment
- Planning ahead for suitable environment (Sutton et al. 2020).

However, in some cases a warm average temperature is a sign of earlier rainy seasons. A contrary climate of a longer above average warm seasons, like short winters and hotter summers, predict a likely flooding events. Such conditions are suitable for the breeding of mosquitoes and other diseases carrying vectors. Therefore, in such circumstances, the MDAs should prepare their communities on mitigation practices, by focusing on insect control, importance of water drainage and provision of weather information among others (CDC 2022b; Sutton et al. 2020).

Partnerships

Mitigating the effects of climate change on health is a multi-sector approach. Partnerships can be formed between the public, business and not for-profit sectors. Partnership are required for effective climate change and health hazards mitigation, as no sector can effectively address environmental degrading issues in silo. This implies that public sector alone cannot effectively handled the growing climate change challenges and the associated health hazards, as such partnerships among the governance sectors is inevitable (Carola 2019, Gogoi et al. 2017). Therefore, climate change mitigation practices can be better achieved through partnerships between public, civil society, and private sectors (Gogoi et al. 2017). Partnership for climate change mitigation should be coordinated and lead by the public sector. Partnerships built on good will, will address the need-do gaps, through capacity development of the bureaucrats by the civil society sector, on how to better manage climate change related health hazards, and also on how to examine the strength, weakness, and resilience status of the existing policies (OECD 2009). Consequently, implementation of climate change mitigation policies can be done through partnerships between the MDAs and private sector in form of Public Private Partnerships (PPPs), MDAs and civil society, and vice versa as highlighted below.

Private Sector

Private sectors are the for-profit institutions. For profit sectors are involved in development activities within the recent past decades; their involvement is mostly in form of infrastructure provision, especially in low-income countries. Private sectors have of recent support environmental services, in the areas of water, sanitation and waste management, through Public Private Partnerships' (PPPs) arrangements (DANIDA 2020).

Public Private Partnership (PPP) refers to the arrangement that ensure joint financing task, between government and the private (for profit) sector, among other stipulations. PPP is formed mainly to benefit from the private sector provision of public utilities in the areas of needs, such as climate change mitigation practices. For example, cost of sanitation and waste management can be supported through PPPs (DANIDA 2020).

To date PPP's success depends on the private sector commitment and political will of the public sector. There should be commitment from both the private and public sector to serve the interest upon which the partnership arrangement is made (Stern and Fankhauser 2016). In this regard, private sectors have substantially contribute in the reduction of greenhouse gas emission (NASEM 2014), as part of their Corporate Social Responsibilities (CSR).

Social responsibility is a role expected to be played by the private sectors in support of other sectors through partnership or independently. For effective CSR functions, governments set certain objectives for private sectors to meet, as part of their. For example, in Germany, the government introduced the concept of social responsibility in public sectors processes. Also, government in the Netherlands, introduced responsible practice, as role model for private sectors to imbibe. Therefore, public sectors should set example for responsible practices that will ensue environmental friendly practices which will in turn mitigates the effects of climate change on human health (NASEM 2014).

Civil Society Organizations

Civil Society Organizations (CSOs) are the not-for-profit institutes; they includes, trade unions, faith based organizations, non-governmental organizations, and community based organizations among others (Umar and Wan-Puteh 2018). Partnerships sometimes occurs between public sector and civil society organization for activities that could mitigate climate change effect on health. CSOs in partnerships with government share information that can mitigate the effect of climate change. For example, in Paraguay, CSOs advocates to end illegal deforestation, which leads to a withdrawal of a decree that previously allowed native forest deforestation (Carola 2019). Public sector should coordinate the mitigation activities during partnership with CSOs. Coordination roles during the partnership will avoid duplication and multiplication of duties between CSOs and the public sectors. Several partnerships have taken place, between the CSOs and public sector across the globe. For example, in the year 2016, the Netherlands launched partnerships with 25 consortia of CSOs, the partnerships focus on lobbying and advocacy in various spaces that includes sustainable use of natural resources among others. Similarly, in Pakistan, CSOs activities prompt the appointment of a Climate change Focal Person by the Prime Minister and launched of Green Pakistani Program, a flagship project within the forestry agency (Qazlbash et al. 2021). Also, in India and Pakistan, legal activism enabled the used of courts by the civil society organization, to make the government accountable to climate change related issues (Qazlbash et al. 2021; Gogoi et al. 2017). Additionally, CSOs allocate financial resources for climate sensitive infrastructure, capacity building and technical assistance among others, to the public sector (OECD 2009).

Environmental Sustainability Role of the Public Sector

Environmental sustainability roles of public sector significantly enhanced sustainable use of energy. The main objective of environmental sustainability in the public sector is to minimize energy consumption toward the reduction in carbon emissions. Adoption of sustainable technology such as the conduct of online meeting, confer solutions that reduce carbon emissions, through decease in the local and internal travels (OECD 2009). Environmental sustainability activities make use of climate change vulnerability information. The information help in managing the climate change associated hazards mitigations like afforestation and reduction in fossil fuel use.

Public sector environmental sustainability activities include:

- Implementation of environmental friendly activities in the power, oil and gas industry, transport and agriculture sectors and proper land use
- Enhanced commitment to environmental change actions at local levels
- Political will and strengthening of environmental change legislation
- Funds mobilization through private sector investment for climate change mitigation
- Provides innovative mechanisms for climate change mitigation practices
- Engaged MDAs, private sectors and stakeholders through partnerships (OECD 2009).

Environmental sustainability as concepts require outcome analysis. Sustainability as concepts discuss how the needed data can be obtain and analyzed for improvement and strategic sustainability plans. Sustainability data are used in addressing potential environmental burden, by bureaucrats (NASEM 2014). Indeed, sustainability concepts agreed that environment protection and sustainability is government's prerogatives (Kulin and Sevä 2019).

Climate Change and Health Hazards Mitigation Measures

Climate change mitigation measures are not made in a vacuum. Mitigation measures were influenced by series of actions and arrangements. For example, water supplying firms may prepared for the likely drought effects through planning of what to do and not to do. Mitigation measures are guided by three criteria, thus; effectiveness, efficiency and equity (Cimato and Mullan 2010). Measures to be taken should considered the three criteria for optimization.

Therefore, every mitigation measures should be based on the criteria that are effective, efficient and equitable and the use of specific measures for specific condition, doing so will prevent the choice and use of wrong measures. Right measures if taken ensure effective mitigation outcome.

Challenges Confronting Climate Change Roles by the MDAs

Climate change effects mitigation are not without challenges. These challenges may be generic or specific depending on the situation and/or issues in contention. This chapter looked at the challenges that are likely to be confronted by the public sectors in their effort to mitigate climate change and health hazards.

Inadequate financial resources constraint public sector's capacity to conduct effective mitigation activities. For example, financing climate change mitigation project like reduction of carbon emission among others is difficult to accomplish (Jackson and Lynch 2011). Financial capacities of MDAs should be strengthen through budget adjustment of the areas of lesser priority, for the areas that could mitigate the effects of climate change (Jackson and Lynch 2011).

Lack of evidence based data is among the barriers to effective mitigation practices. Lack of data pose a challenge of evidence based mitigation techniques. For example, reliable data is required when preparing for behavior change awareness program. Data is also required in addressing complex mitigation decision issues, for example, a decision of where to place climate change and health hazards mitigation hub; is it in the ministry of health or ministry of environment? Decision making in such circumstances have a relatively long-term effects, with additional costs, as such data is required for right decision.

Lack of human resources in terms of expertise and quantity inhibits mitigation practices. Competent and adequate human capital is required to respond to the climate change and health hazards mitigation need; lack of it, is a barrier to the successful mitigation practice. The public sector systems in the global south often lack the needed human capacity that can provide resilience antidotes to the frequent environmental degradation issues. MDAs roles in climate change mitigation require manpower and expertise for optimal outcome (Cimato and Mullan 2010).

Way Forward

Achieving the aims of climate change and health hazards mitigation requires strategic approaches. Climate change health related issues can be mitigated through partnerships between the stakeholders. MDAs should make use of available health data to plan, design, formulate and implement mitigation policies, focusing on the criteria that advance climate change and health hazards mitigation measures.

To cut the real world emission, campaign on greenhouse gas reduction should be intensified. Mitigation and reduction in fossils emissions are inseparable, therefore, MDAs should support innovations that reduce greenhouse gas related pollutants such as ground level ozone. This require commitment and adherence to the global mitigation guidelines like the World Health Organization (WHO) Air Pollution Guidelines (Willetts et al. 2022; WHO 2015).

Climate change governance in form of transdisciplinary collaborations focusing on the international agreements on climate change mitigation would protect both human and the environment health. Therefore, engagement between actors and sectors with stake on health and environment as one, on mitigation policy-making, should be formed (Willetts et al. 2022).

Furthermore, five phased process framework called "Building Resilience against Climate Effects (BRACE)", designed by the United State CDC, guides communities' preparation on the effects of climate change on health mitigation by the MDAs, thus;

1. Predict and evaluate potential climate change effects and vulnerabilities: Identify potential climate change effects related health hazards, at risk communities and locations.
2. Identify associated disease threat: Approximate and/or predict the possible threat to health outcomes related to climate change.
3. Evaluate public health actions: Identify the most appropriate health actions for the identified health effects of greatest priority.
4. Prepare and implement climate change mitigation plan: Prepare a documented mitigation plan that is updated time to time. Share and supervise implementation plan.
5. Assess effects and enhanced the standard of the interventions: Assess the processes to decide the quality of the obtained data and implemented interventions (CDC 2022b).

Consequently, the following key roles will guide the MDAs conduct of climate change mitigation practices (WHO 2015);

1. Provision of evidence-based data on health hazards associated with climate change through operational and empirical research, to develop tools and guidance through which burden of disease will be understood and prepare for interventions.
2. Public sector should provide leadership during partnerships/collaboration. Provision of leadership by public sector, during a multi-sector and/or multi-stakeholder partnerships, for climate change mitigation is a requisite and mandatory.
3. Strengthening the capacity of public sectors' functions and competency, to;

a. Use decision tools and support instrument like Health Impact Assessment and Environmental Impact Assessment, to harness co-benefits and other opportunities for climate change mitigation practices.
b. Monitor, evaluate and report climate change mitigation policies.
c. Advocate for and engage stakeholders on issues that promotes climate change mitigation.

4. Climate change mitigation practices should be included in national health development plans (Fox et al. 2019; WHO 2015).

Conclusions

Mainstreaming climate change mitigation practices thrives on the integration of policy development, implementation, monitoring and evaluation processes, through partnership/collaborations between public, private and civil society organization and other players in the development field, with stake in health and environment as one. This indicates that, MDAs mitigation roles are multi-stakeholder's engagement, through which sustainable development can be attained (UNDP-UNEP 2011). In this regard, roles of public sector in climate change and health hazards mitigation, can considerably reduce related morbidity and mortality rates. Succinctly, the roles of public sector in mitigating climate change and health hazards is critical; the roles could elucidate and address all of the identified related health hazards through multi-sectors engagement, partnership and/or collaborations. It is therefore, imperative for the public sector, to lead and coordinate the provision of effective policies and frameworks that when implemented, could ultimately protect the planet from the effect of climate change, and promote the health and well-being of the population in question, simultaneously.

References

Abdeltawab KM, Essafi S, Moran K, Wardrope A, Pétrin-Desrosiers C, Fiolis D (2021) Relationship between Climate Change and the enjoyment of human right to health. In: IFMSA Contribution to the UN OHCHR study on the Climate Change and human rights to health, in the road to COP21

Carol van R (2019) Environmental challenges: too big for government and business to tackle alone

CDC (2022a) Climate effects on health. Available from https://www.cdc.gov/climateandhealth/effects/default.htm

CDC (2022b) Public health adaptation strategies for climate change. Available from https://www.epa.gov/arc-x/public-health-adaptation-strategies-climate-change

Cimato F, Mullan M (2010) Adapting to climate change: analysing the role of government. Defra evidence and analysis series, paper no. 1

Control, Centers for Disease, and Prevention (2016) Climate change decreases the quality of the air we breathe. Retrieved from www.cdc.gov/climateandhealth/pubs/air-quality-final_508.pdf

DANIDA, Workshop Papers (2020) Allocation of key responsibilities for private participation options Improving the Urban Environment and Reducing Poverty. Copenhagen, Denmark

Fawzy S, Osman AI, Doran J, Rooney DW (2020) Strategies for mitigation of climate change: a review. Environ Chem Lett 18(6):2069–2094
Fox M, Zuidema C, Bauman B, Burke T, Sheehan M (2019) Integrating public health into climate change policy and planning: state of practice update. Int J Environ Res Public Health 16(18):3232
Georgeson L, Maslin M, Poessinouw M (2017) Global disparity in the supply of commercial weather and climate information services. Sci Adv 3(5):e1602632
Gogoi E, Bahadur AV, del Rio CR (2017) Mainstreaming adaptation to climate change within governance systems in South Asia: an analytical framework and examples from practice. ACT learning paper
Hallegatte S, Henriet F, Corfee-Morlot J (2011) The economics of climate change impacts and policy benefits at city scale: a conceptual framework. Clim Change 104(1):51–87
Harris JM, Roach B, Codur A (2017) The economics of Global Climate Change. Global Development and Environment Institute. Tufts University
Hepburn C (2010) Environmental policy, government, and the market. Oxf Rev Econ Policy 26(2):117–136
Jackson T, Lynch W (2011) Public sector responses to climate change: evaluating the role of Scottish local government in implementing the climate change (Scotland) Act 2009. Commonwealth Journal of Local Governance (8/9):112–135
Kulin J, Sevä IJ (2019) The role of government in protecting the environment: quality of government and the translation of normative views about government responsibility into spending preferences. Int J Sociol 49(2):110–129
Mordecai EA, Ryan SJ, Caldwell JM, Shah MM, LaBeaud AD (2020) Climate change could shift disease burden from malaria to arboviruses in Africa. Lancet Planet Health 4(9):e416–e423
NASEM, National Academic of Science Engineering, and Medicine (2014) Sustainability concepts in decision-making: tools and approaches for the US environmental Protection agency. The National Academies Press, Washington, DC
OECD Publishing (2009) Integrating climate change adaptation into development co-operation: policy guidance. Organisation for Economic Co-operation and Development
Patrick R, Capetola T, Townsend M, Nuttman S (2012) Health promotion and climate change: exploring the core competencies required for action. Health Promot Int 27(4):475–485
Qazlbash SK, Zubair M, Manzoor SA, ul Haq A, Baloch MS (2021) Socioeconomic determinants of climate change adaptations in the flood-prone rural community of Indus Basin, Pakistan. Environmental Development (37):100603
Sinha A, Gupta M, Shahbaz M, Sengupta T (2019) Impact of corruption in public sector on environmental quality: implications for sustainability in BRICS and next 11 countries. J Clean Prod 232:1379–1393
Stern N, Fankhauser S (2016) Climate change, development, poverty and economics
Sutton B, Mulvenna V, Voronoff D, Humphrys T (2020) Acting on climate change and health in Victoria. Med J Aust 212(8):345
Tol RSJ (2009) The economic effects of climate change. J Econ Perspect 23(2):29–51
Umar I, Wan-Puteh SE (2018) An overview of civil society organizations' roles in health project sustainability in Bauchi State, Nigeria. Pan Afr Med J (30)
UNDP-UNEP (2011) Mainstreaming climate change adaptation into development planning: a guide for practitioners. In: United Nations Development Program - United Nations Environment Program
UNHCHR (2016) Analytical study on the relationship between climate change and the human right. Human Rights Council Thirty-second session. In: Annual report of the office of the United Nations High Commissioner for. Human Rights. Advance Edited Version, A/HRC/32/23
WHO (World Health Organization) (2015) Promoting health while mitigating climate change. Author, Geneva
Willems S, Baumert K (2003) Institutional capacity and climate actions. Citeseer
Willetts L, Bansard J, Kohler P, Bettelli P, Rosen T, Schröder M, Grant L (2022) Health in the Global Environmental Agenda: a policy guide

World Bank (n.d.) Social dimensions of climate change. Available from https://www.worldbank.org/en/topic/social-dimensions-of-climate-change#1
World Food, Summit (1996) Rome declaration on world food security. In: Plan of action to UNCLOS

Umar Ibrahim I was borne and brought up in Nigeria. I have several years of progressive experience in both clinical and preventive health services in Africa, Asia and Pacific. My career focus evolved around environmental change and public health. I obtained a Ph.D. in Global Health Governance & Community Health; Masers in Public Health (MPH); Post Graduate Diploma in Public Health (PGDPH), Post Graduate Diploma in Public Administration (PGDPA), Graduate Diploma in Accident & Emergency Nursing (A&E), BSC Nursing; Registered Nurse (RN); and Community Health Extension Work (CHEW) respectively, in addition to several off-line and on-line self-development courses. I have also worked in public, private, civil society and academic sectors, thus; Federal University Dutse, Jigawa State; Health and Human Services Secretariat, FCTA—Abuja, Nigeria; International Health Management Services Limited, Lagos, Nigeria; Nigerian Navy; and Bauchi State Ministry of Health and Human Services; as well as in Abubuwa Societal Development Initiative, a non-governmental organization; International Institute for Global Health, Malaysia; and Institute for Sustainable Futures, Sydney, Australia. The highlighted work experiences and academic background enables me an understanding of planetary health facilitation, research and community service prerogative and processes.

Chapter 20
Changing Fertility, Child Mortality and Contraceptive Prevalence Rates in Bangladesh: Effects of Disaster Risk Reduction and Family Planning Programs

Khandaker Jafor Ahmed, Yan Tan, and Dianne Rudd

Abstract Bangladesh is densely populated and vulnerable to climatic, natural, human-made, and biological disasters. Bangladesh is particularly vulnerable to cyclones, floods, drought, and temperature-related events. Disasters affect the well-being of the population of the country in many ways, while the effects are unevenly distributed across regions and population subgroups. Child mortality, total fertility, and contraceptive prevalence rates are not equally distributed across different regions. Existing population policy and family planning programs have made efforts to address the regional disparities. However, the role of disaster risk reduction and family planning programs in changing these rates remains unexplored. This research uses a mixed methods approach (quantitative and qualitative) to collect diverse information to understand how disaster risk reduction and family planning programs can help change geographical disparities in fertility, child mortality, and contraceptive prevalence rates. Primary data were collected through household surveys ($N = 544$) from married women aged 18–49 years with at least one child living with their husbands, and key-informant interviews with stakeholders ($N = 8$) in 2019–2020. This research found that disaster risk reduction and family planning programs can change fertility, improve child mortality, and contraceptive prevalence rates. Moreover, disaster risk reduction and family planning programs were more active and effective in the cyclone-hit village, where fertility and child mortality were lower, along with higher contraceptive access and use. It is crucial to ensure that these programs actively function in the targeted resource-constrained areas, and they are able to collaborate effectively to reduce regional disparities in those rates.

K. J. Ahmed (✉)
Walsh School of Foreign Service, Georgetown University, 37th and O St NW, Washington 20057, District of Columbia, USA
e-mail: ka906@georgetown.edu

Y. Tan · D. Rudd
Department of Geography, Environment and Population, School of Social Sciences, The University of Adelaide, Adelaide 5005, South Australia, Australia

W. Leal Filho et al. (eds.), *Climate Change and Health Hazards*, Climate Change Management, https://doi.org/10.1007/978-3-031-26592-1_20

Introduction

Bangladesh's unique geographical conditions and location make it one of the most vulnerable countries in the world and will become even more so due to the increasing frequency and intensity of climate change and extreme weather events. Climate scientists confirm and warn that extreme climate events will increase over time in frequency, intensity, and severity of impact on the environment and population dynamics (e.g., mortality, migration, and fertility) (Fischer et al. 2021). Between 2000 and 2019, more than 11,000 extreme weather events in the world caused the loss of over 475,000 lives and US$ 2.56 trillion (Eckstein et al. 2021). According to the long-term Climate Risk Index (CRI) 2021, Bangladesh is ranked seventh among the top ten countries most affected between 2000 and 2019. This ranking is based on Bangladesh having experienced 185 climate-related events that caused losses of US$ 1,860.04 (in a million US$ in purchasing power parity [PPP]) during this period (Eckstein et al. 2021). Bangladesh is most likely to suffer adverse effects from droughts, heavy rainfall, floods, cyclones, and sea-level rise associated with anthropogenic climate change (Huq et al. 2019).

Bangladesh is undergoing significant changes in population size, density, and distribution in response to changing climate and associated extreme events (M. R. Khan et al. 2021). Disaster risk reduction and family planning are the two most crucial programs that can play an important role in tackling the adverse impacts of extreme climate events on the population. The vision of the disaster management of Bangladesh is "to reduce the risk to people, especially the poor and the disadvantaged, from the effects of natural, environment and human induced hazards to a manageable and acceptable humanitarian level and to have in place an efficient emergency response management system" (GoB 2020, p. 1). The Government of Bangladesh (GOB) has developed structural frameworks and comprehensive disaster response strategies to reduce disaster losses to human life and property (Ahmad 2019). The development of disaster risk reduction programs, such as the National Plan for Disaster Management (NPDM) of 2010–2015, 2016–2021, and 2021–2025, Standing Orders on Disasters (SOD) of 2010, Disaster Management Act (DMA) of 2012, National Disaster Management Policy (NDMP) of 2015, saw excellent progress in disaster management in Bangladesh (Ahmad 2019; Hasan et al. 2019). Spontaneous community participation is the primary aim of NPDM, SOD, NDMP and DMA (Ahmad 2019; Hassan et al. 2019). NDMP and SOD primarily aim to provide complete guidance and clarify the responsibilities of each operational unit engaged in disaster risk reduction activities. Union Disaster Management Committee (UDMC), led by the Union Parishad (UP), is the first responding government unit to help local people before, during, and after disasters. UDMC has shown noticeable success in rural Bangladesh, facilitating village-level disaster risk reduction training sessions for members and volunteers and preparing risk-reduction action plans, evacuation processes and hazard vulnerability indicators (Paul et al. 2022).

Bangladesh is unique in the sense that many government and non-governmental organisations (NGOs) seriously focus on family planning programs. Notably, the

International Center for Diarrheal Diseases Research, Bangladesh (ICDDRB), BRAC Bangladesh, and the World Vision Bangladesh collaborated regularly with the government of Bangladesh, and have played a significant role in formulating health programs and providing extensive reproductive health services in rural and remote villages in Bangladesh over the years (Streatfield and Kamal 2013). Since its inception, the National Family Planning Program (FPP) in Bangladesh have been supported in developing research practice and the proper implementation of plans by many local and international donor agencies. The Directorate of Family Planning in Bangladesh has employed female field workers called Family Welfare Assistants (FWAs) since the inception of FPP in the 1970s, and they are an important part of the outreach services they offer to married couples. Family welfare assistants visit the same family or the same couple every two months, record new marriages, entice them to adopt family planning methods, distribute contraceptives and refer clients to family planning facilities for antenatal care, postnatal care and normal delivery (Schuler et al. 1995).

Bangladesh's contraceptive prevalence rate (CPR) has increased steadily since its introduction in 1975. CPR increased by 54 percent between 1975 and 2014 and has maintained this rate. According to the latest statistics, 62 percent of currently married women of reproductive age are using a contraceptive method, with 52 percent of them using modern methods. The public sector provides 40 percent of modern contraceptive users, while family welfare assistants are the leading source, serving 16 percent of users. Non-governmental organisations provide services to 5 percent of modern method users (NIPORT et al. 2019).

This research examines how these programs can change fertility, child mortality, and contraceptive prevalence rates in areas vulnerable to cyclones and floods in Bangladesh and contrasts the findings between areas where these programs are successful. Although both programs differ in scope and function, in areas vulnerable to extreme climate events, both are able to collaborate and implement their tasks effectively. This research has significance in considering what policy interventions should be taken to address regional disparities in fertility, child mortality and contraceptive use in Bangladesh.

This chapter begins with an introduction to this research, followed by describing the variations in fertility, child mortality, and contraceptive prevalence rates in administrative divisions and districts in Bangladesh. The following section summarises the methods and materials that have been used for this research. The results section outlines the socio-economic and demographic characteristics of the two study villages and relates them to differences in fertility, child mortality, and contraceptive use. The final section discusses the role of disaster risk reduction and family planning programs in changing those rates and concludes the chapter.

Regional Variations in Fertility, Child Mortality and Contraceptive Prevalence Rates in Bangladesh

There are uneven geographical distributions in under-five mortality, total fertility, and contraceptive use across Bangladesh's administrative divisions and districts (NIPORT et al. 2019). Bangladesh has eight administrative divisions: Barishal, Chattogram, Dhaka, Khulna, Mymensingh, Rajshahi, Rangpur, and Sylhet. Table 20.1 presents the divisional rates of contraceptive prevalence for any method, under-five mortality, and total fertility. Three divisions—Dhaka, Khulna, and Rajshahi—have achieved replacement-level fertility (2.1 or fewer children). Divisions with the highest fertility rates are Sylhet (2.7 births) and Chattogram (2.5 births). Contraceptive prevalence rates were highest in Rangpur (68.1%), followed by Rajshahi (66%), and lowest in Sylhet (54.3%) and Chattogram (55.3%). These differences in contraceptive use were clearly established even after controlling for socio-demographic and cultural factors between them (Alam et al. 2018; Islam et al. 2020). The highest under-five mortality rate was in the Sylhet division (65%) and the lowest in Dhaka (39%) and Khulna (40%). These differences in under-five mortality have been attributed to high-risk fertility behaviour and differentials in access and use of maternal healthcare and family planning services (M. A. Khan et al. 2021). Moreover, parts of the divisions are vulnerable to climate extremes such as cyclones, floods and drought. In addressing the regional disparities in child mortality, fertility, and contraceptive prevalence there must be a strengthening of service delivery and improved coordination between disaster risk reduction and family planning programs.

In 2016, 14 districts in Bangladesh had achieved a below-replacement-fertility level (TFR 2.0 or less). However, there were 9 districts with a TFR of 3.0 and more (NIPORT et al. 2019). Although these fertility differentials are primarily attributed to socio-economic conditions and educational attainment (Bora et al. 2022; Roy and

Table 20.1 Regional disparities in child mortality, fertility, and contraceptive prevalence in eight administrative divisions of Bangladesh

Divisions	Contraceptive use	Under-five mortality	Total fertility rate (TFR)
Barishal Division	64.7	45	2.4
Chattrogram Division	*55.3*	47	**2.53**
Dhaka Division	64.7	*39*	*2.01*
Khulna Division	64.7	*40*	*1.97*
Mymensingh Division	63.4	53	2.49
Rajshahi Division	**66**	48	*2.03*
Rangpur Division	**68.1**	53	2.17
Sylhet Division	*54.3*	**65**	**2.69**
Bangladesh	**62.8**	**47**	**2.23**

Data source NIPORT et al. (2019)
Bold indicates the highest values and italic for the lowest values

Hossain 2017), the relative importance of family planning programs in Bangladesh should be acknowledged (Bora et al. 2022). There is also a need to recognise the importance of considering disasters in fertility studies (Ahmed and Tan 2022), and the degree to which the fertility decisions of women vary according to whether areas are hit by a disaster or not (Haq and Ahmed 2019). It was found that the likelihood of disaster risk reduction considerations was more active and more focused in those regions most vulnerable to extreme climate events.

Bangladesh has 19 districts where contraceptive use and access are still inadequate and poorly practised. The lowest rate of contraceptive use was in Cumilla (47.8), and the highest in Rangamati (75.1). In relation to child mortality which tends to be closely related to a country's socio-economic development, standard of living and appropriate health and nutrition programs. In 2016, only four districts (namely, Munshiganj, Meherpur, Rajshahi, and Cumilla) attained the 4th Health Nutrition and Population Sector Program (HNPSP) 2022 target for the under-five mortality rate ($\leq$34 under-five deaths per 1,000 live births). Bangladesh has 11 districts with higher under-five mortality, notably more than 55 deaths per 1,000 live births (NIPORT et al. 2019). It has been found that above average under-five mortality was concentrated in the east of Bangladesh, associated with the region's poor socio-economic status, antenatal care, sanitation, and increased malaria prevalence (Gruebner et al. 2017).

Methodology

Study Areas

Based on the literature and statistics of loss and damage from recent cyclones and floods in 2019, and in consultation with climate change experts in Bangladesh, two districts of Bangladesh were selected for this study: *Satkhira*, in the high-cyclone-risk region, and *Sunamganj*, in extreme flood–prone region. The Sunamganj district is in the northeast of Bangladesh and Satkhira district is located in the southwest. Table 20.2 presents the socio-demographic and health care utilisation in these two districts. Since both districts are geographically and socio-culturally distinct, there are also significant differences in their socio-demographic and other health indicators. Sunamganj has a higher poverty rate (48.8%) and a lower number of educated women (25.9%) than those in Satkhira (14% in poverty and 47.9% with schooling). More than half of the younger women aged 15 to 19 years in Satkhira were married at the time of data collection, and had fewer children, compared to women in Sunamganj with more than three children.

Child mortality indicators suggest that women in Sunamganj were more likely (almost 2 times) to witness the death of a child compared to their counterparts in Satkhira. Women in Satkhira were more likely to access contraceptives (64%) than those in Sunamganj (54.7%), and were also more likely (almost 2 times) to have the deliveries of their babies attended by a skilled birth attendant and at a health

Table 20.2 Differences in socio–economic, fertility, health, and child mortality in Satkhira (*Cyclones*) and Sunamganj (*Floods*) districts

Indicators	*Satkhira*	*Sunamganj*
Socio–economic (%)		
Households in poverty	14	48.8
Females (aged 15–49) with secondary or higher education	47.9	25.9
Marriage and fertility (%)		
Female adolescents (aged 15–19) currently married	52.4	31.6
Total fertility rate (TFR.)	1.79	3.64
CPR for any method	64	54.7
CPR for a modern method	56.9	47.1
Maternal health (%)		
Deliveries attended by a skilled birth attendant (SBA)	56	28.2
Deliveries at a health facility	51.7	26.3
Childhood mortality (%)		
Neonatal mortality	28	45
Post neonatal mortality	6	16
Infant mortality	33	60
Child mortality	6	12
Under-five mortality	39	72

Note Two villages have been selected from the districts: Chadnimukha from Satkhira and Lamagaon from Sunamganj
Data source NIPORT et al. (2019)

facility than their counterparts in Sunamganj. While these underlying differences in socio-demographic and health indicators could explain the variations in fertility, child mortality, and contraceptive prevalence rates between the selected districts, the relative importance of disaster risk reduction and access to family planning programs and the availability of health resources in changing these rates are undeniable, and merit further investigation.

After the two districts were selected for study, a sub-district (*upazila)* from each district was also selected, based on the available literature on recent extreme climate events that had occurred in the districts. Tahirpur was the sub-district of Sunamganj deemed highly vulnerable to floods (Bhowmik et al. 2021; Raihan and Hossain 2021). Shyamnagar, a sub-district of Satkhira, was found to have been severely affected by cyclones (Hossain et al. 2018; Jakariya et al. 2016; Martin et al. 2014; Tajrin and Hossain 2017).

During the fieldwork, consultation with experts was undertaken to identify the most vulnerable areas. After meeting with the Upazila Nirbahi Officers (UNO) of Shyamnagar and Tahirpur to find out the *union* (the layer of government below *upazila*) within each of the two sub-districts, it was possible to establish the most vulnerable areas to recent extreme climate events in 2019. UNO know the socio-economic and demographic conditions as well as the climatic hazards of their areas. Following the consultation with the local UNO, the Union Chairman from that area, who serves as the Administrative Chief of a union and the Chairperson of the Union Disaster Management Committee (UDMC), was consulted to identify the most vulnerable village from within the designated union as well as to inform them about the research. Thirty-six individuals make up UDMC. Depending on regional conditions, the committee chairperson may appoint up to three more members and establish groups and subgroups. With a mandate to act as a rural disaster management agency, the UDMC is expected to be instrumental in disaster planning, mitigation, emergency response and post-disaster rehabilitation.

Ultimately, two unions[1] were selected: *Gabura* of Shyamnagar Upazila[2] and *Sreepur South* of Tahirpur Upazila. A number of studies had previously selected Gabura Union as a cyclone-affected study area (Garai 2014; Mallick et al. 2011; Martin et al. 2014). *Sreepur South* Union had been selected as a severe flood-affected study area by Haq and Ahmed (2019). Ultimately, one village[3] of each union was selected based on the information provided by the Upazila Nirbahi Officer and the Union Chairman of each area: Chadnimukha of Gabura and Lamagaon of Sreepur South (Fig. 20.1).

Data Sources and Methods

This research applied a simple random sampling technique to select married women from the village of reproductive aged 18 to 49 years, with at least one or more children and currently living with their husbands, to complete a face-to-face household survey. Household surveys were distributed to eligible women in each village. The representative sample size[4] was drawn from the population with a 99 percent confidence level, a standard error of 0.01941 with a relative standard error of 3.88, and a proportion of 0.5 (upper and lower limits are 0.55 and 0.45). From the two villages, 544 women met the stated criteria. The overall survey sample included 318 women from Chadnimukha and 231 from Lamagaon, out of populations of 611 and 355,

[1] *Unions* are the smallest rural administrative and local government units and represent the lowest tier of local government.

[2] The second-lowest tier of regional administration in Bangladesh, the upazila, is administratively similar to a district and plays the most crucial role. The core functions of an upazila involve the coordination of the development and administrative activities at the division, district, and upazila levels, the maintenance of general and revenue administration, and the performance of magistracy.

[3] In Bangladesh, a village is the smallest territorial and social unit for administrative and representative purposes. Each union is made up of nine villages, one of which is designated as a ward.

[4] This research used the online sample size calculator as provided at the website of Australian Bureau of Statistics. The link is https://www.abs.gov.au/websitedbs/D3310114.nsf/home/Sample+Size+Calculator.

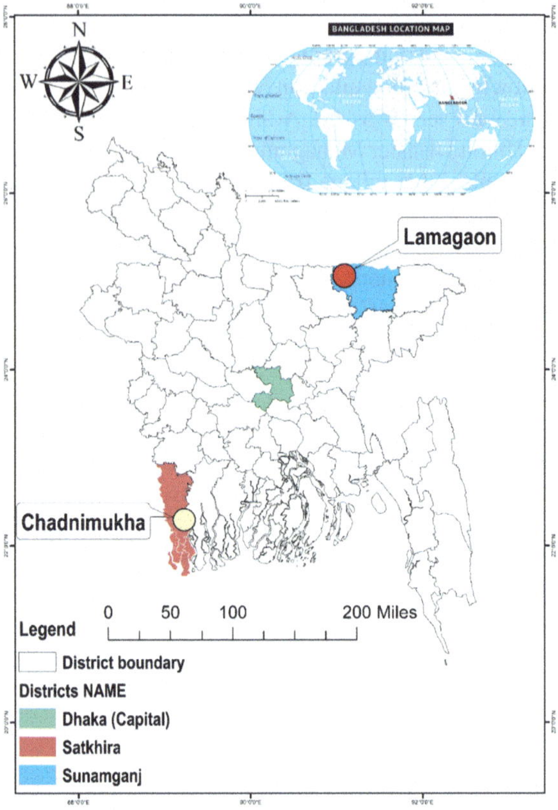

Fig. 20.1 Location of study areas in Bangladesh (*Source* Map produced by using ArcMap version 10.6.1)

respectively. The married women who participated were chosen randomly through a random number generator software, "Research Randomizer".[5] The household survey asked for information about all household members, including fertility intentions, birth histories, and child mortality, and targeted the female of the household to provide the most accurate information.

A household-level survey was conducted face-to-face through a questionnaire. Women were asked to provide demographic and health information about the householders, family planning facilities, contraceptive use, and birth histories of all children. Based on the inclusion criteria, women were selected from the villages using family planning registration records from the local family planning office. The interviewers—four female research assistants who resided in the selected areas—asked the respondents the questions orally and recorded their responses according to the given response codes.

The research began by conducting four key informant interviews (KIIs) in each of the selected upazilas of Shyamnagar and Tahirpur (Table 20.3), a total of eight.

[5] Urbaniak, G. C., and Plous, S. (2019). Research Randomizer (Version 4.0) [Computer software]. Retrieved from http://www.randomizer.org/.

This interview provided an overview of the selected upazilas, their vulnerability to extreme climate events, and the identification of the most highly vulnerable unions and subsequent villages. The KII 1 with each local Upazila Nirbahi Officer helped select the unions and villages most highly vulnerable to floods and cyclones. The KII 2, with an Upazila Family Planning Officer (UFPO), aimed to explore the region's family planning issues. The third KII, with one of the leading non-governmental organisations in each village, focused on women's reproductive health and child nutrition. The last KII was with each village's local family planning worker (FWA) to discuss their family planning services and their work with local married women.

Descriptive statistics were used to establish differentials in fertility, child mortality, contraceptive use, and access to health and family planning facilities in the two areas. Qualitative data from field notes and audio recordings were transcribed into protocols and transcripts and encoded via NVivo quality data management software. Protocols and transcripts were coded and one or more distinguished summaries of each transcript were developed to keep the content of a sentence or paragraph consistent. The coding process is considered part of the qualitative analysis (Miles and Huberman 1994) because it helps organise data into meaningful groups or patterns (Tuckett 2005). Interview quotes were used to develop a section of findings to complement quantitative data.

Table 20.3 Participants of key informant interviews

KII. No.	Name of key informant interviews	Targeted expert
KII 1	KII with Upazila Nirbahi Officer (UNO) or Project Implementation Officer (PIO) to gain an overview of the upazila and its vulnerability to extreme climate events and to select highly vulnerable areas for study	Upazila Nirbahi Officer (UNO) or Project Implementation Officer (PIO)
KII 2	KII with Upazila Family Planning Officer (UFPO) to identify the family planning issues of the upazila and study area	Upazila Family Planning Officer (UFPO)
KII 3	KII with World Vision NGO officer to explore local issues associated with women's reproductive health and child nutrition	Branch Manager
KII 4	KII with Family Welfare Assistant (FWA) to discuss family planning services	Family Welfare Assistant (FWA)

Results

Demographic, Economic and Social Differences in the Study Villages

Table 20.4 presents the socio-demographic characteristics of the women of cyclone-prone Chadnimukha and flood-prone Lamagaon, as reported in the household survey questionnaires undertaken in both villages. Comparing the age of respondents, the cyclone-affected village had more younger females aged between 18 and 24 years while the flood-hit one had a higher proportion of older mothers, but in both villages, the majority were aged between 25 and 34 years. Around one in four respondents in both study areas had not received formal schooling; notably, slightly more respondents in the cyclone-hit village had been through secondary and higher secondary schooling than those in the flood-hit village. It was also found that respondents in the flood-hit village were significantly more likely to be married before the legal age at marriage (18 years for girls), at 83 percent, compared to their counterparts in the cyclone-hit village, at 59 percent. This finding is consistent with a previous study in Bangladesh (Ahmed et al. 2019) that showed that early marriage of girls was higher in a flood-hit area (86%) than in a cyclone-hit area (62%).

According to these descriptive statistics, two-thirds of the households surveyed in Chadnimukha (cyclone-affected area) rely on wage labour, either agriculture- or non-agriculture-based, as their primary source of income. It is difficult to make a living farming or homestead gardening, because agricultural production in the region is sensitive to soil and water salinity: the growing number of storms in the region are associated with increased damage to dams, which then allow salt water to enter the farmland and affect crop production, particularly the rice varieties that are the primary food of the community. In Lamagaon (the flood-affected area), more than one-third of households depend on agriculture and two-fifths on fishing for their primary source of income. The region is a wetland area with diverse fish species, and the land often benefits from floods, becoming more fertile for crop production. Landless families in the flood-hit village can depend on agricultural wages from harvesting rice paddies in the dry season.

Notably, it took more time (an average of 53 min using the most common mode of transportation) for cyclone-affected dwellers to reach their closest hospital than it did for those from the flood-hit village (17 min). These differences between the study areas can help explain the differences in fertility, child mortality, and contraceptive use; however, it is also crucial to consider how disaster risk reduction and family planning programs play their roles in this matter.

Table 20.4 Demographic, economic and social characteristics of women in villages hit by cyclones (*Chadnimukha*) and floods (*Lamagaon*)

Characteristics	*Cyclones*	*Floods*
Demographic		
Age (**Mean/Std. Dev**)	**30.35 (6.80)**	**30.66 (7.08)**
18–24	23.4	18.2
25–34	45.0	49.8
35–49	31.6	32.0
Education (%)		
No years of schooling	26.8	23.4
Primary school [levels 1–5]	26.2	34.6
Secondary school [levels 6–10]	33.5	31.2
Higher secondary school [levels 11–12]	13.5	10.8
Age at marriage (**Mean/Std. Dev**)	**17.37 (1.33)**	**16.26 (1.39)**
Less than 18 years	59.1	82.7
18 years or over	40.9	17.3
Economic		
Income sources (%)		
Farming	16.9	35.5
Fishing	16.0	42.4
Agricultural wage labour	40.6	13.9
Non–agricultural wage labour	26.5	8.2
Social		
Times to the closest hospital (minutes)	53.16 (30.62)	16.68 (3.25)

Source Household Survey, 2019–2020

Fertility, Child Mortality, and Contraceptive Use in the Study Areas

As shown in Table 20.5, the mean number of children ever born as well as the number of those who survived was significantly higher in the flood-hit village than in the cyclone-hit village. The mean number of children ever born to household survey respondents was 3.75 in the flood-hit village and 2.06 in the cyclone-hit village. It was found that women of different age groups in the cyclone-hit village had more children than those living in the flood–hit village. Notably, younger respondents aged 18–24 years in the cyclone-hit village had lower numbers of children ever born (mean = 1.32) than their counterparts in the flood-hit village (mean = 1.79).

Table 20.5 shows that the number of child deaths was comparatively higher in the flood-hit village than in the cyclone-hit village. Differences in fertility and child

Table 20.5 Age-specific fertility of women, child mortality, and contraceptive use in villages hit by cyclones (*Chadnimukha*) and floods (*Lamagaon*)

Age-specific fertility	*Cyclones*	*Floods*
Number of children ever born (Mean/SD)		
18–24	1.32 (0.52)	1.79 (0.84)
25–34	2.10 (0.88)	3.53 (1.40)
35–49	2.55 (0.99)	5.20 (1.79)
Total	**2.06 (0.96)**	**3.75 (1.88)**
Number of children alive (Mean/SD)		
18–24	1.21 (0.40)	1.62 (0.69)
25–34	1.90 (0.74)	3.05 (1.13)
35–49	2.34 (0.84)	4.39 (1.47)
Total	**1.88 (0.82)**	**3.22 (1.52)**
Experience of child mortality (%)		
Yes	16.6	36.8
No	83.4	63.2
Contraceptive use (%)		
No	22.4	31.2
Yes	77.6	68.8

Source Household Survey, 2019–2020

mortality between the areas are likely to be associated with active participation in disaster risk reduction and family planning. For example, more than one-third of respondents in Lamagaon reported the death of at least one child during their reproductive span, whereas less than one-fifth of respondents in Chadnimukha had done so. Research has found that women may tend to bear more children as insurance against the adverse effects of extreme climate events or hazardous environments (Cain 1981, 1983, 1986; Frankenberg et al. 2015) or to replace the lost ones (Finlay 2009). This appears to be the case for women living in the flood-hit village.

About 10 percent more respondents in the cyclone-hit area were using contraceptives at the time of the survey in 2019 and 2020, which could be due to the interventions of non-government organisations (e.g., World Vision) working in collaboration with the existing government-facilitated family planning programs.

Disaster Risk Reduction and Family Planning Programs in Study Areas: Experts' Views

This section utilises qualitative information derived from interviews with stakeholders and experts. It was found that disaster risk reduction and family planning programs were more effectively functioning in the village hit by cyclones than in floods. Members of UDMCs in the cyclone-hit village were more active in creating public awareness about cyclone disasters and providing evacuation procedures and training. Regular meetings regarding formulating flood disaster action plans were often absent in practice in the flood-hit village (Lamagaon). The Upazila Nirbahi Officer of Tahirpur Upazila said:

> *UDMCs in Lamagaon are not fully effective due to various reasons such as lack of awareness, resources, proper training, budget, decision-making authority, and proper monitoring. In addition, we do not receive adequate support from government and non-government organisations. And, the remote villages in our sub-district rarely receive attention from media. (Upazila Nirbahee Officer, Tahirpur Upazila, 2019)*

Relief activities and programs to rehabilitate the affected population are inadequate, and flood shelters are poorly equipped and maintained in Lamagaon. UDMC of Gabura union is comparatively more effective than Sreepur South Union. The results of field observations and consultations with officials indicated various institutional capacities at different stages of a cyclone. During the pre-cyclone period, Union Parishads (UP), Red Crescent Society (RCS), and Non-Governmental Organisations provided services more efficiently in promoting cyclone warning and evacuation procedures. Cyclone Preparedness Program (CPP) volunteers are more efficient than any other organisation during the pre-disaster period in disseminating early warning messages and influencing people to decide to evacuate to cyclone shelters during disasters. A government official stated that:

> *UDMC is active in Gabura Union, providing services more efficiently before, during and after cyclones. They disseminate cyclone warning information and evacuate vulnerable people according to their evacuation plans. During a cyclone, they organise emergency rescue operations using locally available facilities and resources and help people move the resources they need to safer places. After a cyclone, they coordinate all relief operations provided by the government and non-government organisations. In this union, many national and international non-governmental organisations collaborate with UDMC to build adaptive capacity and increase household resilience. (Upazila Nirbahee Officer, Shyamnagar Upazila, 2019)*

World Vision Bangladesh was more actively engaged in disaster risk reduction activities in the cyclone-hit village than in the flood-hit village. According to the official of World Vision Bangladesh:

> *World Vision Bangladesh works with various services related to disaster management in the Gabura union, such as early warning operations during disasters, evacuation assistance, and immediate needs such as food and non-food items, health, water, sanitation and shelter, and enabling communities to resettle, build recovery, and increase adaptive capacity and resilience. (Manager, World Vision Bangladesh, Shyamanagr, 2019)*

Government and non-government organisations have a significant role to play in family planning and reproductive decisions. These organisations can raise awareness among women about contraceptive use and mother-and-child health, empower them through financial activities, facilitate their contribution to food security through homestead gardening, and enhance the well-being of households. During fieldwork, it was observed that some non-governmental organisations actively supported women's reproductive health, particularly in the cyclone-hit village of Chadnimukha. An internationally reputed NGO (World Vision) was actively functioning in the cyclone-hit village, focusing on a project called *Nobo Jatra* (which means *New Beginning*), which addressed women's reproductive health and child nutrition. This initiative, led by World Vision Bangladesh and in collaboration with the Ministry of Disaster Management and Relief (MoDMR) of the Government of Bangladesh, collaborated with the World Food Program and Winrock International. According to the Family Welfare Assistant in Chadnimukha:

> *Nobo Jatra provided assistance to pregnant and nursing mothers who were below the lower poverty line and faced chronic food insecurity. Selected individuals were enrolled in a 15-month nutritional security net program and received BDT 2240 (USD 27.5) each month. The 15-month window for cash transfers had been identified to coincide with the period beginning in the second and third trimesters of pregnancy and ending when the child is 9 months old to cover the most critical transition period, from exclusive breastfeeding to the consumption of a combination of breast milk and formula. (Family Welfare Assistant, Chadnimukha, 2019)*

In Chadnimukha, the program aimed to enhance women's involvement in nutrition and health decision-making by helping improve women's capacity to accept cash, open savings accounts, and manage their mobile banking information, and strengthen the government frontline staff's responsibility to assess participants' gender sensitivity, nutrition and health status, and decision-making skills.

> *World Vision Bangladesh focuses on various activities like health, nutrition, washing, livelihood and food security in 28 administrative districts of Bangladesh. We collaborate with government-facilitated community clinics in rural and remote areas of the Satkhira district to increase awareness of services available at community clinics, increase knowledge and access to maternal and child services, and build the capacity of health facilities to meet the needs of all patients. (Manager, World Vision Bangladesh, Shyamanagr, 2019)*

Floods cause colossal damage to health infrastructure, medicine, and contraceptive supplies. The flood-hit village (Lamagaon) and its upazila have inadequate health facilities. It is difficult for family welfare assistants to visit community clinics and service holders in Lamagaon. The Upazila Family Planning Assistant articulated that:

> *There are Union Health and Family Welfare Centers in 6 out of 7 unions of Tahirpur Upazila. The [UHFW] Centre of Sreepur South Union collapsed ten years ago due to floods. Moreover, there is no Centre in Borodal North. So we have five Centres in the upazila. Out of these five Centres, family planning camps are held in four; no camp is held in the Borodal South due to a lack of human resources. There is no Mother and Child Welfare Centre in our upazila. Pregnant women can give birth only in four centres where their deliveries can be attended by skilled birth attendants (S.B.A.). The road transport systems are poor. There is no ambulance service here, as it is a disaster-prone area. Every year, the water level rises 3 to 4 feet, [flooding the] Balijuri Union Health and Family Welfare Center, and medicines get wasted. (Upazila Family Planning Assistant, Tahirpur Upazila, 2019)*

Health Facilities in the Study Areas

Access to health facilities can be a major determinant of fertility and child mortality in areas affected by extreme climate events. Table 20.6 shows the five nearest health facilities visited by women in the study villages, listed from closest to farthest away. A community clinic is the nearest-available health facility. Community clinics actively provide free contraceptives and other incentives to help women of reproductive ages plan their pregnancies. The family welfare assistants from the community clinics visit the households to maintain family planning registration records. After a few weeks of marriage, family welfare assistants usually reach out to newly married women to discuss family planning issues and provide contraceptives and other services. The family welfare assistants also discuss the mother's health and the community clinic's antenatal care. In addition, the family welfare assistants offer free contraceptives to women who want to stop bearing children.

The Health Inspector (HI) and FWA usually reside in the village and live adjacent to the community clinic. Local pharmacies and doctors can offer prescription, over-the-counter medication, general health advice, and assistance if available in a village. Of the other health services listed in Table 20.6 of most interest is the Union Health and Family Welfare Centre (UHFWC), which is in proximity to rural populations and provides essential services such as family planning, menstrual regulation, vaccinations, and in general, reproductive, and maternal health services.

In the cyclone-hit village of Chadnimukha, the closest available health facility is the community clinic; villagers can access their services on foot taking about ten to 20 min. According to the village's family welfare assistants, the community clinic operates regularly and effectively. The next-closest health facility for these villagers is the Union Health and Family Welfare Centre, about a 30 min walk from the village, except during the flood season, when villagers use either a boat or walk, which takes around 50 min. Other health facilities available, listed according to distance from the village, are the Upazila Health Complex (UHC), NGO health centre, and private facilities in Shyamnagar Upazila; these facilities provide primary health care, outpatient care, family planning services, and other preventive health care, which are used less often, particularly when cyclones and their effects increase travel time to as much as 150 min.

The flood-hit village of Lamagaon does not have a Union Health and Family Welfare Centre, but two pharmacies and unqualified doctors in the local market provide the closest available services. The villagers also do not have a community clinic in their village; they can go to Mohjompur, the village next door, a 15-min walk, for the services provided by the community clinic. For essential treatment, the majority of the villagers of Lamagaon go to the nearest sub-district, Kalmakanda, because the health facilities are better. Respondents reported that the Union Health and Family Welfare Centre of Sreepur South, which these villagers had used in the past, was damaged by floods and has been out of action for years.

Table 20.6 Nearest health facilities in villages hit by cyclones (*Chadnimukha*) and floods (*Lamagaon*)

	Five nearest health facilities in this village	Average time to travel (minutes)		Primary modes of transportation	
		Throughout the year (except for disasters)	During disasters	Throughout the year (except for disasters)	During disasters
Cyclones					
1	Community clinic	10	20	Walk	Walk
2	Union Health and Family Welfare Centre[a]	30	50	Walk	Walk/Boat
3	Upazila Health Complex	90	120	Motorbike	Motorbike
4	NGO Health Centre	110	130	Motorbike	Motorbike
5	Private clinic/hospital/doctor	130	150	Motorbike	Motorbike
Floods					
1	Local pharmacy/doctor	10	15	Walk	Boat
2	Community clinic	15	20	Walk	Boat
3	Union Health and Family Welfare Centre[b]	30	60	Walk	Boat
4	Upazila Health Complex	60	120	Motorbike	Boat
5	Kalmakanda[6]	120	240	Motorbike	Boat

Notes [a]Though the community clinic was the closest health facility for the cyclone-hit village dwellers in terms of its distance and time to travel, they typically chose the Union Health and Family Welfare Centre as their nearest hospital-like facility for normal consultations, [b] Like their counterparts from the cyclone-hit village, people from the flood–hit village typically go to other area's Union Health and Family Welfare Centre for consultations

Source Household Survey, 2019–2020

The Health Inspector (HI) and FWA live far from the Lamagaon community clinic, and sometimes, they cannot reach the clinic due to adverse weather conditions. It is also time-consuming for them to travel, either by two-wheeler motorbikes (outside flood season) or by boat (in flood season), to the community clinic from the Tahirpur Upazila. The floodwater comes in May; sometimes, a flash flood happens in early April, and the floodwater levels remain high until October, and are typically at exceedingly dangerous levels in June, July, and August.

A critical difference between the ability of the villagers from the cyclone-hit and the flood-hit villages to access essential health facilities is that, even though Gabura

[6] Kalmakanda is an upazila of Netrokona District in the newly formed Mymensingh Division, Bangladesh.

Union, and, therefore, Chadnimukha, is surrounded by water, the road transportation system across the river is sound. As soon as the villagers cross the Kholpetua river, which typically takes 3 min, and go through Nildumur ghat [*landing place*], their primary means of transport to Shyamnagar Upazila are motorbikes, easy bikes, and buses. In the flood-hit village, the only mode of transportation in winter on deteriorated roads is a two-wheel motorbike. During the flood season, a manual or engine-driven boat has to take a slow, circular route to reach the hospital at Tahirpur or the nearby Kalmakanda sub-district. During the monsoon season, boats are needed to go in any direction. Consequently, villagers living in remote areas rely heavily on untrained medical assistants and unqualified so-called 'doctors'.

Family Planning Facilities and Access

The experience of an extreme climate event and its aftermath can significantly impact the use of contraceptives and, in turn, fertility rates. Table 20.7 presents the different types of contraceptives that were being used by women at the time of the survey in the study areas. In the cyclone-hit village of Chadnimukha 77.6 percent of women were using any contraceptive method compared to 68.8 percent in the flood-hit village of Lamagaon. In the cyclone-hit village, two thirds of respondents used modern methods compared to about half in the flood-hit village. It is interesting to note that not only did more respondents in the flood-hit village rely more on conventional methods than those in the cyclone-hit village, but the use of the oral contraceptive pill among respondents from the cyclone-hit village was nearly twice that of those from the flood-hit village. The oral contraceptive pill was by far the most popular modern approach in both villages, followed by less favoured injectables and female sterilisation. The withdrawal method was one of the most common traditional practices of respondents from both villages, and was notably more widespread in the flood-hit village (10.8%) than in the cyclone-hit village (6.4%). The least popular methods of contraception in both villages were male sterilisation, IUDs, implants, and condoms. Nearly a quarter of respondents in the cyclone-hit village and about a third in the flood-hit village were not using any method of contraception, which can be linked to the older age structure of women in the latter, with many having three or more children.

Respondents to the household questionnaire survey were also asked to identify the most accessible source of modern contraceptives in their area. Table 20.8 shows that the most contraceptives were available in the public sector, in community clinics (CCs), union health and family welfare centres (UHFWCs), Upazila (sub–district) Health Complexes (UHCs) and District Health Complexes (DHCs). The survey found that neither the respondents nor their partners typically depended on non–governmental organisations or other private sources (e.g., grocery shops) to provide modern contraceptives. Ultimately, the preferred sources of modern contraceptives varied according to their type. To access the pill, most participants (90.6% in the cyclone-hit village of Chadnimukha and 77.9 percent in the flood-hit village of Lamagaon) visited the community clinic, a location that is the quickest and easiest for them

Table 20.7 Current method of contraception of married women aged 18–49, in villages hit by cyclones (*Chadnimukha*) and floods (*Lamagaon*)

Methods	*Cyclones* %	*Floods* %
Any method (modern or traditional)	77.6	68.8
Any modern method	65.8	51.9
Modern methods		
Oral contraceptive pill	46.6	25.1
Injectables	9.9	12.6
Condoms	0.6	2.6
Female sterilisation	5.1	6.5
Male sterilisation	1.0	0.9
IUD	/	1.7
Implants	0.6	2.6
LAM	1.9	/
Any traditional method	11.8	16.9
Traditional methods		
Rhythm	3.5	3.5
Withdrawal	6.4	10.8
Others (any method)	1.9	2.6
No method (includes pregnant women)	22.4	32.2

Notes If more than one method was used, only the most effective method was considered in this tabulation; IUD denotes "Intrauterine device"; LAM stands for "Lactational amenorrhea method"
Source Household Survey, 2019–2020

to reach (even in the flood-hit village). However, more respondents in the flood-hit village relied on the private sector, using, for example, a local pharmacy to access pills (20.7%) and injectables (43.8%). This is likely because it is difficult for women in the flood-hit village to travel to the community clinic during the flood season.

Respondents also mentioned that the family welfare assistants could not often visit them due to extreme weather conditions and heavy flooding. Community clinics and union health and family welfare centres were the most accessible source for injectables in the cyclone-hit village. Many respondents (17.4%) in the cyclone-hit village reported having visited an NGO health centre to access injectables, while none did so in the flood-hit village. Community clinics were the common source for condoms in both areas. However, a considerable number of women (12.4% in the cyclone-hit village and 18.8% in the flood-hit village) reported that they purchased condoms in grocery shops, likely because these women or their partners depended on condoms during or immediately after extreme events, when public health infrastructure was damaged, or it was difficult to access other familiar sources. Women visited the Upazila Health Complexes for other modern methods such as male or female sterilisation, IUDs, and implants.

Table 20.8 Current use of modern contraceptive methods among women respondents aged 18–49 by source of a method in villages hit by cyclones (*Chadnimukha*) and floods (*Lamagaon*)

	Oral pill		Injectable		Condoms		Female sterilisation		Male sterilisation		IUDs		Implants	
	Cyclones	*Floods*	*Cyclones*	*Floods*	*Cyclones*	*Floods*	*Cyclones*	*Floods*	*Cyclones*	*Floods*	*Cyclones*	*Floods*	*Cyclones*	*Floods*
Public sector	**97.3**	**79.3**	**81.5**	**56.2**	**76.4**	**75.0**	**100.0**	**100.0**	**100.0**	**100.0**		**100.0**	**100.0**	**100.0**
Community clinic	90.6	77.9	43.5	43.8	69.7	75.0		5.6						
Union Health and Family Welfare Centre	6.7	1.4	30.4	1.5	6.7									25
Upazila (sub–district) Health Complex			7.6	10.9			100.0	94.4	100.0	100.0		100.0	100.0	75
District hospital														
Private medical sector	**2.7**	**20.7**	**1.1**	**43.8**	**10.1**	**6.2 asis>**								
Private clinic/hospital/doctor			1.1											
Local pharmacy	2.7	20.7		43.8	10.1	6.2								
NGO sector			**17.4**		**1.1**									
NGO health centre			17.4		1.1									
Other private sources					**12.4**	**18.8**								
Grocery shops					12.4	18.8								
Total	100.0	100.0	100.0	100.0	100.0	100.0	100.0	100.0	100.0	100.0		100.0	100.0	100.0
Number of women	255	145	92	64	89	16	17	36	3	2		7	1	8

Note Women or their husbands may use more than one method
Source Household Survey, 2019–2020

Discussion

The study found that fertility and child mortality were higher and contraceptive use lower in the village hit by floods than in the one hit by cyclones. Findings showed that disaster risk reduction and family planning programs were not actively functioning and collaborating with non-governmental organisations in the village hit by floods in northeast Bangladesh. The family planning facilities were poor, in respect to resources and practice, with very few non-governmental organisations focused on women's reproductive health and child health operating in that village.

Findings of this research suggest that active participation of government, non-government and private sectors is imperative to enable changes in fertility, child mortality, and contraceptive use. Findings showed that government (e.g., community clinics), non-government (e.g., World Vision Bangladesh), and private organisations actively focus on women's reproductive health and child malnutrition in the cyclone-hit area, but less so in the flood-hit region. Active participation of both public and private sectors is imperative in achieving health coverage and ensuring the sustainable development of community health.

The Government of Bangladesh has targeted regional disparities in fertility, child mortality, and contraceptive prevalence rates in order to reduce them across the country. This requires changes to be made to relevant policies, or indeed, a new raft of policies needs to be developed. It is crucial to ensure that any emerging programs are flexible and can function actively in resource-constrained areas, particularly those severely impacted by extreme climate events.

Different districts of Bangladesh are uniquely and unevenly vulnerable to climatic disasters, including cyclones, floods, droughts, and other temperature-related events (Ahmed and Tan 2021; Barua et al. 2016). These findings support the notion that the uneven and variable rates in fertility, child mortality, and contraceptive use across these districts are the result in part to the different climate disasters each experience, which have typically been largely attributed to demographic, economic, social, and cultural determinants (Islam et al. 2003, 2010; Alam et al. 2018). To address the regional disparities in these rates in Bangladesh and lower rates overall, existing population policy and family planning programs would benefit substantially from strong collaboration between disaster risk reduction practices and family planning programs. These programs should focus on improving health infrastructure, designated health support areas and storage facilities in shelters, increasing public awareness about climate risks and an understanding of likely household vulnerabilities to make informed family planning decisions.

Community clinics offer mother and newborn health care services, reproductive health and family planning, the Expanded Program on Immunisation (EPI), nutrition knowledge and micronutrient supplementation, and health care. These clinics are operated using a participatory model, and are easily accessible to all residents. Therefore, these clinics should be at the front line of the government's health initiatives, particularly those addressing fertility, child mortality, and contraceptive use. This is strongly recommended for the region hit by floods in this study, which is currently underserved by government community health services. Moreover, non-governmental organisations focused on the health and well-being of populations should facilitate their interventions to target deprived people who are also subject to extreme climate events and make special efforts to reach those living in remote and flood-hit areas.

This research suggests that multisectoral engagement of disaster management and family planning programs is critical to changing fertility, child mortality, and contraceptive use rates. Climate change adaptation or disaster risk reduction programs cannot enhance adaptive capacity or resilience to extreme climate events alone. Enhancing an individual's or community's adaptive capacity depends on social and human capital. Developing social and human capital has not been the key focus of either climate change adaptation or disaster risk reduction practices. Similarly, any changes in fertility, child mortality, and contraceptive use depend on an understanding of determinants such as education, age at marriage, labour force participation, and wealth status, that need to be considered in conjunction with locational disadvantages and exposure to extreme climate events. Therefore, in addressing regional variations in those rates there is a need to engage a multisector approach within the public sector and between the public and private sectors.

Community clinics should be at the forefront of the government's health development efforts in Bangladesh. One of Bangladesh's groundbreaking health initiatives in the 1970s was the establishment of a community-based healthcare program (Schuler et al. 1995). Since then, the program has expanded in scope, with an increasing number of skilled community health workers performing a critical role in improving the general health and socio-economic status of the Bangladeshi people. Community clinics are now the government's primary approach to achieving universal health coverage (UHC) and are essential in the current battle against COVID-19. The community clinic offers services such as mother and newborn health care, Integrated Management of Childhood Illness (IMCI), reproductive health and family planning, the Expanded Program on Immunisation (EPI), nutrition knowledge and micronutrient supplementation, and health care. Education and counselling, screening for chronic noncommunicable diseases, treatment of minor illnesses, general ailments, and first aid and other amenities are all available at community clinics. In Bangladesh, a community clinic is a unique example of public–private partnerships in providing primary health care that can play an important role in the sustainable development of community health through providing quality health care (Riaz et al. 2020). Since community clinics are run through community participation and are easily accessible to the population, they should remain at the forefront of the government's health improvement efforts in Bangladesh.

Media plays a significant role in the disaster risk reduction process: informing, educating, and empowering communities with relevant knowledge to influence public action, disaster preparedness, and mitigation policy (Shaw et al. 2021). Media coverage heavily influences disaster relief, and the extent of media coverage depends on the type of climate extremes. Before a disaster, the media focus primarily on the importance of preparedness in the event of disasters, which helps practitioners and vulnerable communities understand what and when appropriate action should be taken to tackle future disasters. At this point, the media explicitly analyses the sources and type of risk, disseminates information to the public, broadcasts early warning and scientific forecasts of the hazards, and promotes precautionary measures (Hong et al. 2019). In times of disaster, the media focus on protecting life and property as much as possible and simply avoiding the risk of secondary disasters. In doing so, the media provides the public with accurate information from time to time, about what to do, and what actions are being taken by authorities and donors (Norris et al. 2008). In the aftermath of a disaster, integrating disaster risk reduction into the rehabilitation and reconstruction process is integral to reducing future risks and minimising potential loss of life and damage to properties. After a disaster, the media focus on rehabilitating and rebuilding damaged infrastructure, livelihood activities, the environment, and economic and social systems. The media has a unique role in attracting the attention of various national and international donors and organisations to help disaster-affected people with relief, food, medicine, cash, and house-building materials (Garai 2018).

Media coverage is commonly high for cyclones, even when associated with minor loss and damage. In Bangladesh, cyclones get far more attention, and affected populations receive far more aid than if an area has experienced floods. The discrepancy is even more significant for more minor floods, which can account for considerable crop loss and damage but do not make the headlines and hence receive little attention from donor agencies. Flood occurrence tends to be regular and has long-term impacts on the community's welfare, but is largely ignored by the media. Bangladesh's mainstream media ignores local disasters like floods—usually more minor in loss and damage than significant and more dramatic disasters like tropical cyclones. Reporting local disasters by local journalists is important to gain national media attention and the notice of central authorities, increasing pressure on them to implement measures to reduce loss and damage (Kakuchi and Shaw 2021).

Conclusion

This study provides an understanding of the importance of programs in dealing with disaster risk reduction and family planning in making changes to rates of fertility, child mortality, and contraceptive use in Bangladesh. This research examined the effects of these two programs, and found that these programs had the potential to change fertility, child mortality, and contraceptive availability in the area hit by cyclones in the southwest more than in the area hit by floods in the northeast. Given the undeniable effects of extreme climate events on fertility, child mortality, and contraceptive access, it can be argued that these effects must be incorporated into disaster risk reduction and population policies. Strengthening the synergy and coordination of service delivery between disaster management and family planning programs is important to address the regional disparities in child mortality, fertility, and contraceptive prevalence rates.

Expanding the disaster risk reduction and family planning coverage to hard-to-reach settings should be a priority and require policies to strengthen partnerships with other government entities, non-governmental organisations and the private sector. Areas hardest hit by extreme climate events are mostly hard-to-reach settings. The flood-hit village was far away from the centre of administrative offices and primary health facilities, consequently the population depended on private sources to access ealth facilities. It is also challenging for family planning workers to reach those women residing in hard-to-reach settings and provide essential resources to reduce fertility and child mortality. Priority should be given to expanding family planning coverage and increasing disaster risk reduction activities in hard-to-reach areas in Bangladesh. Very few national and international non-governmental organisations, which focus on disaster risk reduction, are present in the flood-hit village, and simultaneously, the local level institution plays a minor role due to a lack of resources and media coverage. Policies can be developed to enhance collaborations and strengthen partnerships between disaster risk reduction and family planning programs, as well as with other government entities, non-governmental organisations and private sectors.

Disaster risk reduction deals with a range of activities—such as building resilient infrastructure, providing early warning systems, taking early action, increasing employment, reducing poverty, ensuring food security, and sustainable agriculture—that can significantly increase the socio-economic development of communities, increase their adaptive capacities and resilience to climate change and extreme weather or climate events. This development, together with ensuring the functioning of effective and properly resourced family planning programs, can help reduce fertility and child mortality and increase contraceptive access and use, particularly in crises as a result of recurring extreme climate events.

Conflict of Interest The authors declare no conflict of interest.

References

Ahmad H (2019) Bangladesh coastal zone management status and future trends. J Coast Zone Manag 22(1):1–7

Ahmed KJ, Haq SMA, Bartiaux F (2019) The nexus between extreme weather events, sexual violence, and early marriage: a study of vulnerable populations in Bangladesh. Popul Environ 40(3):303–324

Ahmed KJ, Tan Y (2021) Assessing and mapping spatial variations in climate change and climatic hazards in Bangladesh. In: Alam GMM, Erdiaw-Kwasie MO, Nagy GJ, Filho WL (eds) Climate vulnerability and resilience in the Global South. Springer, Cham, pp 465–486

Ahmed KJ, Tan Y (2022) The disaster-fertility nexus in Sub-Saharan African countries. In: Odimegwu C, Adewoyin Y (eds), Handbook of African demography. Routledge, United Kingdom, pp 363–379

Alam MA, Chamroonsawasdi K, Chansatitporn N, Munsawaengsub C, Islam MS (2018) Regional variations of fertility control behavior among rural reproductive women in Bangladesh: a hierarchical analysis. Behav Sci 8(8)

Barua U, Akhter MS, Ansary MA (2016) District-wise multi-hazard zoning of Bangladesh. Nat Hazard 82(3):1895–1918

Bhowmik J, Irfanullah HM, Selim SA (2021) Empirical evidence from Bangladesh of assessing climate hazard–related loss and damage and state of adaptive capacity to address them. Clim Risk Manag 31:100273

Bora JK, Saikia N, Kebede EB, Lutz W (2022) Revisiting the causes of fertility decline in Bangladesh: the relative importance of female education and family planning programs. Asian Popul Stud 1–24

Cain M (1981) Risk and insurance: perspectives on fertility and agrarian change in India and Bangladesh. Popul Dev Rev 7(3):435–474

Cain M (1983) Fertility as an adjustment to risk. Popul Dev Rev 9(4):688–702

Cain M (1986) Risk and fertility: a reply to Robinson. Popul Stud 40(2):299–304

Eckstein D, Künzel V, Schäfer L (2021) Global climate risk index 2021: who suffers most from extreme weather events? Weather–related loss events in 2019 and 2000 to 2019. Germanwatch, Berlin. https://germanwatch.org/en/19777

Finlay JE (2009) Fertility response to natural disasters: the case of three high mortality earthquakes. World Bank Policy Research Working Paper No. 4883. https://doi.org/10.1596/1813-9450-4883

Fischer EM, Sippel S, Knutti R (2021) Increasing probability of record–shattering climate extreme climates. Nat Clim Chang 11(8):689–695

Frankenberg E, Laurito M, Thomas D (2015) Demographic impact of disasters. In: Wright JD (ed) International encyclopedia of the social and behavioral sciences. Elsevier, pp 101–108

Garai J (2014) The impacts of climate change on the livelihoods of coastal people in Bangladesh: a sociological study. In: Filho WL, Alves F, Azeiteiro UM, Caeiro S (eds) International perspectives on climate change: Latin America and beyond. Springer, pp 151–163

Garai J (2018) Mass media and climate change induced disaster risk reduction and mitigation in coastal area of Bangladesh: a sociological study. In: Filho WL, Manolas E, Azul A, Azeiteiro U, McGhie H (eds) Handbook of climate change communication. Springer, Cham, pp 273–292

GoB (2020) National plan for disaster management (2021–2025): action for disaster risk management towards resilient nation. Ministry of Disaster Management and Relief, Government of Bangladesh (GoB). https://modmr.portal.gov.bd/sites/default/files/files/modmr.portal.gov.bd/page/a7c2be1_6c9d_4ecf_bb53_ec74653e6d05/NPDM2021-25%20DraftVer5_23032020.pdf

Gruebner O, Khan MMH, Burkart K, Lautenbach S, Lakes T, Krämer A, Subramanian SV, Galea S (2017) Spatial variations and determinants of infant and under-five mortality in Bangladesh. Health Place 47:156–164

Haq SA, Ahmed KJ (2019) Is fertility preference related to perception of the risk of child mortality, changes in landholding, and type of family? A comparative study on populations vulnerable and not vulnerable to extreme weather events in Bangladesh. Popul Rev 58(2):61–99

Hasan MR, Nasreen M, Chowdhury MA (2019) Gender-inclusive disaster management policy in Bangladesh: a content analysis of national and international regulatory frameworks. Int J Disaster Risk Reduc 41(101324)

Hong YX, Kim JS, Xiong LH (2019) Media exposure and individuals' emergency preparedness behaviors for coping with natural and human-made disasters. J Environ Psychol 63:82–91

Hossain MAR, Ahmed M, Ojea E, Fernandes JA (2018) Impacts and responses to environmental change in coastal livelihoods of south–west Bangladesh. Sci Total Environ 637–638:954–970

Huq S, Chow J, Fenton A, Stott C, Taub J, Wright H (2019). Confronting climate change in Bangladesh: policy strategies for adaptation and resilience. Springer

Islam M, Rob U, Chakroborty N (2003) Regional variations in fertility in Bangladesh. Genus 59(3/4):103–145

Islam MK, Haque MR, Hema PS (2020) Regional variations of contraceptive use in Bangladesh: a disaggregate analysis by place of residence. PLOS One 15(3): e0230143

Islam S, Islam MA, Padmadas SS (2010) High fertility regions in Bangladesh: a marriage cohort analysis. J Biosoc Sci 42(6):705–719

Jakariya M, Ahmed MF, Sikder MT (2016) Vulnerability analysis of the cyclone Aila affected community in the southwest belt of Bangladesh. J Health Environ Res 2(2):5–12
Kakuchi S, Shaw R (2021) Importance of reporting local disasters: lessons from Varanasi, India. In: Shaw R, Kakuchi S, Yamaji M (eds) Media and disaster risk reduction. Springer, Singapore, pp 179–187
Khan MR, Huq S, Risha AN, Alam SS (2021) High-density population and displacement in Bangladesh. Science 372(6548):1290–1293
Khan MA, Khan N, Rahman O, Mustagir G, Hossain K, Islam R, Khan HT (2021) Trends and projections of under-5 mortality in Bangladesh including the effects of maternal high-risk fertility behaviours and use of healthcare services. PLOS One 16(2):e0246210
Mallick B, Rahaman KR, Vogt J (2011) Coastal livelihood and physical infrastructure in Bangladesh after cyclone Aila. Mitig Adapt Strat Glob Change 16(6):629–648
Martin M, Billah M, Siddiqui T, Abrar C, Black R, Kniveton D (2014) Climate–related migration in rural Bangladesh: a behavioural model. Popul Environ 36(1):85–110
Miles MB, Huberman AM (1994) Qualitative data analysis: an expanded sourcebook, 2nd edn. Sage
National Institute of Population Research and Training (NIPORT), International Centre for Diarrhoeal Disease Research, Bangladesh (ICDDRB), MEASURE Evaluation (2019) Bangladesh district level socio–demographic and health care utilisation indicators. Dhaka, Bangladesh, and Chapel Hill, NC, USA: NIPORT, ICDDRB, and MEASURE Evaluation
Norris FH, Stevens SP, Pfefferbaum B, Wyche KF, Pfefferbaum RL (2008) Community resilience as a metaphor, theory, set of capacities, and strategy for disaster readiness. Am J Community Psychol 41(1–2):127–150
Paul SK, Naim J, Mou MA (2022) Assessing efficiency of disaster management institutions at local level in the coastal region of Bangladesh. Nat-Based Solut 2:100024
Raihan F, Hossain MM (2021) Livelihood vulnerability assessments and adaptation strategies to climate change: a case study in Tanguar haor, Sylhet. J Water Clim Chang 12(7):3448–3463
Riaz BK, Ali L, Ahmad SA, Islam MZ, Ahmed KR, Hossain S (2020) Community clinics in Bangladesh: a unique example of public-private partnership. Heliyon 6(5):e03950
Roy S, Hossain SMI (2017) Fertility differential of women in Bangladesh demographic and health survey 2014. Fertil Res and Pract 3(1):1–5
Schuler SR, Hashemi SM, Jenkins AH (1995) Bangladesh's family planning success story: a gender perspective. Int Fam Plan Perspect 21(132)
Shaw R, Kakuchi S, Yamaji M (2021) Media and disaster risk reduction. In Shaw R, Kakuchi S, Yamaji M (eds) Media and disaster risk reduction. Springer, Singapore, pp 3–12
Streatfield PK, Kamal N (2013) Population and family planning in Bangladesh. The J Pakistan Med Ass 63(4 Suppl 3):S73–S81
Tajrin S, Hossain B (2017) The socio–sconomic impact due to cyclone Aila in the coastal zone of Bangladesh. Int J Law Humanit Soc Sci 1:60–67
Tuckett AG (2005) Applying thematic analysis theory to practice: a researcher's experience. Contemp Nurse 19(1–2):75–87

Dr.Khandaker Jafor Ahmed is a Postdoctoral Fellow in the Walsh School of Foreign Service at Georgetown University, Washington DC. In this position, he is working on a project focused on the population and environment in Bangladesh, with a multidisciplinary team of demographers and environmental scientists who have been conducting research in Bangladesh for more than a decade. He received his doctoral degree in Environmental Studies from the Department of Geography, Environment and Population at the University of Adelaide (UoA), Australia. His research interest falls within the broader field of population and environment, with a special focus on human fertility. Given his academic qualification (Bachelor's and Master's degrees in Sociology, and PhD in Environmental Studies) and research experience (as a Research Associate in the Centre for Climate Change and Environmental Research at BRAC University, Bangladesh), his research seeks to explain the impact of climate change and extreme climate events on population dynamics (mortality, migration, and fertility).

Dr. Yan Tan is an Associate Professor at the Department of Geography, Environment and Population, The University of Adelaide. Yan's research lies at the interface of population and environmental studies, focusing on mobility (migration, displacement, resettlement) and on evolving high-end demographic measures and econometric models for climate change–migration–development nexus. Her most recent projects explore the interplay of environmental, demo-graphic, socio-economic and institutional factors in climate (environmental) change, popula-tion mobility, and adaptation. Her work has focused on China, Australia, and some other parts of the Asia-Pacific region.

Dr. Dianne Rudd is an Honorary Research Fellow in the Department of Geography, Environment and Popula-tion. She has extensive expertise in the areas of Geography and Demography, with a specific focus on population ageing and health; future population growth and international migration to Australia and within the southeast Asia region; gender, family change and fertility, envi-ronmental change and sustainable communities; climate change impacts on populations. She has extensive experience in quantitative and qualitative research methods, survey and inter-view techniques, analysis of census data to establish population trends and social change, fu-ture projected populations taking account of fertility, mortality and migration. She has supervised more than 50 Ph.D. students that have successfully completed and is currently supervis-ing 6 Ph.D. students with projects on population and a range of related environmental issues in countries outside of Australia and also with local indigenous populations in Australia. During a long teaching career, she has taught courses: Geography, Population and Environment, Pop-ulation and Health, Global International Migration, Social Science Techniques, Applied Popu-lation Analysis, Research Methods, and Globalisation.

Chapter 21
Climate Change and Humanitarian Responses: A Proposal of Education for Health Hazards Preparedness

João Casqueira Cardoso, Mahfuza Parveen, Mafalda S. A. Fonseca, and Laurine Grzelak

Abstract This contribution consists in an analysis of the recent developments and proposals for changes in humanitarian practices, in order to improve the responses, and in particular education and awareness-raising, to health hazards linked to climate change. The first part explores the existing global context, and the growing need for more sensitization and education to health hazards related to climate change among humanitarian workers and the populations in general. The position of the United Nations and of other specialized international organizations is explained, as well as the proposals recently made by the major humanitarian NGO networks, such as the ALNAP and the Sphere Project. The second part presents examples of health hazards linked to climate change, and the responses given in the field. This way, this intent is to compare the "by the book" recommendations, and the reality in the field. This section of text uses the methodology of the case study analysis, with a focus on climate change and health hazards situations in Bangladesh and in France (both in French Guiana and in metropolitan France). The cases studied, including a rather complete list of climate change caused problems (heatwaves, drought, cyclones and floods) and their consequences on health hazards. Such situations are contrasted and analysed, in order to underline the role that specific preventive and preparedness measures can have to improve the response to health hazards. The conclusions underline the

---

J. C. Cardoso (✉)
Department of Political and Humanitarian Studies, University Fernando Pessoa/CEPESE, Porto, Portugal
e-mail: jcasq@ufp.edu.pt

M. Parveen
Department of Environmental Science and Disaster Management, Daffodil International University, Dhaka, Bangladesh
e-mail: mahfuza.esdm@diu.edu.bd

M. S. A. Fonseca · L. Grzelak
Master in Humanitarian Action, Cooperation and Development, University Fernando Pessoa, Porto, Portugal
e-mail: 42160@ufp.edu.pt

L. Grzelak
e-mail: 42177@ufp.edu.pt

W. Leal Filho et al. (eds.), *Climate Change and Health Hazards*, Climate Change Management, https://doi.org/10.1007/978-3-031-26592-1_21

relative lack of wide educational or awareness raising measures, and the need to dedicate more attention to this area, for both professionals and the population in general.

Wherever you go,
Whether you are consumed by a forest fire,
I remain high as the ashes
Poem by Amirul Arham

Introduction

Climate change affects mostly health, leading to several problems including injuries and deaths as a result of extreme events, but also infectious diseases, and food and water insecurity. Global warming, in particular, has many negative effects on human health, such as the inability of the human body to regulate its temperature, provoking heat exhaustion and heat stroke, and even intensifying some diseases like heart and lung diseases. The emission of toxic pollutants is dangerous for human health and can cause asthma, infections such as pneumonia and cancer, leading to about seven million deaths every year. Extreme weather events such as floods, droughts and wildfires can destroy infrastructures and lead to the inability of the population to access health care facilities, causing deaths and are also affecting, in turn, water security, as people are unable to access clean water for drinking, cooking, and hygiene. In addition, the increasing temperatures and water temperatures are causing a spread of infectious diseases such as Lyme diseases, which affects the skin, heart, and brain of the population, as well as Vibrio which can cause gastrointestinal illness and infections (Ebi et al. 2017).

In sum, climate change is the cause of multiple variants of humanitarian crisis, stating with the health dimension. The United Nations High Commission for Refugees reminds that climate changes are destroying livelihoods, driving displacement of millions of people every year, and increasing conflicts, and that the adverse effects of climate change contribute to the vulnerability and pose increased risk to the human rights of people affected (UNHCR 2022). Furthermore, climate changes as a factor of humanitarian crisis are exacerbating current inequalities, affecting those who are already most vulnerable and usually with less capabilities (Sen 1993). Taking the multiple challenges put on populations, health structures, and social systems by climate changes, this contribution looks, first, at the recommendations of international agencies as concerns education or awareness good practices for preparedness to health hazards linked to climate change. The second part of the article illustrates the specific case studies Bangladesh and France, in order to look at the reality in the field, as the field reality may be different from what international institutions actually advocate.

The Vision of International Agencies

For international agencies linked to humanitarian action, the implementation of awareness and education among humanitarian workers and the population in general is seen as increasingly necessary. The implementation of awareness can bring numerous positive results such as: reducing people's vulnerability and exposure to climate-related hazards, protecting livelihoods by safeguarding the natural resources of the population, improving communities' health and safety by reducing pollution and waste. It can also promote good practices among the communities, such as the fight against deforestation, desertification, wildfire or other forms of pollution. As such, it also educates the local communities in order to be more resilient, biodiversity sensitive; and it also enhances food security, as well as social and economic development (EHA 2019). It is also crucial that the population are informed and educated about the topic, in order to be aware of how to be less exposed to climate-related hazards. Indeed the example of some countries, like Japan, is a paradigmatic one on this question (Pulhin et al. 2021), but other countries are also developing good examples (*cf.* our case studies). The challenge ahead seems to specify the good practices, taking into account the variety of health hazards; and to generalize and diffuse the best practices among the potentially affected populations.

The World Health Organization (WHO) has been developing a programme on climate change and health, taking on board the impacts of climate change on vulnerable groups, and advocating the implementation of protective measures. In the conference on Health and Climate (WHO 2021), it underlined the importance of reducing the impacts of climate change, meeting the health needs of the population, their resilience and underlining the need to strengthen the health system. The organization has shared some possible interventions, namely: defining adaptation plan based on assessed population vulnerabilities, building response plans in case of new disease outbreak, together with emergency response plans, evacuation procedures and emergency management coordination measures to anticipate and respond to events affecting public health. It also recommends the development and implementation of national all-hazards health emergency and disaster risk management programs, based on risk and capacity assessments (which account for shorthand long-term climate trends). The WHO promotes the integration of climate risks and management measures in the health curriculum. Lastly, it recommends the implementation of a communication strategy to raise awareness about climate change and health (WHO 2021).

Another United Nations agency has been decisive in the initiatives to promote good practices. It is the Food and Agriculture Organization (FAO), which shares several publications about the impacts of climate change on various food safety hazards. The organization highlights the importance of rise awareness for sustainable agriculture reducing emissions. The focus of its studies is on the important to help farmers to avoid health hazards, and to build climate-resilient systems at national level. To address climate risks, the FAO is "developing climate change impact and vulnerability assessments for crops, livestock, fisheries and forestry" (FAO 2017: 10).

More specifically, it promotes an "approach to strengthening resilience and reducing poverty in rural areas through risk-informed and shock-responsive social protection systems" (FAO 2017: 10). Moreover, it recommends the development of disaster risk reduction management capabilities, early warning systems and rapid reaction mechanisms to support the population to adapt and mitigate climate change (FAO 2017).

As regards private international bodies, it is worth mentioning the Sphere network. The Sphere standards are considered a reference tool for international NGOs, volunteers, UN agencies and others. Considering the importance of consequences that climate change have on health hazards, the Sphere movement has taken an important position concerning the quality of the humanitarian response as concerns sustainable development and the mitigation of the impact of climate change. The role of humanitarian workers is not forgotten, as they should also focus their work on reducing the environmental impact of humanitarian activities, otherwise they may also contribute to health hazards for vulnerable populations. On the Sphere thematic sheet "Reducing environmental impact in humanitarian response" based on the 2018 Sphere Handbook, various visions about the positive or negative impacts of the humanitarian response on the environment are underlined. In fact, the degradation of the environment has a really negative impact on the resilience and the capabilities of the population, affecting their wellbeing by reducing their health capacity or their access to education for example. The importance of considering and integrating the environment in every humanitarian response, is seen as necessary in order to increase sustainable development, but also in order to create a solid foundation for conflict resolution and peace-building. But, on top of that, it is also important for the Sphere movement to understand why it is so essential to promote education and awareness on health hazards caused by climate change (Sphere 2019).

For Sphere, the most important is to include the environmental issues at an early stage of a crisis response. Indeed, this methodology will have a sustainable and positive impact for the rest of the response. For example, it will reduce the risk of recurrence, safeguarding the natural resource, and by this way protected livelihood, their health and safety. It will also reduce the risk of local conflict for resources. Through mitigation activities, including education and awareness-raising, it will help to prevent the population and the environment from future health hazard, and contribute to the "[s]lowing or reversing trends that lead to deforestation, desertification and pollution and thereby supporting community resilience, biodiversity, food security and economic development" (Sphere 2019). This early-stage incorporation of the environment in the management of humanitarian actions is also a key to increase the quality and accountability of humanitarian action.

Concluding on the private international actors, it is important to mention ALNAP (Active Learning Network for Accountability and Performance). ALNAP is a "global network of NGOs, UN agencies, members of the Red Cross/ Crescent Movement, donors, networks, academics and consultants dedicated to learning how to improve response to humanitarian crises" (ALNAP s.d.). ALNAP is aware of the fact that humanitarian actors must be well prepared to deal with the effects of hazards related to climate change. It considers that humanitarian workers and, more generally, humanitarian actors, ought to improve capabilities of communities to plan and implement a full range of resilience measures, including DRRR (Disaster Risk Reduction and Response), to ensure the quickest response possible. It also considers that it is necessary to develop more knowledge among the humanitarian workers as regards the complexity of climate change impact. Most interestingly, ALNAP recommends as an important method that humanitarian actors "co-design climate change programmes with vulnerable people and groups, ensuring that they understand and can discuss climate change, its effects and the potential impact on their lives and livelihoods" (De Geoffroy et al. 2021: 33). This aspect is also completed by a recommendation of focusing "support to communities and socio-economic groups that are particularly exposed" (De Geoffroy et al. 2021: 51), giving the example of one of the recurrent climate change issue, namely heatwaves. According to ALNAP, partnerships between humanitarian organizations and governments could also help to set up social protection to anticipate the impact of extreme climate phenomena, This can be completed by partnerships between the humanitarian workers and climate-related actors (e.g. Civil Protection), using "joint analysis and common standards" (De Geoffroy et al. 2021: 32).

Case Study on Bangladesh

Bangladesh, a South Asian country located beside the Bay of Bengal with a coastline of 580 km. The country is one of the utmost susceptible countries in the world to climate change due to its unique flat and low-lying topography and geographical location. The population density, extreme poverty, lack of institutional proficiency and the livelihoods dependence of climate sensitive sectors, such as agriculture and fisheries, make Bangladesh as one of the most vulnerable countries in the world to climate change (Huq 2001; Huq et al. 2004; Huq and Ayers 2007). The sixth report of The Intergovernmental Panel on Climate Change (IPCC)[1] stated that "the low elevation, high population and weak infrastructure make Bangladesh highly vulnerable to climate change". According to the Global Climate Risk Index 2021 Bangladesh ranked 7th most vulnerable country among the 180 countries in the world, and from 2000 to 2019 the country faced 185 extreme weather events due to climate change. Increasing temperatures, rising sea levels, floods, drought and cyclones are the important climate change impacts suffered by Bangladesh.

The changing climate has not only an impact on the environment but also impact on the human life. All the climatic disasters degrade the human activities and ultimately forced populations to migrate from one place to another place. Such migrations tend to reduce the quality of life, especially in rural areas due to the lack of social support,

[1] The Intergovernmental Panel on Climate Change (IPCC) was created in 1988 by the United Nations, and is a clustering of 195 States. It works on the causes and consequences of climate change by assessing scientists, technical, economic, and social knowledge.

standard household shelters, standard employment, poor healthcare systems, mental problems derived from the uncertainty about the future (Naheed and Hort 2015). To overcome these situations, three (humanitarian) approaches have been traditionally used, namely: resilience, disaster risk reduction, and early warning systems (Marin and Naess 2017). They can be useful to define the targets for the adaptation to climate change. Yet, to analyse the adaptation to climate change from the standpoint of health hazards implies a dynamic approach, due to the complexity of climate related problems existing in Bangladesh. In particular, it requires to identify the different types of disasters and the different types of problems they create, both in nature and in human activities.

Cyclones

Cyclones are tropical storms, which formed over the South Pacific and Indian Ocean. In the last 2–3 decades the frequency and intensity of the tropical cyclone/hurricane/tornedo has been increased worldwide due to the global climate change. Increasing sea surface temperature is one of the most favourable criteria for the genesis of cyclone along with other climatic criteria (Nasher et al. 2022). The intensity and frequency of tropical cyclone has been increased due to the geographical location and funnel-shaped coastline of Bangladesh. Every year about 5% of global tropical cyclones are formed over the Bay of Bengal, and the cyclones caused 53% of the world death and 80% of the casualties (Rahman et al. 2022a, b, c; Fakhruddin et al. 2022).

In Bangladesh, most affected and high risk zones by tropical cyclone associated phenomena (wind, storm surge) are Barguna, Patuakhali and Bhola districts (southern part), Cox's Bazar and Chottagram districts (the eastern part), and Satkhira, Khulna and Bagerhat districts (south western part) (Hoque et al. 2018) (Fig. 21.1A). Since the independence of Bangladesh (1971), a number of cyclones have hit all over the country. For example, in November 1970 (Bhola Cyclone), in November 1971, in November 1974, in May 1975, in May 1977, in November 1983, in May 1984, in November 1988, in December 1990, in April 1991, in May 1997, in September 1997, in May 1998, in November 2007 (cyclone Sidr), in October 2008 (cyclone Rashmi), in May 2009 (cyclone Aila), in May 2013 (cyclone Mahasen), in July 2015 (cyclone Komen), in May 2016 (cyclone Roanu), in May 2017 (Cyclone Mora), in May 2019 (cyclone Fani), in November 2019 (Cyclones Matmo and Bulbul), in May 2020 (cyclone Amphan), and lastly the cyclone Yaas in 2021.

The coastal areas of Bangladesh suffer as a consequence of strong wind, heavy rain, large storm surges near landfall. Agricultural land and livestocks are damaged due to inundation, salt-water intrusion, among other effects. Due to these extreme events the human settlements face severe casualties, leading to migration movements after large portions of the local populations lose their shelters, agricultural lands, livestocks and also due to the shortage of fresh water. For example, large proportions of the population affected changed their occupation and moved other cities after the

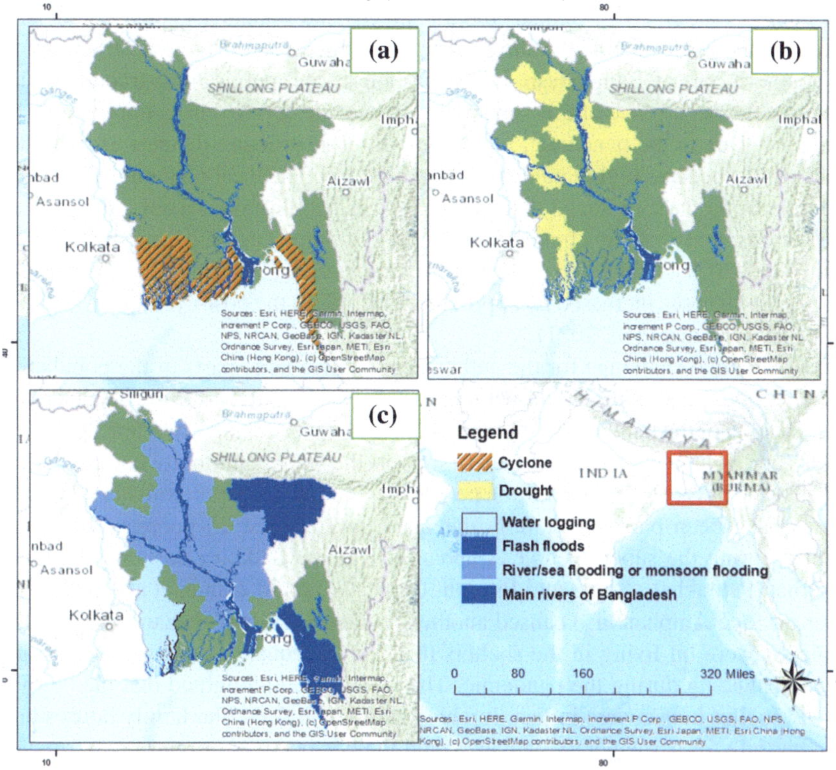

Fig. 21.1 Map of disaster prone areas in Bangladesh. Cyclone (A), drought (B) and flood (C)

cyclone Sidr (2007) and the cyclone Aila (2009). Approximately 350,000 people were displaced and not able to return their previous home, according to the United Nations (UN) Office for the Coordination of Humanitarian Affairs report (Rahman et al. 2022a, b, c). The cyclone Aila affected people obliged to migrate to the nearby places or other big cities because of their loss of houses and physical properties, and damage of agricultural lands due to salinity. As a result, what was observed was an increase in the unemployment problem, an increase in the financial and life insecurity, and most specifically health impacts due to the lack of proper medication, lack of transport access to the local public services, and scarcity of drinkable water (Islam and Hasan 2016). The educational system is also damaged along the affected area after the strike of each cyclone. A study revealed that 90% of the schools were severely impaired due to Cyclone Aila in Shyamnagar Upazila (sub-district), Satkhira District (Parvin et al. 2022).

A case study was conducted on the effect of cyclone displacement by the Department of Environmental Science and Disaster Management, at Daffodil International University (Bangladesh), at the end of 2020, just after the attack of the cyclone Amphan. In reality a "super cyclone", Amphan made landfall on 20th May 2020 on Jammu Island, in West Bengal (India), and crossed Bangladesh on the same day with a wind speed of 60–90 kph. The super cyclone caused huge destructions to the southern part of Bangladesh, including houses and public infrastructures. The effects of the occurrence of this super cyclone became severe during the COVID-19 pandemic. Due to the pandemic, rather than the infrastructure damage as such, the health sector became highly vulnerable and unstable. The local community suffered a lot due to the lack of social distancing, lack of health professionals, shortage of medications, and problems linked to the supply of pure water and sanitation facility (Islam et al. 2021). Compared to the pre-cyclonic period (19 May 2020), the number of COVID-19 cases increased by approximately 70% in the post-cyclonic period (29 May 2020) (Kumar et al. 2021).

It was a huge challenge for the authorities to move large parts of the population in shelter house and maintain proper hygiene and health safety during the pandemic time. Despite of the pandemic, a number of government and non-government organizations and volunteers were involved to support the local community before, during and after the havoc. Over one million people were inundated, approximately 220,000 houses have been partially or fully damaged, and at least 26 people died in the country during the super cyclone Amphan in Khulna, Satkhira, Jashore, Barguna, Bagerhat, Patuakhali, Bhola and Pirojpur districts (Tauhid Zaman et al. 2020). The super cyclone Amphan also caused another "collateral" heakth hazard: the persons displaced were bit living in the shelters that were adequate to maintain necessary social distancing during the pandemic. The study also highlighted that the "provisional" shelter house were in fact durable solutions. Indeed, the family houses had been partially or permanently damaged, and the perspectives of return to normality were low, which also elevated mental pressure and stress as a new health problem. Food scarcity and the almost total destruction of the sanitation system also added to the health crisis and social insecurity, especially for women and children.

Drought

Drought is a most widespread natural hazard today. It develops slowly and can remain for prolonged duration. It is affecting around 55 million people worldwide in every year (Mishra et al. 2022). Bangladesh is a drought-prone country with an index score of 3 out of 5 according to Global Drought Risk Index 2020 (Eckstein et al. 2020). In last few decades, Bangladesh suffered by several climate change related incidents, such as increasing temperature, prolonged Summer season, high variability in rainfall, low annual and seasonal rainfall, among other problems. The high population density and poverty rates, the high reliance on agricultural production, the low adaptive capacity and the lack of awareness greatly influence the impact of droughts in Bangladesh. Even though the pre-monsoon drought intensity has been reduced through sufficient irrigation practices in many areas of the country, the problem is recurrent (Brammer 1987; Sarker et al. 2020).

In terms of severity and occurrence, the North and North-Western, Western, South-western and Central parts of Bangladesh were and are still the most drought-prone areas (Habiba et al. 2012; Rahman & Lateh 2016) (Fig. 21.1B). A number of studies illustrate the recurrent severity of drought along the years (Rahman and Lateh 2016; Rahman et al. 2022a, b, c). Drought may sometimes turned into famine and cause adverse effects on hydrology (surface water and ground water), affecting both human health and biodiversity. Aspects such as food security are impacted, and so is livestock and livelihood, the soil and water pollution, but also the hydroelectric power generation. Here again, the lack of health conditions, and the risk or reality of human diseases may lead to forced migration (Miyan 2015; Rahman and Lateh 2016). Building proper awareness, developing early awareness system, acquiring more capacity for a correct monitorisation and prediction, may be the most urgent response. However, other solutions may be more sustainable, such as to introduce drought resistant crop variety, the conservation of rain water and the adoption of drought-tolerant crop varieties, in order to reduce the actual impacts of the drought (Miyan 2015; Rahman et al. 2022a, b, c; Xenarios et al. 2016). The populations of the drought prone areas should be informed, and trained, in order to have a growing their awareness and know what precautions to adopt in case of drought.

Floods

The geographical location of Bangladesh makes the country one of the most flood-prone areas in the world. Historical records show that at least five major floods occurred in the twentieth century in 1955, 1974, 1987, 1988, 1998, 2004, 2007 and 2010. All of these floods affected at least one-third of the country, and initiated enormous physical damage and fatalities. In the history of Bangladesh, the most severe flood happened in 1998, when it inundated approximately two-thirds of the country as well as the agricultural sectors, fisheries sectors, livestock sectors and damaged severely rural infrastructures (Khandker 2007).

In Bangladesh, floods are usually classified as flash floods, water logging and river/sea flooding or monsoon flooding. Flash floods commonly occurr from April to July and from September to October, and the most affected areas are Sylhet, Sunamganj, Moulvibazar, Habiganj, and Netrakona Districts, Chittagong, Cox's Bazar and Bandarban Districts (Fig. 21.1C). The southwest costal belt areas, namely Satkhira, Jessore, Jhenaidah and Khulna Districts, are examples areas of water logging (Fig. 21.1C).

The government of Bangladesh developed and introduced several processes over the years that helped the population to handle these disasters better. For example, it improved forecasting and warning systems, introduced improved policy for the distribution and handling of disaster relief, created special emergency funds, introduced flood insurance, among other measures. Some lessons can be extracted for such initiatives. A study shows that an insurance scheme can be designed on the basis of the socio-economic conditions. Another aspect evidenced is the risk perceptions of the flood by affected populations, and the fact that their attitudes should be considered carefully to adapt better the educational response and the training practices (Hossain et al. 2022).

Case Study on French Guiana

Since the beginning of the 21th century, the Intergovernmental Panel on Climate Change (IPCC) has been warning about the tipping point of the entire ecosystem that is supposed to regulate climate change. The tipping point is when these ecosystems: The Amazonian Forest, the coral reef, the Siberian permafrost, and the ice cap become also victims of climate change and are not fulfilling their role. The Amazonian Forest also recognized and called the carbon sink and with a role in absorbing CO_2 from human activity, is nowadays on the tipping point, mainly due to deforestation becoming a close future gas emitter as well. It is also essential to underline that the Amazonian Forest is producing a part of the rain in the region, rain that is pushed by the wind in the rest of Southern America. In case of destruction of the forest due to human activity, this weathering activity will be considerably reduced and will risk creating dewatering of the Southern American countries. In order to reduce the risk, the IPCC made Brazil and other Amazonian countries promise to stop deforestation by 2030. This is the first step, but the question is what will happen during all those years. Experts estimated that before 2030, the Northern part of the Amazonian Forest will become savannah and will be more sensitive to fire and destruction of the rest of the forest (Lama 2021).

One of the countries of the Amazonian region is the French Guiana, an overseas department of France. The country is covered by 90% of the less impacted tropical forest in the world: The Amazonian Forest. Only 3% of the whole territory is protected (Stier et al. 2020) and the country hosts many endemic species that need protection.

The country is known for its European aero spatial activity where we can find different actors like the European Spatial Agency (ESA), the Ariane Space Group, the Italian group Advances Vision Into Orbit (AVIO), but also the National Space Study Centre (CNES) and around forty entities that collaborate for the launch of rockets (Centre Spatial Guyanais 2022). The aero spatial activity implies a pollution created by all the products used for the rocket. In the case of the Guiana Space Center (GSP), it is in a 700 km^2 of littoral savannah where about fifty terrestrial species, over 400 bird species and over hundred fishes are living. But we can also find endemic species such as the giant taboo, the jaguar, the great anteater, the ocelot, the broad-billed grebe, the popsicle and the red ibis (Joussen 2017). The GSP is aware of the risk on the wildlife population due to the pollution, and is doing research about rocket rejection and the possibilities of contamination. For example, rockets are releasing paraffin, liquid oxygen, some rockets like the Ariane 5 may release chloridric acid and alumina in very small quantities (Joussen 2017). But the measure reveals that the risk is at maximum around 1 km around the launching zone; the perimeter around it is actually burned due to the warm gas from launching (Joussen 2017).

To be sure the pollution doesn't disturb the wildlife population, some research is on going on the animals behaviours with the National Office of the Hunting and Wildlife Authorities (ONCFS). The first study has been with a jaguar and did not reveal any abnormal behaviour either within 2 to 10 km of the launch area; and the results have been the same with the peccary, a small mammalian (Joussen 2017). Another initiative to mention is the one of the association Kwata, which works on marine species and especially with marine's turtles. Here again, the result of the studies did not show any disturbance. The ornithologist Olivier Tostain and the ECOBIOS association have also been studying the long-term consequences on bird shells, however after only four years of observation it is too early to give an opinion. Lastly, the Hydreco laboratory is also doing research on the fish and aquatic wildlife, without any result for the moment. (Joussen 2017). Even if the place is a protected area with hunting restriction and a reintroducing species area, there are still some issues because they needed to clear a new space for the Ariane 6 rocket launch, where there was a protected fern species. But in order to preserve the environment after each clearing, the CGS must give back land with an ecological value equivalent to that lost (Joussen 2017).

Another aspect that has been affected by climate change is the fishing industry and the conservation of the mangroves, that represent 55,000 hectares overall. About 60% of the French mangroves ecosystem is in French Guiana. The surface water weather increased by +1 °C in 30 years and is affecting the shrimp fishing industry, which has been decreasing drastically in the last 10 years (Guyaweb 2019). As concerns the mangrove, it decreased from 1 to 2% per year, while they were representing three quarters of the French Guiana coastline. In the last 50 years, they have been decreasing from 20 to 35%. The biggest challenges are the pollution created by the urbanisation of the coast, the land-use conversion to agriculture but also the aquaculture farms that are too intensive and not sustainable (Scemama et al. 2022).

In addition, other specific aspect is dangerous for the environment of the Amazonian Forest: it is the gold mining in French Guiana. There is an estimated over 15,000 *Garimperios* (Brazilian landowners) that are mining in over 1,000 clandestine mines. In addition to clearing areas of forest, creating an increased risk of forest fires, illegal mining creates risks for rivers, due to the mercury used to separate the gold. The French authorities cannot do anything to stop it, and limits itself to send the minors back to Brazil before they came again (Bellos 2007).

The consciousness about the issues referred above is rising, but it seems still difficult to show, in the immediate, the effects on health of climate change. This, in a way, is a result of the fact that France and French Guiana are developed territories, where natural disasters are compensated by social and effective emergency measures already. The vice-president of the association Citizen for Climate Guiana, Eric Bourdin, emphasizes that French Guiana is "one of the departments most affected by global warming because the increase in temperature in a climate with a hydrometry of over 80° would make it difficult to evacuate our body temperature from 35° and we are almost there, we are at 34° (…)" (Lama 2021). Things can get worse and worse, over time. He explained that, indeed, in the worse case scenario, in 2050 the country will be uninhabitable, in a context where there has been an increase of +1.36 °C in the last 50 years. During the G26 meeting, the Amazonian region was focused, pinpointing the progressive drying and rising temperatures in this area of over 5 million km (Lama 2021).

The Amazonian region, including the French Guiana, is currently facing the ENSO cycle, which created important inter-annual rainfall, due to El Niño weather pattern, which created a rainfall deficit and La Niña, which created periods of excessive rainfall (Lama 2021). The last climatic study of 2015 already showed risks linked to climate change, like longest and harsh dry seasons and excessively long rainfall seasons. But there are also important risks of vegetation fire, floods because of the increasing of 3.5 mm per year of the ocean that caused marine submersion and lack of rainfall evacuation, and also the spontaneous urbanisation of the savannah and of the forest (Lama 2021). The governments, both local and central (metropolitan), are attempting to protect the population by encouraging good practices at their level, like waste recycling, changing food practices by consuming more locally, energy saving, developing and using renewable energies (Lama 2021). But it remains to be seen if this is enough. In fact, France, as some other countries in the world, are facing a greater challenge and greater responsibility: to prevent health hazards caused by climate chance in both specific locations like the Amazonian area, but also in Europe.

Case Study on Metropolitan France

Heatwaves, floods, desertification, and extreme cold weather are all consequences of climate change, and have affected strongly the European continent. Focusing on the case of metropolitan France, it is relevant to state that climate change was being discussed all around the country as being one of the most important points for the Presidential election of March 2022, as well as in parallel in other forums, such as the Intergovernmental Panel on Climate Change, which has been trying to alert the French population by using interactive posters in the street with a QR code linked to its last report (Mendy 2022; IPCC 2022). The new government was assumed to be ready to answer and to take action against environmental issues to reduce the consequences of the climate change in the country first and in the rest of the world *also*.

More specifically, one health hazard has been recurrent in metropolitan France in the last years: heatwaves. It is important to remember that a heatwave is defined by a period between the month of June and September while the weather during daytime and nightime is higher than the department threshold for at least three days (Adélaïde et al. 2022). Connected to this problem, the City of Paris, in the report *Paris in the face of climate change* gives a definition of the term "climate hazard", stating: "How likely is it that a natural phenomenon will occur? How intense will it be? Where will it occur? Is it foreseeable? The diagnosis highlights climate hazard trends, with regard to the average rainfall, for example, severe climate hazards such

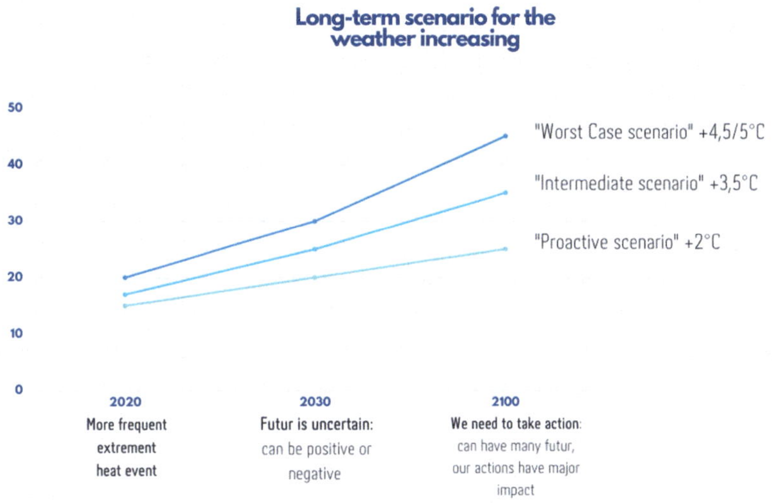

Fig. 21.2 Long-term scenario for the weather increasing. Based on Roussel (2021)

as heatwaves (…)" (Roussel 2021: 5). In the report, the City of Paris indicates that it is facing different difficulties that occur in the capital as heatwaves, also floods, and risks to citizens' health. It used figures to show the different evolution of the climate hazard within a period of one hundred years. Using the data of the IPCC, it shows a proactive scenario where the weather gain +2 °C, the intermediate scenario with + 3,5 °C, and the worst-case scenario with an augmentation of +4,5/+5 °C in the year 2100 (Roussel 2021) (Fig. 21.2).

The City of Paris report shows clearly the expected increase of days under heatwaves – from 7,2 days over 30 °C and 0,2 tropical nights over 20 °C in 2010 to 13,6 days and 5 tropical nights. It also gives an approximation of what it would the future. In 2030, metropolitan France should have almost 20 days of heatwaves, almost 22 days in 2050 and we will have more than 34 days of heatwaves in 2085. For the tropical night, it should rise to almost 18 nights in 2030, more than 20 nights in 2050, and almost 35 tropical nights in 2085. We can see that we will have more tropical nights than a heatwave in 2085. If compared with the result of 2010, it is observed that in 75 years, there will be 20 more days of heatwaves and 30 more tropical nights (Roussel 2021).

These increases of extreme weather will have a considerable bad effect on human health. In their article: *Des impacts sanitaires du changement climatique déjà bien visibles: l'exemple des canicules* (Adélaïde et al. 2022) the authors refer various health impacts due to climate change that already appears in the population, in metropolitan France. One of the most important for them is the loss of well-being. In fact, they considered this loss to the restriction of the practice of activities that could create important symptoms. These symptoms include tiredness, fainted, cramps, decreased alertness, and problems with cognitive functions (Adélaïde et al. 2022).

To overview the victims of the heatwave, a comparative study based in the Greater City of Lyon indicates that the most vulnerable persons during heatwaves are children under 10 and the adults over 75 years old (Alonso and Renard 2020). Yet, the previous health condition must be taken into consideration, as this could change the results of studies. This study shows that the psychiatric status of a person can also affect the capacity to face heatwaves. Additional factors also include gender, and show that women are more affected by higher temperatures (Alonso and Renard 2020). The study shows the percentage of death by age in France: "At the national level, in France, excess mortality [due to heatwaves] was estimated at 20% for the 45–74-year-olds, 70% for the 75–94-year-old age group, and 120% for people over 94 years old" (Alonso and Renard 2020). The heatwave of 2003 caused 19,490 deaths, the one of 2006 caused 1388 deaths, and the one of 2015 caused 3275 deaths (Alonso and Renard 2020). For Chanel and Pascal, since 1970 France counts more than 37,000 deaths caused by extreme weather, and the most important one was the heatwave of 2003, with more than 15,000 deaths in less than two weeks (Adélaïde et al. 2022).

The latter study, in Lyon, connects with the study of the City of Paris, demonstrating that even in developed countries the public health system's vulnerability as far as healthcare resources are concerned, when health hazards occur as a result of climate changes. In fact, the existing risk for the healthcare system is almost always increasing. It is foreseeable that in the next decades metropolitan France will face less extreme cold, less snow, and less freezing rain, that are important for the ecosystem. Resources like water, biodiversity, energy, air, and food will be affected. Such resources are important for the public health system itself. One of the biggest challenges of the City of Paris, for example, is to be able to provide drinkable water all around the city, both in public and private places.

Studies estimate that drinking water will be certain and accessible until 2050. However, due to the increase in the weather and the heatwaves, the population will need more water to stay hydrated, and it is estimated that the consumption of water will increase by 2% every year from 2030. In Paris, the highest level of evaporation of the river Seine is already visible, and the increase of the temperature of 2 or 3 °C by 2100 will lead to issues on the biological and chemical quality of water. Those changes will lead to challenges to access to drinkable water in the city and create risks of disease due to the change in the water quality (Roussel 2021). Moreover, the quality of air in Paris is already one of the biggest issues as regards the health of the resident population, and the climate change with for example the ozone depletion and fine particles contribute to the decrease of the air quality, and will impact more people's health. Cardiovascular or respiratory diseases are causing premature death, and with pandemics like COVID-19, the population's health is getter worsen (Roussel 2021).

As it is possible to observe, heatwaves are not the only hazard of climate change in metropolitan France. Flood is also going to be more common phenomenon, and will lead to health issues. The report of the City of Paris shows that the decennial flood flows should increase from 20% and should be at over 40% in a hundred years.

Colombe Brossel, Deputy Mayor of Paris in charge of the cleanliness of public spaces, sorting and reducing waste, recycling and reuse (Ville de Paris s.d.), suggests an interesting alternative, indicating that: "[i]n the face of climate change and in anticipation of violent rainfall events, we have initiated the ParisPluie Plan, which is designed to make use of rainwater for cooling the city and greening projects and limiting risks of spillage into the Seine and flooding" (Roussel 2021: 14). This way, it seems possible to plan an anticipation of the climate hazard, turning them an opportunity for sustainable and greener development. One of the key issues is to identify the institutions with major risks of being affected by the climate change. Here again, the City of Paris plans a Flood Risk Prevention Plan (PPRI) for hospitals of the city. Over forty health facilities and hospitals have been identified, that are in floodplain areas. Another complementary point is to plan, and educate for, drastic conditions linked to the impact of driving during a flood. Alternative ways must be planned, as the lack of mobility could also slow emergency health operations (Roussel 2021).

In addition to the flood affecting hospitals, another issue is that "each hospital bed 'consumes' an average of 750 L of water a day, roughly the equivalent of the daily quantity consumed by six Parisians" (Roussel 2021: 20). Every hospital will need to reduce its water consumption in order to cope with shortage. In fact, parallel to what was observed with the COVID-19 crisis, which has obliged to shift to emergency development mode in many settings, the health work will have to be reprogrammed in order to respond to the reduction of resources (as well as to the public health needs) as a result to climate change.

Conclusions

Since 2020, the Covid-19 pandemic demonstrated in a drastic manner the way the earth is suffering from the impact of human activity. Due to the fact that most economic and social exchanges stopped during several months, especially from March 2020 to the late months of that same year, humanity as a whole has empirically observed how the various natural elements have improved (e.g. the quality of the air, just to mention this aspect). At the same time, it is now clear that climate change impacts health hazards.

Both in developing countries and in developed countries, as the case studies illustrate, there is a growing need for a better understanding of the way human communities can prepare better for health hazards resulting from climate change. In Bangladesh, disasters force people to switch their occupations and adapt to a life with less earnings, and also to drastically changing the lifestyle and reduce many aspects of the daily consumptions to adapt to the situation. The issue of forced internal migrations is a recurrent one, causing instability and turning the health standards more difficult to implement in such a unstable context. Despite the help coming from the Governmental and the Non-Governmental Organizations, the fates are not changing much, although the effects of disasters also depend on the disaster type. So, it is high time to focus on the humanitarian actions before/during/after the disasters focusing on disaster types.

In France too, both in French Guiana and in metropolitan France, the severe impact of climate change on highly organized public health systems is already visible. For example, despite the attempts and the alerts of NGOs and international bodies to stop deforestation in the Amazonian Forest by 2030, it is likely that well before 2030 the northern part of the Amazonian Forest will become savannah, and will be more sensitive to fire and destruction. In metropolitan France, also, the challenges resulting from recurrent heatwaves have to be combined with the growing need to reduce water consumption. In a way, the fact that developed and modern equipment already exist to respond to humanitarian crisis, as well as social aid support, does not mean that measures should not be taken more intensively. The lack of hygiene in big cities, illustrated by the case of the capital Paris, is illustrative of this. The degradation of environments in complex urban metropolis or city centre, resulting form a combination of bureaucratic, social and cultural phenomena (e.g. strikes), causes new and unusual health risks in developed countries (e.g. associated to the grows of homeless persons). Besides, the awareness of populations tends to diminish in the context of developed countries, as most of the population rely on public services, and the impact of health hazards can be much higher than it should actually be.

As a result, the analysis done above concludes to the need of more education for health hazards preparedness resulting of new health risks potentiated by climate changes, as an essential issue, both for workers in the field of humanitarian action and for populations in general.

References

Adélaïde L, Chanel O, Pascal M (2022) Des impacts sanitaires du changement climatique déjà bien visibles : l'exemple des canicules. Annales des Mines – Responsabilité et environnement, April, N°106. Available at: https://www.annales.org/re/2022/re106/2022-04-10.pdf. Accessed 28 Apr 2022

Alonso L, Renard F (2020) A Comparative study of the physiological and socio-economic vulnerabilities to heat waves of the population of the metropolis of Lyon (France) in a climate change context. Int J Environ Res Public Health 17(3):1004. https://doi.org/10.3390/ijerph17031004

ALNAP (s.d.) 2022–2023 Work Plan and Budget. Available at: https://www.alnap.org/system/files/content/resource/files/main/ALNAP%20work%20plan%202022-2023%20web.pdf. Accessed 8 Jun 2022

Bellos A (2007) Illegal, polluting and dangerous: the gold rush in French Guiana. [online] The Guardian, December, 17. Available at: https://www.theguardian.com/environment/2007/dec/17/pollution.mining. Accessed 22 May 2022

Brammer H (1987) Drought in Bangladesh: lessons for planners and administrators. Disasters 11(1):21–29

Centre Spatial Guyanais (2022) Acteurs. [online] Available at: https://centrespatialguyanais.cnes.fr/en/acteurs. Accessed 28 May 2022

De Geoffroy V, Knox Clarke P, Bhatt M, Grunewald F, Doherty J (2021) Adapting humanitarian action to the effects of climate change An ALNAP Lessons Paper. Available at: https://www.alnap.org/help-library/alnap-lessons-paper-adapting-humanitarian-action-to-the-effects-of-climate-change

Ebi K, Ogden NH, Semenza JC, Woodward A (2017) Detecting and attributing health burdens to climate change. Environ Health Perspect 125(8):085004. https://doi.org/10.1289/EHP1509

Eckstein D, Künzel V, Schäfer L, Winges M (2020) Global climate risk index. Bonn: Germanwatch

EHA (2019) Climate risk, connecting environment and humanitarian action. Available at: https://ehaconnect.org/themes/climate-risk/

FAO (2017) United Nations Climate Change Conference 2017, FAO's work on climate change. Available at: https://www.fao.org/3/i8037e/i8037e.pdf. Accessed 22 May 2022

Fakhruddin B, Kintada K, Hassan Q (2022) Understanding hazards: probabilistic cyclone modelling for disaster risk to the Eastern Coast in Bangladesh. Progress in Disaster Science 13:100216

Guyaweb (2019) L'impact du changement climatique sur la pêche en Guyane, Guyaweb, site d'information et d'investigation en Guyane. [online]. Available at: https://www.guyaweb.com/actualites/news/sciences-et-environnement/limpact-du-changement-climatique-sur-la-peche-en-guyane/. Accessed 22 May 2022

Habiba U, Shaw R, Takeuchi Y (2012) Farmer's perception and adaptation practices to cope with drought: Perspectives from Northwestern Bangladesh. Int J Disaster Risk Reduct 1:72–84. https://doi.org/10.1016/j.ijdrr.2012.05.004

Hoque MA-A, Phinn S, Roelfsema C, Childs I (2018) Assessing tropical cyclone risks using geospatial techniques. Appl Geogr 98:22–33

Hossain MS, Alam GMM, Fahad S, Sarker T, Moniruzzaman M, Rabbany MG (2022) Farmers' willingness to pay for flood insurance as climate change adaptation strategy in northern Bangladesh. J Cleaner Product 130584

Huq S (2001) Climate change and Bangladesh. In Science (Vol. 294, Issue 5547, p. 1617). American Association for the Advancement of Science

Huq S, Ayers J (2007) Critical list: the 100 nations most vulnerable to climate change. IIED Sustainable Development Opinion

Huq S, Reid H, Konate M, Rahman A, Sokona Y, Crick F (2004) Mainstreaming adaptation to climate change in least developed countries (LDCs). Climate Policy 4(1):25–43

IPCC (2022) Français – IPCC. [online] Available at: https://www.ipcc.ch/languages-2/francais/. Accessed 28 May 2022

Islam MR, Hasan M (2016) Climate-induced human displacement: a case study of Cyclone Aila in the south-west coastal region of Bangladesh. Nat Hazards 81(2):1051–1071

Islam SMD-U, Mondal PK, Ojong N, Bodrud-Doza M, Siddique MAB, Hossain M, Mamun MA (2021) Water, sanitation, hygiene and waste disposal practices as COVID-19 response strategy: insights from Bangladesh. Environ Dev Sustain 23(8):11953–11974. https://doi.org/10.1007/s10668-020-01151-9

Joussen I (2017) L'impact de l'activité spatiale en Guyane sur l'écosystème. [online] Sciences et Avenir. Available at: https://www.sciencesetavenir.fr/nature-environnement/impact-des-lancements-de-fusees-sur-l-environnement_109420. Accessed 22 May 2022

Khandker SR (2007) Coping with flood: role of institutions in Bangladesh. Agric Econ 36(2):169–180

Kumar S, Lal P, Kumar A (2021) Influence of super cyclone "Amphan" in the Indian subcontinent amid COVID-19 pandemic. Remote Sens Earth Syst Sci 4(1):96–103. https://doi.org/10.1007/s41976-021-00048-z

Lama C (2021) Dérèglement climatique : les questions qui se posent en Guyane – Guyane la 1ère. [online] Guyane la 1ère. Available at: https://la1ere.francetvinfo.fr/guyane/ouest-guyanais/guyane/dereglement-climatique-les-questions-qui-se-posent-en-guyane-1161202.html. Accessed 22 May 2022

Marin AF, Naess LO (2017) Climate change adaptation through humanitarian aid? Promises, perils and potentials of thenew humanitarianism? Brighton: Institute of Development Studies

Mendy A (2022) IMAGES. Quand le rapport du GIEC s'affiche en géant dans le métro à Paris. [online] Actu Paris, 17 May. Available at: https://actu.fr/ile-de-france/paris_75056/images-quand-le-rapport-du-giec-s-affiche-en-geant-dans-le-metro-a-paris_51007033.html. Accessed 28 May 2022

Mishra D, Goswami S, Matin S, Sarup J (2022) Analyzing the extent of drought in the Rajasthan state of India using vegetation condition index and standardized precipitation index. Model Earth Syst Environ 8:601–610. https://doi.org/10.1007/s40808-021-01102-x

Miyan MA (2015) Droughts in Asian least developed countries: vulnerability and sustainability. Weather Climate Extremes 7:8–23. https://doi.org/10.1016/j.wace.2014.06.003

Naheed A, Hort K (2015) Bangladesh health system review, health systems in transition, Vol. 5, No. 3. World Health Organization, Geneva

Nasher NMR, Karim KHR, Islam MY (2022) Spatio-temporal variation of cyclone intensity over the coastal region of Bangladesh using 134 years track analysis. Trop Cyclone Res Rev 11(1):16–25

Parvin GA, Takashino N, Islam MS, Rahman MH, Abedin MA, Ahsan R (2022) Disaster-induced damage to primary schools and subsequent knowledge gain: case study of the Cyclone Aila-Affected community in Bangladesh. Int J Disaster Risk Reduct 72:102838

Pulhin JM, Inoue M, Shaw R (2021) Climate change, disaster risks, and human security – Asian experience and perspectives. Springer, Singapore

Rahman M, Lateh H (2016) Meteorological drought in Bangladesh: assessing, analysing and hazard mapping using SPI, GIS and monthly rainfall data. Environ Earth Sci 75(12):1–20

Rahman M, Hossain M, Ali M, Ahmed Z, Hedayutul Islam AHM (2022a) Assessing vulnerability and adaptation strategy of the cyclone affected coastal area of Bangladesh. Geoenviron Disasters 9(1):1–20

Rahman MS, Sujan MHK, Acharjee DC, Rasha RK, Rahman M (2022b) Intensity of adoption and welfare impacts of drought-tolerant rice varieties cultivation in Bangladesh. Heliyon, e09490

Rahman SM, Faruk M, Rahman M, Rahman SM (2022c) Drought index for the region experiencing low seasonal rainfall: an application to northwestern Bangladesh. Arab J Geosci 15(3):1–15

Roussel J (coord.) (2021) Paris in the face of climate change – update of the study on Paris' vulnerabilities and strengths in the face of climate change and diminishing resources. Ville de Paris, Paris. Available at: https://ramboll.com/-/media/files/reh/synthseparis-in-the-face-of-climate-changeen.pdf?la=en. Accessed 28 Apr 2022

Sarker M, Latifur R, Janet N, Mansor SA, Ahmad BB, Shahid S, Chung E-S, Reid JS, Siswanto E (2020) An integrated method for identifying present status and risk of drought in Bangladesh. Remote Sensing 12(17):2686

Scemama P, Regnier E, Blanchard F, Thébaud O (2022) Ecosystem services assessment for the conservation of Mangroves in French Guiana using fuzzy cognitive mapping. [online] Front Forests Global Change 4:769182. https://doi.org/10.3389/ffgc.2021.769182. Accessed 22 May 2022

Sen, A (1993) Capability and well-being. In Nussbaum and Sen (eds), The quality of life, pp. 30–53. Clarendon Press, Oxford

Sphere (2019) Reducing environmental impact in humanitarian response

UNHCR (2022) Climate change, displacement and human rights, UNHCR. Available at: https://www.unhcr.org/publications/brochures/6242ea7c4/climate-change-displacement-human-rights.html

Stier A, de Carvalho WD, Rostain S, Catzeflis F, Claessens O, Dewynter M, McKey D, Mustin K, Palisse M, de Thoisy B (2020) The Amazonian Savannas of French Guiana: cultural and social importance, biodiversity, and conservation challenges, tropical conservation science 13:1–21. Accessed 24 May 2022

Tauhid Zaman J, Arif Mostafa P, Asaduzzaman Sardar S, Nazmul Huda Nasim B, Bijoy Dhar R, Aminul Islam Rana S, Ibrahim Rony C, Dulal Abdullah R (2020) Cyclone Amphan: death toll rises to 26. DhakaTribune. 21 May 2020. Available at: https://archive.dhakatribune.com/bangladesh/2020/05/21/cyclone-amphan-slows-down-moves-towards-rajshahi-region. Accessed 14 July 2022

Ville de Paris (s.d.) Colombe Brossel. Available at: https://www.paris.fr/pages/brossel-colombe-2259. Accessed 10 June 2022

WHO (2021) Climate Change and Health, World Health Organization. Available at: https://www.who.int/news-room/fact-sheets/detail/climate-change-and-health

Xenarios S, Nemes A, Sarker GW, Sekhar NU (2016) Assessing vulnerability to climate change: are communities in flood-prone areas in Bangladesh more vulnerable than those in drought-prone areas? Water Resources Rural Develop 7:1–19. https://doi.org/10.1016/j.wrr.2015.11.001

João Casqueira Cardoso is Associate Professor and coordinator of the Master course in Humanitarian Action, Cooperation and Development, University Fernando Pessoa. He holds a Master in International Studies (Reading University) and a Ph.D. in Law (European University Institute), and has a long experience in international relations, including in the field of health. He has worked for the Directorate General for Studies of the European Parliament, for the Council of Europe, and for the International Labour Organization. He is a full member of the CEPESE (Research Centre for the study of Population, Economy and Society), Porto (Portugal).

Mahfuza Parveen is currently working as Associate Professor and Research Coordinator in the Department of Environmental Science and Disaster Management, Daffodil International University, Bangladesh. She has completed her first Masters degree in Geological Sciences from Bangladesh and later earned Master of Engineering degree majoring in Environmental Sciences followed by a Ph.D. degree from Saitama University, Japan. In the beginning of her career she experienced to work with renowned environmental NGO in Bangladesh and also gathered experience in environment and educational sectors while studying in Japan.

Mafalda S. A. Fonseca originates from Portugal. She is currently a researcher-student in the Master course in Humanitarian Action, Cooperation and Development, University Fernando Pessoa.

Laurine Grzelak originates from France. She is currently a researcher-student in the Master course in Humanitarian Action, Cooperation and Development, University Fernando Pessoa.

Chapter 22
Effectiveness of Public Policy in the Face of Climate Change: The Case of Coffee Growing in the State of Chiapas: Mexico

Tanya Méndez-Matus and Gabriela Narcizo de Lima

Abstract The costs of inaction in the face of climate change are higher than those of being preventive and proactive. In the state of Chiapas—Mexico, an increase in the frequency and intensity of extreme hydrometeorological events have been observed in recent decades, which have caused economic, social and environmental damages. In this sense, this research sought to analyze how vulnerable rainfed agriculture, particularly coffee farming, in the state of Chiapas is to the effects of climate change. After this analysis, we sought to understand the perspective of key stakeholders in decision-making regarding the impacts of climate change, agricultural vulnerability, and the effectiveness of public policy to reduce this vulnerability. Even though Chiapas is in third place nationally in the implementation of strategies to face climate change, the results of this work require a much greater effort so that tools for adaptation to climate change are effective in all sectors, and in a particular way, in the agricultural sector, since it represents an aspect of food security in the country. The correct articulation between the laws, programs, and actions against climate change in Chiapas are essential to reduce the vulnerability of the population to extreme climatic phenomena, for which changes are required that lead to practical governmental and social actions that generate economic strategies, as well as environmental and sustainable technologies, as a basis for adaptation to climate change.

Introduction

In recent decades, climate change has been placed on the public agenda of many countries and organizations as one of the main challenges facing humanity today (Borràs 2017; López Cruz et al. 2021; Abeldaño Zuñiga et al. 2021).

T. Méndez-Matus
Research Institute in Risk Management and Climate Change, University of Sciences and Arts of Chiapas, Tuxtla Gutiérrez, Mexico

G. N. de Lima (✉)
Geography Department—Faculty of Letters, Porto University, Porto, Portugal
e-mail: gabrielalima@letras.up.pt; gabrielanl@gmail.com

W. Leal Filho et al. (eds.), *Climate Change and Health Hazards*, Climate Change Management, https://doi.org/10.1007/978-3-031-26592-1_22

This phenomenon is usually considered a global problem with regional consequences that have impacted and modified the natural environment, thus creating imbalances in ecosystems and natural resources that, in turn, imply alterations in human systems in both the social and economic spheres (Rockström et al. 2009; Economic Commission for Latin America and the Caribbean 2015; Abeldaño Zuñiga et al. 2021).

These variations in climate have impacted different production systems, such as agriculture, which is a highly sensitive activity to changes related mainly to temperature and precipitation, which represents a risk for the economy and food security of populations in several countries (McMichael and Lindgren 2011; Harvey et al. 2018; Murray-Tortarolo and Jaramillo 2019; Abeldaño Zuñiga et al. 2021).

The risks related to changes in climate variables have been addressed using impact scenarios. Although these models roughly simulate temperature trends, they are far from being forecasts (Cardona et al. 2012; Neri and Magaña 2016; Pei et al. 2020).

According to Magaña (2013), in the case of Mexico, some of these projections indicate a warming pattern due to the country's climatic and hydrological conditions. In addition to a geographical location that is highly vulnerable to changes in climate and hydrometeorological events such as droughts and floods (Martínez-Austria 2020; Prieto-González 2020).

The country's high vulnerability to such changes is because many still consider them unpredictable phenomena, among other things, and therefore only respond to an event of this type, thus dominating the naturalistic paradigm in which disasters are explained only by the occurrence of the natural phenomenon. In other words, the risk is not analyzed considering the system's vulnerability, but only the hazard and its relationship with disasters (Neri and Magaña 2016).

By analyzing the problem from a risk perspective, it is possible to identify the factors that create it and thus prepare to avoid, or at least reduce, the magnitude of the impact (Cardona et al. 2012; Magaña 2013; Neri and Magaña 2016). Actions before, during, and after the phenomenon are based on recognizing and reducing the most important vulnerability factors.

Because of this situation, it is essential to pay attention to the country's growing areas. According to the Ministry of Agriculture and Rural Development (SAGARPA—acronym in Spanish) (Secretaría de Agricultura Ganadería Desarrollo Rural Pesca y Alimentación 2017), there are 109.3 million productive hectares in Mexico, of which 645 thousand are destined for coffee production, with the state of Chiapas accounting for the most, with 35.4% of national production.

In Mexico, coffee growing is considered a fundamental strategic activity since it allows the integration of productive chains, the creation of foreign exchange and jobs, and the livelihood of many small producers. This activity includes around 30 indigenous groups, in addition to having an enormous ecological relevance, since more than 90% of the area cultivated with coffee is under diversified shade, which contributes to the conservation of biodiversity and the provision of ecosystem services (Fideicomisos Instituidos en Relección con la Agricultura 2016).

For the state of Chiapas, coffee production has represented one of the main sources of income generation while at the same time developing an important social, environmental, and economic function based on agroforestry production with wide territorial coverage (Centro Agronómico Tropical de Investigación y Enseñanza 2011).

The possibility of coffee ceasing to play an important role in the state's local economy due to the effects of climate change would have unprecedented impacts, which could further aggravate the already difficult economic situation of coffee-producing communities (Centro Agronómico Tropical de Investigación y Enseñanza 2011).

Some projections indicate that there will be an increase in the average temperature in the coffee-growing zone of the Sierra Madre de Chiapas (one of the most important for the state) of a magnitude of 2.1–2.2 °C in 30 years. On the other hand, it is also expected, in approximate terms, a reduction of 80–85 mm in average rainfall for the same region (Gobierno del Estado de Chiapas 2011).

If such scenarios are fulfilled, the best areas for coffee cultivation would undergo important changes in zones below 1700 m above sea level (Centro Agronómico Tropical de Investigación y Enseñanza 2011). The Sierra Madre de Chiapas, which today has 265,400 hectares of optimal areas for the production of Arabica coffee, would only have 23% of this usable area (Food and Agriculture Organization of the United Nations and Secretaría de Agricultura y Desarrollo Rural 2012a; Secretaría de Agricultura Ganadería Desarrollo Rural Pesca y Alimentación 2017).

Climate scenarios developed by the Climate Change Action Program for the State of Chiapas (PACCCH—acronym in Spanish) (Gobierno del Estado de Chiapas 2011) indicate increasing probabilities of the occurrence of natural phenomena of greater intensity. Since climate change affects each socioeconomic region in different ways, some populations are more susceptible to the same hazard, and their vulnerability may increase or decrease depending on the specific actions taken in the region.

Chiapas is one of Mexico's states with the greatest physical and social vulnerability to climate events. In addition to the latent risk created by hydrometeorological phenomena such as extreme rains, landslides, and floods that have been occurring with greater frequency and intensity in recent years, with significant environmental, social, and economic impacts (Gobierno del Estado de Chiapas 2011).

It is very important to consider that agricultural production may be affected by changes in temperature and precipitation patterns, especially in those areas that do not have the necessary technology to cope with these changes. An increase of 1.5 °C in the average temperature would decrease coffee crop yields in some of the main coffee-growing areas of Chiapas (Centro Agronómico Tropical de Investigación y Enseñanza 2011; Food and Agriculture Organization of the United Nations and Secretaría de Agricultura y Desarrollo Rural 2012a; Fideicomisos Instituidos en Relección con la Agricultura 2016; Secretaría de Agricultura Ganadería Desarrollo Rural Pesca y Alimentación 2017), creating a strong economic and social impact on the state. In this sense, this research sought to analyze how vulnerable rainfed agriculture, particularly coffee farming, in the state of Chiapas is to the effects of climate change. After this analysis, we sought to understand the perspective of key stakeholders in decision-making regarding the impacts of climate change, agricultural vulnerability, and the effectiveness of public policy to reduce this vulnerability.

Methodological Procedures

Initially, documentary information was obtained to analyze the content of international, national, and state policies on climate change and the agricultural sector. Likewise, the existing participatory bodies in the State of Chiapas for decision-making on agriculture and climate change were identified.

For a more specific and concrete approach, background information was sought on research based on public policies and the agricultural sector, and personal interviews were conducted with some of the authors of this research.

Based on the information obtained from the documentary review and the interviews, it was determined that the best way to analyze the perception of agricultural vulnerability and the effectiveness of the regulatory framework in the state would be through the completion of two questionnaires. One with open and empirical questions and the other with multiple-choice questions, to which LOW, MEDIUM, and HIGH values were assigned as possible answers.

The first questionnaire was directed to 107 stakeholders related to coffee growing in Chiapas and its vulnerability to climate change; it had 11 simple questions, which made it possible to visualize this sector's changes. The second questionnaire was directed to stakeholders directly related to the institutions responsible for decision-making in the field of community development, ecosystem conservation, agricultural development, and climate change in the state. It was designed to examine the perception of risk due to climate change, agricultural vulnerability, and institutional vulnerability.

Once the answers were obtained, the LOW, MEDIUM, or HIGH values of those questions that could be correlated were analyzed and plotted. Likewise, a search of climatic data was carried out in the databases of the National Meteorological Service (SMN—acronym in Spanish) (Servicio Meteorológico Nacional 2021) for the state of Chiapas, specifically for the variables temperature (minimum, average, and maximum) and average precipitation. Seasonal changes in climatic variables were evaluated, among which the observed values for minimum temperatures and average precipitation recorded in the spring season were highlighted in all data series for the period 1985–2019 to perform a statistical analysis of the data.

From the data collected, the following two groups were considered: 1985–2000 and 2001–2019, to which statistical analyses were applied, such as calculation of means, standard deviation, percentiles, variance, and amplitude, among others. The Shapiro Wilk (Shapiro and Wilk 1965) and Kruskal Wallis (Kruskal and Wallis 1952) analysis were applied[1] to the minimum temperature and mean precipitation data for each group of data to determine the normality of the series ($p < 0.05$). For the paired samples T-test, the normality test with the Shapiro Wilk method (Shapiro and Wilk 1965) was used since it is perfectly suited to small samples with less than 50 values.

The results obtained from these two studies and the basic statistical analysis were compared to determine the data's anomaly. In the same way, graphs Box Plot[2] were elaborated on the data to present a visual aid of the results obtained.

Subsequently, an analysis was done using data from the Agri-food and Fisheries Information Service (SIAP—acronym in Spanish) (Servicio de Información Agroalimentaria y Pesquera 2022) and the National Institute of Statistics and Geography (INEGI—acronym in Spanish) (Instituto Nacional de Estadística y Geografía 2021) of the harvest of all crops in the State of Chiapas from 1985 to 2019. Cherry coffee was chosen as the most representative of Chiapas agriculture due to its importance in the state's economy. From the data collected, the annual harvests per hectare from 1980 to 2018 were calculated.

Subsequently, on the coffee production data in the state, a basic statistical analysis and a Shapiro Wilk (Shapiro and Wilk 1965) and Kruskal Wallis (Kruskal and Wallis 1952) analysis were performed to evaluate the difference between planting and harvesting of cherry coffee in the last 38 years. For this analysis, the data were divided into two groups 1980–2000 and 2001–2018. A *Box Plot* graph was elaborated with the results obtained from the previous analysis to visualize the data. These graphs allow us to quickly deduce the existence of changes in the variables studied. Similarly, using different statistical studies allows us to reach a more accurate conclusion for the hypothesis of this study.

Results and Discussion

Quantitative Analysis

There is scientific evidence that seasonal temperatures and precipitation have changed over the last 30 years, mainly concerning minimum and average temperatures and annual precipitation accumulation (Lima and Magaña Rueda 2018; Lima et al. 2018). In the case of this study, spring minimum temperatures were the most prominent in terms of seasonal change trends. As seen in the Box Plot (Fig. 22.1) and the Shapiro Wilk study (Table 22.1), the data referring to the mean minimum spring temperatures in the state of Chiapas for the periods 1985–2000 and 2001–2019 differ in the means

[1] The Kruskal–Wallis rank test is a nonparametric method for testing whether samples originate from the same distribution. It is used to compare two or more independent samples of equal or different sample sizes.

[2] *Box Plot* is a form of statistical representation intended primarily to highlight aspects of the distribution of observations in one or more quantitative data series.

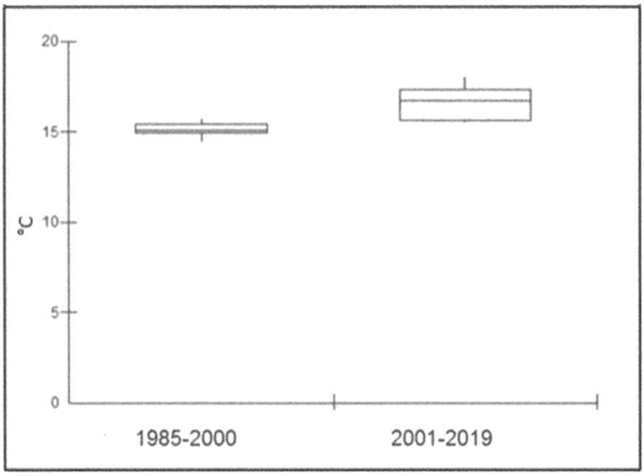

Fig. 22.1 Box Plot minimum temperatures (°C)—Springs 1985–2000 and 2001–2019 (*Source* Own elaboration based on data from Servicio Meteorológico Nacional 2021)

Table 22.1 Shapiro Wilk study for minimum temperatures—Springs 1985–2000 and 2001–2019

Shapiro Wilk Study—Springs		
Results	Group 1: 1985–2000	Group 2: 2001–2019
Sample Size	16	19
Average	22.49	16.54
Deviation Pattern	0.366	0.865
W	0.845	0.885
P	0.0107	0.0293

Source Own elaboration based on data from Servicio Meteorológico Nacional (2021)

by 5.8 °C. The *P*-value in the Shapiro Wilk study was less than 0.05 in both cases, thus ruling out a similarity between the means in the periods analyzed.

The overall statistical study for the mean spring minimum temperature data for the periods 1985–2000 and 2001–2019 (Table 22.2) indicated a difference between the median and arithmetic mean with a coefficient of variation of 2.52% and 5.22%, respectively. The graph of minimum temperatures in springs from 1985 to 2019 (Fig. 22.2) showed an increasing trend, especially from 2005 to 2019.

Finally, the Kruskal Wallis statistical study (Table 22.3) showed a *P*-value of less than 0.05, indicating the discrepancy between the means in the average minimum temperatures of both periods. It can be concluded that the average minimum temperatures of the springs in the period 2001–2019 are different from that found in the period 1985–2000, with an increase observed mainly from 2005.

In this sense, the analysis results of rainfall data from the 1985 to 2019 spring seasons were also alarming. According to the *Box Plot* graph (Fig. 22.3) results and the Shapiro Wilk study (Table 22.4), there is a significant variation in the averages of both periods analyzed (1985–2000 and 2001–2019), with a *P*-value as low as 0.0083. In summary, a considerable decrease in precipitation was found for the 2001–2019 period compared to the 1985–2000 period.

Table 22.2 General statistical study of springs by periods 1985–2000 and 2001–2019

Springs		
Minimum temperatures	1985–2000	2001–2019
Sample size	16	19
Minimum	14.4	15.5
Maximum	15.7	18
Total amplitude	1.3	2.5
Median	15.1	16.7
First quartile	14.95	15.65
Third quartile	15.4	17.35
Interquartile deviation	0.45	1.7
Arithmetic average	15.11	16.594
Variance	0.145	0.749
Deviation pattern	0.381	0.865
Error of pattern	0.095	0.198
Variation coefficient	2.52%	5.22%
Asymmetry	– 0.55	– 0.052
Cuartize	– 0.54	– 1.62
Harmonic average	15.1	16.55
N(ma)	16	19
Geometric average	15.114	16.57
N(mg)	16	19
Geometric variance	1	1
Deviation pattern	1.025	1.0537

Source Own elaboration based on data from Servicio Meteorológico Nacional (2021)

The general statistical study of the precipitation data shows that the coefficient of variation is 71.40% and 85.05%, respectively. The median and arithmetic mean also vary in both groups, with scores higher than 4 l/m^2. The following graph (Fig. 22.4) shows that, from 2003 onwards, there was a decrease in total precipitation. However, the *P*-value in the Kruskal Wallis study (Table 22.5) indicates a value greater than 0.05, revealing that the total precipitation in the springs of the two periods compared (1985–2000 and 2001–2019) does not present a statistically significant difference. This may be due to the precipitation peak that occurred between 2001 and 2003 (Fig. 22.4).

According to different analyses, there is evidence of an alarming decrease in cherry coffee planting and harvesting in the last 30 years (Centro Agronómico Tropical de Investigación y Enseñanza 2011; Fideicomisos Instituidos en Relección con la Agricultura 2016; Secretaría de Agricultura Ganadería Desarrollo Rural Pesca y Alimentación 2017). In this sense, the *Box Plot* (Figs. 22.5 and 22.6) of the cherry coffee planting and harvesting in the state of Chiapas show a considerable decrease

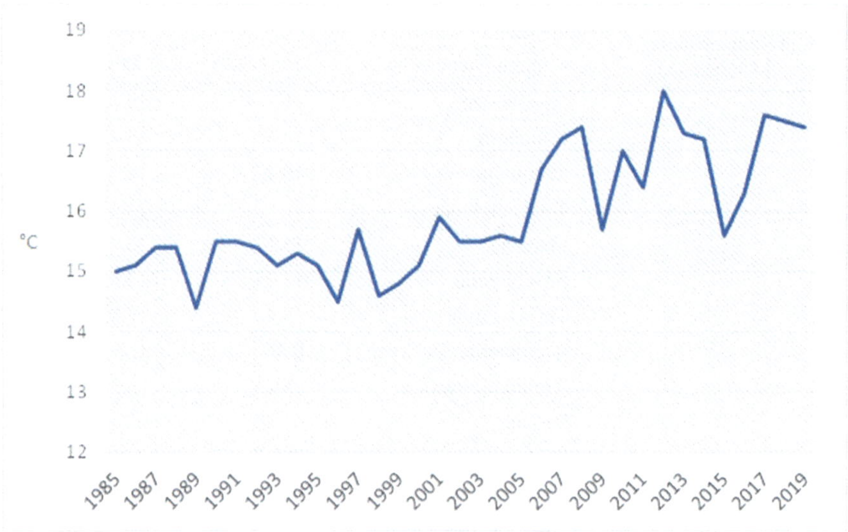

Fig. 22.2 Minimum temperatures (°C) recorded for the state of Chiapas—Springs 1985–2019. (*Source* Own elaboration based on data from Servicio Meteorológico Nacional [2021])

Table 22.3 Kruskal Wallis study for minimum temperatures—Springs 1985–2000 and 2001–2019

Minimum temperature—Springs	
Kruskal Wallis	Results
H	22.7
Degrees of freedom	1
P	<0.0001

Source Own elaboration based on data from Servicio Meteorológico Nacional (2021)

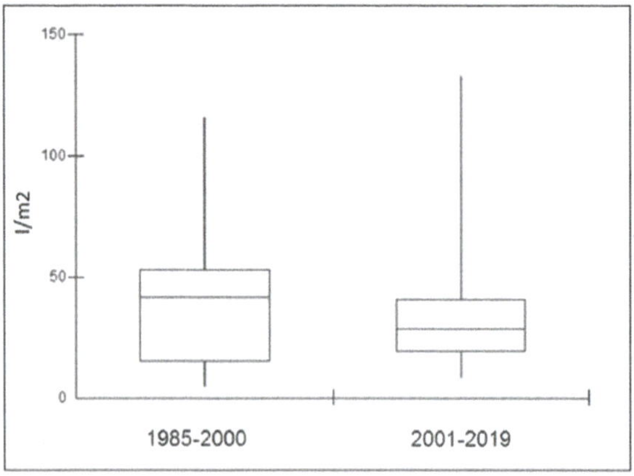

Fig. 22.3 Box Plot average precipitation (l/m^2)—Springs 1985–2000 and 2001–2019 (*Source* Own elaboration based on data from Servicio Meteorológico Nacional [2021])

Table 22.4 Shapiro Wilk Study for average precipitation—Springs 1985–2000 and 2001–2019

Shapiro Wilk Study—Precipitation Springs	
Results	Precipitation
Sample Size	19
Average	37.25
Deviation pattern	31.68
W	0.715
P	0.0083

Source Own elaboration based on data from Servicio Meteorológico Nacional (2021)

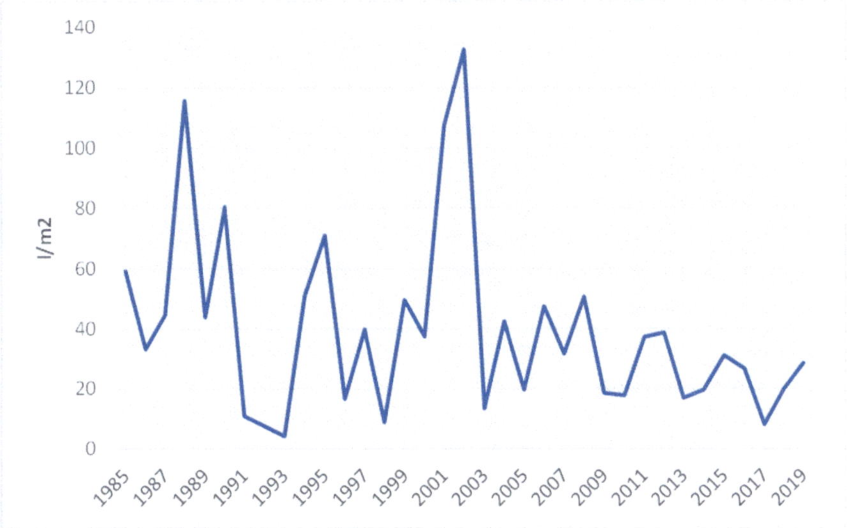

Fig. 22.4 Average precipitation (l/m^2) recorded for the state of Chiapas—Springs 1985–2019 (*Source* Own elaboration based on data from Servicio Meteorológico Nacional [2021])

Table 22.5 Kruskal Wallis study for total precipitation—Springs 1985–2000 and 2001–2019

Precipitation—Springs	
Kruskal Wallis	Results
H	0.58
Degrees of Freedom	1
P	0.446

Source Own elaboration based on data from Servicio Meteorológico Nacional (2021)

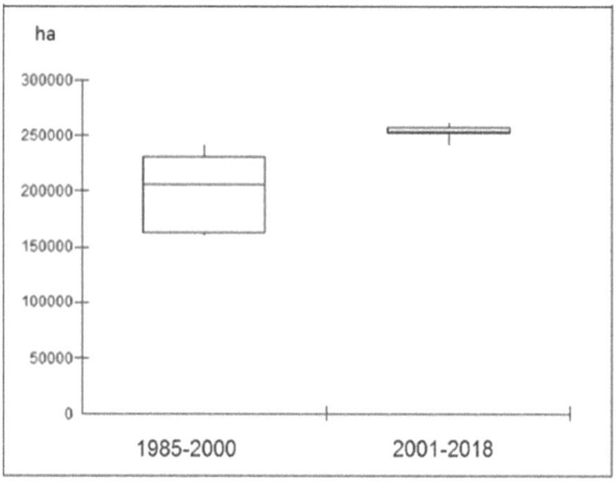

Fig. 22.5 Planting of Cherry Coffee in the state of Chiapas—Comparison Periods 1985–2000 and 2001–2019 (*Source* Own elaboration based on data from Instituto Nacional de Estadística y Geografía [2021] and Servicio de Información Agroalimentaria y Pesquera [2022])

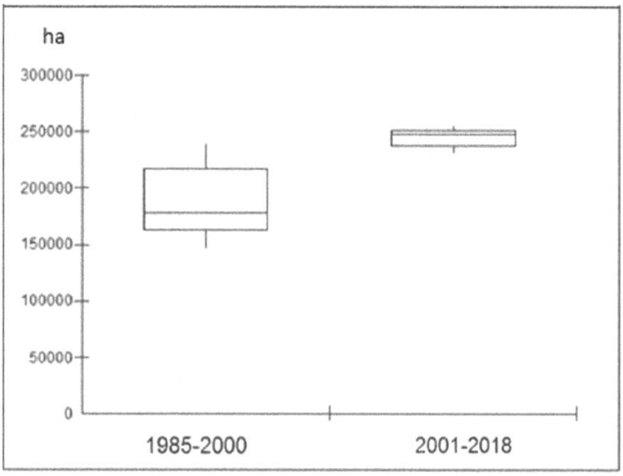

Fig. 22.6 Harvesting of Cherry Coffee in the state of Chiapas–Comparison Periods 1985–2000 and 2001–2019 *Source* Own elaboration based on data from Instituto Nacional de Estadística y Geografía [2021] and Servicio de Información Agroalimentaria y Pesquera [2022])

in both planting and harvesting from 2001 to 2018. The Shapiro Wilk study for the sowing and harvesting of the compared periods 1985–2000 and 2001–2018 yielded values less than 0.05, indicating statistical variation between the means.

Qualitative Analysis

There is national agreement among the social, political, and economic sectors on the importance of agriculture for Mexico's economy, the well-being of its population, and its unquestionable role in combating the adverse effects of climate change (Centro Agronómico Tropical de Investigación y Enseñanza 2011; Food and Agriculture Organization of the United Nations and Secretaría de Agricultura y Desarrollo Rural 2012a; Fideicomisos Instituidos en Relección con la Agricultura 2016; Secretaría de Agricultura Ganadería Desarrollo Rural Pesca y Alimentación 2017). According to the Ministry of Agriculture, Livestock, Rural Development, Fisheries and Food (SAGARPA—acronym in Spanish) (Secretaría de Agricultura Ganadería Desarrollo Rural Pesca y Alimentación 2017), this sector represents a potential for the articulation of policies and the consequent fulfillment of the international goals acquired by the country, especially concerning Sustainable Development Goals (SDGs) 2, 7, 12, and 13 of the United Nations 2030 Agenda (United Nations and Department of Economic and Social Affairs 2015).

Based on the review of the legal and programmatic framework of Mexico and the state of Chiapas, it was found that the National Climate Change Policy Strategy aims to promote better agricultural and forestry practices to increase and preserve natural carbon sinks. Likewise, it was found that to achieve this objective, there are two specific participation bodies in the country, the SAGARPA and the Ministry of Agrarian, Territorial and Urban Development (SEDATU—acronym in Spanish) (Instituto Nacional de Ecología y Cambio Climático 2018).

At the state level, within the Law for Adaptation and Mitigation of Climate Change in the State of Chiapas (Honorable Congreso del Estado de Chiapas 2010), the objective of reducing vulnerability to climate change in the state was found in Article 2. Likewise, in Article 3, it was found that one of the objectives is the development of criteria and indicators in terms of adaptation and mitigation to climate change. Another important objective of this article is to establish the bases for institutional coordination between the State Government agencies, the Municipalities, and the Federal Government regarding vulnerability, risk, adaptation, and mitigation of climate change.

To guarantee technical and financial support for climate change, mitigation, and adaptation actions, the state of Chiapas operates the State Environmental Fund, which is in charge of capturing and channeling resources. But, despite these legal advances, the perception of implementing effective strategies to address the impacts of climate change on coffee farming in the state depends on the space and social group in which one intervenes. The people directly involved in agriculture or decision-making in this sector are the ones who perceive the true fruits of these advances through the results they obtain. This reality was confirmed through the responses obtained from questionnaire number 1 (questionnaire given to the stakeholders involved in the planting and harvesting of coffee).

According to the interviews with 107 stakeholders directly related to coffee growing in Chiapas, this sector still has problems which are mainly related to the following: the change in land use, changing to lower-quality seeds, reduction of coffee growing areas partially replaced by corn crops and livestock, and use of fast-growing varieties that need more sun exposure, among other problems.

The results of the interviews conducted show that the most serious current problems for coffee cultivation in the state are:

a. Deforestation due to using sun varieties to replace the traditionally used Arabica varieties.
b. Migration of Robusta varieties to the highlands.

c. Beans and processed coffee of much lower quality.
d. Support from the governmental agricultural authority is directed and conditioned to using *Catimor* varieties, originating in regions with different climatic conditions than those of Chiapas.
e. Deforestation in high altitude areas to establish new Arabica coffee plantations.
f. The suppliers of nursery plants contracted by the government deliver very poor-quality plants to the producers, which does not guarantee the viability of the planting.
g. Producers are undercapitalized, so they cannot invest too much in planting, resulting in low crop quality and not reaching optimum production levels.
h. There is no access to good quality seeds for producers to have their own nurseries, as used to be done in past decades, since coffee repopulation was done gradually and with local plants.
i. The most serious problem is the aging of coffee plantations. No stakeholders, small or large producers, cooperatives, buyers, or governmental institutions of the productive sector, have invested in the timely renovation of coffee plantations since 1990. This lack of investment is the origin of most of the problems of pests, diseases, and low coffee productivity in the state.
j. A serious problem is the substitution of shade coffee plantations, with Arabica varieties, for sun coffee plantations, with varieties that appear to be resistant to rust.[3] This change was due to a misguided response to the impact of the dramatic expansion of rust beginning in 2012.
k. Loss of coffee value. Organic coffee has a higher price in beans, roasted, and ground, but the producer receives very little of the premium.
l. Official data on the coffee sector in Chiapas are not reliable.

Regarding the responses obtained in the second questionnaire, directed to key stakeholders in different ministries and governmental agencies related to the agricultural sector in the state, it is observed that there is a common perception among them that this is still a very vulnerable sector in the country and in Chiapas. Still, only 50% of the respondents think this vulnerability is directly related to climate change. Many of these stakeholders expressed that the information and data created by scientific studies are seldom used in decision-making. In addition, they stated that the effect of public policy is almost nil, and there is no adequate investment in agriculture to cope with the adverse effects of a changing climate. Finally, most respondents agree that, despite adequate involvement and the existence of some future plans for climate change adaptation, institutional coordination to address these challenges is of very low quality.

Some of the most relevant responses in this survey were:

a. Perception of climate change vs. Perception of agricultural exposure to climate hazards:

[3] Rust is a phytoparasitic fungus that is exclusive to coffee plantations. It is considered one of the most catastrophic diseases that can affect crops in history and is one of the diseases that has caused the most losses in the last 100 years (CropLife Latin America).

More than 66% of respondents agree that the exposure of the agricultural sector to climate hazards is high, as drought and rising temperatures have significantly affected production. However, it is interesting to note that only 37% believe that changes in climate variables are extreme and, in many cases, unpredictable. In fact, more than 38% believe that changes in climatic variables have been recorded in the area but that they are not extreme. It should be noted that close to 29% believe that drought and the increase in temperature have insufficiently affected coffee production and 22% state that there have been no changes in climatic patterns in the last 30 years.

b. Existence of measurement mechanisms vs. Use of scientific data for decision making:

Nearly 57% of those surveyed believe that the mechanisms for measuring, reporting, monitoring, verifying, and evaluating for decision-making on climate change and agriculture are scarce, and 37% believe that these mechanisms are non-existent. Likewise, 57% of respondents concluded that scientific data (temperature, precipitation, risk, vulnerability) are rarely taken into account in decision making, and 30% expressed that these data are taken into account, but only to a limited extent.

c. Effect of public policy on reducing the vulnerability of the agricultural sector to climate change vs. Local, regional, and international investment for agriculture applied to climate change in the state:

About 58% of the people who responded to the surveys believed that public policy had no positive effect since agriculture is still vulnerable. On the other hand, 38% said that public policy had benefited somewhat, but agriculture continues to be vulnerable. On the other hand, 65% said that the funds allocated for agriculture in the area of climate change are non-existent and about 32% said that some funds are allocated, but they are limited. It can be noted that, in both cases, the perception of low public policy impact and low investment considerably exceeded the medium and high levels.

d. Institutional coordination on vulnerability, risk, adaptation, and mitigation to climate change vs. Involvement of key stakeholders in decision-making and the definition of present and future plans for mitigation and adaptation to climate change:

51% of respondents stated that institutional coordination on climate change is non-existent or of very poor quality, and 50% stated that the involvement of key stakeholders in climate change decision-making is insufficient. Regarding the definition of present and future plans for mitigation and adaptation to climate change, the responses were divided between 46% who said they had an idea of some mitigation and adaptation plans that are being implemented and/or being developed in their agencies and 45% who said they did not know if their agency was working on the development of any plan to address this issue. In all cases, the responses were divided into almost 50% of those who believe that the definition of plans, stakeholder involvement, and institutional coordination is of medium quality and those who believe it is of low quality.

A summary of the results observed in this study can be seen in the following scheme (Fig. 22.7):

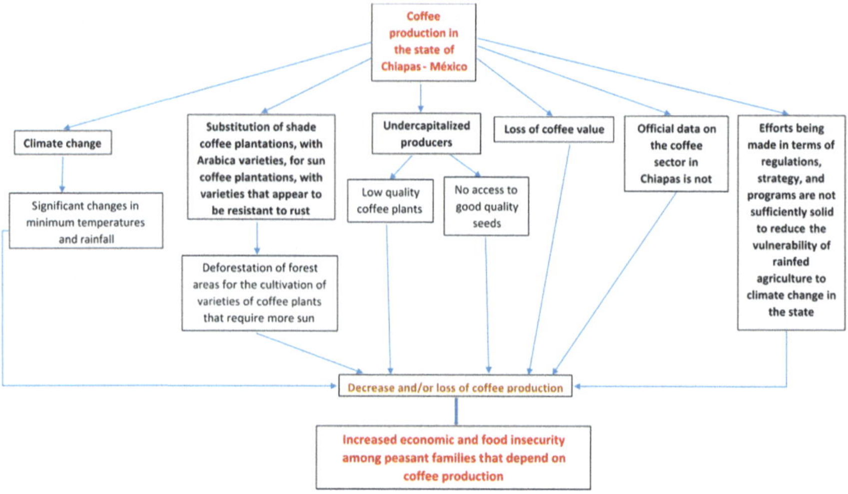

Fig. 22.7 Outline of research results (*Source* Own elaboration)

Conclusions

The state of Chiapas is located in the southeastern region of Mexico and is formed by 27 Protected Natural Areas, which add up to 11,461,244 hectares. Despite the wealth of Its natural resources, 83% of households in Chiapas present some level of food insecurity, of which 85% are in rural areas (Food and Agriculture Organization of the United Nations and Secretaría de Agricultura y Desarrollo Rural 2012b; Consejo Nacional de Población 2018).

For these poorest populations in rural areas, coffee cultivation represents the basis of their income, and the impacts generated by climate change in the region may cause the levels of food insecurity among peasant families to worsen, so it is important to build the adaptive capacity of these groups.

According to the data obtained in this study, there is evidence of significant changes in climatic variables, especially in minimum temperatures and rainfall, in the state of Chiapas. These changes are affecting productivity and creating losses for coffee growing in the state. These impacts have created changes in farmers' incomes, land use, and agricultural practices implemented in the region. This scenario will likely increase the vulnerability of both the producer and the ecosystems, which further distances the country from the process of adapting to climate change.

At the national and state level, there is a legal framework for climate change that is very similar to the international one. This framework allows for the development of public policies that can address the adverse effects of climate change in the agricultural sector; however, there is a perception that these policies are not visualized in the programmatic structure of the state of Chiapas.

Although there are areas of work at the institutional level focused on climate change mitigation and adaptation, it is still not possible to visualize their impact on the territory, and the policies applied.

It is observed that the key decision-makers in the state's agricultural sector agree that coffee growing is a vulnerable sector in the region, but this perception is vague and confusing. A relationship between climate change and agricultural vulnerability is not yet perceived. Likewise, these key stakeholders have not yet identified how this situation should be addressed, nor who is responsible for it.

On the other hand, the perspective of the stakeholders involved in the planting and harvesting of coffee in Chiapas is that the efforts being made in terms of regulations, strategy, and programs are not sufficiently solid to reduce the vulnerability of rainfed agriculture to climate change in the state, which makes it impossible to maintain a constant production of coffee in the region and harms the economic and food security of the peasant families that depend on it.

Prospects and limitations of the study: The study addressed the entire state of Chiapas, which, as mentioned, is one of the poorest in Mexico and presents very high levels of food insecurity, especially among peasant families that depend on coffee production for their subsistence. Despite this, as Chiapas is a large state, marked differences can be observed between the different regions, both concerning the impacts of climate change on agriculture and the socioeconomic conditions of the producers, therefore, an analysis that takes into account these regional differences could open new perspectives on the situation of coffee production in the state in the face of the impacts of climate change.

References

Abeldaño Zuñiga RA, Lima GN, González Villoria AM (2021) Impact of slow-onset events related to climate change on food security in Latin America and the Caribbean. Curr Opin Environ Sustain 50:215–224. https://doi.org/10.1016/j.cosust.2021.04.011

Borràs S (2017) Movimientos para la justicia climática global: replanteando el escenario internacional del cambio climático [Movements for global climate justice: reframing the international scenario of climate change]. Relac Int 0:97–119

Cardona OD, van Aalst MK, Birkmann J (2012) Determinants of risk: exposure and vulnerability. Managing the risks of extreme events and disasters to advance climate change adaptation. Cambridge University Press, Cambridge, Inglaterra, pp 65–108

Centro Agronómico Tropical de Investigación y Enseñanza (2011) Estrategia del sector cafetalero para la adaptación, mitigación y reducción de la vulnerabilidad ante el cambio climático en la Sierra Madre de Chiapas [Strategy of the coffee sector for the adaptation, mitigation and reduction of vulnerability to climate change in the Sierra Madre de Chiapas]

Consejo Nacional de Población (2018) Proyecciones de la Población de México y de las Entidades Federativas, 2016–2050 y Conciliación Demográfica de México, 1950 -2015 [Projections of the Population of Mexico and of the Federal Entities, 2016–2050 and Demographic Reconciliation of Mexico, 1950 -2015]. CropLife Latin America Roya del cafeto. https://www.croplifela.org/es/plagas/listado-de-plagas/roya-del-cafeto. Accessed 7 Jul 2022

Economic Commission for Latin America and the Caribbean (2015) The economics of climate change in Latin America and the Caribbean: paradoxes and challenges of sustainable development. Santiago, Chile

Fideicomisos Instituidos en Relección con la Agricultura (2016) Panorama Agroalimentario (Café 2016) [Agrifood Panorama (Coffee 2016)]

Food and Agriculture Organization of the United Nations, Secretaría de Agricultura y Desarrollo Rural (2012a) Agricultura familiar con potencial productivo en México [Family farming with productive potential in Mexico]

Food and Agriculture Organization of the United Nations, Secretaría de Agricultura y Desarrollo Rural (2012b) México: el sector agropecuario ante el desafío del cambio climático [Mexico: the agricultural sector facing the challenge of climate change]

Gobierno del Estado de Chiapas (2011) Programa de Acción ante el Cambio Climático del Estado de Chiapas [Chiapas State Climate Change Action Program]

Harvey CA, Saborio-Rodríguez M, Martinez-Rodríguez MR et al (2018) Climate change impacts and adaptation among smallholder farmers in Central America. Agric Food Secur 7:57. https://doi.org/10.1186/s40066-018-0209-x

Honorable Congreso del Estado de Chiapas (2010) Ley para la Adaptación y Mitigación ante el Cambio Climático en el Estado de Chiapas [Law for Adaptation and Mitigation to Climate Change in the State of Chiapas]

Instituto Nacional de Ecología y Cambio Climático (2018) Evaluación de la política nacional de cambio climático [Evaluation of the national climate change policy]. https://www.gob.mx/inecc/acciones-y-programas/evaluacion-de-la-politica-nacional-de-cambio-climatico. Accessed 7 Jul 2022

Instituto Nacional de Estadística y Geografía (2021) Datos – Oferta de información estadística a nivel nacional y por entidad federativa [Data – Offer of statistical information at the national level and by federal entity]. https://www.inegi.org.mx/datos/. Accessed 28 Aug 2021

Kruskal WH, Wallis WA (1952) Use of ranks in one-criterion variance analysis. J Am Stat Assoc 47:621. https://doi.org/10.2307/2280779

de Lima GN, Lombardo MA, Magaña V (2018) Urban water supply and the changes in the precipitation patterns in the metropolitan area of São Paulo – Brazil. Appl Geogr 94:223–229. https://doi.org/10.1016/j.apgeog.2018.03.010

de Lima GN, Magaña Rueda VO (2018) The urban growth of the metropolitan area of Sao Paulo and its impact on the climate. Weather Clim Extrem 21:17–26. https://doi.org/10.1016/j.wace.2018.05.002

López Cruz DB, Abeldaño Zuñiga RA, de Lima GN (2021) Analysis of international climate change agreements in the context of Latin America and the Caribbean. In: Leal Filho W (ed) Handbook of Climate Change Management, 1st edn. Springer International Publishing, Cham, pp 1–26

Magaña VO (2013) Guía metodológica para la evaluación de la vulnerabilidad ante cambio climático [Methodological guide for the assessment of vulnerability to climate change]. INE/PNUMA., Ciudad de México, México

Martínez-Austria PF (2020) Tendencias observadas en las temperaturas máximas en el noroeste de México [Observed trends in maximum temperatures in northwestern Mexico]. In: Martínez-Austria PF, Corona-Vásquez B, Ptiño-Gómez C (eds) Cambio climático y riesgos hidrometeorológicos, 1st edn. Fundación Universidad de las Américas, Puebla, pp 65–86

McMichael AJ, Lindgren E (2011) Climate change: present and future risks to health, and necessary responses. J Intern Med 270:401–413. https://doi.org/10.1111/j.1365-2796.2011.02415.x

Murray-Tortarolo GN, Jaramillo VJ (2019) The impact of extreme weather events on livestock populations: the case of the 2011 drought in Mexico. Clim Change 153:79–89. https://doi.org/10.1007/s10584-019-02373-1

Neri C, Magaña V (2016) Estimation of vulnerability and risk to meteorological drought in Mexico. Weather Clim Soc 8:95–110. https://doi.org/10.1175/WCAS-D-15-0005.1

Pei S, Dahl KA, Yamana TK, et al (2020) Compound risks of hurricane evacuation amid the COVID-19 pandemic in the United States. medRxiv. https://doi.org/10.1101/2020.08.07.20170555

Prieto-González R (2020) Riesgos hidrometeorológicos en México [Hydrometeorological risks in Mexico]. Cambio climático y riesgos hidrometeorológicos, 1st edn. Fundación Universidad de las Américas, Puebla, Puebla, pp 31–44

Rockström J, Steffen W, Noone K et al (2009) Planetary boundaries: exploring the safe operating space for humanity. Nature 461:472–475. https://doi.org/10.1038/461472a

Secretaría de Agricultura Ganadería Desarrollo Rural Pesca y Alimentación (2017) Planeación agrícola nacional 2017–2030 (Café Mexicano) [National agricultural planning 2017–2030 (Mexican Coffee)]

Servicio de Información Agroalimentaria y Pesquera (2022) Producción Agrícola [Agricultural production] . In: Serv. Inf. Agroaliment. y Pesq. https://www.gob.mx/siap/acciones-y-programas/produccion-agricola-33119. Accessed 7 Jul 2022

Servicio Meteorológico Nacional (2021) Servicio Meteorológico Nacional [National Metereological Service] In: Serv. Meteorológico Nac. https://smn.conagua.gob.mx/es/. Accessed 28 Aug 2021

Shapiro SS, Wilk MB (1965) An analysis of variance test for normality (complete samples). Biometrika 52:611. https://doi.org/10.2307/2333709

United Nations, Department of Economic and Social Affairs (2015) Sustainable development goals. https://sdgs.un.org/goals. Accessed 20 Mar 2022

Tanya Méndez Matus is originally from Chiapas, Mexico, and currently lives in Idaho, USA. She has a degree in Biology in the area of Natural Services from Brigham Young University—Idaho/USA. MSc Tanya Mendez worked as a volunteer in many projects such as crocodile population assessment and sea turtle reintroduction, both led by the National Commission of Natural Protected Areas of the Government of Mexico. She obtained her master's degree in Risk Management and Climate Change from the Research Institute in Risk Management and Climate Change of the University of Sciences and Arts of Chiapas—Mexico.

Gabriela Narcizo de Lima Assistant Professor at the Department of Geography of the Faculty of Letters—University of Porto—Portugal. Doctor of Science (Physical Geography) from the University of Sao Paulo (USP)—Brazil, with a master's and bachelor's degree in Geography from the São Paulo State University "Julio de Mesquita Filho" (UNESP)—Brazil. Between 2016 and 2018, Professor Gabriela Narcizo de Lima carried out two postdoctoral research stays in the Department of Territorial Planning and Geoprocessing of the Institute of Geosciences and Exact Sciences of Rio Claro/Sao Paulo (UNESP) and the Institute of Geography of the National Autonomous University of Mexico (UNAM). Her main lines of research are related to the impacts of anthropic activities on climate at various scales and issues associated with managing water resources and land use planning.

Chapter 23
The Role of Education and Research in Human and Planetary Health

Walter Leal Filho, Juliane Boenecke, Derya Taser, and Johanna Heimfarth

Abstract This chapter outlines the role of education and research in fostering human and planetary health and describes the example of the project "KlimaGESUND", which attempts to foster awareness of and preparedness for the health impacts of climate change in health sciences and public health higher education in Germany.

Introduction

The year 2015 was special in the field of human and planetary health. In that year, the report produced by the Rockefeller Foundation and the journal The Lancet called "Safeguarding human health in the Anthropocene epoch: report of The Rockefeller Foundation-Lancet Commission on planetary health" was launched. Also in 2015, the World Health Organization and the Secretariat of the Convention on Biological Diversity published the report "Connecting global priorities: biodiversity and human health: a state of knowledge review" with over 100 contributors, meant to guide future joint actions. Both documents comprehensively address the need for a better understanding of the connections between human health and ecosystems, and the

W. Leal Filho (✉) · J. Boenecke · D. Taser
Research and Transfer Centre "Sustainable Development and Climate Change Management", Hamburg University of Applied Sciences, Ulmenliet 20, 21033 Hamburg, Germany
e-mail: walter.leal2@haw-hamburg.de

J. Boenecke
e-mail: juliane.boenecke@haw-hamburg.de

D. Taser
e-mail: derya.taser@haw-hamburg.de

W. Leal Filho · J. Boenecke
Department Health Sciences, Hamburg University of Applied Sciences, Ulmenliet 20, 21033 Hamburg, Germany

J. Heimfarth
School of Public Health, Charité - Universitätsmedizin Berlin, Seestraße 73, 13347 Berlin, Germany

W. Leal Filho et al. (eds.), *Climate Change and Health Hazards*, Climate Change Management, https://doi.org/10.1007/978-3-031-26592-1_23

risks associated with damages to the integrity of the planet. The period in which humanity finds itself right now, the Anthropocene, is a risky one since mankind is putting the planet under considerable pressure. These elements have led to the emergence of a new field of research, namely planetary health.

Planetary health refers to human health and the state of natural systems surrounding humans. It identifies that both groups are linked because human civilization and health are dependent on natural systems and the preservation of natural resources. Recent decades have seen the degradation of natural systems and resources whose cumulative effects now threaten the existence of future generations (UNFCCC 2022). The debate on planetary health has been expanded to the vaster term of "planetary wellbeing". This is used to describe the highest attainable standard of well-being for human and non-human beings and their social and natural systems. An example of what is being done in this area is a scheme titled "Planetary Wellbeing Initiative", which was implemented by Pompeu Fabra University in 2018. This is a long-term institutional plan that also accounts for the UN Sustainable Development Goals (SDG). This programme recognizes the need for research into developing concepts as well as inter and transdisciplinary collaborations, that can engage students and encourage them to work together towards a healthier planet (Antó et al. 2021).

Planetary Health is now being driven as an effort to deal with the question of sustainability and human life on the planet, under an increasingly integrative, transdisciplinary, and global perspective, since the problems of this planetary crisis cross geopolitical borders, and academic boundaries and affect humanity as a whole. Along these lines, the most decisive issue seems to be the choice of the adjective "planetary" to express the imperative of relating human health to ecosystem health. This has a meaning deeper than the expression "global health" and seems more appropriate since the adjective "planetary" means the health of the biosphere. The concepts of Global Health and Planetary Health are closely associated with general perspectives, especially the SDGs in the context of global crises, and in particular the COVID-19 pandemic. Indeed, the many problems and changes in lifestyles by the pandemic shed light on the challenges related to tackling infectious diseases, food insecurity, and the climate emergency at the same time. Planetary health seeks to address a very concrete contemporary problem, namely the need to understand, quantify and act in order to reverse the effects of human population growth and the acceleration of socio-economic activities on the environment and, inter alia, the disturbances on the Earth's natural ecosystems. And how these, in turn, impact human health and well-being. Anthropic disturbances in natural ecosystems are characterised by changes in climate, land use, changes in the nitrogen and phosphorus cycle, chemical pollution of soil, water, and air, reduction in the availability of drinking water, loss of biodiversity, destruction of the ozone layer, and ocean acidification, among others. In all these areas, there is a perceived need to document and promote examples of initiatives and good practices, which may change current trends. In this context, education has a key role to play, as outlined in the next section.

The Role of Education and Research in Human and Planetary Health

One of the best ways to improve planetary health is by introducing the subject into education. This can be implemented at various stages of education and thus has numerous benefits. With regard to the earlier levels of education such as primary and secondary, basic education for planetary health can help to shape the mindsets of young individuals. Early schooling years are important for building a strong foundation for responsible individuals at a late stage. By introducing planetary health education to young kids, mindsets can be changed to ensure that children are climate-conscious, and are in a better position to understand the consequences of their actions towards them, and their environment (von Borries et al. 2020). Changes triggered by education processes may be attributed to children understanding the linkages between nature, animals, and humans and adopting behaviours that protect the health of the environment and inter alia, of humans. Some studies indicate that children who were taught about climate and environmental issues and how these related to human health, showed an increased adaptive capacity and reduced vulnerability to climate change in their adult life (Lutz et al. 2014).

In tertiary education, the incorporation of planetary health studies is often associated with a drive toward transdisciplinary collaboration. This serves the purpose of increasing transformative action, that is backed by a diverse array of knowledge and moves towards improving health and achieving the SDGs (Guzmán et al. 2021). More recently, planetary health education has been discussed, quite frequently, in connection with health profession degrees. All healthcare professions have a unique role to play and the responsibility of educating current and future professionals but also their patients and clients, providing them with the required knowledge to properly respond to the environmental, social, and health problems faced today, especially in respect of achieving the SDGs (Watts et al. 2021). Many countries have agreed to reduce carbon emissions by at least 45% within the next 10–15 years, and achieve carbon neutrality by 2050, including in the field of healthcare.

However, despite these encouraging trends, healthcare graduates are not yet fully equipped to handle matters related to planetary health. Therefore, healthcare education in tertiary institutions needs to provide students with the vital information, tools, skills, and mindsets they require to implement sustainable healthcare and foster the well-being of all generations while reiterating the need to protect the health of the planet (Shaw et al. 2021). Aside from its incorporation into the curriculum, research for planetary health should also be prioritised This is important since innovative solutions and strategies need to be developed to improve well-being and health on earth. A recent workshop organised by the "Future Earth Health Knowledge-Action Network" (Health KAN) highlighted a research agenda that focused on understanding the management of public health in connection to environmental changes being experienced, in four primary areas:

1. risk and management;
2. improving climate-resilient health systems;
3. monitoring surveillance and evaluation; and
4. risk communication.

These four areas looked at major problems such as climate change, extreme weather events, food production, and consumption and waste management. The event aimed to influence research worldwide, especially at higher education institutions, by identifying current knowledge gaps and providing a basis for policy design through transdisciplinary efforts. Therefore, research as part of education is a key component in the promotion of planetary health (Ebi et al. 2020). In order to achieve this, planetary health needs to be well integrated into the university curriculum. An example of an attempt to achieve this is the project "KlimaGESUND", which aims at strengthening higher education and capacity building on climate change and health in the field of Health Sciences and Public Health in Germany.

Climate Change in Health Sciences and Public Health Higher Education in Germany

Climate change poses a major threat to the health of humans, animals, and the environment, with Germany being no exception: Increasing numbers of heat-related deaths during the summer months; newly emerging infectious diseases; the expansion of pollen season in length and intensity; compromised health care performance due to severe adverse effects caused by extreme weather; psychological stress and trauma following extreme weathers and biodiversity losses—those, among others, pose the diverse direct and indirect effects of climate change on health that our health system must face today and in the near future (Gu et al. 2020; Schulz et al. 2019; Heuson and Traidl-Hoffmann 2018). The development of effective climate adaptation strategies as well as climate mitigation are pivotal to securing and promoting health and well-being at the individual and population level against the impacts of climate change. Hence, there is an urgent need to raise awareness, build profound knowledge and capacity, and increase literacy about the health risks and challenges of climate change already starting in professional training and higher education, with medical and health-related stakeholders playing an essential role (Yang et al. 2018).

Against this background, strengthening the teaching and capacity building of health professions on the health challenges of climate change in Germany is critical (Lancet Countdown on Health and Climate Change 2021). The recent climatic extremes in Germany, including heavy rain, flooding, storms, and extreme heat, add to this urgency and are already posing new challenges in health care and disaster response. While medical science and nursing have set promising examples in higher education, a stronger effort is needed in the disciplines of Health Sciences and Public Health. For example, the German Alliance on Climate Change and Health (KLUG e.V.) published a report on the existing teaching initiatives on Planetary Health in

Table 23.1 Participatory evaluation of KlimaGESUND during the design and pilot phase (N = 79)

How threatening do you perceive global climatic and environmental changes to be for your health?					
Highly threatening	Somewhat threatening	Neither nor	Barely threatening	Not at all threatening	I don't know
35 (44%)	**40 (51%)**	2 (3%)	2 (3%)	0 (0%)	0 (0%)
How threatening do you perceive the health effects of climate change in Germany?					
5 (6%)	**64 (81%)**	7 (9%)	2 (3%)	0 (0%)	1 (1%)
In your professional role, how prepared do you feel for the health challenges of climate change?					
Very well prepared	Well prepared	Somewhat prepared	Rather poorly prepared	Not at all prepared	I don't know
2 (3%)	9 (11%)	**44 (56%)**	**18 (23%)**	5 (6%)	1 (1%)

higher health education in Germany and outlined shortcomings in the respective fields compared to medical sciences (Gepp et al. 2022). In the KlimaGESUND project, we could gain similar insights, while at the same time an increased interest and demand among students and lecturers could be observed in strengthening and promoting climate education in vocational and postgraduate training.

To gain a better understanding, we asked 79 students of Health Sciences and Public Health from different universities in Hamburg, Berlin, and Baden-Wuerttemberg how threatening they perceive climate change to be, more specifically in the geographic setting of Germany, whether they feel sufficiently prepared for the climate change-induced challenges associated with their health professionals, and whether curriculum content on climate change and health is already taught in their study or training programme (cf. Table 23.1).

Overall, all participants perceived climate change as a health threat, although less so in their own geographic context, with more than half feeling only somewhat or rather poorly prepared for the challenges ahead. In addition, more than half of the participants stated that little or no content on climate change and health had been taught in their study or training programme (66%). This formed the basis of the idea and motivation for the educational project KlimaGESUND.

Insights Into KlimaGESUND

The future healthcare and Public Health actors are part of a generation that will be significantly involved in the development of climate change mitigation and adaptation strategies in Germany and beyond. Climate change curricula are pivotal in Health Sciences and Public Health to prepare future actors in various fields of health care, health services, and health research for practice in a drastically changing world in order to ensure sustainable health care for all (SDG No. 3). To address this gap, the project KlimaGESUND developed and piloted teaching and training materials on different topics of climate change, climate adaptation, and climate change mitigation from a Health Sciences and Public Health perspective, aiming for easy access and flexible implementation in higher education and vocational training programmes. Across the various resources provided by KlimaGESUND, pre-knowledge will be strengthened and basic scientific facts on climate change as a driver of health will be explained from multiple perspectives (Toolbox 1). The consequences, risks, and opportunities for population health will be explored, with relevant adaptation strategies in the field of Health Sciences and Public Health being discussed and future fields of action in the regional context being identified (Toolbox 2). Interactive case studies and table-top exercises on selected climate change and health challenges supplement the materials (Toolbox 3).

To offer diverse learning experiences suitable for different settings, KlimaGESUND is based on the didactic concept of "blended learning" that combines the strengths and well-established methods of traditional teaching with interactive, web-based approaches, with the learning objectives of increasing awareness and knowledge about climate change and health among the future health actors and strengthening associated competencies in an interactive way. Moreover, a modular design was adopted to enable the use of single materials as well as to combine resources for more extensive sessions. This way, individual teaching units of 30–45 min up to longer-term student projects could be created. By using this approach, we aim to offer students and lecturers from Health Sciences and Public Health but also associated health professions a variety of learning and teaching resources that combine offline and online materials, including classic slide sets, reading materials, and recommendations, a scientific literature database, interactive web-based learning tools, audio, and video features, as well as case study and table-top exercises for more hands-on and team-oriented training. The materials cover the fundamentals of Health Sciences and Public Health, selected climate change-associated health challenges in Germany (e.g., extreme heat, weather extremes, newly emerging diseases, hazards in the air like pollen and air pollution, supply of water and food, changes in regional ecology and biodiversity), and facilitate the transfer into the practice of both disciplines, primarily addressing aspects of climate adaptation.

Transfer into Practice

Health Sciences and Public Health can serve a very special mission in health professions education and transfer into practice. Both disciplines represent a unique cross-section between individual, population, and global health, between research and practice, and between the individual disciplines of health care, services, and research. At the same time, climate change, as an extremely complex crisis for health and society, has diverse impacts on the individual, the local, the national, the regional, the global, and even the planetary level. Important fields of action also exemplified in the KlimaGESUND framework (cf. Fig. 23.1), can be found in the areas of

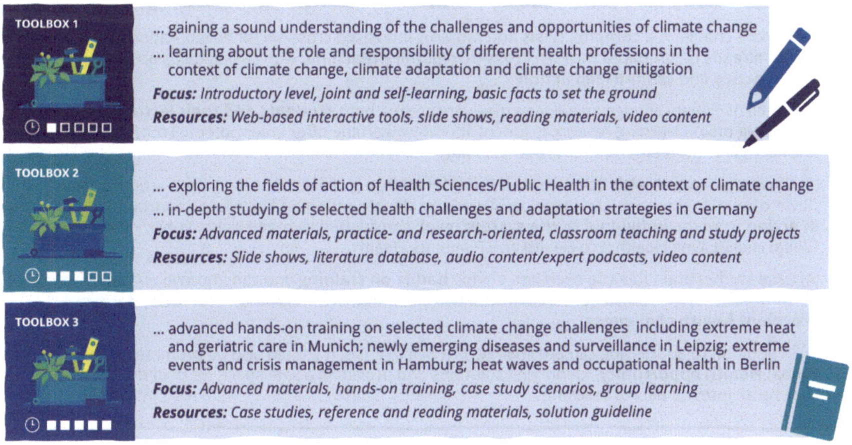

Fig. 23.1 KlimaGESUND framework (*Source* Authors)

action established in Public Health, including surveillance, disaster management, occupational health, health promotion, prevention, governance, health economics, risk communication, among others (Kurth and Matusall 2018; Schaeffer et al. 2021), with the overarching scheme of Health Sciences research linking to all fields of action. As a complex emergency similar to the COVID-19 pandemic, climate change in the context of health and well-being requires cross-sectoral solutions, transfer of established methodologies to new domains of application, and innovative concepts of risk monitoring, early warning, and response. Moreover, stronger investments are needed in disaster preparedness, public communication and education, and policy decision support, contributing to both climate adaptation and climate change mitigation. This also reflects the unique potential of KlimaGESUND, which is primarily designed for these disciplines. We can therefore share the following recommendations for teaching and training derived from the project (Fig. 23.2).

The didactic concept of the educational module KlimaGESUND has proven highly valuable in both national and international teaching and has been able to facilitate international exchange in addition to interdisciplinary collaboration. However, to enable successful transfer to other geographic settings, we strongly recommend integrating local and regional climatic changes and challenges into teaching activities rather than addressing global concepts and engaging students in developing their own climate change strategies. The provision and promotion of basic climate change knowledge (cf. Fig. 23.1, "Toolbox 1") can provide a sound starting point, which should then be complemented by data, literature, and other materials related to the students' geographic and environmental context. Finally, we would like to encourage open access and sharing of educational initiatives and materials in order to provide lecturers and students with an easily deployable resource to support them in integrating climate change and health into their teaching and curricula.

(1) Climate change is a **complex crisis** and requires increased cross-sector collaboration; this should already be prepared through **cross-disciplinary teaching**, with strong focus on **transfer into practice** and **team-based action**

(2) States of climate change knowledge may vary among both **students and their lecturers;** interactive online learning resources for collaborative learning offer great potential for **joint knowledge enhancement** and further exchange

(3) **Concepts from Health Sciences and Public Health**, e.g., fields of public health action or the „River of Live" metaphore, can meaningfully inform climate action and highlight important **fields of action** from the **individual to the system level**; considering Human Health, Public Health, Global Health, One Health, Eco Health and Planetary Health

(4) Case studies and table-top exercises enable **hands-on training** and can improve preparedness for future challenges; these become particularly valuable when integrating **local facts, data and prevalent health challenges**

(5) In addition to addressing climate change as a health crisis, it should also be promoted as a **Global Health opportunity**; identifying solutions and providing a safe space for peer discussion shall be an integral part of teaching

Fig. 23.2 Lessons learnt from the KlimaGESUND project (*Source* Authors)

Conclusions

Young academics are demanding that universities start implementing changes towards the implementation of profound planetary health education and transformative action in the field of medical sciences, health care, public health, and health research. Firstly, they urge universities to promote research that bridges the gap between sound scientific findings and actual positive changes in society. Secondly, they emphasise the need to implement a so-called "planetary health curricula". This can be done by the integration into existing modules, as pursued by the KlimaGESUND project, or the development of new, standalone courses. Thirdly, they require universities to more actively engage in communicating scientific findings to the general public, so as to increase their levels of information and awareness. Such changes are vital to equip young people with the necessary skills and knowledge to improve planetary health while engaging the general community. Furthermore, it makes it easier to address planetary health challenges from a social equity lens (Wabnitz et al. 2020).

But to yield the expected benefits, education for planetary health needs to consider a variety of factors, as outlined in Fig. 23.3.

Despite the need for planetary health education, very few guidelines have been provided as to how it may be implemented under an international consensus. To provide a sense of direction for institutions, educators, and students, the Planetary Health Alliance created the "Planetary Health Education Framework", which is now being developed. A special task force was created with experts on planetary health to design a functional framework that has common foundational languages across multiple disciplines, sectors, and regions (Guzmán et al. 2021). The framework aims to work on twelve cross-cutting principles for planetary health education that will lead to the promotion of diverse education strategies to protect and restore planetary health

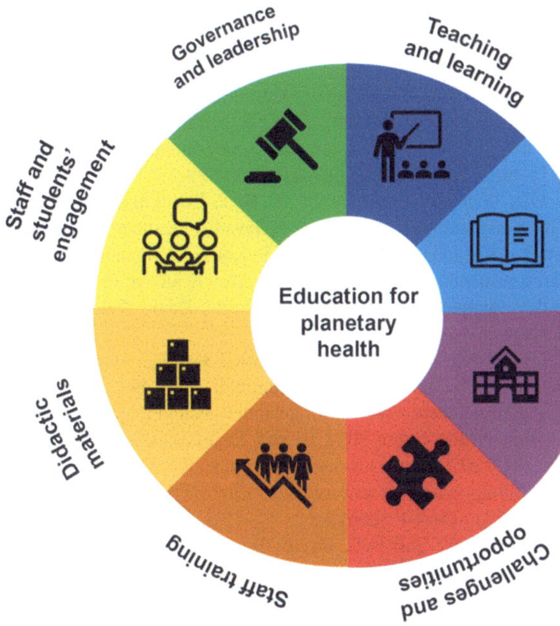

Fig. 23.3 Some of the factors to be considered in implementing education for planetary health (*Source* Authors)

(Guzmán et al. 2021; Stone et al. 2018). In this context, the framework developed for Germany by Gepp et al. (2022) shall also be highlighted. It is hoped that the "Planetary Health Education Framework" will provide a sound basis, upon which the education potential of planetary health may be realised, also supported by research that may shed light on some of the areas where improvements are needed.

Acknowledgements The KlimaGESUND project was carried out at the *Research and Transfer Centre "Sustainable Development and Climate Change Management"* (FTZ-NK) at the Hamburg University of Applied Sciences. We thank all Bachelor's and Master's students for their contributions and engagement with the project and are grateful for collaborating with highly committed partners in the field of health research and education in Hamburg, Stuttgart, Munich, Berlin, and beyond. The project was funded by the Federal Ministry for the Environment, Nature Conservation, Nuclear Safety, and Consumer Protection (ID 67DAS199) from 01/2020 to 03/2022.

Conflict of Interest The authors declare no conflict of interest.

Ethical Statement No ethical clearance was required for undertaking the project KlimaGESUND.

References

Antó JM, Martí JL, Casals J, Bou-Habib P, Casal P, Fleurbaey M, Frumkin H, Jiménez-Morales M, Jordana J, Lancelotti C, Llavador H, Mélon L, Solé R, Subirada F, Williams A (2021) The planetary wellbeing initiative: pursuing the sustainable development goals in higher education. Sustainability 13(6):3372. https://doi.org/10.3390/su13063372

Ebi KL, Harris F, Sioen GB, Wannous C, Anyamba A, Bi P, Boeckmann M, Bowen K, Cissé G, Dasgupta P, Dida GO, Gasparatos A, Gatzweiler F, Javadi F, Kanbara S, Kone B, Maycock B, Morse A, Murakami T, Mustapha A, Pongsiri M, Suzán G, Watanabe C, Capon A (2020) Transdisciplinary research priorities for human and planetary health in the context of the 2030 agenda for sustainable development. Int J Environ Res Public Health 17(23):8890. https://doi.org/10.3390/ijerph17238890

Gepp S, Hammelbeck J, Hornschuch M, Jung L, Manderschied S, Regler H, Saß R, Schwienhorst-Stich E-M, Wabnitz K (2022). Klima. Umwelt. Gesundheit. Ein Leitfaden für Lehrangebote zu planetarer Gesundheit. Online. https://www.klimawandel-gesundheit.de/wp-content/uploads/2022/01/Leitfaden-Planetary-Health-Lehre-2022_01.pdf. Last Accessed 18 Mar 2022

Gu H, Yan W, Elahi E, Yuxia C (2020) Air pollution risks human mental health: an implication of two-stages least squares estimation of interaction effects. Environ Sci Pollut Res Int 27(2):2036–2043. https://doi.org/10.1007/s11356-019-06612-x

Guzmán CAF, Aguirre AA, Astle B, Barros E, Bayles B, Chimbari M, El-Abbadi N, Evert J, Hackett F, Howard C (2021) A framework to guide planetary health education. Lancet Planet Health 5(5):e253–e255. https://doi.org/10.1016/S2542-5196(21)00110-8

Heuson C, Traidl-Hoffmann C (2018) The significance of climate and environment protection for health under special consideration of skin barrier damages and allergic sequelae. Bundesgesundheitsblatt—Gesundheitsforschung—Gesundheitsschutz 61(6):684–696. https://doi.org/10.1007/s00103-018-2742-y

Kurth BM, Matusall S (2018). Das Zukunftsforum public health. Public Health Forum 26(3):205–208. https://doi.org/10.1515/pubhef-2018-0070

Lancet Countdown on Health and Climate Change (2021) Policy brief for Germany. Online. https://klimagesund.de/wp-content/uploads/2020/12/Lancet-Countdown-Policy-Brief-Germany_ENG.pdf. Last Accessed 18 Mar 2022

Lutz W, Muttarak R, Striessnig E (2014) Universal education is key to enhanced climate adaptation. Science 346(6213):1061–1062. https://doi.org/10.1126/science.1257975

Schaeffer D, Gille S, Vogt D, Hurrelmann K (2021) National action plan health literacy in Germany origin, development, and structure. J Public Health. https://doi.org/10.1007/s10389-021-01616-9

Schulz H, Karrasch S, Bölke G, Cyrys J, Hornberg C, Pickford R, Schneider A, Witt C, Hoffmann B (2019). Atmen: Luftschadstoffe und Gesundheit—Teil I. Pneumologie 73(05):288–305. https://doi.org/10.1055/a-0882-9366

Shaw E, Walpole S, McLean M, Alvarez-Nieto C, Barna S, Bazin K, Behrens G, Chase H, Duane B, El Omrani O (2021) AMEE consensus statement: planetary health and education for sustainable healthcare. Med Teach 43(3):272–286. https://doi.org/10.1080/0142159X.2020.1860207

Stone SB, Myers SS, Golden CD (2018) Cross-cutting principles for planetary health education. Lancet Planetary Health 2(5):e192–e193. https://doi.org/10.1016/S2542-5196(18)30022-6

UNFCCC. (2022). Planetary health. https://unfccc.int/climate-action/un-global-climate-action-awards/planetary-health#:~:text=One%20well%2Dknown%20impact%20of,frequent%20incidence%20of%20severe%20storms

Von Borries R, Guinto R, Thomson DJ, Abia WA, Lowe R (2020) Planting sustainable seeds in young minds: the need to teach planetary health to children. Lancet Planet Health 4(11):e501–e502. https://doi.org/10.1016/S2542-5196(20)30241-2

Wabnitz K-J, Guzman V, Haldane V, Ante-Testard PA, Shan Y, Blom IM (2020) Planetary health: young academics ask universities to act. Lancet Planet Health 4(7):e257–e258. https://doi.org/10.1016/S2542-5196(20)30142-X

Watts N, Amann M, Arnell N, Ayeb-Karlsson S, Beagley J, Belesova K, Boykoff M, Byass P, Cai W, Campbell-Lendrum D (2021) The 2020 report of the Lancet Countdown on health and climate change: responding to converging crises. Lancet 397(10269):129–170. https://doi.org/10.1016/S0140-6736(20)32290-X

Yang L, Liao W, Liu C, Zhang N, Zhong S, Huang C (2018) Associations between knowledge of the causes and perceived impacts of climate change: a cross-sectional survey of medical, public health and nursing students in universities in China. Int J Environ Res Public Health 15(12):2650. https://doi.org/10.3390/ijerph15122650

Walter Leal Filho holds the chairs of Climate Change Management at the Hamburg University of Applied Sciences (Germany), and Environment and Technology at Manchester Metropolitan University (UK). He directs the Research and Transfer Centre "Sustainability Development and Climate Change Management." His main research interests are in the fields of sustainable development and climate change, also including aspects of climate change and health. He has over 30 years' experience on climate change projects and directs the International Climate Change Information and Research Programme.

Juliane Boenecke is a Project Manager at the Department of Health Sciences at the Hamburg University of Applied Sciences (Germany).

Derya Taser is a Project Manager at the Research and Transfer Centre "Sustainability Development and Climate Change Management" at the Hamburg University of Applied Sciences (Germany).

Johanna Heimfarth is a Research Associate at the Research and Transfer Centre "Sustainability Development and Climate Change Management" at the Hamburg University of Applied Sciences (Germany).

The manufacturer's authorised representative in the EU is Springer Nature Customer Service Centre GmbH, Europaplatz 3, 69115 Heidelberg, Germany. If you have any concerns regarding our products, please contact ProductSafety@springernature.com

Printed and bound by CPI Group (UK) Ltd, Croydon, CR0 4YY
06/10/2026
02225252-0014